MODELLING OF MECHANICAL SYSTEMS VOLUME 2

MODELLING OF MECHANICAL SYSTEMS VOLUME 2

Structural Elements

François Axisa and Philippe Trompette

AMSTERDAM • BOSTON • HEIDELBERG • LONDON • NEW YORK • OXFORD
PARIS • SAN DIEGO • SAN FRANCISCO • SINGAPORE • SYDNEY • TOKYO

Elsevier Butterworth-Heinemann
Linacre House, Jordan Hill, Oxford OX2 8DP
30 Corporate Drive, Burlington, MA 01803

First published in France 2001 by Hermes Science, entitled '*Modélisation des systèmes mécaniques, systèmes continus, Tome 2*'

First published in Great Britain 2005

British Library Cataloguing in Publication Data
A catalogue record for this book is available from the British Library

Library of Congress Cataloguing in Publication Data
A catalogue record for this book is available from the Library of Congress

ISBN 0 7506 6846 6

For information on all Elsevier Butterworth-Heinemann publications visit our website at http://books.elsevier.com

Typeset by Newgen Imaging Systems (P) Ltd., Chennai, India
Printed and bound by CPI Group (UK) Ltd, Croydon, CR0 4YY
Transferred to Digital Print 2011

Contents

Preface .... **xvii**

Introduction .... **xix**

Chapter 1. Solid mechanics **1**
1.1. Introduction 2
1.2. Equilibrium equations of a continuum 3
1.2.1. Displacements and strains 3
1.2.2. Indicial and symbolic notations 9
1.2.3. Stresses 11
1.2.4. Equations of dynamical equilibrium 13
1.2.5. Stress–strain relationships for an isotropic elastic material 16
1.2.6. Equations of elastic vibrations (Navier's equations) 17
1.3. Hamilton's principle 18
1.3.1. General presentation of the formalism 19
1.3.2. Application to a three-dimensional solid 20
1.3.2.1. Hamilton's principle 20
1.3.2.2. Hilbert functional vector space 20
1.3.2.3. Variation of the kinetic energy 21
1.3.2.4. Variation of the strain energy 21
1.3.2.5. Variation of the external load work 23
1.3.2.6. Equilibrium equations and boundary conditions 23
1.3.2.7. Stress tensor and Lagrange's multipliers 24
1.3.2.8. Variation of the elastic strain energy 25
1.3.2.9. Equation of elastic vibrations 27
1.3.2.10. Conservation of mechanical energy 28
1.3.2.11. Uniqueness of solution of motion equations 29
1.4. Elastic waves in three-dimensional media 31
1.4.1. Material oscillations in a continuous medium interpreted as waves 31
1.4.2. Harmonic solutions of Navier's equations 32

1.4.3. Dilatation and shear elastic waves ... 32
1.4.3.1. Irrotational, or potential motion ... 33
1.4.3.2. Equivoluminal, or shear motion ... 33
1.4.3.3. Irrotational harmonic waves (dilatation or pressure waves) ... 33
1.4.3.4. Shear waves (equivoluminal or rotational waves) ... 38
1.4.4. Phase and group velocities ... 38
1.4.5. Wave reflection at the boundary of a semi-infinite medium ... 40
1.4.5.1. Complex amplitude of harmonic and plane waves at oblique incidence ... 41
1.4.5.2. Reflection of (SH) waves on a free boundary ... 43
1.4.5.3. Reflection of (P) waves on a free boundary ... 44
1.4.6. Guided waves ... 48
1.4.6.1. Guided (SH) waves in a plane layer ... 48
1.4.6.2. Physical interpretation ... 51
1.4.6.3. Waves in an infinite elastic rod of circular cross-section ... 53
1.4.7. Standing waves and natural modes of vibration ... 53
1.4.7.1. Dilatation plane modes of vibration ... 54
1.4.7.2. Dilatation modes of vibration in three dimensions ... 55
1.4.7.3. Shear plane modes of vibration ... 58
1.5. From solids to structural elements ... 59
1.5.1. Saint-Venant's principle ... 59
1.5.2. Shape criterion to reduce the dimension of a problem ... 61
1.5.2.1. Compression of a solid body shaped as a slender parallelepiped ... 61
1.5.2.2. Shearing of a slender parallelepiped ... 62
1.5.2.3. Validity of the simplification for a dynamic loading ... 63
1.5.2.4. Structural elements in engineering ... 64

Chapter 2. Straight beam models: Newtonian approach ... 66
2.1. Simplified representation of a 3D continuous medium by an equivalent 1D model ... 67
2.1.1. Beam geometry ... 67
2.1.2. Global and local displacements ... 67
2.1.3. Local and global strains ... 70
2.1.4. Local and global stresses ... 72
2.1.5. Elastic stresses ... 74
2.1.6. Equilibrium in terms of generalized stresses ... 75
2.1.6.1. Equilibrium of forces ... 75
2.1.6.2. Equilibrium of the moments ... 77
2.2. Small elastic motion ... 78
2.2.1. Longitudinal mode of deformation ... 78
2.2.1.1. Local equilibrium ... 78

2.2.1.2. General solution of the static equilibrium without external loading ... 79
2.2.1.3. Elastic boundary conditions ... 79
2.2.1.4. Concentrated loads ... 82
2.2.1.5. Intermediate supports ... 84
2.2.2. Shear mode of deformation ... 86
2.2.2.1. Local equilibrium ... 86
2.2.2.2. General solution without external loading ... 88
2.2.2.3. Elastic boundary conditions ... 88
2.2.2.4. Concentrated loads ... 88
2.2.2.5. Intermediate supports ... 89
2.2.3. Torsion mode of deformation ... 89
2.2.3.1. Torsion without warping ... 89
2.2.3.2. Local equilibrium ... 89
2.2.3.3. General solution without loading ... 90
2.2.3.4. Elastic boundary conditions ... 90
2.2.3.5. Concentrated loads ... 90
2.2.3.6. Intermediate supports ... 90
2.2.3.7. Torsion with warping: Saint Venant's theory ... 91
2.2.4. Pure bending mode of deformation ... 99
2.2.4.1. Simplifying hypotheses of the Bernoulli–Euler model 99
2.2.4.2. Local equilibrium ... 100
2.2.4.3. Elastic boundary conditions ... 102
2.2.4.4. Intermediate supports ... 103
2.2.4.5. Concentrated loads ... 103
2.2.4.6. General solution of the static and homogeneous equation ... 104
2.2.4.7. Application to some problems of practical interest ... 104
2.2.5. Formulation of the boundary conditions ... 114
2.2.5.1. Elastic impedances ... 114
2.2.5.2. Generalized mechanical impedances ... 116
2.2.5.3. Homogeneous and inhomogeneous conditions ... 116
2.2.6. More about transverse shear stresses and straight beam models ... 116
2.2.6.1. Asymmetrical cross-sections and shear (or twist) centre ... 117
2.2.6.2. Slenderness ratio and lateral deflection ... 118
2.3. Thermoelastic behaviour of a straight beam ... 118
2.3.1. 3D law of thermal expansion ... 118
2.3.2. Thermoelastic axial response ... 119
2.3.3. Thermoelastic bending of a beam ... 121
2.4. Elastic-plastic beam ... 123
2.4.1. Elastic-plastic behaviour under uniform traction ... 124
2.4.2. Elastic-plastic behaviour under bending ... 124
2.4.2.1. Skin stress ... 125

2.4.2.2. Moment-curvature law and failure load 126
2.4.2.3. Elastic-plastic bending: global constitutive law 127
2.4.2.4. Superposition of several modes of deformation 128

Chapter 3. Straight beam models: Hamilton's principle 130
3.1. Introduction 131
3.2. Variational formulation of the straight beam equations 132
3.2.1. Longitudinal motion 132
3.2.1.1. Model neglecting the Poisson effect 132
3.2.1.2. Model including the Poisson effect (Love–Rayleigh model) 133
3.2.2. Bending and transverse shear motion 135
3.2.2.1. Bending without shear: Bernoulli–Euler model 135
3.2.2.2. Bending including transverse shear: the Timoshenko model in statics 136
3.2.2.3. The Rayleigh–Timoshenko dynamic model 139
3.2.3. Bending of a beam prestressed by an axial force 141
3.2.3.1. Strain energy and Lagrangian 142
3.2.3.2. Vibration equation and boundary conditions 143
3.2.3.3. Static response to a transverse force and buckling instability 145
3.2.3.4. Follower loads 148
3.3. Weighted integral formulations 149
3.3.1. Introduction 149
3.3.2. Weighted equations of motion 151
3.3.3. Concentrated loads expressed in terms of distributions 151
3.3.3.1. External loads 152
3.3.3.2. Intermediate supports 155
3.3.3.3. A comment on the use of distributions in mechanics 156
3.3.4. Adjoint and self-adjoint operators 156
3.3.5. Generic properties of conservative operators 162
3.4. Finite element discretization 163
3.4.1. Introduction 163
3.4.2. Beam in traction-compression 167
3.4.2.1. Mesh 168
3.4.2.2. Shape functions 169
3.4.2.3. Element mass and stiffness matrices 169
3.4.2.4. Equivalent nodal external loading 171
3.4.2.5. Assembling the finite element model 171
3.4.2.6. Boundary conditions 172
3.4.2.7. Elastic supports and penalty method 173

3.4.3. Assembling non-coaxial beams 174
3.4.3.1. The stiffness and mass matrices of a beam element for bending 174
3.4.3.2. Stiffness matrix combining bending and axial modes of deformation.. 177
3.4.3.3. Assembling the finite element model of the whole structure 177
3.4.3.4. Transverse load resisted by string and bending stresses in a roof truss.. 180
3.4.4. Saving DOF when modelling deformable solids 186

Chapter 4. Vibration modes of straight beams and modal analysis methods **188**
4.1. Introduction 189
4.2. Natural modes of vibration of straight beams 190
4.2.1. Travelling waves of simplified models 190
4.2.1.1. Longitudinal waves 190
4.2.1.2. Flexure waves 193
4.2.2. Standing waves, or natural modes of vibration 196
4.2.2.1. Longitudinal modes..................................... 196
4.2.2.2. Torsion modes.. 200
4.2.2.3. Flexure (or bending) modes 200
4.2.2.4. Bending coupled with shear modes 205
4.2.3. Rayleigh's quotient 207
4.2.3.1. Bending of a beam with an attached concentrated mass 207
4.2.3.2. Beam on elastic foundation 209
4.2.4. Finite element approximation 210
4.2.4.1. Longitudinal modes..................................... 210
4.2.4.2. Bending modes... 211
4.2.5. Bending modes of an axially preloaded beam 213
4.2.5.1. Natural modes of vibration 213
4.2.5.2. Static buckling analysis................................. 214
4.3. Modal projection methods 217
4.3.1. Equations of motion projected onto a modal basis 218
4.3.2. Deterministic excitations.. 220
4.3.2.1. Separable space and time excitation 220
4.3.2.2. Non-separable space and time excitation 221
4.3.3. Truncation of the modal basis 222
4.3.3.1. Criterion based on the mode shapes 222
4.3.3.2. Spectral criterion.. 224
4.3.4. Stresses and convergence rate of modal series................... 229
4.4. Substructuring method 231
4.4.1. Additional stiffnesses 231
4.4.1.1. Beam in traction-compression with an end spring 232

4.4.1.2. Truncation stiffness for a free-free modal basis 235
4.4.1.3. Bending modes of an axially prestressed beam 237
4.4.2. Additional inertia 238
4.4.3. Substructures by using modal projection........ 240
4.4.3.1. Basic ideas of the method 240
4.4.3.2. Vibration modes of an assembly of two beams linked by a spring 243
4.4.3.3. Multispan beams........ 245
4.4.4. Nonlinear connecting elements........ 247
4.4.4.1. Axial impact of a beam on a rigid wall 248
4.4.4.2. Beam motion initiated by a local impulse followed by an impact on a rigid wall........ 254
4.4.4.3. Elastic collision between two beams 256

Chapter 5. Plates: in-plane motion **259**
5.1. Introduction........ 260
5.1.1. Plate geometry........ 260
5.1.2. Incidence of plate geometry on the mechanical response........ 260
5.2. Kirchhoff–Love model........ 262
5.2.1. Love simplifications........ 262
5.2.2. Degrees of freedom and global displacements........ 262
5.2.3. Membrane displacements, strains and stresses........ 263
5.2.3.1. Global and local displacements 263
5.2.3.2. Global and local strains........ 263
5.2.3.3. Membrane stresses........ 265
5.3. Membrane equilibrium of rectangular plates........ 265
5.3.1. Equilibrium in terms of generalized stresses........ 265
5.3.1.1. Local balance of forces 266
5.3.1.2. Hamilton's principle 267
5.3.1.3. Homogeneous boundary conditions 270
5.3.1.4. Concentrated loads........ 270
5.3.2. Elastic stresses........ 272
5.3.3. Equations and boundary conditions in terms of displacements........ 273
5.3.4. Examples of application in elastostatics........ 275
5.3.4.1. Sliding plate subject to a uniform longitudinal load at the free edge 275
5.3.4.2. Fixed instead of sliding condition at the supported edge 277
5.3.4.3. Three sliding edges: plate in uniaxial strain configuration 278
5.3.4.4. Uniform plate stretching........ 278

5.3.4.5. In-plane uniform shear loading 279
5.3.4.6. In-plane shear and bending 280
5.3.5. Examples of application in thermoelasticity 283
5.3.5.1. Thermoelastic law 283
5.3.5.2. Thermal stresses 284
5.3.5.3. Expansion joints 285
5.3.5.4. Uniaxial plate expansion 286
5.3.6. In-plane, or membrane, natural modes of vibration 289
5.3.6.1. Solutions of the modal equations by variable separation 289
5.3.6.2. Natural modes of vibration for a plate on sliding supports 290
5.3.6.3. Semi-analytical approximations: Rayleigh–Ritz and Galerkin discretization methods 293
5.3.6.4. Plate loaded by a concentrated in-plane force: spatial attenuation of the local response 299
5.4. Curvilinear coordinates 303
5.4.1. Linear strain tensor 304
5.4.2. Equilibrium equations and boundary conditions 305
5.4.3. Elastic law in curvilinear coordinates 307
5.4.4. Circular cylinder loaded by a radial pressure 307

Chapter 6. Plates: out-of-plane motion **311**
6.1. Kirchhoff–Love hypotheses 312
6.1.1. Local displacements 312
6.1.2. Local and global strains 313
6.1.2.1. Local strains 313
6.1.2.2. Global flexure and torsional strains 313
6.1.3. Local and global stresses: bending and torsion 314
6.2. Bending equations 316
6.2.1. Formulation in terms of stresses 316
6.2.1.1. Variation of the inertia terms 316
6.2.1.2. Variation of the strain energy 317
6.2.1.3. Local equilibrium without external loads 318
6.2.2. Boundary conditions 319
6.2.2.1. Kirchhoff effective shear forces and corner forces 319
6.2.2.2. Elastic boundary conditions 322
6.2.2.3. External loading of the edges and inhomogeneous boundary conditions 322
6.2.3. Surface and concentrated loadings 324
6.2.3.1. Loading distributed over the midplane surface 324
6.2.3.2. Load distributed along a straight line parallel to an edge 325
6.2.3.3. Point loads 326

6.2.4. Elastic vibrations 327
6.2.4.1. Global stresses 327
6.2.4.2. Vibration equations 327
6.2.4.3. Elastic boundary conditions 328
6.2.5. Application to a few problems in statics 329
6.2.5.1. Bending of a plate loaded by edge moments 329
6.2.5.2. Torsion by corner forces 331
6.3. Modal analysis 332
6.3.1. Natural modes of vibration 332
6.3.1.1. Flexure equation of a plate prestressed in its own plane 332
6.3.1.2. Natural modes of vibration and buckling load 335
6.3.1.3. Modal density and forced vibrations near resonance 338
6.3.1.4. Natural modes of vibration of a stretched plate 340
6.3.1.5. Warping of a beam cross-section: membrane analogy 347
6.4. Curvilinear coordinates 348
6.4.1. Bending and torsion displacements and strains 348
6.4.2. Equations of motion 349
6.4.3. Boundary conditions 350
6.4.4. Circular plate loaded by a uniform pressure 350

Chapter 7. Arches and shells: string and membrane forces 354
7.1. Introduction: why curved structures? 355
7.1.1. Resistance of beams to transverse loads 355
7.1.2. Resistance of shells and plates to transverse loads 356
7.2. Arches and circular rings 358
7.2.1. Geometry and curvilinear metric tensor 358
7.2.2. Local and global displacements 359
7.2.3. Local and global strains 360
7.2.4. Equilibrium equations along the neutral line 361
7.2.5. Application to a circular ring 364
7.2.5.1. Simplifications inherent in axisymmetric structures 364
7.2.5.2. Breathing mode of vibration of a circular ring 365
7.2.5.3. Translational modes of vibration 365
7.2.5.4. Cable stressed by its own weight 366
7.3. Shells 367
7.3.1. Geometry and curvilinear metrics 367
7.3.2. Local and global displacements 369
7.3.3. Local and global strains 369
7.3.4. Global membrane stresses 369
7.3.5. Membrane equilibrium 370

7.3.6. Axisymmetric shells ... 371
7.3.6.1. Geometry and metric tensor ... 371
7.3.6.2. Curvature tensor ... 372
7.3.7. Applications in elastostatics ... 375
7.3.7.1. Spherical shell loaded by uniform pressure ... 375
7.3.7.2. Cylindrical shell closed by hemispherical ends ... 376
7.3.7.3. Pressurized toroidal shell ... 378
7.3.7.4. Spherical cap loaded by its own weight ... 382
7.3.7.5. Conical shell of revolution loaded by its own weight ... 386
7.3.7.6. Conical container ... 388

Chapter 8. Bent and twisted arches and shells ... 391
8.1. Arches and circular rings ... 392
8.1.1. Local and global displacement fields ... 392
8.1.2. Tensor of small local strains ... 393
8.1.3. Pure bending in the arch plane ... 394
8.1.3.1. Equilibrium equations ... 394
8.1.3.2. Vibration modes of a circular ring ... 396
8.1.4. Model coupling in-plane bending and axial vibrations ... 398
8.1.4.1. Coupled equations ... 398
8.1.4.2. Vibration modes of a circular ring ... 400
8.1.4.3. Arch loaded by its own weight ... 402
8.1.5. Model coupling torsion and out-of-plane bending ... 407
8.1.5.1. Coupled equations of vibration ... 407
8.1.5.2. Natural modes of vibration of a circular ring ... 410
8.2. Thin shells ... 412
8.2.1. Local and global tensor of small strains ... 412
8.2.1.1. Local displacement field ... 412
8.2.1.2. Expression of the local and global strain components ... 412
8.2.2. Love's equations of equilibrium ... 414
8.3. Circular cylindrical shells ... 415
8.3.1. Equilibrium equations ... 415
8.3.1.1. Love's equations in cylindrical coordinates ... 415
8.3.1.2. Boundary conditions ... 416
8.3.2. Elastic vibrations ... 418
8.3.2.1. Small elastic strain and stress fields ... 418
8.3.2.2. Equations of vibrations ... 419
8.3.2.3. Pure bending model ... 420
8.3.2.4. Constriction of a circular cylindrical shell ... 421
8.3.2.5. Bending about the meridian lines ... 425
8.3.2.6. Natural modes of vibration $n = 0$ ... 426

8.3.3. Bending coupled in z and θ 428
8.3.3.1. Simplified model neglecting the hoop and shear stresses 428
8.3.3.2. Membrane and bending-torsion terms of elastic energy 430
8.3.3.3. Point-wise punching of a circular cylindrical shell 433
8.3.3.4. Natural modes of vibration 434
8.3.3.5. Donnel–Mushtari–Vlasov model 435
8.3.4. Modal analysis of Love's equations 436
8.3.5. Axial loading: global and local responses 438

Appendices **441**
A.1. Vector and tensor calculus 441
A.1.1. Definition and notations of scalar, vector and tensor fields 441
A.1.2. Tensor algebra 443
A.1.2.1. Contracted product 443
A.1.2.2. Non-contracted product 445
A.1.2.3. Cross-product of two vectors in indicial notation 445
A.2. Differential operators 446
A.2.1. The Nabla differential operator 446
A.2.2. The divergence operator 446
A.2.3. The gradient operator 447
A.2.4. The curl operator 448
A.2.5. The Laplace operator 449
A.2.6. Other useful formulas 449
A.3. Differential operators in curvilinear and orthonormal coordinates 449
A.3.1. Metrics 449
A.3.2. Differential operators in curvilinear and orthogonal coordinates 452
A.3.2.1. Gradient of a scalar and the Nabla operator 452
A.3.2.2. Gradient of a vector 452
A.3.2.3. Divergence of a vector 453
A.3.2.4. Divergence of a tensor of the second rank 453
A.3.2.5. Curl of a vector 454
A.3.2.6. Laplacian of a scalar 454
A.3.2.7. Polar coordinates 454
A.3.2.8. Cylindrical coordinates 455
A.4. Plate bending in curvilinear coordinates 457
A.4.1. Formulation of Hamilton's principle 457
A.4.2. Equation of local equilibrium in terms of shear forces 459
A.4.3. Boundary conditions: effective Kirchhoff's shear forces and corner forces 460
A.5. Static equilibrium of a sagging cable loaded by its own weight 461
A.5.1. Newtonian approach 462

A.5.2. Constrained Lagrange's equations, invariance of the cable length 463
A.5.3. Constrained Lagrange's equations: length invariance of a cable element 465
A.6. Mechanical properties of some solids in common use 466

References **468**

Index **472**

Preface

In mechanical engineering, the needs for design analyses increase and diversify very fast. Our capacity for industrial renewal means we must face profound issues concerning efficiency, safety, reliability and life of mechanical components. At the same time, powerful software systems are now available to the designer for tackling incredibly complex problems using computers. As a consequence, computational mechanics is now a central tool for the practising engineer and is used at every step of the designing process. However, it cannot be emphasized enough that to make a proper use of the possibilities offered by computational mechanics, it is of crucial importance to gain first a thorough background in theoretical mechanics. As the computational process by itself has become largely an automatic task, the engineer, or scientist, must concentrate primarily in producing a tractable model of the physical problem to be analysed. The use of any software system either in a University laboratory, or in a Research department of an industrial company, requires that meaningful results be produced. This is only the case if sufficient effort was devoted to build an appropriate model, based on a sound theoretical analysis of the problem at hand. This often proves to be an intellectually demanding task, in which theoretical and pragmatic knowledge must be skilfully interwoven. To be successful in modelling, it is essential to resort to physical reasoning, in close relationship with the information of practical relevance.

This series of four volumes is written as a self-contained textbook for engineering and physical science students who are studying structural mechanics and fluid–structure coupled systems at a graduate level. It should also appeal to engineers and researchers in applied mechanics. The four volumes, already available in French, deal respectively with Discrete Systems, Basic Structural Elements (beams, plates and shells), Fluid–Structure Interaction in the absence of permanent flow, and finally, Flow-Induced Vibrations. The purpose of the series is to equip the reader with a good understanding of a large variety of mechanical systems, based on a unifying theoretical framework. As the subject is obviously too vast to cover in an exhaustive way, presentation is deliberately restricted to those fundamental physical aspects and to the basic mathematical methods which constitute the backbone of any large software system currently used in mechanical engineering. Based on the experience gained as a research engineer in nuclear engineering at the French Atomic Commission, and on course notes offered to

2nd and 3rd year engineer students from ECOLE NATIONALE SUPERIEURE DES TECHNIQUES AVANCEES, Paris and to the graduate students of Paris VI University, the style of presentation is to convey the main physical ideas and mathematical tools, in a progressive and comprehensible manner. The necessary mathematics is treated as an invaluable tool, but not as an end in itself. Considerable effort has been taken to include a large number of worked exercises, especially selected for their relative simplicity and practical interest. They are discussed in some depth as enlightening illustrations of the basic ideas and concepts conveyed in the book. In this way, the text incorporates in a self-contained manner, introductory material on the mathematical theory, which can be understood even by students without in-depth mathematical training. Furthermore, many of the worked exercises are well suited for numerical simulations by using software like MATLAB, which was utilised by the author for the numerous calculations and figures incorporated in the text. Such exercises provide an invaluable training to familiarize the reader with the task of modelling a physical problem and of interpreting the results of numerical simulations. Finally, though not exhaustive the references included in the book are believed to be sufficient for directing the reader towards the more specialized and advanced literature concerning the specific subjects introduced in the book.

To complete this work I largely benefited from the input and help of many people. Unfortunately, it is impossible to properly acknowledge here all of them individually. However, I wish to express my gratitude to Alain Hoffmann head of the Department of Mechanics and Technology at the Centre of Nuclear Studies of Saclay and to Pierre Sintes, Director of ENSTA who provided me with the opportunity to be Professor at ENSTA. A special word of thanks goes to my colleagues at ENSTA and at Saclay – Ziad Moumni, Laurent Rota, Emanuel de Langre, Ianis Politopoulos and Alain Millard – who assisted me very efficiently in teaching mechanics to the ENSTA students and who contributed significantly to the present book by pertinent suggestions and long discussions. Acknowledgements also go to the students themselves whose comments were also very stimulating and useful. I am also especially grateful to Professor Michael Païdoussis from McGill University Montreal, who encouraged me to produce an English edition of my book, which I found quite a challenging task afterwards! Finally, without the loving support and constant encouragement of my wife Françoise this book would not have materialized.

François Axisa
August 2003

Introduction

To understand what is meant by structural elements, it is convenient to start by considering a whole structure made of various components assembled together with the aim to satisfy various functional and cost criteria. Depending on the domain of application, the terminology used to designate such assemblies varies; they are referred to as buildings, civil engineering works, machines and devices, vehicles etc. In most cases, the shapes of such structures are so complicated that the appropriate way to make a mathematical model feasible, is to identify simpler structural elements, defined according to a few generic response properties. Such a theoretical approach closely follows the common engineering practice of selecting a few appropriate generic shapes to build complex structures. Since the architects and engineers of the Roman Empire, two geometrical features have been recognized as key factors to save material and weight in a structure. The first one is to design slender components, that is, at least one dimension of the body is much less than the others. From the analyst standpoint this allows to model the actual 3D solid by using an equivalent solid of reduced dimension. Accordingly, one is led to distinguish first between 1D and 2D structural elements. The second geometrical property of paramount importance to optimise the mechanical resistance of structural elements is the curvature of the equivalent solid. Based on these two properties structural elements can be identified as:

1. Straight beams, modelled as a one-dimensional and rectilinear equivalent solid.
2. Plates, modelled as a two-dimensional and planar equivalent solid.
3. Curved beams, modelled as a one-dimensional and curved equivalent solid.
4. Shells, modelled as a two-dimensional and curved equivalent solid.

The second volume of this series deals with modelling and analysis of the mechanical responses of such structural elements. However, this vast subject is restricted here, essentially, to the linear elastodynamic domain, which constitute the cornerstone of mathematical modelling in structural mechanics. Moving on from discrete systems to deformable solids, as material is assumed to be continuously distributed over a bounded domain defined in a 3D Euclidean space, two new salient points arise. First, motion must be described in terms of continuous functions of space and then appropriate boundary conditions have to be specified in order to describe the mechanical equilibrium of the solid boundary. That mastering the consequences of these two features in structural modelling is by far not a simple task can be amply asserted by recalling that it progressed, along with the necessary mathematics, step by step over a long period lasting essentially from the eighteenth to the first half of the twentieth century. Apart from the concepts and methods inherent to the continuous nature of the problem, those already described in Volume 1, to deal with discrete systems keep all their interest, in particular the concept of natural modes of vibration and the methods of modal analysis. Actually, in practice, to analyse most of the engineering structures, it is necessary to build first a finite element model, according to which the structure is discretized into a finite number of parts, leading to a finite set of time differential equations. The latter can be solved numerically on the computer, either by using a spectral or a time stepping method.

Chapter 1 reviews the fundamental concepts and results of continuum mechanics used as a necessary background for the rest of the book. Major points concern the concepts of strain and stress tensors, the formulation of equilibrium equations, using the Newtonian approach and Hamilton's principle, successively. Then, they are particularized to the case of linear elastodynamics, producing the Navier's equation which govern the elastic waves in a solid. The concept of natural modes of vibration in a solid is introduced by solving the Navier's equations in terms of harmonic waves and accounting for the reflection conditions at the solid boundary. Finally, the Saint-Venant's principle is used as a guiding line to model a solid as a structural element.

Chapter 2 presents the basic ideas to model beam-like structures as a 1D solid; the starting point is to assume that the beam cross-sections behave as rigid bodies. Here, modelling is restricted to the case of straight beams and the 1D equilibrium equations, including boundary conditions, are derived by using the Newtonian approach, i.e. by balancing directly the forces and moments acting on a beam element of infinitesimal length. Study is further particularized to the case of linear elastodynamics producing the so called vibration equations. Presented here in their simplest and less refined form, they comprise three uncoupled equations which govern stretching, torsion and bending, respectively. The lateral contraction induced by stretching, due to the Poisson ratio is neglected, which is a realistic assumption in most engineering applications. According to the Bernoulli–Euler model, coupling of bending with transverse shear strains is negligible, which is a reasonable assumption if the beam is slender enough. Concerning torsion, in the case of noncircular cross-sections they are found to warp in such a way that torsion rigidity can be considerably lowered with respect to the value given by a pure torsion model. Warping induced by torsion is classically described based on the Saint-Venant model. The chapter is concluded by presenting a few problems of thermoelasticity and plasticity to illustrate further the modelling process required to approximate a 3D solid as an equivalent 1D solid.

In Chapter 3, the problem of modelling straight beams is revisited and completed by presenting a few distinct topics of theoretical and practical importance. At first, Hamilton's principle is used to improve the basic beam models established in Chapter 2, by accounting for the deformation of the cross-sections and the effect of axial preloads on beam bending. Then, the weighted integral equations of motion are introduced as a starting point to introduce various mathematical concepts and techniques. They are used first together with the singular Dirac distribution, already introduced in Volume 1, to express the equilibrium equations in a unified manner, independently from the continuous or discrete nature of the physical quantities involved in the system. As a second application of the weighted integral equations, the symmetry properties of the stiffness and mass operators are demonstrated, based on the beam operators. Finally, weighted integral equations together with Hamilton's principle give us a good opportunity to present an introductory description of the finite element method.

Chapter 4 is devoted to the modal analysis method, which is a particularly elegant and efficient tool for modelling a large variety of problems in mechanics, independently of their discrete or continuous nature. At first, the natural modes of vibration of straight beams are described. Then they are used as convenient structural examples to present several aspects of modal analysis, focusing on those specific to the case of continuous systems. In particular, the criteria to truncate suitably the modal series are established and illustrated by several examples. Finally, the substructuring method using truncated modal bases for describing each substructure is introduced and illustrated by solving a few linear and nonlinear problems involving intermittent contacts.

Chapters 5 and 6 deal with thin plates described as 2D solids by assuming that strains in the thickness direction can be neglected. Plates are characterized by a plane geometry bounded by edges comprising straight and/or curved lines. Chapter 5 is concerned with the in-plane solicitations and responses, where the part is played by the so called membrane components solely. Chapter 6 is concerned with the out-of-plane, or transverse, solicitations and responses, where the part is played by the flexure and torsion components and the in-plane preloads. Modelling is based on the so called Kirchhoff–Love hypotheses which extend to the 2D case the Bernoulli–Euler model of straight beam bending. Solution of a few problems help to concretize the major features of plate responses to various load conditions. Amongst others, enlightening results concerning the Saint-Venant principle invoked in Chapter 1, are obtained by using the modal analysis method to the response of a rectangular plate to an in-plane point load. On the other hand, the Rayleigh–Ritz discretization method is described and applied to the semi-analytical calculation of the natural modes of vibration of rectangular plates.

Chapters 7 and 8 are devoted to curved structures, namely arches and thin shells. In curved beams and shells, tensile or compressive stresses can resist transverse loads, even in the absence of a prestress field. This can be conveniently emphasized by considering first simplified arch and shell models where bending and torsion terms are entirely discarded, which is the object of Chapter 7. Though the range of validity of the equilibrium equations obtained by using such a simplifying assumption, is clearly limited to certain load conditions, it is believed appropriate to present and discuss them in a rather detailed manner before embarking on the more elaborate models presented in Chapter 8, which account for string or membrane stresses as well as for bending and torsion stresses. Solution of a few problems concerning circular arches or rings and then shells of revolution, brings out that transverse loads cannot be exactly balanced by tensile or compressive stresses in the case of beams but they can in the case of shells. In any case, to deal with general loading conditions, it is necessary to include bending and torsion into the equilibrium equations of arches and shells which is the object of Chapter 8, the last of this volume. As illustrated by the solution of a few problems, the relative importance of the various coupling terms arising in the arch and shell equations, largely depend on the geometry of the structure and on the space distribution of the loads.

The content of the English version of the present volume is basically the same as that of the first edition in French. However, it benefited from various significant improvements and complements, concerning in particular the reflection and the guided propagation of elastic waves and the presentation of the finite element method. Finally, a special word of thanks goes again to Philip Kogan, for checking and rechecking every part of the manuscript. His professional attitude has contributed significantly to the quality of this book. Any remaining errors and inaccuracies are purely the author's own.

François Axisa and Philippe Trompette
November 2004

Chapter 1

Solid mechanics

Real mechanical systems generally comprise an assembly of deformable solids, which must be modelled within the framework of the theory of continuum mechanics. Accordingly, material is assumed to be distributed continuously in a 3D domain. However, in most instances, the engineer deals with *structural elements* endowed with geometrical particularities which allow for further simplification in modelling, based on the concept of 2D or even 1D equivalent continuous media. Before embarking on the presentation of such models, which is the central object of this book, it is appropriate to review first a few fundamental concepts, definitions and laws of continuum mechanics. This vast subject is restricted here to a few important aspects of linear elasticity and elastodynamics of solids.

1.1. Introduction

The mechanics of solid bodies is concerned with the motion of deformable media, in which solid matter is continuously distributed over a domain of the three-dimensional Euclidean space, bounded in every direction. Accordingly, it extends the mechanics of discrete particles, dealing with the new following aspects:

1. To describe the motion, use is made of an infinite, and even more significant, an uncountable set of degrees of freedom (DOF). The three displacement components of all material points define a vector field $\vec{X}(\vec{r};t)$, which is a continuous and differentiable function of the position vector $\vec{r}$ and depends on time t. A priori, $\vec{X}(\vec{r};t)$ and $\vec{r}$ are defined in a 3D Euclidean space. Assumption of continuity of $\vec{X}(\vec{r};t)$ implies that occurrence of any cracks or holes during deformation is precluded.
2. $\vec{r}$ may specify either the position of the points of the space in which the motion takes place (*Eulerian description*) or the position of material points during motion (*Lagrangian description*).
3. The mechanical properties of the continuum are described by scalar, vector, and tensor fields, which can be Eulerian or Lagrangian in nature. Nevertheless, so long as the theory is restricted to small displacements, as it is the case in linear elasticity, the Eulerian and Lagrangian descriptions become equivalent to each other.
4. A body made of one or several continua fills a finite volume $(\mathcal{V})$ limited by a closed surface $(\mathcal{S})$, termed the *boundary*. To formulate the equilibrium equations of the volume $(\mathcal{V})$, one is led to distinguish between forces which are distributed over either a volume (force per unit volume), or a surface (force per unit area) or a line (force per unit length), or even concentrated at some discrete points.
5. The boundary may be constrained by various types of relations, involving kinematical fields (displacement, velocity and acceleration) and/or dynamical components (internal and external forces), which define the *boundary conditions*. Furthermore, distinct boundary conditions may hold at distinct positions of the boundary domain; for instance, a displacement field is prescribed over a part $(\mathcal{S}_1)$ of $(\mathcal{S})$ whereas a pressure field is applied to the complementary part $(\mathcal{S}_2) = (\mathcal{S}) \cap (\mathcal{S}_1)$.

In Section 1.2 a few basic notions of continuum mechanics theory are reviewed which are needed subsequently throughout this book. Here, a Newtonian approach is chosen, i.e. the equilibrium conditions are derived directly by writing down the balances of forces and torques. The reader is referred to more specialized books such as [FUN 68], [FUN 01], [SAL 01] for a more thorough and advanced study of this vast subject.

In Section 1.3, *Hamilton's principle of least action* is used to extend Lagrange's formalism to deformable bodies. This analytical approach will be used abundantly (but not exclusively) in the subsequent chapters to model the basic structural components of common use in mechanical engineering as equivalent 2D and 1D continuous media.

In Section 1.4, a few notions needed to analyse the propagation of material waves in elastic solids are presented. Here, interest is focused on wave reflection at the boundaries and on standing waves. The latter are identified with the natural modes of vibration of the elastic solid, provided with conservative (elastic or inertial) boundary conditions. Modal frequencies and wavelengths provide suitable scaling factors to validate the simplifying assumptions adopted in structural modelling to analyse dynamical problems.

Finally, in Section 1.5, Saint-Venant's principle, which allows one to distinguish between local and global effects in the response of solids, is discussed in the context of the simplifying assumptions which allow one to model a 3D solid as an equivalent 2D or 1D solid.

1.2. Equilibrium equations of a continuum

1.2.1 *Displacements and strains*

When loaded, the solid body is deformed, but, in most cases of practical interest, very slightly in comparison with the deformations experienced by fluids. So, in a solid, material points which are initially very close together remain close together during deformation and the Lagrangian description is well adapted to formulate the equations of mechanical equilibrium. The motion is described by a displacement vector field which is referenced to the initial (non-deformed) configuration (Figure 1.1). If the body is deformed during motion, the distance between two material points is changed. So, the deformation rate has to be related in some suitable

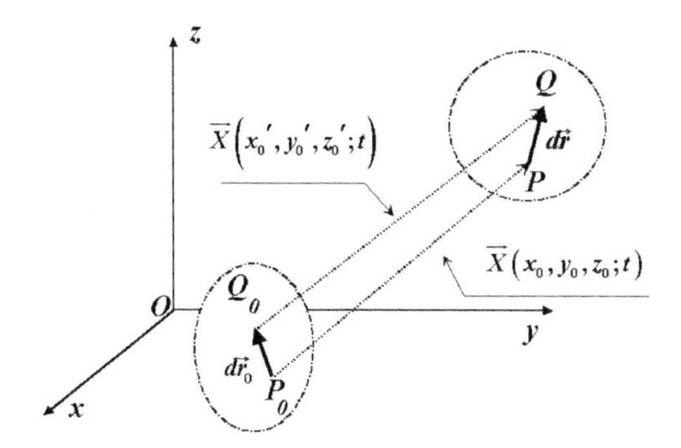

Figure 1.1. *Lagrangian displacement and strain fields of two closely spaced points*

manner to the relative change of length of an infinitesimal segment, giving rise to the concept of *strain tensor*, denoted in symbolic notation $\overline{\overline{\varepsilon}}$. The tensor nature of $\overline{\overline{\varepsilon}}$ arises as a consequence of the fact that the change of length generally depends upon the direction, but not upon the coordinate system. See Appendix A.1 for a brief presentation of vector and tensor calculus.

Let P_0 and Q_0 be two infinitely neighbouring material points of the initial configuration (time $t = 0$). Their position is defined in a Cartesian coordinate system of unit vectors $\vec{i}, \vec{j}, \vec{k}$ as:

$$\begin{aligned} P_0: \quad \vec{r}_0 &= x_0\vec{i} + y_0\vec{j} + z_0\vec{k} \\ Q_0: \quad \vec{r}'_0 &= x'_0\vec{i} + y'_0\vec{j} + z'_0\vec{k} \end{aligned} \qquad [1.1]$$

At a later time t, P_0 and Q_0 are mapped into the slightly displaced points P and Q respectively:

$$\begin{aligned} P: \quad \vec{r} &= x\vec{i} + y\vec{j} + z\vec{k} \\ Q: \quad \vec{r}' &= x'\vec{i} + y'\vec{j} + z'\vec{k} \end{aligned} \qquad [1.2]$$

where the coordinates are referred to the initial configuration and described by functions of space and time of the type:

$$\begin{aligned} &x(x_o, y_o, z_o; t); \quad x'(x'_o, y'_o, z'_o; t) \\ &y(x_o, y_o, z_o; t); \quad y'(x'_o, y'_o, z'_o; t) \\ &z(x_o, y_o, z_o; t); \quad z'(x'_o, y'_o, z'_o; t) \end{aligned} \qquad [1.3]$$

The displacement vectors of P and Q are defined as:

$$\begin{aligned} \vec{X}(\vec{r}_0; t) &= \overrightarrow{OP} - \overrightarrow{OP_0} = \overrightarrow{P_0P} = \vec{r} - \vec{r}_0 \\ \vec{X}(\vec{r}'_0; t) &= \overrightarrow{OQ} - \overrightarrow{OQ_0} = \overrightarrow{Q_0Q} = \vec{r}' - \vec{r}'_0 \end{aligned} \qquad [1.4]$$

Let $d\vec{r}_0$ denote the infinitesimal vector $\overrightarrow{P_0Q_0}$ of Cartesian components dx_0, dy_0, dz_0, and $d\vec{r}$ the infinitesimal vector $\overrightarrow{PQ}$, of Cartesian components dx, dy, dz. It is easily shown that:

$$\vec{X}(\vec{r}'_0; t) = \overrightarrow{OQ} - \overrightarrow{OQ_0} = \vec{r} + d\vec{r} - (\vec{r}_0 + d\vec{r}_0) = \vec{X}(\vec{r}_0; t) + d\vec{r} - d\vec{r}_0 \qquad [1.5]$$

As the coordinates [1.3] are continuous functions of space which can be differentiated at least to the first order, the chain derivation rule can be used to obtain:

$$\begin{aligned} dx &= \frac{\partial x}{\partial x_0}dx_0 + \frac{\partial x}{\partial y_0}dy_0 + \frac{\partial x}{\partial z_0}dz_0 \\ dy &= \frac{\partial y}{\partial x_0}dx_0 + \frac{\partial y}{\partial y_0}dy_0 + \frac{\partial y}{\partial z_0}dz_0 \\ dz &= \frac{\partial z}{\partial x_0}dx_0 + \frac{\partial z}{\partial y_0}dy_0 + \frac{\partial z}{\partial z_0}dz_0 \end{aligned} \qquad [1.6]$$

[1.6] can be written in symbolic notation as:

$$d\vec{r} = \overline{\overline{\operatorname{grad}\vec{r}}} \cdot d\vec{r}_0 \qquad [1.7]$$

$\overline{\overline{\operatorname{grad}\vec{r}}}$ is a tensor of the second rank called the *gradient* of the position vector $\vec{r}$. Its Cartesian components identify with the partial derivatives appearing in [1.6]. A form like [1.7] is termed *intrinsic* as it makes no specific reference to a coordinate system.

In the same manner, the components of $\vec{X}(\vec{r}_0;t)$ can be differentiated to give:

$$\begin{aligned} dX &= \frac{\partial X}{\partial x_0}dx_0 + \frac{\partial X}{\partial y_0}dy_0 + \frac{\partial X}{\partial z_0}dz_0 \\ dY &= \frac{\partial Y}{\partial x_0}dx_0 + \frac{\partial Y}{\partial y_0}dy_0 + \frac{\partial Y}{\partial z_0}dz_0 \\ dZ &= \frac{\partial Z}{\partial x_0}dx_0 + \frac{\partial Z}{\partial y_0}dy_0 + \frac{\partial Z}{\partial z_0}dz_0 \end{aligned} \qquad [1.8]$$

written in intrinsic form as:

$$d\vec{X} = \overline{\overline{\operatorname{grad}\vec{X}}} \cdot d\vec{r}_0$$

On the other hand, from [1.5] and [1.7] it follows that:

$$\vec{X}\left(\vec{r}_0';t\right) - \vec{X}(\vec{r}_0;t) = d\vec{X} = d\vec{r} - d\vec{r}_0 = \left(\overline{\overline{\operatorname{grad}\vec{r}}} - \overline{\overline{I}}\right) \cdot d\vec{r}_0 \qquad [1.9]$$

where $\overline{\overline{I}}$ is the identity tensor such that $d\vec{r}_0 = \overline{\overline{I}} \cdot d\vec{r}_0$

Then, from [1.8] and [1.9], it follows that:

$$\overline{\overline{\operatorname{grad}\vec{X}}} = \overline{\overline{\operatorname{grad}\vec{r}}} - \overline{\overline{I}} \qquad [1.10]$$

At this step, the mathematical manipulations which are necessary to proceed in the definition of the strain tensor are carried out more easily by shifting either to a matrix or to an indicial notation, rather than by using directly the symbolic vector and tensor notation. This is because matrix and indicial notations deal with the scalar components of vectors and tensors, as defined in a specific coordinate system. In this way, the rules of algebra with scalars are immediately applicable. It is important to get well trained in the symbolic, matrix and indicial notations because all of them are used with equal frequency in the literature on continuum and structural mechanics. As we have to deal here with Cartesian tensors of the first rank (vectors) and of the second rank only, the matrix notation is particularly convenient for dealing with the present problem. The results will be converted into the two other kinds of notation afterwards. Accordingly, the linear system [1.6] is rewritten in matrix form as:

$$\begin{bmatrix} dx \\ dy \\ dz \end{bmatrix} = \begin{bmatrix} \partial x/\partial x_o & \partial x/\partial y_o & \partial x/\partial z_o \\ \partial y/\partial x_o & \partial y/\partial y_o & \partial y/\partial z_o \\ \partial z/\partial x_o & \partial z/\partial y_o & \partial z/\partial z_o \end{bmatrix} \begin{bmatrix} dx_o \\ dy_o \\ dz_o \end{bmatrix} \qquad [1.11]$$

that is, in concise form as:

$$[dr] = [J][dr_0] \qquad [1.12]$$

$[dr]$ and $[dr_0]$ are the column vectors built with the Cartesian components of $d\vec{r}$ and $d\vec{r}_0$, respectively. $[J]$ is the *gradient transformation matrix*, also called the *Jacobian matrix* of transformation, which is built with the Cartesian components of $\overline{\overline{\mathrm{grad}\,\vec{r}}}$. If the material is deformed when passing from the initial to the actual configuration, the length of $[dr]$ differs from that of $[dr_0]$. Using [1.12], it follows that:

$$\|d\vec{r}\|^2 = [dr]^T[dr] = [dr_0]^T[J]^T[J][dr_0] = [dr_0]^T[C][dr_0] \qquad [1.13]$$

where the upper script $(^T)$ stands for a matrix transposition.

This quadratic form is independent of the coordinate system and is used to define the *Cauchy kinematic tensor* $\overline{\overline{C}}$, written as the symmetric matrix:

$$[C] = [J]^T[J] \qquad [1.14]$$

In the particular case of a rigid body motion, $\|d\vec{r}\| = \|d\vec{r}_0\|$, hence $[C]$ reduces necessarily to the identity matrix $[I]$ (diagonal elements equal to one, and non-diagonal elements equal to zero). Furthermore, using [1.10], it is possible to express $\overline{\overline{C}}$ in terms of $\overline{\overline{\mathrm{grad}\,\vec{X}}}$. First, [1.10] is rewritten in matrix notation as:

$$[J] = [I] + [\mathrm{grad}[X]] \qquad [1.15]$$

Then, substituting [1.15] into [1.14], and applying the rules of matrix product, we arrive at:

$$[C] - [I] = [\text{grad}[X]] + [\text{grad}[X]]^T + [\text{grad}[X]][\text{grad}[X]]^T \qquad [1.16]$$

As in a rigid body motion $[C]$ reduces to $[I]$, it is appropriate to use the right-hand side of [1.16] to define the strain tensor. It turns out that a suitable definition is the so called *Green–Lagrange strain tensor* represented by the matrix:

$$[\varepsilon] = \frac{1}{2}([C] - [I]) = \frac{[\text{grad}[X]] + [\text{grad}[X]]^T + [\text{grad}[X]][\text{grad}[X]]^T}{2} \qquad [1.17]$$

Obviously, $[\varepsilon]$ is symmetric. Physically, the Green–Lagrange strain tensor provides a means to measure the relative change of length of an infinitesimal segment by the relation:

$$[dr_0]^T[\varepsilon][dr_0] = \frac{[dr]^T[dr] - [dr_0]^T[dr_0]}{2} \qquad [1.18]$$

Though [1.18] is suitably independent from the coordinate system, its pertinence to the measurement of deformations is not obvious at first glance. The two following specific cases can be used to understand it better.

1. The so called *engineering strain*, denoted here by ε_E, measures the relative change of length of the straight segment *AB* (see Figure 1.2) by:

$$\varepsilon_E = \frac{dx - dx_0}{dx_0} \qquad [1.19]$$

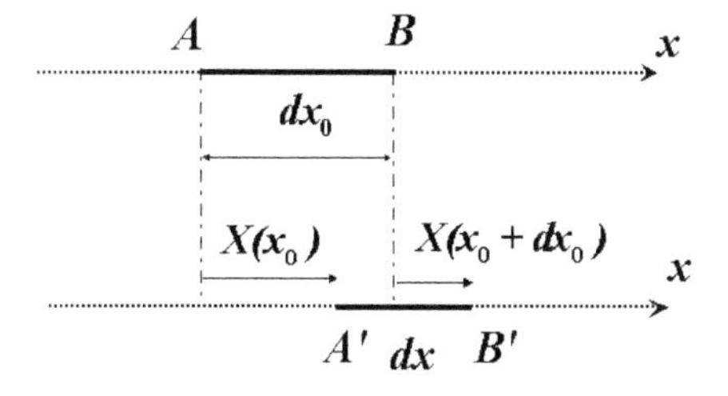

Figure 1.2. *Extension of a straight segment*

It is noticed that the definition [1.19] can also be transformed as follows:

$$\left.\begin{aligned} dx &= (\varepsilon_E + 1)\, dx_0 \\ \varepsilon_E &= \frac{(dx - dx_0)(dx + dx_0)}{dx_0(dx + dx_0)} \end{aligned}\right\} \Rightarrow \varepsilon_E = \frac{(dx^2 - dx_0^2)}{dx_0^2(2 + \varepsilon_E)} \qquad [1.20]$$

If ε_E is sufficiently small, as will be always the case in this book, the relative change of length can also be expressed by the Green–Lagrange strain denoted here ε_G:

$$\varepsilon_E \simeq \varepsilon_G = \frac{dx^2 - dx_0^2}{2dx_0^2} = \frac{\partial X}{\partial x} + \frac{1}{2}\left(\frac{\partial X}{\partial x}\right)^2 \qquad [1.21]$$

2. Motion of a rigid body, for instance a rotation, must induce no strain at all. This can be checked by rotating a straight segment of length L through a finite angle θ, see Figure 1.3. The easiest way is to use the Cauchy kinematic tensor.

   The initial configuration is defined by $A(x_0, y_0)$ and the actual one by $A'(x, y)$. The coordinates are transformed according to the formula:

$$\begin{bmatrix} x \\ y \end{bmatrix} = \begin{bmatrix} \cos\theta & -\sin\theta \\ \sin\theta & \cos\theta \end{bmatrix} \begin{bmatrix} x_0 \\ y_0 \end{bmatrix} \Rightarrow \begin{bmatrix} X \\ Y \end{bmatrix} = \begin{bmatrix} \cos\theta - 1 & -\sin\theta \\ \sin\theta & \cos\theta - 1 \end{bmatrix} \begin{bmatrix} x_0 \\ y_0 \end{bmatrix}$$

The Jacobian matrix identifies with the rotation matrix; consequently, $[C] = [J]^T[J] = [I]$ and $[\varepsilon] = [0]$, as suitable. It is also of interest to express the components of the Green–Lagrange tensor directly; which emphasises that the

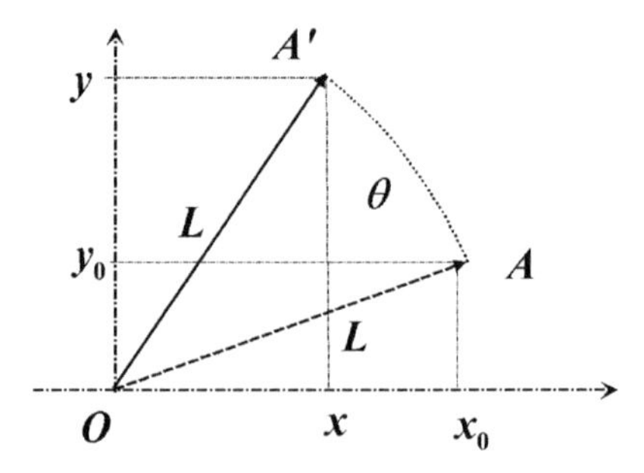

Figure 1.3. *Rotation of a straight segment*

quadratic terms of $\bar{\bar{\varepsilon}}$ must be taken into account to obtain the correct result:

$$
\begin{aligned}
\varepsilon_{xx} &= \frac{\partial X}{\partial x_0} + \frac{1}{2}\left(\left(\frac{\partial X}{\partial x_0}\right)^2 + \left(\frac{\partial Y}{\partial x_0}\right)^2\right) \\
&= \cos\theta - 1 + \frac{(\cos\theta - 1)^2 + \sin^2\theta}{2} = 0 \\
\varepsilon_{yy} &= \frac{\partial Y}{\partial y_0} + \frac{1}{2}\left(\left(\frac{\partial X}{\partial y_0}\right)^2 + \left(\frac{\partial Y}{\partial y_0}\right)^2\right) \\
&= \cos\theta - 1 + \frac{(\cos\theta - 1)^2 + \sin^2\theta}{2} = 0 \\
2\varepsilon_{xy} &= \frac{\partial X}{\partial y_0} + \frac{\partial Y}{\partial x_0} + \frac{\partial X}{\partial y_0}\frac{\partial X}{\partial x_0} + \frac{\partial Y}{\partial y_0}\frac{\partial Y}{\partial x_0} \\
&= -\sin\theta + \sin\theta + -\sin\theta(\cos\theta - 1) + \sin\theta(\cos\theta - 1) = 0
\end{aligned}
$$

Nevertheless, if the displacements and the strains are small enough, the nonlinear terms of the Green–Lagrange strain tensor can be omitted, giving rise to the so called *infinitesimal*, or *small strain tensor*:

$$
[\varepsilon] = \frac{[\mathrm{grad}[X]] + [\mathrm{grad}[X]]^T}{2} \qquad [1.22]
$$

1.2.2 *Indicial and symbolic notations*

As detailed in Appendix A.1, the indicial notation is a way to describe vectors and tensors in any orthogonal coordinate system by their generic component. For instance, the Cartesian coordinates x, y, z of the position vector $\vec{r}$ are denoted collectively as $x_i (i = 1, 2, 3)$, where the index i takes on the values 1, 2, 3. Those of $\overline{\overline{\mathrm{grad}\,\vec{r}}}$ are denoted $\partial x_i/\partial x_{0_j}$, where the indices i and j take on the values 1, 2, 3 independently from each other. Besides the advantage of conciseness, the indicial notation allows one to deal with scalar variables only, avoiding thus the need to worry about the specific operation rules appropriate for scalar, vector and tensor quantities. For instance, using the indicial notation, relation [1.7] is written as:

$$
dx_i = \frac{\partial x_i}{\partial x_{0_j}} dx_{0_j} = J_{ij}\, dx_{0_j} \qquad [1.23]
$$

where use is made of the convention of implicit summation on the repeated index j.

The index notation for the Green–Lagrange strain tensor [1.17] is:

$$\varepsilon_{ij} = \frac{1}{2}\left(\frac{\partial X_i}{\partial x_{0_j}} + \frac{\partial X_j}{\partial x_{0_i}} + \frac{\partial X_k}{\partial x_{0_j}}\frac{\partial X_k}{\partial x_{0_i}}\right) \qquad [1.24]$$

As an exercise, formula [1.24] can be derived directly, starting from:

$$x_i = x_{0_i} + X_i \Rightarrow \frac{\partial x_i}{\partial x_{0_j}} = \delta_{ij} + \frac{\partial X_i}{\partial x_{0_j}}$$

where $\delta_{ij} = 1$ if $i = j$ and zero otherwise.

From [1.14], it follows that $C_{ij} = J_{ki}J_{kj}$, further expressed as:

$$C_{ij} = \frac{\partial x_k}{\partial x_{0_i}}\frac{\partial x_k}{\partial x_{0_j}} = \left(\delta_{ki} + \frac{\partial X_k}{\partial x_{0_i}}\right)\left(\delta_{kj} + \frac{\partial X_k}{\partial x_{0_j}}\right)$$

then,

$$C_{ij} = \delta_{ij} + \frac{\partial X_k}{\partial x_{0_i}}\delta_{kj} + \frac{\partial X_k}{\partial x_{0_j}}\delta_{ki} + \frac{\partial X_k}{\partial x_{0_i}}\frac{\partial X_k}{\partial x_{0_j}}$$

and finally, in agreement with [1.17]:

$$C_{ij} - \delta_{ij} = 2\varepsilon_{ij} = \frac{\partial X_j}{\partial x_{0_i}} + \frac{\partial X_i}{\partial x_{0_j}} + \frac{\partial X_k}{\partial x_{0_i}}\frac{\partial X_k}{\partial x_{0_j}}$$

On the other hand, the small strain tensor is written as:

$$\varepsilon_{ij} = \frac{1}{2}\left(\frac{\partial X_i}{\partial x_j} + \frac{\partial X_j}{\partial x_i}\right) \qquad [1.25]$$

where the subscript ($_0$) in the position coordinates is dropped, since no difference is made between the initial and the actual configurations in the case of infinitesimal motions.

Returning finally to the symbolic notation, the intrinsic expression of the Green–Lagrange strain tensor is written as:

$$\overline{\overline{\varepsilon}} = \frac{1}{2}\left(\overline{\overline{C}} - \overline{\overline{I}}\right) = \frac{\left(\overline{\overline{\text{grad}\,\vec{r}}}\right) + \left(\overline{\overline{\text{grad}\,\vec{r}}}\right)^T + \left(\overline{\overline{\text{grad}\,\vec{r}}}\right)^T \cdot \left(\overline{\overline{\text{grad}\,\vec{r}}}\right)}{2} \qquad [1.26]$$

where the dot product stands for the contracted product, marked in indicial notation by a repeated index.

1.2.3 *Stresses*

The concept of stress extends the notion of internal restoring force in discrete systems and that of pressure, familiar in hydrostatics. The basic idea is to consider the equilibrium of the solid at the local scale of an infinitesimal element. Both the force and moment resultants acting on it must cancel out at static equilibrium. When writing down such balances, it is appropriate to distinguish between the *body forces*, acting on elements of volume of the body, such as the weight or the inertia, and the *surface forces* exerted by the 'exterior' on the boundary $(\mathcal{S})$ of the element, which are proportional to the area of $(\mathcal{S})$. The latter are often termed '*contact forces*'. Thus, if the element is infinitesimal, body forces can be neglected in comparison with contact forces, since they are less by one order of magnitude. Let consider a solid body notionally cut into two portions by a plane. It can be arbitrarily decided that the cross-sectional surface $(\mathcal{S})$ separating the two portions belongs to one of them, forming thus a facet of that portion, see Figure 1.4 in which $(\mathcal{S})$ is assumed to belong to the portion (I). Let $(d\mathcal{S})$ be an infinitesimal surface element of the facet. The contact force exerted by the portion (II), through $(d\mathcal{S})$ of area dS, is $\vec{T} = \vec{t}\,dS$, where $\vec{t}$ is the *stress vector*.

A priori, $\vec{t}$ depends both on the position and on the orientation of $(d\mathcal{S})$. The latter is defined by the unit normal vector $\vec{n}$, conventionally taken as positive when directed outward from the portion (I), as indicated in Figure 1.4. Let us consider an infinitesimal tetrahedron with three facets parallel to the three Cartesian coordinate planes (unit vectors $\vec{i}, \vec{j}, \vec{k}$). Static equilibrium implies that the force $\vec{T}$ exerted by the external medium on the oblique facet of the tetrahedron, defined by the area dS and the unit normal $\vec{n}$ (coordinates n_x, n_y, n_z), has to be balanced by all the forces $\vec{T}_x, \vec{T}_y, \vec{T}_z$ which act on the other tetrahedron facets and which are induced by the

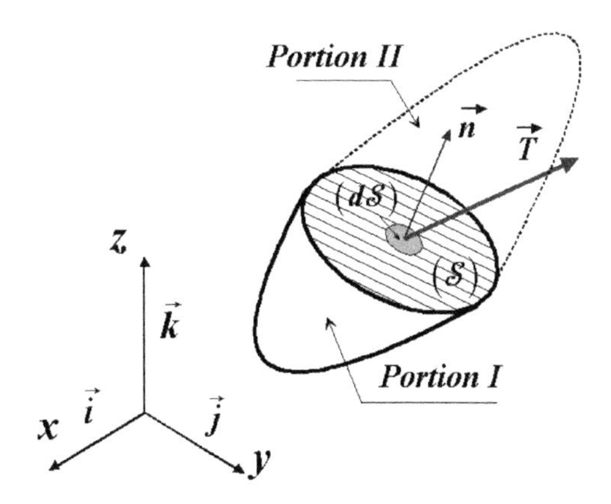

Figure 1.4. *Section of a solid by a facet*

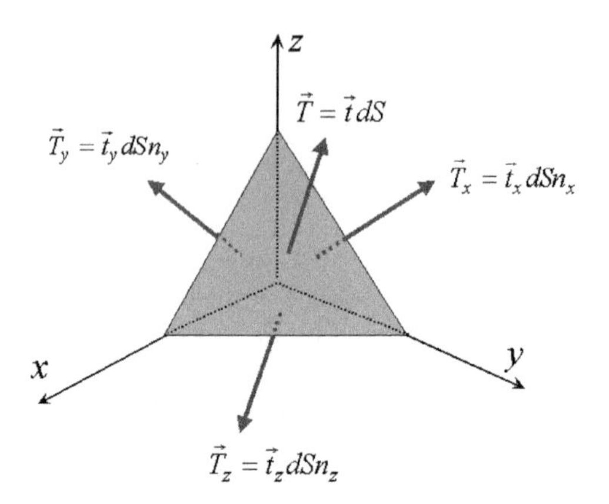

Figure 1.5. *Equilibrium of an infinitesimal tetrahedron*

stress vectors:

$$\vec{t}_x = t_{xx}\vec{i} + t_{xy}\vec{j} + t_{xz}\vec{k} : \text{ facet with normal } \vec{i} \text{ and algebraic area: } -dSn_x$$
$$\vec{t}_y = t_{yx}\vec{i} + t_{yy}\vec{j} + t_{yz}\vec{k} : \text{ facet with normal } \vec{j} \text{ and algebraic area: } -dSn_y$$
$$\vec{t}_z = t_{zx}\vec{i} + t_{zy}\vec{j} + t_{zz}\vec{k} : \text{ facet with normal } \vec{k} \text{ and algebraic area: } -dSn_z$$

The minus sign arises as a consequence of the outward orientation of the normal vectors, see Figure 1.5.

The force balance is thus written as:

$$\vec{T} + \vec{T}_x + \vec{T}_y + \vec{T}_z = (\vec{t} - (\vec{t}_x n_x + \vec{t}_y n_y + \vec{t}_z n_z))\, dS = 0$$

It follows that the Cartesian components of $\vec{t}$ verify the matrix relation:

$$\begin{bmatrix} t_x \\ t_y \\ t_z \end{bmatrix} = \begin{bmatrix} t_{xx} & t_{yx} & t_{zx} \\ t_{xy} & t_{yy} & t_{zy} \\ t_{xz} & t_{yz} & t_{zz} \end{bmatrix} \begin{bmatrix} n_x \\ n_y \\ n_z \end{bmatrix}$$

This matrix defines *the Cauchy stress tensor*, which is usually written $\overline{\overline{\sigma}}(\vec{r})$:

$$\vec{t} = \overline{\overline{\sigma \cdot}}\, \vec{n} \Leftrightarrow \begin{bmatrix} t_x \\ t_y \\ t_z \end{bmatrix} = \begin{bmatrix} \sigma_{xx} & \sigma_{yx} & \sigma_{zx} \\ \sigma_{xy} & \sigma_{yy} & \sigma_{zy} \\ \sigma_{xz} & \sigma_{yz} & \sigma_{zz} \end{bmatrix} \begin{bmatrix} n_x \\ n_y \\ n_z \end{bmatrix} \qquad [1.27]$$

The tensor nature of $\overline{\overline{\sigma}}$ is evidenced by noting that the virtual work $\delta\mathcal{W} = (\delta d)([n]^{\mathrm{T}}[\sigma][n])\, dS$, produced by a virtual displacement $(\delta d)\vec{n}$ of the facet, does

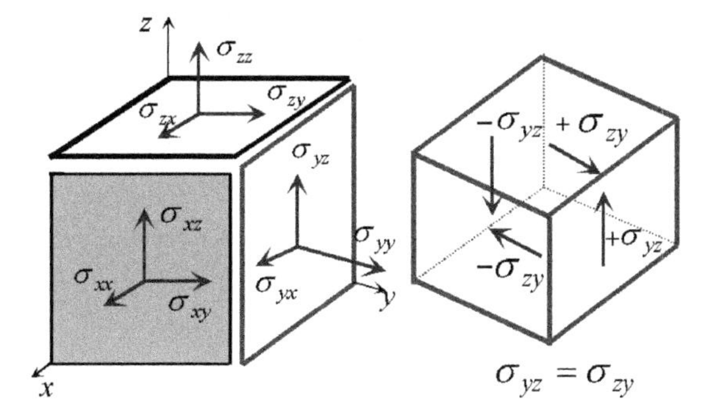

Figure 1.6. *Stress Cartesian components on the facets of a cubical element*

not depend on the coordinate system. Hence $[n]^T[\sigma][n]$ is also invariant. Using Cartesian coordinates and indicial notation, [1.27] can be written as:

$$t_j = \sigma_{ij} n_i \quad i, j = 1, 2, 3 \qquad [1.28]$$

As a definition, σ_{ij} is the j-th component of the stress force per unit area through the facet of normal unit vector n_i.

The positive Cartesian components of the stress tensor are shown in Figure 1.6 (sketch on the left-hand side). The diagonal terms are called *normal stresses*, which may be either tensile or compressive, depending whether the sign is positive or negative. The other components are termed *tangential* or *shear stresses*. Of course, all these components are local quantities, which are defined at each position $\vec{r}$ within the solid and at the boundary. So $\overline{\overline{\sigma}}(\vec{r})$ is a tensor field. The equilibrium of moments leads to the symmetry of $\overline{\overline{\sigma}}$. This very important property can be understood referring to the sketch on the right-hand side of Figure 1.6. The resulting moment of stresses about the origin of the axes must be zero. For instance, the component $d\mathcal{M}_x^{(1)} = -(\sigma_{zy}\, dx\, dy)\, dz$ is balanced by the component $d\mathcal{M}_x^{(2)} = +(\sigma_{yz}\, dx\, dz)\, dy$, consequently $\sigma_{yz} = \sigma_{zy}$, etc. Eigenvectors of the symmetric stress matrix are termed *principal stress directions*, which can be used to define an orthogonal coordinate system, and the eigenvalues are termed *principal stresses*.

1.2.4 *Equations of dynamical equilibrium*

Let us isolate mentally a finite portion of a solid medium, which occupies a volume $(\mathcal{V}(t))$, bounded by a closed surface $(\mathcal{S}(t))$. Such a portion may be viewed as a solid body. Let $\vec{X}(\vec{r};t)$ be the Lagrangian displacement field of all the material particles within $(\mathcal{V}(t))$ and $\rho(\vec{r};t)$ be the mass per unit volume. The portion of material remains in dynamical equilibrium at any time t, when subjected to inertia, stress and external forces. The latter may comprise body forces $\vec{f}^{(e)}(\vec{r};t)$ and contact forces $\vec{t}^{(e)}(\vec{r};t)$ acting respectively in $(\mathcal{V}(t))$ and on $(\mathcal{S}(t))$; that is $\vec{f}^{(e)}$ is assumed to vanish on $(\mathcal{S}(t))$ and $\vec{t}^{(e)}$ within $(\mathcal{V}(t))$. The global force balance over the body is written as:

$$\int_{(\mathcal{V}(t))} -\rho\ddot{\vec{X}}\,d\mathcal{V} + \int_{(\mathcal{S}(t))} \overline{\overline{\sigma}} \cdot \vec{n}\,d\mathcal{S} + \int_{(\mathcal{V}(t))} \vec{f}^{(e)}(\vec{r};t)\,d\mathcal{V} = 0 \quad [1.29]$$

$$\overline{\overline{\sigma}}(\vec{r}) \cdot \vec{n}\,(\vec{r}) - \vec{t}^{(r)}(\vec{r};t) = \vec{t}^{(e)}(\vec{r};t) \quad \forall \vec{r} \in (\mathcal{S}(t)) \quad [1.30]$$

The contact forces $\vec{t}$ applied to $(\mathcal{S}(t))$ can stand for external loads $\vec{t}^{(e)}(\vec{r};t)$ and/or for reaction forces $\vec{t}^{(r)}(\vec{r};t)$ induced by some support conditions, or prescribed displacements.

Equation [1.29] stands for the balance of the internal and external forces acting on the body itself and equation [1.30] for that of the internal and external forces acting on the boundary. Further, in [1.29] the surface integral can be suitably transformed into a volume integral by using the divergence theorem (cf. Appendix A.2, formula [A.2.5]). The internal terms are then collected on the left-hand side and the external forces on the right-hand side of the equations. [1.29] is thus transformed into:

$$\int_{(\mathcal{V}(t))} (\rho\,\ddot{\vec{X}} - \operatorname{div}\overline{\overline{\sigma}})\,d\mathcal{V} = \int_{(\mathcal{V}(t))} \vec{f}^{(e)} d\mathcal{V} \quad [1.31]$$

As $(\mathcal{V}(t))$ can be chosen arbitrarily, the 'global' equilibrium is equivalent to the 'local' equilibrium defined by the two local equations:

$$\begin{aligned} \rho\ddot{\vec{X}} - \operatorname{div}\overline{\overline{\sigma}} &= \vec{f}^{(e)}(\vec{r};t); \quad \forall \vec{r} \in (\mathcal{V}(t)) \\ \overline{\overline{\sigma}}(\vec{r}) \cdot \vec{n}\,(\vec{r}) &= \vec{t}(\vec{r};t); \quad \forall \vec{r} \in (\mathcal{S}(t)) \end{aligned} \quad [1.32]$$

The first equation of this system asserts that any point of the volume $(\mathcal{V}(t))$ is in dynamical equilibrium, and the second gives the boundary conditions which must be fulfilled at each point of $(\mathcal{S}(t))$. As a general case, they may include support reactions, external contact forces $\vec{t}^{(e)}$ and prescribed motions $\vec{D}(\vec{r};t)$, as sketched in Figure 1.7. The latter can be interpreted as a particular type of external contact forces, not given explicitly but expressible in terms of constraint reactions (cf. [AXI04], Chapter 4). As a consequence, it is illegal to assign the values of a prescribed motion and of an external contact force at the same position; that is $\vec{t}^{(e)}$ and $\vec{D}$ must be specified on two complementary parts of $(\mathcal{S})$, denoted $(\mathcal{S}_1)$ and $(\mathcal{S}_2) = (\mathcal{S}) \cap (\mathcal{S}_1)$ respectively, as already mentioned in the introduction. On the other hand, a boundary condition applied to a subdomain $(\mathcal{S}_3)$ is said to be *homogeneous* if it does not prescribe any external loading (neither

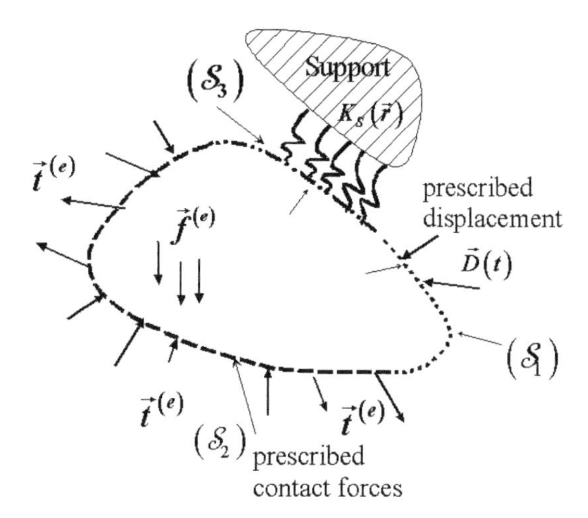

Figure 1.7. *Solid with homogeneous and inhomogeneous boundary conditions*

prescribed forces nor motions); otherwise, the boundary condition is said to be *inhomogeneous*.

As the theory is restricted here to conservative systems, we will consider only elastic supports. In linear elasticity, they are represented by a homogeneous relation of the type:

$$\overline{\overline{\sigma}}(\vec{r}) \cdot \vec{n}(\vec{r}) - K_S[\vec{X}] = 0 \quad \forall \vec{r} \in (\mathcal{S}_3) \qquad [1.33]$$

K_S is a linear stiffness operator defined $\forall \vec{r} \subset (\mathcal{S})$, which can be a scalar (stiffness coefficient of springs), or a differential operator (stiffness operator of another solid used as a supporting device). Furthermore, the relation [1.33] can also describe the condition for a given degree of freedom to be free or, alternatively, to be fixed, by connecting it to a spring and letting the stiffness coefficient tend either to zero, or to infinity.

It is appropriate to conclude the present subsection by writing down the equations of dynamical equilibrium in the following general form which includes the particular cases discussed just above:

$$\begin{aligned} \rho\ddot{\vec{X}} - \operatorname{div}\overline{\overline{\sigma}} + \vec{f}^{(i)} &= \vec{f}^{(e)}(\vec{r};t); \quad \forall \vec{r} \in (\mathcal{V}(t)) \\ \vec{X}(\vec{r};t) &= \vec{D}(\vec{r};t); \quad \forall \vec{r} \subset (\mathcal{S}_1) \\ \overline{\overline{\sigma}}(\vec{r}) \cdot \vec{n}(\vec{r}) - K_S[\vec{X}] &= \vec{t}^{(e)}(\vec{r};t); \quad \forall \vec{r} \subset (\mathcal{S}_2) \end{aligned} \qquad [1.34]$$

where

$$(\mathcal{S}_2) = (\mathcal{S}) \cap (\mathcal{S}_1)$$

In the system [1.34], an additional body force density $\vec{f}^{(i)}$ is included, which depends on the problem studied. For instance, $\vec{f}^{(i)}$ can stand for a component such as a gravity force $\rho\vec{g}$, where $\vec{g}$ is the acceleration vector of gravity, or for centrifugal and Coriolis forces. On the other hand, if K_S does not vanish on $(\mathcal{S}_1)$ the constraint force induced by the prescribed displacement depends on K_S. Finally, the system of equations [1.34] is said to be *mixed*, because it is formulated partly in terms of kinematical variables $\vec{X}, \ddot{\vec{X}}$, and partly in terms of forces. The formulation involves three scalar equations and nine unknowns, namely the six stress and the three displacement components. The system is thus underdetermined, except if use can be made of six additional independent relationships, which are given by the material laws. This is precisely the case if the material behaves elastically, as detailed in the next subsection.

1.2.5 *Stress–strain relationships for an isotropic elastic material*

This study is restricted here to materials in which the stresses depend on strain components $\overline{\overline{\sigma}} = B(\overline{\overline{\varepsilon}})$ solely. The simplest law of this kind is the generalized Hooke's law, which defines the so called *linear elastic* material law. It is written both in symbolic and indicial notations as:

$$\overline{\overline{\overline{\overline{h}}}} : \overline{\overline{\varepsilon}} = \overline{\overline{\sigma}} \Longleftrightarrow \sigma_{ij} = h_{ijkl}\varepsilon_{kl} \quad i,j,k,l = 1,2,3 \qquad [1.35]$$

where $\overline{\overline{\varepsilon}}$ is the small strain tensor and $\overline{\overline{\overline{\overline{h}}}}$ is the Hooke *elasticity tensor*. The symbol (:) indicates that the product is contracted twice, as evidenced in the index notation by the presence of the repeated indices k and l.

In a 1D medium, Hooke's law states that the stress is proportional to the strain, the coefficient of proportionality being the elasticity constant. A priori, in a 3D medium a law of the type [1.35] would lead to the definition of $3^4 = 81$ elastic constants for describing the elastic properties of the material. However, since $\overline{\overline{\varepsilon}}$ and $\overline{\overline{\sigma}}$ are both symmetric, $\overline{\overline{\overline{\overline{h}}}}$ is symmetric with respect to i, j and k, l respectively, which leads to at most $9 \times 4 = 36$ independent elasticity constants. Furthermore, if the material is isotropic, the number of independent elasticity constants reduces to only two, which are defined either as the Lamé parameters, λ and μ (μ is the shear modulus, often denoted G in structural engineering), or as the Young's modulus E and Poisson ratio ν. The relations between these parameters are:

$$\lambda = \frac{\nu E}{(1+\nu)(1-2\nu)}; \quad \mu = G = \frac{E}{2(1+\nu)} \qquad [1.36]$$

In Appendix A.6, numerical values are given for a few materials of common use. Hooke's law for an isotropic material can be written as:

$$\overline{\overline{\sigma}} = \lambda \operatorname{Tr}\left(\overline{\overline{\varepsilon}}\right) \overline{\overline{I}} + 2G\overline{\overline{\varepsilon}} = (\lambda \operatorname{div} \vec{X})\overline{\overline{I}} + 2G\overline{\overline{\varepsilon}} \qquad [1.37]$$

$Tr\left(\overline{\overline{\varepsilon}}\right) = Tr[\varepsilon] = \varepsilon_{ii}$ is the trace of the matrix $[\varepsilon]$; equal to the divergence of the displacement vector, it measures the relative variation of volume induced by the strains, see Appendix A.2, formula [A.2.3]. Inversion of [1.37] results in:

$$\overline{\overline{\varepsilon}} = \frac{1+\nu}{E}\overline{\overline{\sigma}} - \frac{\nu}{E}\operatorname{Tr}\left(\overline{\overline{\sigma}}\right)\overline{\overline{I}} \qquad [1.38]$$

Relation [1.38] is derived by transforming first [1.37] into:

$$\overline{\overline{\varepsilon}} = \frac{\overline{\overline{\sigma}}}{2G} - \frac{\lambda}{2G}\operatorname{Tr}\left(\overline{\overline{\varepsilon}}\right)\overline{\overline{I}} \Rightarrow \operatorname{Tr}\left(\overline{\overline{\varepsilon}}\right) = \frac{1}{2G}\operatorname{Tr}\left(\overline{\overline{\sigma}}\right) - \frac{3\lambda}{2G}\operatorname{Tr}\left(\overline{\overline{\varepsilon}}\right)$$

$$\Rightarrow \operatorname{Tr}\left(\overline{\overline{\varepsilon}}\right) = \frac{1}{2G+3\lambda}\operatorname{Tr}\left(\overline{\overline{\sigma}}\right)$$

then,

$$\overline{\overline{\varepsilon}} = \frac{\overline{\overline{\sigma}}}{2G} - \frac{\lambda}{2G(2G+3\lambda)}\operatorname{Tr}\left(\overline{\overline{\sigma}}\right)\overline{\overline{I}}$$

To write down the final result [1.38], use is made of the relations [1.36] between λ, G and E, ν.

1.2.6 *Equations of elastic vibrations (Navier's equations)*

In linear elasticity, the equations of dynamical equilibrium are often called *vibration equations* since they describe small oscillations of the elastic material in the neighbourhood of a permanent and stable state of equilibrium, chosen as the configuration of reference. Here, the latter is chosen as a static and stable state of equilibrium, in which the stresses and strains are identically zero. These equations are expressed in terms of displacements. Using Hooke's law [1.37], and the small strain tensor [1.26], the equations of dynamical equilibrium [1.34] are written as:

$$\rho\ddot{\vec{X}} - \left[G\,\Delta\vec{X} + (\lambda + G)\overrightarrow{\mathrm{grad}(\mathrm{div}\,\vec{X})}\right] = \vec{f}^{\,e)}(\vec{r};t) \quad \forall\vec{r} \in (\mathcal{V})$$

$$\lambda\,\mathrm{div}\,\vec{X}\bar{\bar{I}}\cdot\vec{n} + G\left(\overline{\overline{\mathrm{grad}\,\vec{X}}} + \left(\overline{\overline{\mathrm{grad}\,\vec{X}}}\right)^T\right)\cdot\vec{n} - K_S[\vec{X}] = \vec{t}^{\,e)}(\vec{r};t) \quad \forall\vec{r} \in (\mathcal{S})$$

$$(\mathcal{V}(t)) \equiv (\mathcal{V}(0)) = (\mathcal{V}); \quad (\mathcal{S}(t)) \equiv (\mathcal{S}(0)) = (\mathcal{S}) \qquad [1.39]$$

Though vectors may be considered as being merely tensors of first rank, it is preferred to mark the gradient of a scalar quantity by an upper arrow instead of a double bar in order to stress that the result is a vector. The first equation governs the local equilibrium at time t of a material particle located at $\vec{r}$ and the second equation stands for elastic boundary conditions. No prescribed motion has been assumed, as it would bring nothing new to the formalism, at least at this step. Finally, the system [1.39], taken as a whole, is said to be homogeneous if no external loading of any kind is applied either to $(\mathcal{V})$, or to $(\mathcal{S})$, even as non-zero initial conditions. Otherwise, it is said to be inhomogeneous.

1.3. Hamilton's principle

Hamilton's principle has already been introduced and extensively used in [AXI 04] for deriving the Lagrange equations of discrete systems. It is recalled that this variational principle is expressed analytically as:

$$\delta[\mathcal{A}(t_1, t_2)] = \delta\left[\int_{t_1}^{t_2} \mathcal{L}\,dt\right] = 0$$

where $\delta[\,]$ denotes the operator of variation. $\mathcal{A}(t_1, t_2)$ is the action between two arbitrary times t_1 and t_2 of the extended Lagrangian $\mathcal{L}$, defined as:

$$\mathcal{L} = \mathcal{E}_\kappa - \mathcal{E}_p + \mathcal{W}_Q$$

$\mathcal{E}_\kappa([q], [\dot{q}])$ denotes the kinetic energy of the system, $\mathcal{E}_p([q])$ the internal potential energy, expressed in terms of the generalized displacements and velocity vectors $[q]$ and $[\dot{q}]$. $\mathcal{W}_Q$ is the work function of extra external or/and internal generalized force vectors $[Q]$ applied to the system, which are not necessarily conservative. The dimension of all the vectors just mentioned is equal to the number *ND* of the degrees of freedom (DOF) of the system. Here, Hamilton's principle will be extended to continuous media, providing us with a very efficient analytical tool for dealing with:

1. the kinematical constraints,
2. the boundary conditions,
3. various numerical methods for obtaining approximate solutions of the differential equations of static and dynamic equilibrium.

1.3.1 *General presentation of the formalism*

The spatial domain occupied by the body and its boundary are still denoted by $(\mathcal{V})$ and $(\mathcal{S})$ respectively, though, depending on the dimension of the Euclidean space considered, $(\mathcal{V})$ may be either a volume, or a surface, or a line; accordingly $(\mathcal{S})$ may be either a surface, a line, or a finite set of points. To formulate Hamilton's principle in a continuous medium, one proceeds according to the following steps:

1. A continuous displacement field and its associated strain tensor is suitably defined in $(\mathcal{V})$. The components of the displacement field are functions of space and time. They can be defined by using the coordinate system which is the most suitable in relation to the geometry of $(\mathcal{V})$. The displacement field $\vec{X}(\vec{r};t)$, which is a continuous vector function of space, extends the independent generalized displacements used in the discrete systems to the continuous case.
2. The Lagrangian $\mathcal{L}$ is defined again as the difference between the kinetic energy $\mathcal{E}_\kappa$ and the internal potential energy $\mathcal{E}_p$, plus the work $\mathcal{W}$ of extra external or internal forces which are eventually applied within $(\mathcal{V})$ and/or on $(\mathcal{S})$. Calculation involves a spatial integration over $(\mathcal{V})$ and/or $(\mathcal{S})$ of the corresponding energy and work densities, denoted e_κ, e_p and w respectively. Fortunately, the actual calculation of $\mathcal{W}$ can be avoided. Indeed, because of the variational nature of Hamilton's principle, only the virtual variation $\delta[\mathcal{W}]$ is needed. $\delta[\mathcal{W}]$ is far more easily expressed than $\mathcal{W}$ itself, when one has to deal with internal forces which are neither inertial nor potential in nature.
3. Hamilton's principle is applied, according to which the action integral of the Lagrangian between two arbitrary times t_1 and t_2 is stationary with respect to any admissible variation $\delta[\vec{X}]$. To be admissible $\delta[\vec{X}]$ must comply with the boundary conditions of the problem and must vanish at t_1 and t_2. In most cases, integrations by parts are needed to formulate such variations explicitly in terms of the components of the displacement field. Boundary terms arising from the spatial integrals contribute to define the boundary conditions of the problem. At this final step, the equilibrium equations are obtained in terms of generalized forces. Obviously, they are necessarily identical to the equilibrium equations which would result from the Newtonian approach.
4. As in the case of discrete systems, the kinematical constraints which can be eventually prescribed on the body may be conveniently introduced by using Lagrange's multipliers (cf. [AXI 04], Chapter 4).

5. Finally, by specifying the stress-strain relationships, the equilibrium equations can be expressed in terms of displacement variables only.

Summarizing briefly the above procedure, it can be said that the major differences between the mechanics of discrete and continuous systems originate from the replacement of a countable set of independent displacement variables by a continuous displacement field, which is a function of the position vector $\vec{r}$ in $(\mathcal{V})$, and from the emergence of boundary conditions, which specify the contact forces and/or the kinematical conditions prescribed on the boundary $(\mathcal{S})$.

1.3.2 *Application to a three-dimensional solid*

For the sake of simplicity, we consider here a 3D solid with either free or fixed boundary conditions, though extension to more general elastic conditions would not lead to particular difficulties.

1.3.2.1 Hamilton's principle

According to the considerations of the preceding subsection, the Lagrangian of 3D bodies is written as:

$$\mathcal{L} = \int_{\mathcal{V}(t)} (e_k - e_s + w_F)\, d\mathcal{V} + \int_{\mathcal{S}(t)} w_T\, d\mathcal{S} \qquad [1.40]$$

where e_κ is the kinetic and e_s the strain energy densities per unit volume. Here, w_F stands for the work density of an external force field acting within the volume of the body and w_T stands for the work density of an external force field acting on the boundary. In what follows, the problem is restricted to the linear domain. Accordingly, the magnitude of the displacements is infinitesimal, and the differences between the initial and the deformed geometries are negligible. Thus, Hamilton's principle takes the form:

$$\delta[\mathcal{A}] = \int_{t_1}^{t_2} \left\{ \int_{(\mathcal{V})} (\delta[e_k] - \delta[e_s] + \delta[w_F])\, d\mathcal{V} + \int_{(\mathcal{S})} \delta[W_T]\, d\mathcal{S} \right\} dt = 0 \qquad [1.41]$$

where $(\mathcal{V})$ and $(\mathcal{S})$ are time independent.

1.3.2.2 Hilbert functional vector space

The displacement field is a vector of the three-dimensional Euclidean space. Using a Cartesian coordinate system, it is written in symbolic notation as:

$$\vec{X}(\vec{r};t) = X(x,y,z;t)\vec{i} + Y(x,y,z;t)\vec{j} + Z(x,y,z;t)\vec{k} \qquad [1.42]$$

As X, Y, Z are real functions of the Cartesian components x, y, z of $\vec{r}$, $\vec{X}$ belongs also to a *functional vector space* provided with the functional scalar product:

$$\langle U, V \rangle_{(\mathcal{V})} = \int_{(\mathcal{V})} \vec{U} \cdot \vec{V} \, d\mathcal{V} = \int_{(\mathcal{V})} (U_x V_x + U_y V_y + U_z V_z) \, d\mathcal{V} \qquad [1.43]$$

where $\vec{U} \cdot \vec{V}$ is the usual notation for the scalar product in the Euclidean space and $\langle U, V \rangle_{(\mathcal{V})}$ is the notation for the scalar product in the functional space.

In contrast to the case of discrete systems, the dimension of the functional vector space is infinite, and even uncountable. Here, it will be asserted, without performing the mathematical proof, that it is complete, which means that any Cauchy sequence of functional vectors is convergent to a functional vector within the same space. This space is thus an *Hilbert space*. The definition holds independently from the dimension of the Euclidean space in which $(\mathcal{V})$ is embedded. The reader interested in a more formal and detailed presentation of the functional vector spaces is referred to the specialized literature, for instance [STA 70].

1.3.2.3 Variation of the kinetic energy

The density of kinetic energy is defined as the kinetic energy per unit volume of a fictitious material particle of infinitesimal mass $\rho \, d\mathcal{V}$:

$$e_\kappa(\vec{r};t) = \frac{1}{2}\rho \left(\dot{\vec{X}}(\vec{r};t) \cdot \dot{\vec{X}}(\vec{r};t) \right) \qquad [1.44]$$

The total kinetic energy is:

$$\mathcal{E}_\kappa(t) = \int_{(\mathcal{V})} e_\kappa \, d\mathcal{V} = \frac{1}{2} \left\langle \dot{\vec{X}}(\vec{r};t), \rho \dot{\vec{X}}(\vec{r};t) \right\rangle_{(\mathcal{V})} \qquad [1.45]$$

As in the case of discrete systems, [1.45] is a quadratic form of the velocity field, which is symmetric and positive definite. Its variation is:

$$\delta[\mathcal{E}_\kappa(t)] = \left\langle \dot{\vec{X}}(\vec{r};t), \rho \delta \dot{\vec{X}}(\vec{r};t) \right\rangle_{(\mathcal{V})} \qquad [1.46]$$

1.3.2.4 Variation of the strain energy

To obtain an explicit formulation of the strain energy, a *material law* describing the mechanical behaviour of the material must be specified first. Nevertheless, if the problem is limited to infinitesimal strain variations, it is always possible to write

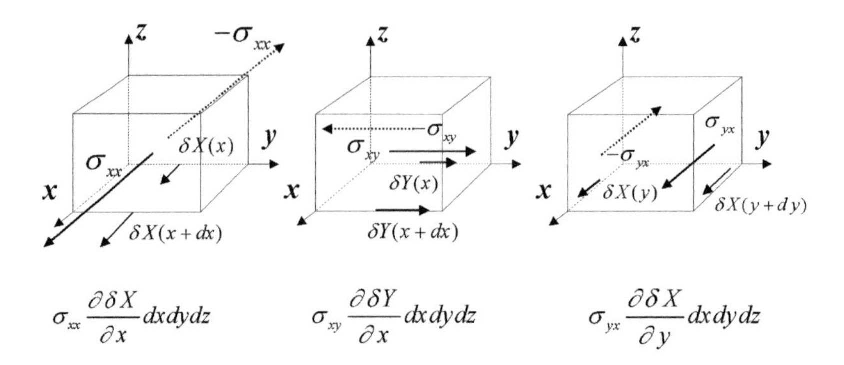

Figure 1.8. *Virtual work of the stresses*

the variation of the strain energy as the contracted tensor product:

$$\delta\left[e_s\right] = \overline{\overline{\sigma}} : \delta\left[\overline{\overline{\varepsilon}}\right] = \sigma_{ij}\delta\left[\varepsilon_{ij}\right] = \sigma_{ij}\delta\left[\frac{\partial X_i}{\partial x_j}\right] = \sigma_{ij}\frac{\partial \delta X_i}{\partial x_j} \qquad [1.47]$$

This result can be found by summing the virtual works induced by each stress component and a related virtual displacement field, as applied to a cubical element, subjected to contact forces. As indicated in Figure 1.8, it is found that:

$$\sigma_{xx}(\delta X(x+dx) - \delta X(x))\,dy\,dz = \sigma_{xx}\frac{\partial \delta X}{\partial x}\,dx\,dy\,dz$$
$$\sigma_{xy}(\delta Y(x+dx) - \delta Y(x))\,dy\,dz = \sigma_{xy}\frac{\partial \delta Y}{\partial x}\,dx\,dy\,dz$$
$$\sigma_{yx}(\delta X(y+dy) - \delta X(y))\,dx\,dz = \sigma_{yx}\frac{\partial \delta X}{\partial y}\,dx\,dy\,dz$$

The calculation rule is the same for the other components. Gathering together all the partial results in a suitable way, the Cartesian form of [1.47] is readily obtained. Furthermore, as a mere consequence of the tensor character of [1.47], the result holds in any coordinate system. On the other hand, it is also worthy of mention that if e_s is a differentiable function of $\overline{\overline{\varepsilon}}$, as in the case of elasticity, $\overline{\overline{\sigma}}$ can be calculated by using the following formula:

$$\delta\left[e_s\right] = \frac{\partial e_s}{\partial \varepsilon_{ij}}\delta\left[\varepsilon_{ij}\right] = \frac{\partial e_s}{\partial \overline{\overline{\varepsilon}}} : \delta\left[\overline{\overline{\varepsilon}}\right] \Rightarrow \overline{\overline{\sigma}} = \frac{\partial e_s}{\partial \overline{\overline{\varepsilon}}} \qquad [1.48]$$

1.3.2.5 Variation of the external load work

Writing the variation of the external work is immediate, leading to the following volume and surface integrals:

$$\int_{(\mathcal{V})} \delta[w_F]\, d\mathcal{V} = \langle \vec{f}^{(e)}, \delta\vec{X}\rangle_{(\mathcal{V})}; \qquad \int_{(\mathcal{S})} \delta\,[w_T]\, d\mathcal{S} = \langle \vec{t}^{(e)}, \delta\vec{X}\rangle_{(\mathcal{S})} \qquad [1.49]$$

1.3.2.6 Equilibrium equations and boundary conditions

By substituting the relations [1.46] to [1.49] into [1.41], Hamilton's principle can be written in indicial notation as:

$$\int_{t_1}^{t_2}\int_{(\mathcal{V})} \left(\rho\dot{X}_i\delta\dot{X}_i - \sigma_{ij}\delta\left(\frac{\partial X_i}{\partial x_j}\right) + f_i^{(e)}\,\delta\, X_i\right) d\mathcal{V} + \int_{t_1}^{t_2}\int_{(\mathcal{S})} t_i^{(e)}\,\delta\, X_i d\mathcal{S} = 0$$
[1.50]

Further, if the internal terms are integrated by parts, the first with respect to time and the second with respect to space, all the variations can be expressed in terms of δX_i exclusively. Gathering together the volume terms in one integral and the surface terms in another one, the equation [1.50] is thus transformed into:

$$\int_{t_1}^{t_2}\int_{(\mathcal{V})} \left(-\rho\ddot{X}_i + \frac{\partial\sigma_{ij}}{\partial x_j} + f_i^{(e)}\right)\delta X_i\, d\mathcal{V} + \int_{t_1}^{t_2}\int_{(\mathcal{S})} \left(-\sigma_{ij}\, n_j + t_i^{(e)}\right)\delta X_i\, d\mathcal{S} = 0$$
[1.51]

Since the δX_i are arbitrary (but admissible) and independent variations, the system [1.51] is satisfied if the two kernels vanish. The kernel within the brackets of the volume integral produces the equations of dynamical equilibrium of the body, whereas the kernel of the surface integral provides the boundary conditions. Of course, it is immediately apparent that such equations are identical to those already established in section 1.2 (cf. system [1.32]). Moreover, if the boundary, or a part $(\mathcal{S}_k)$ of it is free (admissible $\delta X_i \neq 0$) the disappearance of the kernel of the surface integral leads to the disappearance of the stresses on $(\mathcal{S}_k)$. On the contrary, if the displacement is constrained by the condition $X_i = 0$ on $(\mathcal{S}_k)$, a Lagrange multiplier Λ_i is associated with the locking condition and the surface integral becomes:

$$\int_{(\mathcal{S}_k)} (\Lambda_i - \sigma_{ij}\, n_j)\delta X_i\, d\mathcal{S} \quad \text{and} \quad \delta X_i \neq 0 \qquad [1.52]$$

Letting the integral [1.52] vanish produces the reaction force at the fixed boundary:

$$\Lambda_i = \sigma_{ij}(\vec{r})n_j \quad \forall\vec{r} \in (\mathcal{S}_k) \qquad [1.53]$$

1.3.2.7 Stress tensor and Lagrange's multipliers

A comment is in order here concerning the relation between stresses and Lagrange's multipliers in a constrained medium. Indeed, rigidity of a solid may be understood as a particular material law expressed analytically by the vanishing of the strain tensor. Turning now to the problem of determining the stress-strain relationship associated with the law $\overline{\overline{\varepsilon}} \equiv 0$, the relation $\varepsilon_{ij} = 0$ is interpreted as a holonomic constraint with which a Lagrange multiplier Λ_{ij} is associated. Thus, the stress tensor describes the internal reactions of the rigid body to an external loading. The simplest way to prove this important result is to consider the static equilibrium of a rigid body loaded by contact forces only (body forces could be included but are not necessary). The constrained Lagrangian is:

$$\mathcal{L}' = \int_{(\mathcal{V})} (-e_s + \Lambda_{ij}\,\varepsilon_{ij})\,d\mathcal{V} + \int_{(\mathcal{S})} w_T\,d\mathcal{S} \qquad [1.54]$$

Here $e_s = 0$, as there are no strains. Λ_{ij} denotes the generic component of the Lagrange multipliers tensor.

In the same way as in discrete systems, the Lagrange equations are obtained from [1.54] by equating to zero the variations of $\mathcal{L}'$ with respect to the independent functions X_i, which are assumed to be free in the variation process. Using [1.47], the variation of [1.54] is written as:

$$\delta[\mathcal{L}'] = \int_{(\mathcal{V})} (\Lambda_{ij})\delta\,\varepsilon_{ij}\,d\mathcal{V} + \int_{(\mathcal{S})} \delta w_T\,d\mathcal{S} = 0 \qquad [1.55]$$

This form is suitable to identify the Lagrange multipliers tensor as a stress tensor. Therefore, the stress tensor is found to describe the internal efforts via a strain-stress relationship, even if the body is supposed to be rigid. In this limit case, the strain-stress relationship reduces to the condition of vanishing strains and the stresses arise as the reactions of a rigid body to external loading. This point of view is useful, at least conceptually, to define stress components under rigidifying assumptions, for instance the pressure in an incompressible fluid (constraint condition $div(\vec{X}) = 0$), as further detailed in the following example.

EXAMPLE. – *Water column enclosed in a rigid tube*

As shown in Figure 1.9, a rigid tube at rest contains a column of liquid. The fluid is subjected to a normal load $T^{(e)}$ which is applied through a rigid waterproof piston. We are interested in determining the pressure field in the fluid. Obviously, the condition of local and/or global static equilibrium leads immediately to a uniform pressure $P = -T^{(e)}/S$ where S is the tube cross-sectional area (section normal to the piston axis). This result is clearly independent of the material law of the

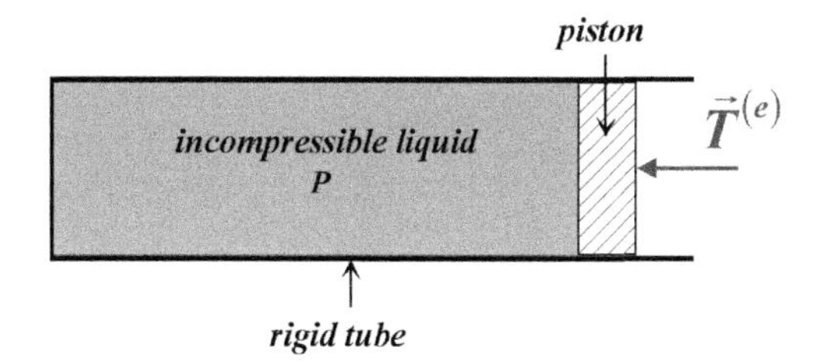

Figure 1.9. *Column of liquid compressed in a rigid tube*

fluid. However, we want to define the pressure in a logical manner starting from the material behaviour of the fluid, which is supposed here to be incompressible. Let us assume that the problem is one-dimensional, as reasonably expected. The law of incompressibility reduces to $\partial X/\partial x = 0$ and the pressure is given by the Lagrange multiplier associated with this condition. The variation of the constrained Lagrangian is:

$$\delta \mathcal{L}' = S \int_0^L \Lambda \frac{\partial(\delta X)}{\partial x} dx + T^{(e)} \delta\, X(L) = 0$$

After integrating by parts,

$$\delta \mathcal{L}' = -S \int_0^L \frac{\partial \Lambda}{\partial x} \delta X\, dx + [\Lambda S \delta X]_0^L + T^{(e)} \delta X(L) = 0$$

where δX is arbitrary, but admissible. Accordingly, at the bottom of the fixed and rigid tube, $\delta X(0) = 0$ and the expected result is obtained:

$$\Lambda = \frac{-T^{(e)}}{S} = P; \quad \frac{\partial \Lambda}{\partial x} = \frac{\partial P}{\partial x} = 0$$

Pressure is positive if $T^{(e)}$ is negative, as suitable.

1.3.2.8 Variation of the elastic strain energy

In the preceding subsections the material law has not yet been specified, except in the limit case of rigidity. It is now particularized to the case of linear elasticity. The virtual variation of the elastic energy density per unit volume is expressed as:

$$\begin{aligned} \delta\, e_e &= \sigma_{ij} \delta \varepsilon_{ij} = h_{ijk\ell}\, \varepsilon_{k\ell} \delta\, \varepsilon_{ij} = h_{ijk\ell}\, \varepsilon_{ij} \delta\, \varepsilon_{k\ell} = h_{ijk\ell} \delta \left[\tfrac{1}{2} \varepsilon_{ij} \varepsilon_{k\ell} \right] \\ &= \tfrac{1}{2} \delta [\varepsilon_{ij} h_{ijk\ell} \varepsilon_{k\ell}] \end{aligned} \qquad [1.56]$$

or in symbolic notation:

$$\delta\, e_e = \tfrac{1}{2}\delta\left[\bar{\bar{\varepsilon}} : \bar{\bar{h}} : \bar{\bar{\varepsilon}}\right]$$

Thus, the elastic energy density is found to be:

$$e_e = \frac{\varepsilon_{ij} h_{ijk\ell}\varepsilon_{k\ell}}{2} = \frac{\sigma_{k\ell}\varepsilon_{k\ell}}{2}$$

or in symbolic notation:

$$e_e = \frac{\bar{\bar{\varepsilon}} : \bar{\bar{h}} : \bar{\bar{\varepsilon}}}{2} = \frac{1}{2}\bar{\bar{\sigma}} : \bar{\bar{\varepsilon}} \qquad [1.57]$$

The result [1.57], known as the Clapeyron formula, shows that e_e is a quadratic form of strains, symmetric and positive. For an isotropic medium e_e can be expressed as:

$$e_e = \tfrac{1}{2}(\lambda(\varepsilon_{ii})^2 + 2\,G(\varepsilon_{ij}\varepsilon_{ij}))$$

or in symbolic notation:

$$e_e = \tfrac{1}{2}\left(\lambda\left(\mathrm{Tr}\left[\bar{\bar{\varepsilon}}\right]\right)^2 + 2\,G\left(\bar{\bar{\varepsilon}} : \bar{\bar{\varepsilon}}\right)\right) \qquad [1.58]$$

Then, using the infinitesimal strain tensor [1.25], e_e can be further written in terms of the displacement field as the quadratic form:

$$\begin{aligned} e_e &= \frac{1}{2}\left(\lambda\left(\frac{\partial X_i}{\partial x_i}\right)^2 + \frac{G}{2}\left(\frac{\partial X_i}{\partial x_j} + \frac{\partial X_j}{\partial x_i}\right)^2\right) \\ &= \frac{1}{2}\left(\lambda\left(\frac{\partial X_i}{\partial x_i}\right)^2 + G\left(\left(\frac{\partial X_i}{\partial x_j}\right)^2 + \frac{\partial X_i}{\partial x_j}\frac{\partial X_j}{\partial x_i}\right)\right) \end{aligned} \qquad [1.59]$$

or in symbolic notation:

$$e_e = \frac{1}{2}\left(\lambda(\mathrm{div}\vec{X})^2 + G\left(\overline{\overline{\mathrm{grad}\,\vec{X}}} : \overline{\overline{\mathrm{grad}\,\vec{X}}} + \overline{\overline{\mathrm{grad}\,\vec{X}}} : \left(\overline{\overline{\mathrm{grad}\,\vec{X}}}\right)^T\right)\right)$$

which is symmetric and positive, or eventually null if displacements of rigid body are included.

1.3.2.9 Equation of elastic vibrations

The material is supposed to be isotropic and linear elastic. The external loads are either contact forces $\vec{t}^{(e)}$ or/and body forces $\vec{f}^{(e)}$. It is recalled that in order to avoid redundancy in the boundary loading by contact and body forces, the latter are assumed to vanish at the boundary. As the contribution of the external loading to the equilibrium equations gives rise to no difficulty, we concentrate here on the variation of internal terms. Retaining the inertial and elastic terms solely, Hamilton's principle is written in indicial notation as:

$$\int_{t_1}^{t_2} dt \int_{(\mathcal{V})} \left(\rho \dot{X}_i \delta \dot{X}_i - \lambda \frac{\partial X_j}{\partial x_j} \delta \frac{\partial X_i}{\partial x_i} - G \frac{\partial \delta X_i}{\partial x_j} \left(\frac{\partial X_i}{\partial x_j} + \frac{\partial X_j}{\partial x_i} \right) \right) d\mathcal{V} = 0$$

One integration by parts of the first term with respect to time gives the inertia force density $-\rho \ddot{X}_i$ per unit volume. A spatial integration by parts of the other terms leads to the elastic force density per unit volume and to boundary terms which are suitable to specify the boundary conditions.

$$\frac{\partial X_j}{\partial x_j} \frac{\partial \delta X_i}{\partial x_i} \Rightarrow \int_{(\mathcal{V})} \frac{\partial X_j}{\partial x_j} \frac{\partial \delta X_i}{\partial x_i} d\mathcal{V} = \int_{(\mathcal{S})} \frac{\partial X_j}{\partial x_j} \delta X_i n_i \, d\mathcal{S} - \int_{(\mathcal{V})} \frac{\partial^2 X_j}{\partial x_i \partial x_j} \delta X_i \, d\mathcal{V}$$

$$\left(\frac{\partial X_i}{\partial x_j} + \frac{\partial X_j}{\partial x_i} \right) \frac{\partial \delta X_i}{\partial x_j} \Rightarrow$$

$$\int_{(\mathcal{V})} \left(\frac{\partial X_i}{\partial x_j} + \frac{\partial X_j}{\partial x_i} \right) \frac{\partial \delta X_i}{\partial x_j} d\mathcal{V} = \int_{(\mathcal{S})} \left(\frac{\partial X_i}{\partial x_j} + \frac{\partial X_j}{\partial x_i} \right) n_i \delta X_i \, d\mathcal{S}$$
$$- \int_{(\mathcal{V})} \left(\frac{\partial^2 X_i}{\partial x_j^2} + \frac{\partial^2 X_j}{\partial x_i \partial x_j} \right) \delta X_i \, d\mathcal{V}$$

Regrouping the volume and the surface terms into two distinct integrals, we arrive at:

$$\int_{t_1}^{t_2} dt \int_{(\mathcal{V})} d\mathcal{V} \left(-\rho \ddot{X}_i + G \frac{\partial^2 X_i}{\partial x_j \partial x_j} + (\lambda + G) \frac{\partial^2 X_i}{\partial x_i \partial x_j} \right) \cdot \delta X_i$$
$$+ \int_{t_1}^{t_2} dt \int_{(\mathcal{S})} \left(\lambda \frac{\partial X_j}{\partial x_j} + G \left(\frac{\partial X_i}{\partial x_j} + \frac{\partial X_j}{\partial x_i} \right) \right) n_i \delta X_{ij} \, d\mathcal{S} = 0$$

Again, as the variation of action must vanish whatever the admissible δX_i may be, the equation of motion is obtained by equating to zero the kernel within the brackets of the volume integral whereas the boundary conditions are given by equating to zero the kernel of the surface integral. In the absence of any external loading, or any elastic support, this reduces to a condition of either a free boundary such that $\delta X_i \neq 0$ and the kernel within the brackets equal to zero, or that of a fixed boundary $\delta X_i = 0$ and the kernel within the brackets not equal to zero.

Finally, it is possible to shift from the indicial to the symbolic notation, by using the following identities:

$$\frac{\partial^2 X_j}{\partial x_i \partial x_j} = \overrightarrow{\text{grad}(\text{div}\vec{X})} \text{ and } \frac{\partial^2 X_i}{\partial x_j \partial x_j} = \text{div}\left(\overline{\overline{\text{grad}\vec{X}}}\right) = \Delta\vec{X} \qquad [1.60]$$

where $\Delta = \text{div grad}()$ is the Laplace operator (see Appendix A.2).

The vibration equations are thus found to agree with the intrinsic form [1.39], as suitable.

1.3.2.10 Conservation of mechanical energy

As above, the solid occupies the finite volume $(\mathcal{V})$ closed by the surface $(\mathcal{S})$. At time $t = 0$ the stable and unstressed configuration of equilibrium is chosen as the state of reference; then the external loads characterized by a volume density $\vec{f}^{(e)}(\vec{r};t)$ and a surface density $\vec{t}^{(e)}(\vec{r};t)$ are applied. The work done by these loads during the infinitesimal time interval $(t, t+dt)$ is:

$$d\mathcal{W} = \left\{\int_{(\mathcal{V})} \vec{f}^{(e)} \cdot \dot{\vec{X}}\, d\mathcal{V} + \int_{(\mathcal{S})} \vec{t}^{(e)} \cdot \dot{\vec{X}}\, d\mathcal{S}\right\} dt \qquad [1.61]$$

The work rate, or instantaneous power $\mathcal{P}(t)$ delivered to the solid is:

$$\mathcal{P}(t) = \frac{d\mathcal{W}}{dt} = \int_{(\mathcal{V})} \vec{f}^{(e)} \cdot \dot{\vec{X}}\, d\mathcal{V} + \int_{(\mathcal{S})} \vec{t}^{(e)} \cdot \dot{\vec{X}}\, d\mathcal{S} \qquad [1.62]$$

By using the equations of equilibrium [1.32], the result [1.62] is expressed in terms of internal forces as:

$$\mathcal{P}(t) = \int_{(\mathcal{V})} \left(\rho\ddot{\vec{X}} - \text{div}\,\overline{\overline{\sigma}}\right) \cdot \dot{\vec{X}}\, d\mathcal{V} + \int_{(\mathcal{S})} \left(\overline{\overline{\sigma}} \cdot \vec{n}\right) \cdot \dot{\vec{X}}\, d\mathcal{S} \qquad [1.63]$$

Using the divergence theorem [A.2.5] of Appendix A.2, this result is transformed into the volume integral:

$$\mathcal{P}(t) = \int_{(\mathcal{V})} \left\{\rho \cdot \ddot{\vec{X}} + \overline{\overline{\sigma}} : \overline{\overline{\text{grad}\,\dot{\vec{X}}}}\right\} d\mathcal{V} \qquad [1.64]$$

The stress term of the kernel in [1.64] can be further transformed into the symmetric form:

$$\overline{\overline{\sigma}} : \overline{\overline{\text{grad}\,\dot{\vec{X}}}} = \frac{1}{2}\left\{\overline{\overline{\sigma}} : \overline{\overline{\text{grad}\,\dot{\vec{X}}}} + \overline{\overline{\sigma}} : \left(\overline{\overline{\text{grad}\,\dot{\vec{X}}}}\right)^T\right\} = \overline{\overline{\sigma}} : \dot{\overline{\overline{\varepsilon}}} \qquad [1.65]$$

To establish this relation, the symmetry of the stress tensor and the hypothesis of small displacements are used. The work rate supplied to the solid may thus be expressed in terms of kinetic and elastic energies (cf Clapeyron formula [1.57]). If the material is elastic $\overline{\overline{\sigma}} : \dot{\overline{\overline{\varepsilon}}} = de_e/dt$, and [1.64] leads to:

$$\mathcal{P}(t) = \int_{(\mathcal{V})} \left\{ \rho \ddot{\vec{X}} \cdot \dot{\vec{X}} + \overset{\Rightarrow}{\sigma} : \dot{\overset{\Rightarrow}{\varepsilon}} \right\} d\mathcal{V} = \frac{d\mathcal{E}_\kappa}{dt} + \frac{d\mathcal{E}_e}{dt} = \frac{d\mathcal{E}_m}{dt} \qquad [1.66]$$

The relation [1.66] formulates the law of energy conservation, according to which the variation rate of mechanical energy stored in the solid is equal to the work rate produced by the external loading acting on it. This energy balance is purely mechanical in nature; thermal exchanges for instance are discarded. On the other hand, the instantaneous variation rate of energy [1.66] can be integrated with respect to time to obtain:

$$\int_0^t \mathcal{P}(\tau) d\tau = \mathcal{E}_\kappa(t) + \mathcal{E}_e(t) = \mathcal{E}_m(t) \qquad [1.67]$$

1.3.2.11 Uniqueness of solution of motion equations

Starting from the conservation law of energy, it is possible to prove the theorem of uniqueness of the solution of any linear elastodynamic problem, which is stated as follows:

The equations of motion of a linear elastic solid, subjected to suitably prescribed loading and/or displacement fields,(including the boundary conditions) have a solution which is unique.

The proof is due to Neumann [NEU 85]. First it is noted that provided the problem is linear, the principle of superposition can be applied. Accordingly, let $\vec{X}_1, \vec{X}_2$ be the respective solutions of the two following problems:

$$\begin{aligned} &\rho \ddot{\vec{X}}_1 - \operatorname{div} \overline{\overline{\sigma_1}} + \vec{f}_1^{(i)} = \vec{f}_1^{(e)}(\vec{r};t); \quad \forall \vec{r} \in (\mathcal{V}) \\ &\overline{\overline{\sigma_1}}(\vec{r}) \cdot \vec{n}(\vec{r}) = \vec{t}_1^{(e)}(\vec{r};t); \quad \forall \vec{r} \in (\mathcal{S}_1) \\ &\overline{\overline{\sigma_1}}(\vec{r}) \cdot \vec{n}(\vec{r}) - K_S[\vec{X}_1] = 0; \quad \forall \vec{r} \in (\mathcal{S}_2) \end{aligned} \qquad [1.68]$$

with

$$(\mathcal{S}_1) \cup (\mathcal{S}_2) = (\mathcal{S})$$

$$\vec{X}_1(\vec{r};0) = \vec{D}_1(\vec{r}); \quad \dot{\vec{X}}_1(\vec{r};0) = \dot{\vec{D}}_1(\vec{r})$$

and

$$\begin{aligned}
\rho\ddot{\vec{X}}_2 - \mathrm{div}\overline{\overline{\sigma_2}} + \vec{f}_2^{(i)} &= \vec{f}_2^{(e)}(\vec{r};t); \quad \forall \vec{r} \in (\mathcal{V}_\mathrm{V}) \\
\overline{\overline{\sigma_2}}(\vec{r}) \cdot \vec{n}(\vec{r}) &= \vec{t}_2^{(e)}(\vec{r};t); \quad \forall \vec{r} \in (\mathcal{S}_1) \\
\overline{\overline{\sigma_2}}(\vec{r}) \cdot \vec{n}(\vec{r}) - K_S[\vec{X}_2] &= 0; \quad \forall \vec{r} \in (\mathcal{S}_2) \\
\vec{X}_2(\vec{r};0) = \vec{D}_2(\vec{r}); &\quad \dot{\vec{X}}_2(\vec{r};0) = \dot{\vec{D}}_2(\vec{r})
\end{aligned} \qquad [1.69]$$

Then $\vec{X} = \alpha\vec{X}_1 + \beta\vec{X}_2$ will be solution of:

$$\begin{aligned}
\rho\ddot{\vec{X}} - \mathrm{div}\overline{\overline{\sigma}} + \vec{f}^{(i)} &= \vec{f}^{(e)}(\vec{r};t); \quad \forall \vec{r} \in (\mathcal{V}) \\
\overline{\overline{\sigma}} \cdot \vec{n}(\vec{r}) &= \vec{t}^{(e)}(\vec{r};t); \quad \forall \vec{r} \in (\mathcal{S}_1) \\
\overline{\overline{\sigma}} \cdot \vec{n}(\vec{r}) - K_S[\vec{X}] &= 0; \quad \forall \vec{r} \in (\mathcal{S}_2) \\
\vec{X}(\vec{r};0) = \vec{D}(\vec{r}); &\quad \dot{\vec{X}}(\vec{r};0) = \dot{\vec{D}}(\vec{r})
\end{aligned} \qquad [1.70]$$

where

$$\begin{aligned}
\vec{f}^{(i)} = \alpha\vec{f}_1^{(i)} + \beta\vec{f}_2^{(i)}; \quad \vec{f}^{(e)} = \alpha\vec{f}_1^{(e)} + \beta\vec{f}_2^{(e)}; \quad \vec{t}^{(e)} = \alpha\vec{t}_1^{(e)} + \beta\vec{t}_2^{(e)} \\
\vec{D}(\vec{r}) = \alpha\vec{D}_1(\vec{r}) + \beta\vec{D}_2(\vec{r}); \quad \dot{\vec{D}}(\vec{r}) = \alpha\dot{\vec{D}}_1(\vec{r}) + \beta\dot{\vec{D}}_2(\vec{r})
\end{aligned} \qquad [1.71]$$

Now, let us assume that two distinct solutions denoted $\vec{X}_1$ and $\vec{X}_2$ do exist for a same problem. Then $\vec{X} = \vec{X}_2 - \vec{X}_1$ must be a solution of the homogeneous system:

$$\begin{aligned}
\rho\ddot{\vec{X}} - \mathrm{div}\,\overline{\overline{\sigma}} &= 0; \quad \forall \vec{r} \in (\mathcal{V}) \\
\overline{\overline{\sigma}}(\vec{r}) \cdot \vec{n}(\vec{r}) &= 0; \quad \forall \vec{r} \in (\mathcal{S}_1) \\
\overline{\overline{\sigma}}(\vec{r}) \cdot \vec{n}(\vec{r}) - K_S[\vec{X}] &= 0; \quad \forall \vec{r} \in (\mathcal{S}_2) \\
\vec{X}(\vec{r};0) = \vec{0}; &\quad \dot{\vec{X}}(\vec{r};0) = \vec{0}
\end{aligned} \qquad [1.72]$$

Because in [1.72], the initial conditions and the external loading are nil, no mechanical energy is provided to the system. Therefore the kinetic and the elastic energy are zero at any time, so the system remains at rest:

$$\vec{X}(\vec{r};t) = \vec{X}_2(\vec{r};t) - \vec{X}_1(\vec{r};t) \equiv 0 \quad \forall \vec{r}, t$$

To conclude this subsection, it is worth mentioning that a similar theorem was obtained by Kirchhoff in the case of statics. Kirchhoff's theorem of uniqueness differs from that of Neumann, since in statics, kinetic energy is discarded. Hence, if the body is free (i.e. not provided with any support) it is always possible to add a uniform, and otherwise arbitrary, displacement field to a given static solution.

1.4. Elastic waves in three-dimensional media

1.4.1 *Material oscillations in a continuous medium interpreted as waves*

When the particles contained in an elastic medium are removed from their position of static equilibrium – assumed here to be stable – the stresses related to the local change of configuration have the tendency to take them back to the position of static equilibrium, but the inertia forces are acting to the opposite, having the tendency to make the particles overshoot it. As a result, they start to oscillate. Due to the principle of action and reaction, the particles lying in the immediate vicinity are also excited and start to oscillate too. In this way, the motion is found to propagate throughout the whole solid. In the absence of inertia, the propagation would have the instantaneous character of the elastic forces. The inertia introduces however a delay in the propagation, in such a way that the speed is finite. Such progressive oscillations are termed *travelling waves*. It is important to point out first that in this "chain reaction" what propagates is not matter but mechanical energy. A discrete version of material waves was already discussed in [AXI 04], Chapters 7 and 8. As schematically illustrated in Figure 1.10, two kinds of waves can be distinguished. They are termed *transverse waves* if the particles oscillate in a direction perpendicular to that of wave propagation and *longitudinal waves* if

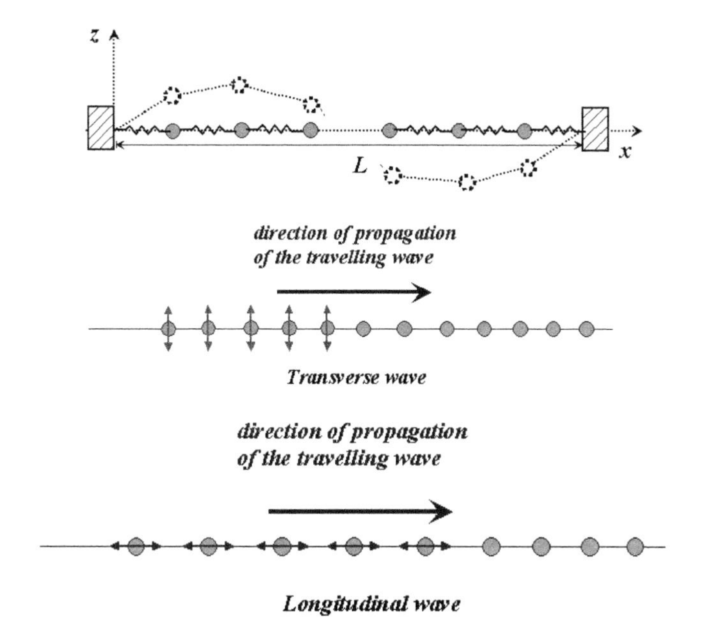

Figure 1.10. *Discrete model of the oscillations of material points in transverse and longitudinal waves*

the particles oscillate in the direction of wave propagation. On the other hand, in an infinite conservative medium, the amount of mechanical energy conveyed by the waves is constant during the propagation. However, in reality nonconservative forces are always present, so the waves are damped out, or alternatively amplified, depending on the sign of the energy transfer to the wave.

1.4.2 *Harmonic solutions of Navier's equations*

In a solid, the material waves are governed by Navier's equations [1.39]. It is appropriate to study first their general properties independently of external loading. Furthermore, it is also suitable to start by assuming a medium extending to infinity in all directions, in such a way that boundary effects can be discarded. Navier's equations are thus reduced to the homogeneous vector equation:

$$\rho\ddot{\vec{X}} - \left[G\Delta\vec{X} + (\lambda + G)\overrightarrow{\mathrm{grad}(\mathrm{div}\,\vec{X})}\right] = 0 \qquad [1.73]$$

In terms of vector components, [1.73] is a linear system of three partial differential equations, whose coefficients are constants if the solid is homogeneous. A well known mathematical technique used to solve this kind of equation is the *method of variables separation.* If the problem is further particularized to harmonic oscillations of pulsation ω, solutions sought can be written as the complex field:

$$\begin{aligned}\vec{X}(x,y,z;t) &= e^{i\omega t}\{X\vec{i} + Y\vec{j} + Z\vec{k}\}\\ X(x,y,z) &= f_x(x)g_x(y)h_x(z)\\ Y(x,y,z) &= f_y(x)g_y(y)h_y(z)\\ Z(x,y,z) &= f_z(x)g_z(y)h_z(z)\end{aligned} \qquad [1.74]$$

Substitution of [1.74] into [1.73] allows one to reduce the problem of solving the partial differential vector equation [1.73] to one of solving nine ordinary differential equations. However, determination of the appropriate nine space functions involved in [1.74] is not a simple task and it is advisable to particularize the problem further, in order to obtain comparatively simple analytical solutions which can be easily discussed from a physical point of view. This is the object of the following subsections.

1.4.3 *Dilatation and shear elastic waves*

The relation $\overrightarrow{\mathrm{grad}(\mathrm{div}\vec{X})} = \Delta\vec{X} + (\mathrm{curl}\,\mathrm{curl}(\vec{X}))$ (see formula [A.2.17] in Appendix A.2) allows a meaningful simplification of [1.73] to be made, by separating the motion into two physically distinct types, namely:

1.4.3.1 Irrotational, or potential motion

$$\rho\ddot{\vec{X}} - \kappa\Delta\vec{X} = 0 \quad \forall\vec{r} \in (\mathcal{V}) \;\; \text{curl}\,\vec{X} = \vec{0}$$

$$\kappa = (\lambda + 2G) = \frac{(1-\nu)E}{(1+\nu)(1-2\nu)} \qquad [1.75]$$

The motions governed by the system [1.75] can also be described by using a scalar displacement potential denoted Φ such as:

$$\vec{X} = \text{grad}\,\Phi \qquad [1.76]$$

which satisfies automatically the condition curl $\vec{X} = 0$.

1.4.3.2 Equivoluminal, or shear motion

$$\rho\ddot{\vec{X}} - G\Delta\vec{X} = 0 \quad \forall\vec{r} \in (\mathcal{V}) \;\; \text{div}\,\vec{X} = 0 \qquad [1.77]$$

The motions governed by the system [1.77] can also be described by using a vector displacement potential denoted $\vec{\Psi}$ such as:

$$\vec{X} = \text{curl}\,\vec{\Psi} \qquad [1.78]$$

which satisfies automatically the condition div$\vec{X} = 0$.

1.4.3.3 Irrotational harmonic waves (dilatation or pressure waves)

The system [1.75] describes waves in which the volume of the medium fluctuates since div $\vec{X} \neq 0$ (otherwise the $\Delta\vec{X}$ term would vanish identically); for this reason such waves are often referred to as 'volume' or dilatation waves. To point out their major features, the easiest way is to study the plane harmonic waves which travel along the Ox axis, of unit vector $\vec{i}$. The displacement field reduces thus to the complex amplitude:

$$\vec{X}(x;t) = X(x)\vec{i}e^{i\omega t} \qquad [1.79]$$

The condition curl $\vec{X} = 0$ is obviously satisfied by [1.79]. If this form is substituted into the first equation [1.75], the following ordinary differential equation is obtained:

$$\kappa \frac{d^2 X}{dx^2} + \omega^2 \rho X = 0 \qquad [1.80]$$

The general solution of [1.80] is written as:

$$X(x;t) = X_+ e^{i\omega(t - x/c_L)} + X_- e^{i\omega(t + x/c_L)} \qquad [1.81]$$

The complex amplitude $X(x;t)$ is the superposition of two plane harmonic waves. Each of them is described by a complex number of modulus $X_\pm$ and argument $\omega(t \mp x/c_L)$. The modulus gives the magnitude of the wave and the argument gives the phase angle referred to $t = 0$ and $x = 0$. If there is a source of plane harmonic waves at $x = 0$, two waves are excited, which travel in two opposite directions along Ox, as sketched in Figure 1.11. The speed of propagation is given by:

$$c_L = \sqrt{\frac{\kappa}{\rho}} = \sqrt{\frac{1-\nu}{(1+\nu)(1-2\nu)} \frac{E}{\rho}} \qquad [1.82]$$

The delay $\tau(x) = -|x|/c_L$ is the time spent by the harmonic waves $X_\pm e^{i\omega t}$ to cover a distance $\pm|x|$. The negative sign agrees with the principle of causality, according to which the response of the medium cannot anticipate the excitation.

In terms of phase angle, $\tau(x)$ is replaced by the phase shift between the oscillations located at the source and at a distance x from the source. It is given by $\psi(x) = -\omega|x|/c_L$. Accordingly, c_L is interpreted as the *phase speed* of the dilatation waves. The wave which travels from left to right ($x > 0$) has the magnitude X_+ and phase shift $-\omega x/c_L$ and the wave which travels from right to left ($x < 0$)

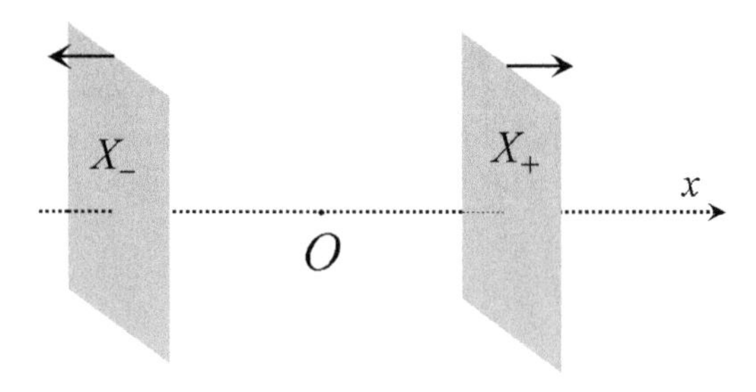

Figure 1.11. *Propagation of plane waves*

has the magnitude X_- and phase shift $-\omega|x|/c_L$. As a general definition, the phase speed of a wave, denoted c_ψ, is such that:

$$\psi(x) = -\frac{\omega|x|}{c_\psi} = -\frac{2\pi|x|}{\lambda} = -k|x| \Rightarrow c_\psi = \frac{\omega}{k} \qquad [1.83]$$

where the *wavelength* λ is the distance travelled by the wave during one period $T = 2\pi/\omega$ of oscillation and $k = 2\pi/\lambda$ is the *wave number*.

If the phase speed is independent of the pulsation, the wave is said *nondispersive*, as is the case of dilatation waves, cf. [1.82], if not it is said to be *dispersive*. To understand the meaning of this terminology, it is appropriate to consider first a compound wave defined as the superposition of two distinct harmonic waves of frequency f_1 and f_2 respectively. Its complex amplitude is written as:

$$X(x;t) = e^{i2\pi f_1(t-x/c_1)} + e^{i2\pi f_2(t-x/c_2)}$$

If $c_1 = c_2 = c$, each component travels at the same speed, so the time profile of the wave is the same from one position to another, and the same holds for the space profile from one time to another. If c_2 differs from c_1, each component travels at is own speed, so the time profile of the wave changes from one position to another, as the space profile does from one time to another. This is illustrated in Figure 1.12, where the real part of $X(x;t)$ is plotted versus time at two distinct positions $x_1 = 0$ and $x_2 = 1.75\lambda_1$ where $\lambda_1 = c_1 f_1$. The spectral components are at $f_1 = 10$ Hz and $f_2 = 20$ Hz, the period of the compound wave is $T = 1/f_1 = 0.1s$.

Figure 1.12a refers to the nondispersive case $c_1 = c_2 = c$. The shape of the wave at x_2 is the same as that at $x_1 = 0$, the time profile being simply translated to the right by the propagation delay $\tau = x_2/c$. Figure 1.12b refers to the dispersive case $c_1 \neq c_2$. The time profile of the wave at x_2 differs from that at $x_1 = 0$. Thus, in the dispersive case, propagation cannot be described simply in terms of propagation delay.

Such elementary considerations can be extended to more complicated waves, such as transients by using the Fourier transformation. Transients are described in the time domain by the displacement field $X(x;t)$ which usually vanish outside a finite time interval $0 \leq t \leq t_1$. It is necessary to stress that here $X(x;t)$ denotes a real valued function, in contrast with the former case where it denoted the complex amplitude of superposed harmonics waves. Shifting to the spectral domain, the transients are described by the Fourier transform of $X(x;t)$, denoted $\widehat{X}(x;\omega)$.

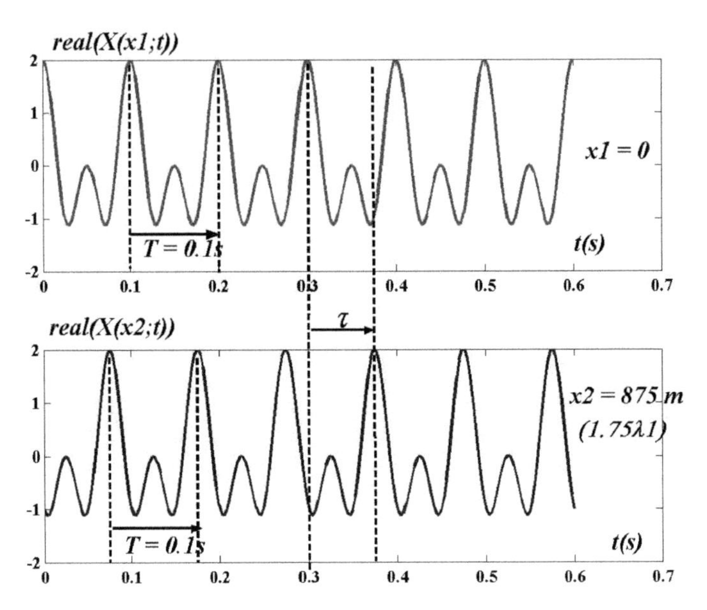

Figure 1.12a. *Time profile of the compound wave, nondispersive case:* $c_1 = c_2 = 5000\,m/s$

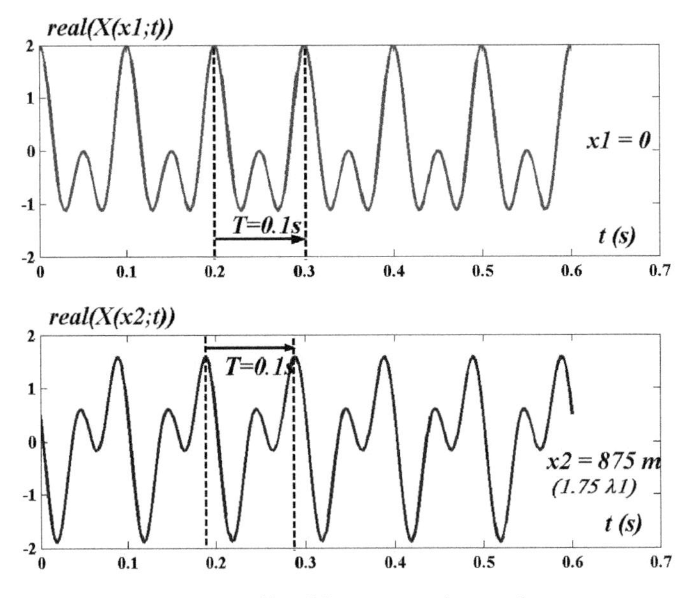

Figure 1.12b. *Time profile of the compound wave, dispersive case:* $c_1 = 5000\,m/s, c_2 = 3000\,m/s$

It is recalled that by definition of the Fourier transform $\widehat{X}(x;\omega)$ and $X(x;t)$ are related to each other by:

$$\widehat{X}(x;\omega) = \int_{-\infty}^{+\infty} X(x;t)e^{-i\omega t}dt \quad \text{and} \quad X(x;t) = \frac{1}{2\pi}\int_{-\infty}^{+\infty} \widehat{X}(x;\omega)e^{+i\omega t}d\omega \tag{1.84}$$

Substituting $t - \tau$ for t, the shift theorem follows:

$$\begin{aligned} &\int_{-\infty}^{+\infty} X(x;t-\tau)e^{-i\omega t}dt = \widehat{X}(x;\omega)e^{-i\omega t} \\ &X(x;t-\tau) = \frac{1}{2\pi}\int_{-\infty}^{+\infty} \widehat{X}(x;\omega)e^{+i\omega(t-\tau)}d\omega \end{aligned} \tag{1.85}$$

where τ is the time delay.

Let $X(t)$ stand for the displacement field of a transient plane wave emitted at $x = 0$, starting from $t = 0$. In the nondispersive case, the wave observed at $x > 0$ is $X_+(x;t) = 0.5X(x;t-\tau)$ where $\tau = x/c_\psi$ as sketched in Figure 1.13 and in the spectral domain $\widehat{X}_+(\omega,x) = 0.5\,\widehat{X}(\omega)e^{-i\omega x/c_\psi}$. Of course, the same result holds for a wave $X_-(t,x)$ travelling in the domain $x < 0$, with $\tau = |x|/c_\psi$. This is precisely the reason why the multiplying factor 0.5 appears in the travelling waves, in such a way that $\widehat{X}_+(\omega,0) + \widehat{X}_-(\omega,0) = \widehat{X}(\omega)$. Contrasting with such simple results, if c_ψ is frequency dependent, the Fourier transform of $X_+(x;t)$

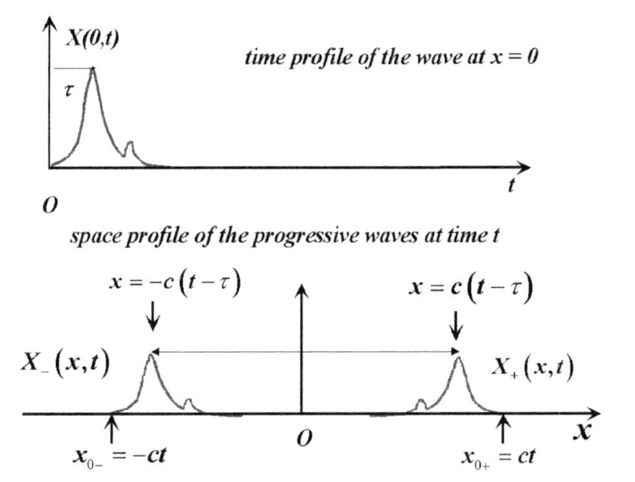

Figure 1.13. *Propagation of nondispersive waves: $c = c_\psi$ is constant*

can be written as $\widehat{X}(x;\omega)e^{-i\omega\tau(\omega)}$. However, as τ is frequency dependent, the shift theorem cannot be used to express $X_+(x;t)$ in terms of $X(t)$. Stated in a qualitative way, the individual spectral components contained in $\widehat{X}(\omega)$ travel at distinct speeds and instead of reaching the same position at the same time, they are dispersed along the axis of propagation. Such waves are thus termed *dispersive waves*. Two examples of travelling dispersive waves will be presented in Chapter 4, see Figures 4.2 and 4.5.

1.4.3.4 Shear waves (equivoluminal or rotational waves)

The system [1.77] describes waves which do not generate any volume variation (div $\vec{X} \equiv 0$) but differential rotations and shear strains in the material. This feature can be evidenced with plane waves propagating in the $\pm x$ directions, which are solutions of [1.77]. As div $\vec{X} \equiv 0$, $\vec{X}$ must be perpendicular to the Ox axis. Hence the harmonic plane waves are necessarily of the general form:

$$\vec{X}(x;t) = (Y_\pm(x)\vec{j} + Z_\pm(x)\vec{k})e^{i\omega(t \pm x/c_s)} \quad \text{where } c_s = (G/\rho)^{1/2} \qquad [1.86]$$

The shear waves are thus found to be transverse and nondispersive. The phase speed is less than that of the dilatation waves by the ratio $c_S/c_L = \sqrt{(1-2\nu)/2(1-\nu)}$. Material oscillates along a direction prescribed by the ratio Z/Y, which defines the polarization state of the wave.

1.4.4 *Phase and group velocities*

As will be shown in several examples introduced later in this book, the phase speed [1.83] of the waves is often found to vary with frequency, leading to propagation features far more complicated than in the nondispersive case; this is because the individual spectral components of the emitted waves interfere with each other in an intricate manner. A first question of interest concerns the propagation of the wave energy. The speed at which wave energy is propagated is known as the *group velocity* and is defined by:

$$c_g = \frac{d\omega}{dk} \qquad [1.87]$$

The physical meaning of this definition can be clarified by superposing a fairly large number of harmonic waves. As a preliminary, it is recalled that the superposition of two sine waves of equal magnitude gives:

$$X(x;t) = X_o\left[\sin(\omega t - kx) + \sin(\omega' t - k'x)\right]$$
$$= 2X_o \cos\left[\frac{(\omega - \omega')t - (k - k')x}{2}\right]\sin\left[\frac{(\omega + \omega')t - (k + k')x}{2}\right] \qquad [1.88]$$

If k is close to k' and ω to ω', the sine function varies much more rapidly than does the cosine. Accordingly, both the time and the space profiles of the compound wave are shaped as a high frequency signal slowly modulated in amplitude, leading to the well known 'beat phenomenon'. This basic result may be extended to the summation of N sine waves slightly detuned from each other. Figure 1.14 shows the graph of the time dependent signal obtained with $N = 20$. The frequencies of the N sine waves vary linearly from 1 Hz to 1.2 Hz. Calculation was carried out using the commercial software MATLAB.

As shown in Figure 1.14, it is observed that as N increases, the oscillations cluster together, giving rise to small wave packets separated by long time intervals during which the wave magnitude – and so energy – remains negligible.

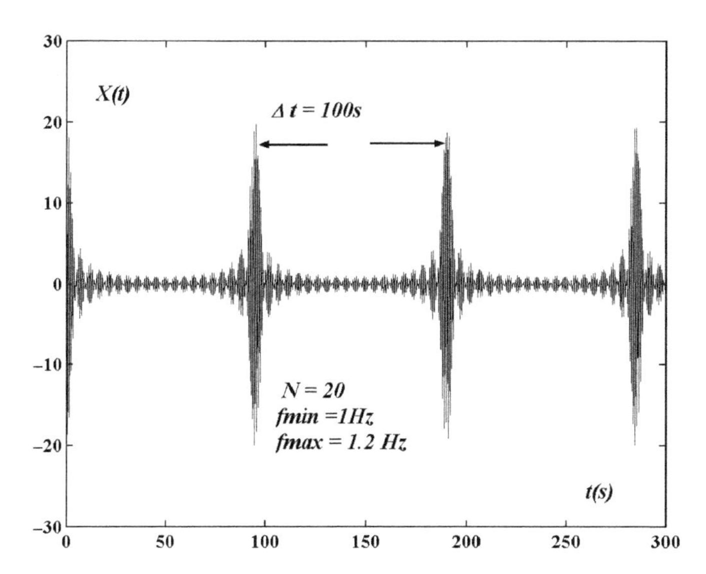

Figure 1.14. *Wave packets resulting from the addition of slightly different harmonic components*

The duration of these intervals can be satisfactorily described by the following formula:

$$\Delta t = \frac{N}{f_N - f_1} \qquad [1.89]$$

where f_1 is the smallest frequency and f_N the largest frequency of the N superposed sine waves.

A similar result is observed if the resulting wave is plotted versus x, at a given time. The distance between two wave packets is given by:

$$\Delta x = \frac{2\pi N}{k_N - k_1} \qquad [1.90]$$

where k_1 and k_N are the smallest and the largest wave numbers, respectively.

Then, it is possible to define the propagation velocity of the wave packets by:

$$\frac{\Delta x}{\Delta t} = \frac{\omega_N - \omega_1}{k_N - k_1} \qquad [1.91]$$

Relation [1.91] is a finite difference approximation of [1.87], which defines the group velocity. It elucidates the physical meaning of c_g and explains why it represents the transport velocity of the mechanical energy of the wave. Furthermore, if a continuous spectrum is considered, defined by a function of dispersion $\omega(k)$, the preceding results can be also interpreted in a slightly different way. The phase shifts between all the sine components of the resulting signal make the individual waves interfere with each other. Additive interferences occur solely over infinitesimal space-time intervals *dt, dx* which make the phase function $\omega t - kx$ stationary:

$$(\omega - \omega')dt - (k - k')dx \cong 0 \Rightarrow d\omega\, dt - dk\, dx \cong 0$$

1.4.5 *Wave reflection at the boundary of a semi-infinite medium*

A thorough and comprehensive study of the propagation of elastic waves in three-dimensional solid bodies requires analytical developments which are beyond the scope of this book. The reader who is interested in such problems is referred for instance to [SOM 50], [ACH 73] and [MIK 78]. However, it is of interest to present here a few basic aspects of the problem which are necessary for the understanding of the physical content of the simplifications made when modelling solid bodies as structural elements. In particular, it will be shown that elastic waves encountered in structural elements differ from those travelling in an unbounded solid, as further discussed in Chapter 4 based on a few examples.

The major difficulty encountered in linear elastodynamics stems from the peculiarities of wave reflection and/or refraction which occur at the boundary of the solid and at the interfaces between two distinct solid media. In contrast with electromagnetic waves and dilatation waves in a fluid, reflection and refraction of solid waves generally give rise to *mode conversion*, which means that the type of the reflected, or refracted waves is not necessarily the same as that of the incident wave, but in most cases a combination comprising a wave of the same type as the incident wave and another wave of a distinct type. Mode conversion arises as a necessary feature to comply with the boundary conditions at the reflecting, or refracting surface. As illustrated below for two reflection cases, it depends on the type of the solid wave considered, the angle of incidence and the boundary conditions.

1.4.5.1 Complex amplitude of harmonic and plane waves at oblique incidence

As sketched in Figure 1.15, the $z = 0$ plane of a Cartesian frame is assumed to be the interface between a semi-infinite elastic solid and vacuum. An elastic plane wave propagates along the direction specified by the unit vector:

$$\vec{\ell} = \sin\theta\vec{i} + \cos\theta\vec{k} \qquad [1.92]$$

where θ is the angle made with the unit vector $\vec{k}$.

If the boundary conditions at $z = 0$ are conservative, the wave is necessarily fully reflected since it cannot propagate in vacuum and mechanical energy is conserved.

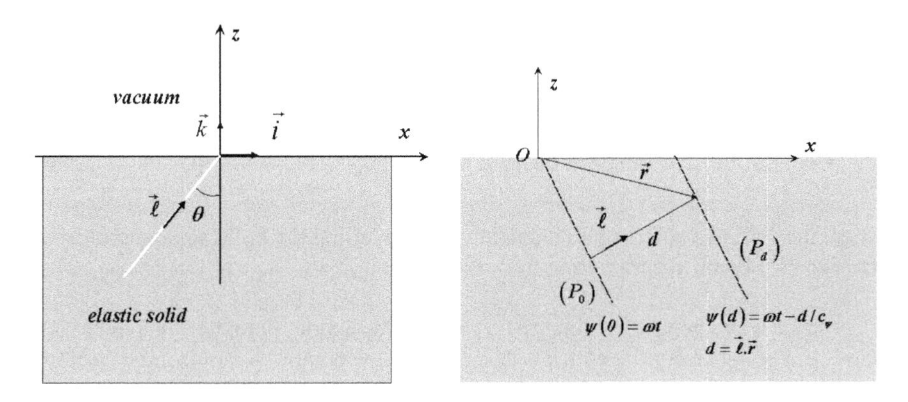

Figure 1.15. *Plane wave incident to the interface between a solid and vacuum: direction of propagation and planes of constant phase*

Material motion is described by the complex displacement vector:

$$\vec{A} = A\vec{d}e^{i\psi} \tag{1.93}$$

where $\vec{d}$ is the unit vector which specifies the direction of material motion and ψ designates the phase shift due to propagation of the wave. At position $\vec{r} = x\vec{i} + z\vec{k}$ and time t, ψ is given by:

$$\psi = \omega\left(t - \frac{\vec{r}\cdot\vec{\ell}}{c_\psi}\right) = \omega\left(t - \frac{x\sin\theta + z\cos\theta}{c_\psi}\right) \tag{1.94}$$

Relation [1.94] extends suitably relation [1.83] to the 3D case since the distance between two planes of constant phase is expressed as $\vec{r}\cdot\vec{\ell}$ instead of x. In the present problem it is noticed however that the properties of the waves are independent of y, since the planes of constant phase are parallel to the Oy direction. They verify the equation $z = -x\tan\theta$. The phase of a wave plane passing through the axes origin O at time t is ωt and the phase at a wave plane separated by the distance d from the former is $t - d/c_\psi$.

On the other hand, at $z = 0$ the Cartesian components X, Y, Z of the material displacements are obtained by superposing the components relative to the incident and reflected waves. The elastic stresses are:

$$\sigma_{zz} = \lambda\left(\frac{\partial Z}{\partial z} + \frac{\partial X}{\partial x}\right) + 2G\left.\frac{\partial Z}{\partial z}\right|_{z=0}; \quad \sigma_{zx} = G\left.\left(\frac{\partial Z}{\partial x} + \frac{\partial X}{\partial z}\right)\right|_{z=0};$$
$$\sigma_{zy} = G\left.\frac{\partial Y}{\partial z}\right|_{z=0} \tag{1.95}$$

To deal with the boundary conditions, it turns out that three distinct types of waves have to be considered, which are sketched in Figure 1.16. A dilatational or pressure wave, hereafter denoted (P) wave, travelling at angle θ with $\vec{k}$ is described by the complex displacement vector:

$$\vec{U} = U\vec{\ell}e^{i\psi} \quad \vec{\ell} = \sin\theta\vec{i} \pm \cos\theta\vec{k} \quad \psi = \omega\left(t - \frac{x\sin\theta \pm z\cos\theta}{c_L}\right) \tag{1.96}$$

where the sign $(+)$ refers to an incident wave travelling towards the boundary and the sign $(-)$ to a wave travelling from the boundary.

A shear wave polarized along the $\vec{j} = \vec{k}\times\vec{i}$ direction, hereafter denoted (SH) wave, travelling at angle α with $\vec{k}$ is described by the complex displacement vector:

$$\vec{W} = W\vec{j}e^{i\psi} \quad \vec{\ell} = \sin\alpha\vec{i} \pm \cos\alpha\vec{k} \quad \psi = \omega\left(t - \frac{x\sin\alpha \pm z\cos\alpha}{c_S}\right) \tag{1.97}$$

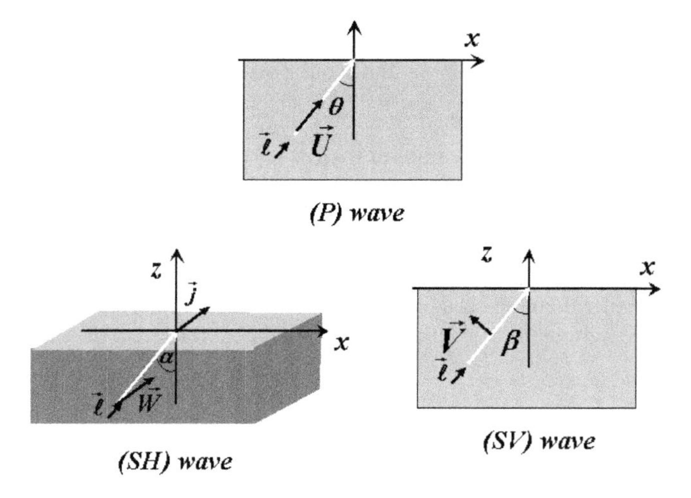

Figure 1.16. *(P), (SH) and (SV) incident waves*

Finally, a shear wave polarized along the $\vec{v} = \vec{\ell} \times \vec{j}$ direction, hereafter denoted *(SV)* wave, travelling at angle β with $\vec{k}$ is described by the vector:

$$\vec{V} = V\vec{v}e^{i\psi} \quad \vec{\ell} = \sin\beta\vec{i} \pm \cos\beta\vec{k} \quad \psi = \omega\left(t - \frac{x\sin\beta \pm z\cos\beta}{c_S}\right) \qquad [1.98]$$

1.4.5.2 Reflection of (SH) waves on a free boundary

Let us consider first the simplest case of (SH) wave reflection. We will show that such waves are reflected as (SH) waves only. No mode conversion occurs, because it is unnecessary to invoke any other waves to comply with the conditions of reflection. In agreement with [1.97], the incident and the reflected (SH) waves are written as:

$$\begin{aligned} &\text{incident wave: } \vec{W}_i = W_i\vec{j}e^{i\psi_i} \quad \psi_i = \omega\left(t - \frac{x\sin\alpha_i + z\cos\alpha_i}{c_S}\right) \\ &\text{reflected wave: } \vec{W}_r = W_r\vec{j}e^{i\psi_r} \quad \psi_r = \omega\left(t - \frac{x\sin\alpha_r - z\cos\alpha_r}{c_S}\right) \end{aligned} \qquad [1.99]$$

Y denoting the displacement at the reflecting plane, the boundary conditions [1.95] are written as:

$$\sigma_{zz} \equiv 0; \quad \sigma_{zx} \equiv 0; \quad \sigma_{zy} = G\left.\frac{\partial Y}{\partial z}\right|_{z=0} = \frac{i\omega GY}{c_S}e^{i\omega t}(-\cos\alpha_i + \cos\alpha_r) = 0 \qquad [1.100]$$

The last condition [1.100] leads to the usual reflection law of optics:

$$\alpha_i = \alpha_r \qquad [1.101]$$

Finally, conservation of energy implies that $W_i = W_r = Y$

1.4.5.3 Reflection of (P) waves on a free boundary

Let us consider then the case of (P) wave reflection. In agreement with [1.96], the Cartesian components of the material displacement related to the incident wave is written as:

$$X_1 = U_1 \sin\theta_1 e^{i\psi_1}; \quad Z_1 = U_1 \cos\theta_1 e^{i\psi_1}; \quad \psi_1 = \omega\left(t - \frac{x\sin\theta_1 + z\cos\theta_1}{c_L}\right) \qquad [1.102]$$

The stress components associated to the incident wave are found to be:

$$\begin{aligned}(\sigma_{zz})_1 &= \frac{-i\omega U_1}{c_L}\left(\lambda + 2G(\cos\theta_1)^2\right)e^{i\psi_1};\\ (\sigma_{zx})_1 &= \frac{-i\omega U_1}{c_L} G\sin 2\theta_1 e^{i\psi_1}; \quad (\sigma_{zy})_1 \equiv 0\end{aligned} \qquad [1.103]$$

In the same way, for the reflected (P) wave we get:

$$\begin{aligned}&X_2 = U_2 \sin\theta_2 e^{i\psi_2}; \quad Z_2 = U_2\cos\theta_2 e^{i\psi_2};\\ &\psi_2 = \omega\left(t - \frac{x\sin\theta_2 - z\cos\theta_2}{c_L}\right)\\ &(\sigma_{zz})_2 = \frac{-i\omega U_2}{c_L}\left(\lambda + 2G(\cos\theta_2)^2\right)e^{i\psi_2};\\ &(\sigma_{zx})_2 = \frac{+i\omega U_2}{c_L} G\sin 2\theta_2 e^{i\psi_2}; \quad (\sigma_{zy})_2 \equiv 0\end{aligned} \qquad [1.104]$$

Now, the attempt to solve the reflection problem in terms of (P) waves only generally fails because the boundary conditions cannot be fulfilled. Disappearance of the normal stress leads to:

$$\begin{aligned}&\sigma_{zz}|_{z=0} = ((\sigma_{zz})_1 + (\sigma_{zz})_2)|_{z=0} = 0 \quad \Rightarrow\\ &-\frac{i\omega}{c_L}\Big\{U_1(\lambda + 2G(\cos\theta_1)^2)e^{i(t-(x\sin\theta_1)/c_L)}\\ &+ U_2\left(\lambda + 2G(\cos\theta_2)^2\right)e^{i(t-(x\sin\theta_2)/c_L)}\Big\} = 0\end{aligned} \qquad [1.105]$$

Disappearance of shear stress leads to:

$$\sigma_{zx}|_{z=0} = ((\sigma_{zx})_1 + (\sigma_{zx})_2)|_{z=0} = 0 \quad \Rightarrow$$
$$+\frac{i\omega G}{c_L}\left\{U_2 \sin 2\theta_2 e^{i(t-(x\sin\theta_2)/c_L)} - U_1 \sin 2\theta_1 e^{i(t-(x\sin\theta_1)/c_L)}\right\} = 0 \qquad [1.106]$$

As the conditions [1.105] and [1.106] must hold independently from the value of the phase angles, we arrive at the following relation between the angles of reflection and incidence:

$$\psi_1|_{z=0} = \psi_2|_{z=0} = \psi \Rightarrow \theta_2 = \theta_1$$

However, when this result is substituted into [1.105] and [1.106], the sole mathematical solution for the wave magnitude is found to be the trivial solution $U_2 \equiv U_1 \equiv 0$, except in the peculiar cases of either normal ($\theta_1 = 0$), or grazing incidence ($\theta_1 = 90°$). Consequently, the sole possibility physically acceptable is to assume that mode conversion occurs in such a way that if θ_1 differs from 0, or 90°, the incident (P) wave is reflected as a pair of (P) and (SV) waves. Obviously, a (SH) wave would be unsuitable here for complying with the boundary conditions. Using [1.98] the reflected (SV) wave is written as:

$$X_3 = V\cos\beta e^{i\psi_3}; \quad Z_3 = V\sin\beta e^{i\psi_3}; \quad \psi_3 = \omega\left(t - \frac{x\sin\beta - z\cos\beta}{c_S}\right) \qquad [1.107]$$

The associated stress components are:

$$(\sigma_{zz})_3 = \frac{i\omega VG}{c_S}\sin 2\beta e^{i\psi_3}; \quad (\sigma_{zx})_3 = \frac{i\omega VG}{2c_S}\cos 2\beta e^{i\psi_3}; \quad (\sigma_{zy})_3 \equiv 0 \qquad [1.108]$$

Disappearance of the resultant normal stress leads to the condition:

$$\sigma_{zz}|_{z=0} = ((\sigma_{zz})_1 + (\sigma_{zz})_2 + (\sigma_{zz})_3)|_{z=0} = 0 \quad \Rightarrow$$
$$\frac{-1}{c_L}\left\{U_1\left(\lambda + 2G(\cos\theta_1)^2\right)e^{i\psi_1} + U_2\left(\lambda + 2G(\cos\theta_2)^2\right)e^{i\psi_2}\right\} \qquad [1.109]$$
$$+\frac{GV}{c_S}\sin 2\beta e^{i\psi_3} = 0$$

Disappearance of the resultant shear stress leads to the condition:

$$\begin{aligned}&\sigma_{zx}|_{z=0} = ((\sigma_{zx})_1 + (\sigma_{zx})_2 + (\sigma_{zx})_3)|_{z=0} = 0 \quad \Rightarrow \\ &\frac{1}{c_L}(-U_1 \sin 2\theta_1 e^{i\psi_1} + U_2 \sin 2\theta_2 e^{i\psi_2}) + \frac{V}{c_S} \cos 2\beta e^{i\psi_3} = 0\end{aligned} \qquad [1.110]$$

Again, the condition must hold independently from the value of the phase angles; whence the new reflection law:

$$(\psi_1 = \psi_2 = \psi_3)|_{z=0} \quad \Rightarrow \frac{\sin\theta_1}{c_L} = \frac{\sin\theta_2}{c_L} = \frac{\sin\beta}{c_S}$$

which can be stated in its final form as:

$$\theta_1 = \theta_2 = \theta = \quad \text{and} \quad \sin\theta = \gamma \sin\beta$$

where

$$\gamma = \frac{c_L}{c_s} = \sqrt{\frac{\lambda + 2G}{G}} = \sqrt{\frac{2(1-\nu)}{1-2\nu}} > 1 \quad \Rightarrow \beta < \theta \qquad [1.111]$$

The incident and reflected waves (α) described by [1.101] and [1.111] are shown in Figure 1.17.

Substituting [1.111] into [1.109] and [1.110] we arrive at the linear system:

$$\begin{aligned}&-(U_2 + U_1)(\lambda + 2G(\cos\theta)^2) + \gamma G V \sin 2\beta = 0 \\ &(U_2 - U_1)\sin 2\theta - \gamma V \cos 2\beta = 0\end{aligned} \qquad [1.112]$$

which can be suitably solved in terms of the displacement reflection coefficient $R = U_2/U_1$ and the displacement conversion coefficient $C = V/U_1$.

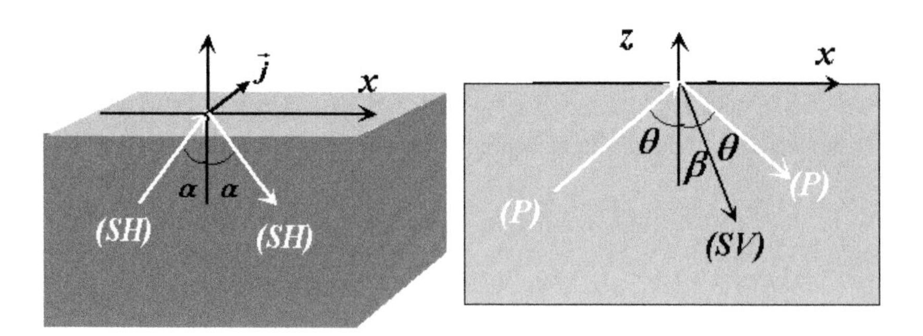

Figure 1.17. *Reflection of a (SH) wave and of a (P) wave on a free boundary*

Using [1.111], it follows that:

$$\frac{\lambda + 2G(\cos\theta)^2}{G} = \frac{\lambda + 2G(1 - (\sin\theta)^2)}{G}$$
$$= \frac{\lambda + 2G(1 - (\gamma\sin\beta)^2)}{G} = \gamma^2(1 - 2(\sin\beta)^2) = \gamma^2\cos 2\beta$$

The system [1.112] is thus transformed into:

$$\begin{bmatrix} \gamma\cos 2\beta & -\sin 2\beta \\ \sin 2\theta & \gamma\cos 2\beta \end{bmatrix}\begin{bmatrix} R \\ C \end{bmatrix} = \begin{bmatrix} -\gamma\cos 2\beta \\ \sin 2\theta \end{bmatrix} \quad [1.113]$$

which has the non-trivial solution:

$$R = \frac{\sin 2\theta \sin 2\beta - (\gamma\cos 2\beta)^2}{\sin 2\theta \sin 2\beta + (\gamma\cos 2\beta)^2}; \quad C = \frac{2\gamma\sin 2\theta\cos 2\beta}{\sin 2\theta\sin 2\beta + (\gamma\cos 2\beta)^2} \quad [1.114]$$

These functions are plotted in Figure 1.18 for a few values of Poisson's ratio.

It is worth noticing that R and C are independent of the wavelength and frequency of the waves. At normal and grazing incidence C vanishes and $R = -1$. The minus sign indicates that at normal incidence, the reflected (P) wave is in phase with the normal incident wave, whereas at grazing incidence it is in phase opposition. Thus in the last case, the superposition of the incident and the reflected wave leads to a vanishing resulting wave; which is surprising a priori. A more detailed study would show that the physical solution is not a plane wave but a *surface wave* confined to the vicinity of the boundary, see for instance [MIK 78]. On the other hand, if ν is sufficiently small, there exist two peculiar incidence angles at which R vanishes, which means total conversion of the (P) wave to a (SV) wave. To conclude on the subject, it is worth mentioning that similar calculations show that mode conversion

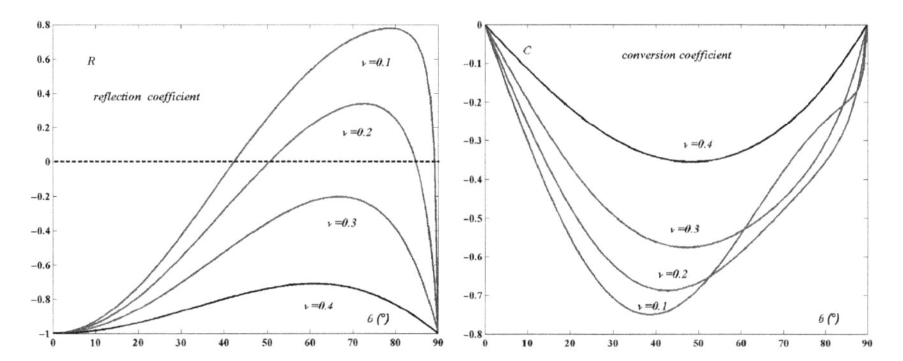

Figure 1.18. *Reflection of (P) waves on a free boundary: reflection and mode conversion coefficients*

of (P) and *(SV)* waves also occurs at a fixed boundary $(X, Z = 0)$, whereas it does not at a sliding boundary $(Z = 0,\ \sigma_{zx} = 0)$.

1.4.6 *Guided waves*

The linear elastic motions of a solid body can be interpreted as elastic waves resulting from the superposition of 'elementary waves'. Even if consideration is restricted to plane waves, it is necessary to distinguish various 'elementary waves' according to the nature (dilatation and shear waves), the polarization of the shear waves, and finally the angle of incidence on the boundaries. A priori, the system of waves one has to deal with seems overwhelmingly complicated to make such an approach practical. This actually turns out to be true in 3D elastodynamics, but the problem proves much easier if the solid body is modelled as a 1D, or 2D structural component, as we shall see later (see in particular Chapter 4). Such a simplification arises because in a structural component we deal with *guided waves* instead of the 3D 'elementary waves'. The concept of guided waves is of major importance to the understanding of the wave systems taking place in structural components. It is introduced here by considering the case of *(SH)* waves which are guided by the free boundaries of a plane solid layer.

1.4.6.1 Guided (SH) waves in a plane layer

The geometry of the problem is shown in Figure 1.19. Starting from the results established in subsection 1.4.5, the wave obtained by superposing the incident and the reflected *(SH)* waves may be written as:

$$\vec{W} = \vec{W}_i + \vec{W}_r = Y\vec{j}e^{i\omega(t-(x\sin\alpha)/c_S)}\left(e^{i\omega((z\cos\alpha)/c_S)} + e^{i\omega((-z\cos\alpha)/c_S)}\right) \quad [1.115]$$

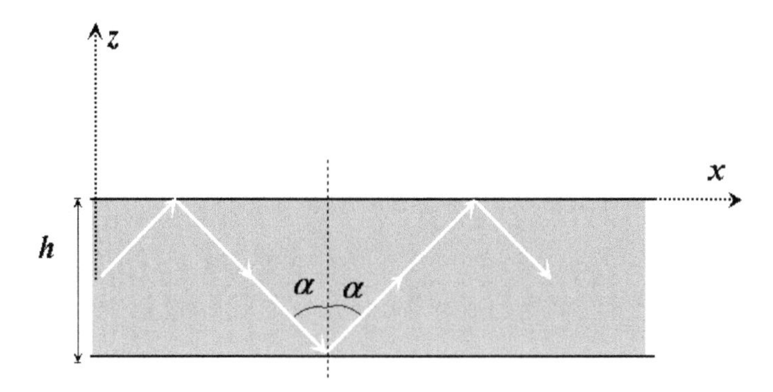

Figure 1.19. *(SH) waves in an infinite plane layer with free boundaries*

The result [1.115] can be transformed into:

$$\vec{W} = 2Y\cos(kz\cos\alpha)\vec{j}e^{ik(c_S t - x\sin\alpha)} = 2Y\cos\left(k^{(a)}z\cot\alpha\right)\vec{j}e^{ik^{(a)}\left(c_S^{(a)}t - x\right)}$$

where

$$k = \frac{\omega}{c_S}; \quad k^{(a)} = k\sin\alpha; \quad c_S^{(a)} = \frac{c_S}{\sin\alpha} \qquad [1.116]$$

The form [1.116] indicates that the resulting wave travels along the Ox direction at the apparent phase speed $c_S^{(a)}$ which is angle dependent. The intrinsic wave number k may be replaced by the apparent wave number $k^{(a)}$. According to such results, it becomes obvious that the elementary (SH) wave travelling at incidence α is guided by the reflecting plane, as a non-planar (SH) wave travelling in the $x > 0$ direction.

Considering now the reflection at the plane $z = -h$, the resulting wave [1.116] must comply also with the condition, which holds at $z = -h$:

$$\sigma_{zy} = G\left.\frac{\partial W}{\partial z}\right|_{z=-h} = 0 \Rightarrow \sin(hk^{(a)}\cot\alpha) = 0 \qquad [1.117]$$

The condition [1.117] leads to the new important feature that only a discrete set of guided waves can travel along the layer, namely those specified by the apparent wave numbers:

$$k_n^{(a)} = \frac{n\pi}{h}\tan\alpha \quad n = 0, 1, 2, \ldots \qquad [1.118]$$

The quantization rule [1.118] means that interferences between incident and reflected waves are generally destructive, except for specific phase relationships. Using the definition of $k^{(a)}$ given in [1.116], the condition [1.118] can be expressed in terms of wavelength as:

$$\frac{\lambda_n^{(a)}}{2} = \frac{h}{n} \quad \text{and} \quad \frac{\lambda_n}{2} = \frac{h}{n}\cos\alpha \qquad [1.119]$$

which means that the apparent half wavelength must be an integer submultiple of the plate thickness.

On the other hand, the z dependency of the guided waves is governed by the stationary wave shapes:

$$\varphi_n(z) = 2Y\cos\left(\frac{n\pi z}{h}\right) \quad n = 0, 1, 2, \ldots \qquad [1.120]$$

The case $n = 0$ corresponds to a plane wave and there are infinite modes of propagation $n > 0$ which are symmetrical or antisymmetrical about the middle plane of the layer, depending whether n is an even or an odd integer.

Another important aspect of the guided waves evidenced just above is the concept of cut-off frequency, which means that the layer cannot let waves propagate if they oscillate below a certain cut-off frequency, independently of the incidence angle. The relations [1.116] and [1.118] can be used to relate the intrinsic wave number to the circular frequency of the waves. This gives:

$$k^2 = \left(\frac{\omega}{c_S}\right)^2 = \left(\frac{k_n^{(a)}}{\sin\alpha}\right)^2 = \left(\frac{n\pi\tan\alpha}{h\sin\alpha}\right)^2 = \left(\frac{n\pi}{h\cos\alpha}\right)^2 \qquad [1.121]$$

An equation such as [1.121], which relates the wave number to the pulsation, is called a *dispersion equation*. It discloses the fact that for each mode of propagation n, there exists a continuous frequency-wave number relationship forming a *branch* of the dispersion equation. By using [1.118] we arrive at:

$$\begin{aligned}(k_n^{(a)})^2 &= \left(\frac{n\pi}{h}\tan\alpha\right)^2 = \left(\frac{n\pi}{h}\right)^2\left(\frac{1-(\cos\alpha)^2}{(\cos\alpha)^2}\right)\\ &= k^2 - \left(\frac{n\pi}{h}\right)^2 = \left(\frac{\omega}{c_S}\right)^2 - \left(\frac{n\pi}{h}\right)^2 \end{aligned} \qquad [1.122]$$

According to [1.122] the apparent wave number becomes imaginary as soon as the wave frequency becomes less than the threshold values:

$$\omega_n^{(c)} \leq \frac{n\pi c_S}{h} \qquad [1.123]$$

which means that, independently of the incidence angle α, waves become spatially *evanescent* if the frequency is less than the threshold value; this because the wave number becomes imaginary, leading to an exponential decay of the wave amplitude. Thus, the guided (*SH*) waves propagate according to the $n = 1$ mode only if the pulsation is higher than $\pi c_S/h$, or according to the $n = 2$ mode if the angular frequency is larger than $2\pi c_S/h$ etc. In contrast, the plane mode $n = 0$ is found to propagate independently of the wave frequency. Furthermore, because for the non-plane modes of propagation k is not proportional to ω, it can be concluded that the non-plane guided waves are dispersive, especially near the cut-off frequency. The dispersion equation [1.122] can be represented graphically as shown in Figure 1.20, which refers to a steel plane layer $h = 1$ m. In this kind of diagram, commonly called *frequency spectrum*, the wave frequency is plotted versus the real part and the imaginary part of the complex wave number k. To help visualization, $\mathrm{Re}(k)$ is plotted as a positive abscissa whereas $\mathrm{Im}(k)$ is plotted as a negative abscissa.

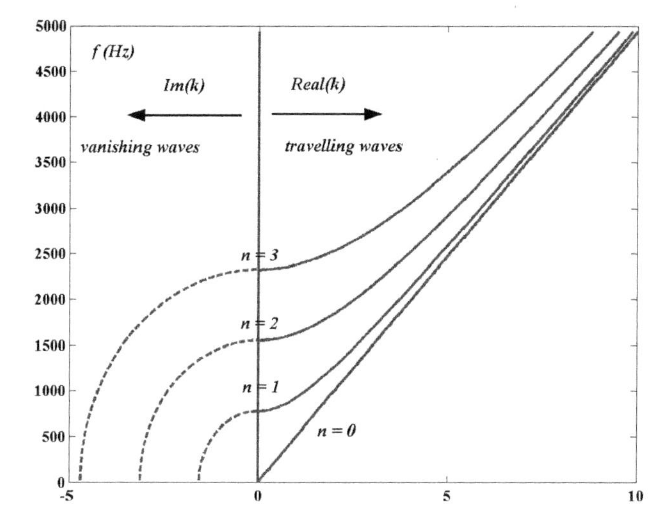

Figure 1.20. *Frequency spectrum of (SH) waves in a steel layer 1 m thick*

Furthermore, the travelling part of the branches which belong to *Re(k)* are in full lines and the vanishing parts belonging to $\text{Im}(k)$ are in dashed lines.

Of course, the complete frequency spectrum of propagation modes in an infinite plane layer is much more complicated than that displayed in Figure 1.20, because it also includes the branches related to the (P) and (SV) waves, which are found to be coupled in a free layer, in agreement with the results of subsection 1.4.5.

1.4.6.2 Physical interpretation

Figure 1.21 shows the *(SH)* wave system in the layer at a given time. It is represented by the traces of a sequence of planes of constant phase and by the direction of propagation. They comprise waves moving upward (full lines) and waves moving downward (dashed lines). The former are incident to the upper boundary $(z = 0)$ and are at the same time the waves reflected by the lower boundary $(z = -h)$ and the reverse occurs concerning downward waves. Direction of propagation is indicated by a white line which makes an angle α with the transverse direction *Oz*. Where a wave plane intersects a boundary, a reflected wave is initiated in such a way that the boundary conditions are satisfied. To understand the physical meaning of the apparent phase speed, we consider two planes of constant phase belonging to the same kind of waves and distant of a wavelength λ. It is immediately seen that the distance along the $0x$ direction is $\lambda_x = \lambda/\sin\alpha$. The time taken by a wave of a given frequency f to sweep this distance is the same whatever α may be. Thus, in

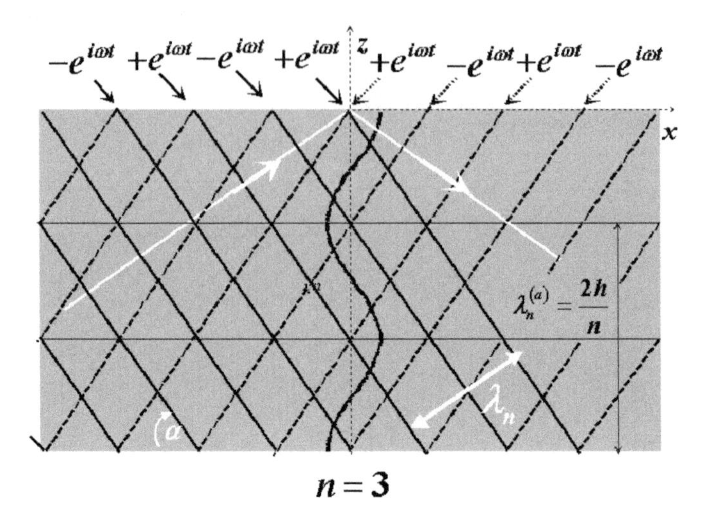

Figure 1.21. *Geometry of the (SH) wave system in the plane layer with free boundaries*

agreement with [1.116], the apparent phase speed is found to be:

$$c_S^{(a)} = \frac{f\lambda}{\sin\alpha} = \frac{c_S}{\sin\alpha}$$

Geometrical reasoning also makes clear why $c_S^{(a)}$ becomes infinite at normal incidence ($\alpha = 0$) and equal to c_S at grazing incidence. It is also possible to understand the major features of the guided waves propagation versus frequency. The wave shapes [1.120] create a stationary transverse pattern, of nodes and antinodes (displacements nulls and of maximum magnitude respectively). The boundaries correspond necessarily to antinodes and the integer n fixes the number of nodes. Let us consider first a guided wave mode $n \neq 1$. Using [1.119], the wavelength λ_n is related to the frequency by:

$$\lambda_n = \frac{2h\cos\alpha}{n} = \frac{c_S}{f_n}$$

As f_n diminishes λ_n increases and α must decrease to comply with the stationary transverse pattern of the n mode. Such an adjustment becomes clearly impossible below the cut-off frequency $f_n^{(c)}$ given by:

$$\frac{c_S}{f_n^{(c)}} = \frac{2h}{n} \Rightarrow f_n^{(c)} = \frac{nc_S}{2h}$$

Furthermore, as the apparent phase speed is frequency dependent, propagation is dispersive, except for the plane wave mode $n = 0$.

$$c_n^{(a)} = \frac{c_S}{\sin\alpha} = \frac{c_S}{\sqrt{1-(\cos\alpha)^2}} = \frac{c_S}{\sqrt{1-((nc_S)/(2hf_n))^2}}$$

$$\text{where } f_n > f_n^{(c)} = nc_S/2h$$

1.4.6.3 Waves in an infinite elastic rod of circular cross-section

As a consequence of the effects briefly described in this and the former subsection, the analytical study of waves in an elastic solid body is intractable, except in very particular cases such as the famous problem treated by Pochhammer in 1876. This author succeeded in working out the exact solution of the axial propagation of the elastic waves guided within a circular cylinder of infinite length. The dispersion equation is marked by the existence of many branch solutions, analogous to those depicted in Figure 1.21, each one characterizing a specific type of guided waves. This theoretical work, described in particular in [MIK 78], is considered rightly to be a major achievement, not only because of the analytical difficulties involved in the task, but also for the more important reason that it furnishes an invaluable reference to check the validity of the simplified analytical models of structural elements used in engineering applications. As further discussed in Chapter 4 in the case of straight beams, the waves arising from such simplified models may be understood as suitable approximations of the lowest branches of the exact dispersion equation arising from the three-dimensional theory.

1.4.7 *Standing waves and natural modes of vibration*

As already pointed out, in a bounded and conservative medium the waves must be fully reflected at the boundaries to comply with the condition of conservation of mechanical energy. As a consequence, when waves are triggered in the medium by some excitation process, as the incident waves are interfering with the reflected waves, a steady oscillatory state is established, giving rise to a system of *standing waves*. The natural modes of vibration of an elastic body identify precisely with these standing waves. The mechanism is basically the same as that already described to establish the guided modes of (*SH*) waves in an infinite plane layer. In an analogous way, it is found that because of the boundary conditions, nontrivial solutions occur only for discrete values of frequency/wave number pairs which provide a discrete spectrum of natural frequencies of the vibration modes. It has to be pointed out that the analytical determination of these modes is tractable only in a relatively few number of geometries and boundary conditions. Again, the difficulty stems in particular from the mode conversion encountered in reflection of elastic waves impinging at oblique incidence on the boundaries of the solid, as illustrated by the few following examples.

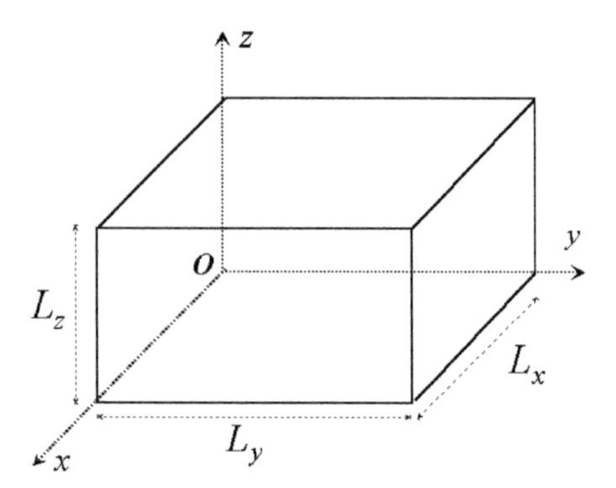

Figure 1.22. *Elastic solid shaped as a cuboid*

1.4.7.1 Dilatation plane modes of vibration

As a first example, we consider an elastic body shaped as a right parallelepiped (or cuboid). Of course, this particular geometry is chosen for convenience in using the Cartesian coordinate system specified in Figure 1.22. The problem is further simplified by assuming that the boundaries parallel to the $0yz$ plane at $x = 0$ and $x = L_x$ are fixed in the normal direction ($O\vec{x}$ axis). Appropriate conditions on the other boundaries will be discussed later.

Starting from the travelling plane waves [1.81], the boundary conditions $X(0) = X(L_x) = 0$ specify both the shapes and the frequencies of the standing waves:

$$X_+ + X_- = 0 \Rightarrow X(x) = X_0 \sin\left(\frac{\omega x}{c_L}\right)$$

$$X(L_x) = 0 \Rightarrow \omega_n = \frac{n\pi c_L}{L_x}; \quad n = 1, 2, \ldots \qquad [1.124]$$

The conditions $X(0) = X(L_x) = 0$ imply a total reflection of the waves with a change of sign at the boundaries $x = 0$ and $x = L_x$, in such a manner that the incident wave is exactly cancelled by the reflected wave at these positions. From [1.124], it is verified that standing waves can exist only for specific wavelengths given by the condition:

$$\lambda_n = \frac{2L_x}{n} \qquad [1.125]$$

which is analogous to the condition [1.119] for the guided waves in a plane layer. The half-length of the standing waves is thus found to be necessarily a submultiple of the length L_x. On the other hand, the amplitude X_0 of this admissible system of standing harmonic waves is arbitrary. Then, it is convenient to define the following sequence of longitudinal modes of vibration given by:

$$\varphi_n^{(L)}(x) = \sin\left(\frac{n\pi x}{L_x}\right); \quad f_n = \frac{nc_L}{2L_x}; \qquad n = 1, 2, \ldots \qquad [1.126]$$

The mode shapes $\varphi_n^{(L)}(x)$ are normalized here to a maximum displacement of unit magnitude. f_n is the natural frequency at which the mode vibrates harmonically. As indicated by the integer index n, there exists an infinite and countable sequence of such modes. However, their actual existence cannot be asserted before having specified suitable conditions at the other faces of the parallelepiped, which have not yet been discussed. Substituting the assumed mode shapes [1.126] into the elastic strain-stress law [1.37], it is readily verified that the shear stresses σ_{xy}, σ_{zy}, σ_{xz} do vanish everywhere. Accordingly, the faces of the body must be free to move in their own plane. In contrast, the normal stresses are found to differ from zero, not only σ_{xx}, but also σ_{yy} and σ_{zz} due to the Poisson effect. Then, the normal displacements must vanish on all the faces. Such conditions, termed '*sliding supports*', were already introduced at the end of subsection 1.4.5, in which it was mentioned that they lead to uncoupled (P) and (SV) waves; which explains why dilatation modes without shear are found to exist if the cubical solid is provided with sliding supports. However, it may be noted that such support conditions are difficult to realize physically.

1.4.7.2 Dilatation modes of vibration in three dimensions

Let us consider the same elastic body as above with all its faces supported in the normal directions, as in the first example. To investigate the three-dimensional waves, it is convenient to use the displacement potential $\Phi(x, y, z)$, defined by [1.76]. In terms of this potential, the harmonic dilatation waves are governed by the following equation:

$$\left(\frac{\omega}{c_L}\right)^2 \overrightarrow{\text{grad}\,\Phi} + \Delta\overrightarrow{\text{grad}\,\Phi} = 0 \iff \overrightarrow{\text{grad}}\left\{\left(\frac{\omega}{c_L}\right)^2 \Phi + \Delta\Phi\right\}$$

$$= 0 \iff \left(\frac{\omega}{c_L}\right)^2 \Phi + \Delta\Phi = C_0 \qquad [1.127]$$

C_0 stands for a constant which can however be set to zero since $\Phi(x, y, z)$ is defined except for an arbitrary additive constant. Accordingly, the homogeneous vector equation [1.127] can be reduced to the scalar equation:

$$\left(\frac{\omega}{c_L}\right)^2 \Phi + \Delta\Phi = 0 \qquad [1.128]$$

which is easily solved by separating the variables. If the trial function $\Phi(x, y, z) = A(x)B(y)C(z)$ is substituted into [1.128], we get:

$$\left(\frac{\omega}{c_L}\right)^2 + \frac{A''}{A} + \frac{B''}{B} + \frac{C''}{C} = 0 \qquad [1.129]$$

Noting that A''/A can be a function of x only, B''/B of y only and C''/C of z only, it follows that all these quantities must be constants. Furthermore, owing to the geometry of the body and the boundary conditions which were assumed, x, y, z are interchangeable. So [1.129] gives the dispersion equation:

$$\left(\frac{\omega}{c_L}\right)^2 - \left(k_x^2 + k_y^2 + k_z^2\right) = 0 \qquad [1.130]$$

The scalars k_x, k_y, k_z can be physically interpreted as wave numbers defining the Cartesian components of a wave vector $\vec{k}$. Starting from the general case $k_x, k_y, k_z \neq 0$, the functions to be determined are solutions of the uncoupled differential system composed of three equations of the same form:

$$\frac{d^2A}{dx^2} + k_x^2 A = 0; \quad \frac{d^2B}{dy^2} + k_y^2 B = 0; \quad \frac{d^2C}{dz^2} + k_z^2 C = 0 \qquad [1.131]$$

which have the general solution:

$$A(x) = a_+ e^{-ik_x x} + a_- e^{ik_x x}; \quad B(y) = b_+ e^{-ik_y y} + b_- e^{ik_y y};$$
$$C(z) = c_+ e^{-ik_z z} + c_- e^{ik_z z} \qquad [1.132]$$

The boundary conditions already specified are written in terms of displacements as:

$$X(0, y, z) = X(L_x, y, z) = 0; \quad Y(x, 0, z) = X(x, L_y, z) = 0;$$
$$Z(x, y, 0) = Z(0, y, L_z) = 0 \qquad [1.133]$$

As $\vec{X} = \overrightarrow{\text{grad}\,\Phi}$, [1.133] is immediately expressed in terms of the displacement potential as:

$$\left.\frac{\partial \Phi}{\partial x}\right|_{x=0,L_x} = \left.\frac{\partial \Phi}{\partial y}\right|_{y=0,L_y} = \left.\frac{\partial \Phi}{\partial z}\right|_{z=0,L_z} = 0 \qquad [1.134]$$

which furnishes infinitely many modal solutions depending of the three integer numbers $l,\ m,\ n$:

$$\Phi_{\ell,m,n}(x, y, z) = \phi_0 \cos\left(\frac{\ell\pi x}{L_x}\right)\cos\left(\frac{m\pi y}{L_y}\right)\cos\left(\frac{n\pi z}{L_z}\right) \quad \ell, m, n = 0, 1, 2, \ldots \qquad [1.135]$$

where ϕ_0 is an arbitrary constant.

The corresponding natural pulsations are:

$$\omega_{\ell,m,n} = \frac{1}{c_L}\sqrt{\left(\frac{\ell\pi}{L_x}\right)^2 + \left(\frac{m\pi}{L_y}\right)^2 + \left(\frac{n\pi}{L_z}\right)^2}; \quad \ell, m, n = 0, 1, 2, \ldots \quad [1.136]$$

In terms of displacements, the mode shapes are written as:

$$\begin{cases} \varphi^{(x)}_{\ell,m,n}(x,y,z) = \dfrac{\partial \Phi_{\ell,m,n}}{\partial x} = \left(-\dfrac{\ell\pi\phi_0}{L_x}\right)\sin\left(\dfrac{\ell\pi x}{L_x}\right)\cos\left(\dfrac{m\pi y}{L_y}\right)\cos\left(\dfrac{n\pi z}{L_z}\right) \\ \varphi^{(y)}_{\ell,m,n}(x,y,z) = \dfrac{\partial \Phi_{\ell,m,n}}{\partial y} = \left(-\dfrac{m\pi\phi_0}{L_y}\right)\cos\left(\dfrac{\ell\pi x}{L_x}\right)\sin\left(\dfrac{m\pi y}{L_y}\right)\cos\left(\dfrac{n\pi z}{L_z}\right) \\ \varphi^{(z)}_{\ell,m,n}(x,y,z) = \dfrac{\partial \Phi_{\ell,m,n}}{\partial z} = \left(-\dfrac{n\pi\phi_0}{L_z}\right)\cos\left(\dfrac{\ell\pi x}{L_x}\right)\cos\left(\dfrac{m\pi y}{L_y}\right)\sin\left(\dfrac{n\pi z}{L_z}\right) \end{cases} \quad [1.137]$$

A few comments about these results are useful. First, the particular case where the three indices are zero is irrelevant to the present problem, since the solution $\omega = k_x = k_y = k_z = 0$ describes free motions of the rigid body without supports. The plane waves correspond to a single index differing from zero. Then, it is noted that the ratios between the relative amplitude of the Cartesian components of the modal vector $\vec{\varphi}_{\ell,m,n}$, have specific values, which indicates that the three directions are coupled together. Finally, the shear modal stresses are found to be:

$$\begin{aligned} \sigma_{xy} &= \frac{G}{2}\left\{\frac{\partial \varphi^{(x)}}{\partial y} + \frac{\partial \varphi^{(y)}}{\partial x}\right\} \\ &= \phi_0 G\left(\frac{\ell\pi}{L_x}\right)\left(\frac{m\pi}{L_y}\right)\sin\left(\frac{\ell\pi x}{L_x}\right)\sin\left(\frac{m\pi y}{L_y}\right)\cos\left(\frac{n\pi z}{L_z}\right) \\ \sigma_{xz} &= \frac{G}{2}\left\{\frac{\partial \varphi^{(x)}}{\partial z} + \frac{\partial \varphi^{(z)}}{\partial x}\right\} \\ &= \phi_0 G\left(\frac{\ell\pi}{L_x}\right)\left(\frac{n\pi}{L_z}\right)\sin\left(\frac{\ell\pi x}{L_x}\right)\cos\left(\frac{m\pi y}{L_y}\right)\sin\left(\frac{n\pi z}{L_z}\right) \\ \sigma_{yz} &= \frac{G}{2}\left\{\frac{\partial \varphi^{(z)}}{\partial y} + \frac{\partial \varphi^{(y)}}{\partial z}\right\} \\ &= \phi_0 G\left(\frac{n\pi}{L_z}\right)\left(\frac{m\pi}{L_y}\right)\cos\left(\frac{\ell\pi x}{L_x}\right)\sin\left(\frac{m\pi y}{L_y}\right)\sin\left(\frac{n\pi z}{L_z}\right) \end{aligned} \quad [1.138]$$

Thus, they vanish on the faces of the parallelepiped. As the tangential displacements are different from zero there, all the faces must be free to move in their own plane. As already indicated in subsection 1.4.5, if such sliding boundary conditions hold, the dilatational (P) waves are reflected without mode conversion. Again, it is stressed that sliding supports would be difficult to manufacture. Incidentally, if the solid is replaced by a non-viscous fluid, i.e. a medium in which the Poisson effect and the shear stresses do not exist, the present calculation takes on its full practical interest. In a fluid, the dilatation waves are called *acoustic waves* and the natural modes defined by the formulas [1.136], [1.137] are the *acoustic* modes of a rectangular enclosure provided with rigid and motionless walls.

1.4.7.3 Shear plane modes of vibration

Starting from the plane shear waves [1.86], it is immediately verifiable that shear stresses $\sigma_{xy} = GdY/dx$ and $\sigma_{xz} = GdZ/dx$ are the only nonzero stress components. They both must vanish at the faces $x = 0$ and $x = L_x$ which are assumed to be free. Whence the boundary conditions:

$$\left.\frac{dY}{dx}\right|_0 = \left.\frac{dY}{dx}\right|_{L_x} = 0; \quad \left.\frac{dZ}{dx}\right|_0 = \left.\frac{dZ}{dx}\right|_{L_x} = 0 \qquad [1.139]$$

From [1.77] it follows that the partial differential equation to be solved is:

$$G\Delta\left[(Y(x)\vec{j} + Z(x)\vec{k})\right] + \omega^2\rho_0((Y(x)\vec{j} + Zx\vec{k})) = 0 \qquad [1.140]$$

Applying the relation [A.2.14] in Appendix A.2 to calculate the Laplacian of the displacement vector, it is readily shown that the vector equation [1.140] can be split into two identical and uncoupled equations of the same type as those already encountered in the last examples:

$$\frac{d^2Y}{dx^2} + \left(\frac{\omega}{c_s}\right)^2 Y = 0; \quad \frac{d^2Z}{dx^2} + \left(\frac{\omega}{c_s}\right)^2 Z = 0 \qquad [1.141]$$

Hence the waves oscillating in the *Oy* direction are found to be identical to those oscillating in the *Oz* direction. Such a result could have been anticipated, since the transverse sections parallel to the *Oyz* plane are not deformed and oscillate as a rigid body in the *Oy* and in the *Oz* directions, independently. As a consequence, the corresponding natural modes of vibration can be split into two families which are identical to each other, except that they vibrate into two orthogonal directions. The common modal quantities are:

$$\varphi_n^{(T)}(x) = \cos\left(\frac{n\pi x}{L_x}\right); \quad \omega_n = \frac{n\pi c_s}{L_x}; \qquad n = 0, 1, 2, \ldots \qquad [1.142]$$

By using the principle of superposition, the physical mode shapes can be written as:

$$\vec{\varphi}_n^{(T)}(x) = (a_n\vec{j} + b_n\vec{k})\varphi_n^{(T)}(x) \qquad [1.143]$$

where the vector $(a_n\vec{j} + b_n\vec{k})$ specifies the direction of vibration, that is the polarization of the wave. Clearly, any transverse direction is possible and equivalent to the others. It can be also noted that modes at zero frequency exist, which are immediately recognized as a rigid displacement of the whole body in an arbitrary transverse direction.

1.5. From solids to structural elements

1.5.1 *Saint-Venant's principle*

The difficulty in obtaining the analytical solutions of elastodynamic problems is also encountered when attempting to solve static problems, as for instance that of determining the stresses in a three-dimensional body induced by forces applied on small portions of it. As an example, let us consider a cylindrical rod inserted in a tensile test machine, see Figure 1.23. If a numerical simulation of the experiment is performed, by using the finite element method (regarding the finite element method see Chapter 3, section 3.4), quite enlightening results are obtained concerning the displacement field and the stress distribution, especially sufficiently near the grips. Major information can be suitably summarized as follows:

1. Near the loaded parts of the rod, in a portion extending no more than a fraction of the rod radius in each direction, the displacement field, and even more conspicuously the stress field, are found to vary in all directions, in connection with the detailed space distribution of the contact forces exerted by the grips. Such fields characterize the so called *local response* of the system rod plus grips.

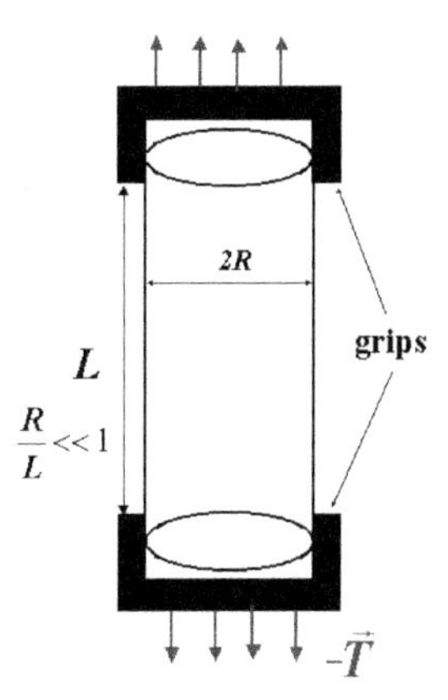

Figure 1.23. *Cylindrical rod set in a tensile test machine*

2. In the current portion of the rod, sufficiently far from the grips, the stress field is found to reduce practically to only one non-vanishing component, namely the uniform axial stress $\sigma_{zz}(r,\theta,z) = T/S$, where cylindrical coordinates r, θ, z are used. Here, T stands for the resultant of the tensile force system exerted by the grips and S denotes the cross-sectional area of the rod. The displacement field is marked by an axial component denoted by W and a radial component denoted by U, which are independent of the polar angle θ. W is observed to increase in proportion to z, starting from the fixed grip, and U varies in proportion to r. All these fields are independent of the distribution of the loading by the grips and characterize the *global response* of the rod.

As can be expected from the simplicity of the response observed sufficiently far from the grips, the analytical solution of the problem becomes straightforward, provided it is formulated in a suitable way. A first idea is to adopt the model of a circular cylinder loaded at its bases by a uniform axial force density whose resultant is T. The problem can be thus solved by using a 2D model, since the θ dependency is removed. Further, if the interest is restricted to determining the axial displacement and the axial stress, a 1D model becomes sufficient. Finally, to determine σ_{zz} only, it becomes unnecessary to model the rod as an elastic body because a single DOF system suffices. This elementary example emphasizes, if necessary, the importance of modelling mechanical systems in close relation to the nature of the information desired. In continuous systems, a major simplification occurs when the study of the *local* and of the *global* responses can be split into two distinct problems. This is often the case in solid mechanics and this kind of simplification has been stated as a principle by Barré de Saint-Venant (1885), which is enunciated as follows:

The elastic response induced by a local force system, whose resultant force and torque are both zero, become negligible far enough from the small loaded portion of the body. In other words, if sufficiently far from the loaded domain, the response depends solely upon the resultant force and torque of the actual loading system.

Here 'local' means a part which is much smaller than the size of the whole body, or of the boundary in the case of contact forces, and 'far enough' means distances substantially greater than the length scales of the loaded part.

As detailed further in the next chapters, the Saint-Venant principle may be considered rightly as the cornerstone for modelling solids as structural elements, even if it is not explicitly invoked in most cases. Nevertheless, care has to be taken when using it, as its validity is not universal. In particular, an important exception is encountered in the case of thin shells of revolution loaded by concentrated forces. This can be easily realized by observing the deformation of a flexible pail filled of water, or sand, and held by its handle. Indeed, it can be observed that the reactions at the handle fasteners induce an ovalization of the pail cross-sections which extends far beyond the vicinity of the fasteners. The colour Plates 1 and 2 illustrate such results which are further discussed in Chapter 8, see also Plates 13 and 14.

1.5.2 *Shape criterion to reduce the dimension of a problem*

In most engineering applications, use is made of structural elements which are characterized by a few generic particularities in their geometry. Such peculiarities can be advantageously used to simplify their mathematical modelling. In particular, most structural elements can be characterized by the fact that one or two dimension(s) is/are much smaller than the other(s). Then, the corresponding strain components can be neglected and finally the dimensions connected to the small scales can be suppressed, as detailed in the rest of the book for several types of structures. A few simple 'order of magnitude' calculations can be used to support this simplifying statement.

1.5.2.1 Compression of a solid body shaped as a slender parallelepiped

An elastic bar shaped as a slender parallelepiped is constrained on the face $x = 0$ and loaded on the opposite face $x = L_x$ by a compressive force $\vec{T}^{(e)} = T_0\vec{i}$ parallel to $0x$, see Figure 1.24. The three-dimensional problem is not simple if the local response is needed. Indeed, it depends upon the stress distribution on the loaded face, and also on the boundary conditions in the vicinity of the opposite face. As already indicated in the example of the rod, the same problem is drastically simplified when restricted to the global response. According to Saint-Venant's principle, the global solution may be expected to be valid in the largest portion of the bar, if L_x is much larger than L_y and L_z. The actual loading exerted on the face $x = L_x$ is described by the force density $t_x^{(e)}(y,z)$. However, for determining the global response, an equivalent load can be defined as the resultant $T_0 = \iint_{(S)} t_x^{(e)}(y,z)\,dy\,dz$ applied to the centre of the free end of the bar. For mathematical convenience, the face $x = 0$ is provided with a sliding support condition, i.e. $(X(0,y,z) = 0)$. The other faces are free. With these hypotheses, the

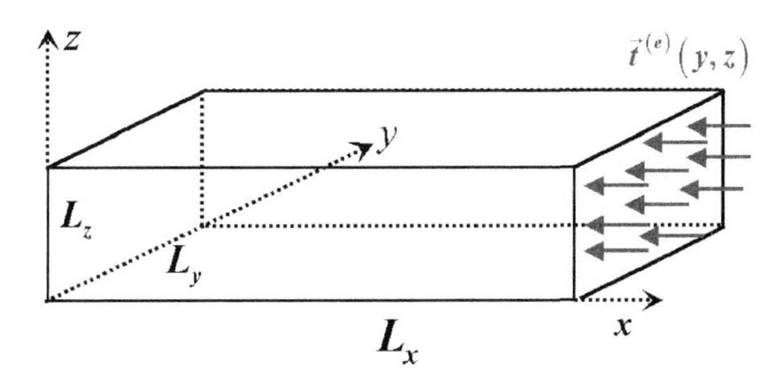

Figure 1.24. *Bar shaped as a slender parallelepiped under compressive load*

equilibrium is obtained from an axial state of stress whose nonzero component is:

$$\sigma_{xx} = \frac{T_0}{L_y L_z} \quad [1.144]$$

Of course, this result fully agrees with the rod example of subsection 1.5.1. However, the interesting point here is that by solving the same problem along the *Oy* and *Oz* axes instead of *Ox*, we find much lower stresses which are,

$$\sigma_{yy} = \frac{T_0}{L_x L_z}; \quad \sigma_{zz} = \frac{T_0}{L_x L_y} \Rightarrow \frac{\sigma_{yy}}{\sigma_{xx}} = \frac{L_z}{L_x} \ll 1; \quad \frac{\sigma_{zz}}{\sigma_{xx}} = \frac{L_y}{L_x} \ll 1 \quad [1.145]$$

On the other hand, as the strains increase with the stresses, so, for any load of same magnitude, the following inequality holds:

$$\varepsilon_{xx} \gg \varepsilon_{yy}, \varepsilon_{zz} \quad [1.146]$$

Thus, the idea which arises naturally is to neglect the transverse strains $\varepsilon_{yy}, \varepsilon_{zz}$ which are found to be much smaller than the axial strain. In other words, if the body extends much more in the axial direction than in the transverse directions, it will be modelled as an equivalent body which is completely rigid in the transverse directions. The corresponding stresses needed to obtain the equilibrium of the solid are given by the Lagrange multipliers associated with the constraint conditions:

$$\varepsilon_{zz} = \varepsilon_{yy} = 0 \quad [1.147]$$

1.5.2.2 Shearing of a slender parallelepiped

What was said just above in the case of a normal loading is also true in the case of a tangential loading. For instance, let us consider the two similar problems sketched in Figure 1.25. In the case (*a*), a load $T_0\vec{i}$ is applied on the face $z = L_z$, the opposite face being fixed in any direction. From the equilibrium conditions it is not difficult to show that normal stresses are identically zero. Hence div $\vec{X} = 0$ and the displacement field is necessarily in the longitudinal direction ($Y = Z = 0$). The shear stress and strain fields are:

$$\sigma_{zx} = \frac{T_0}{L_x L_y}; \quad \varepsilon_{zx} = \frac{1}{2}\frac{\partial X}{\partial z} \quad [1.148]$$

In the case (*b*), the load $T_0\vec{k}$ is applied on $x = L_x$. Of course, the same reasoning as in case (*a*) applies. The shear stress and strain fields are:

$$\sigma_{xz} = \frac{T_0}{L_y L_z}; \quad \varepsilon_{xz} = \frac{1}{2}\frac{\partial Z}{\partial x} \quad [1.149]$$

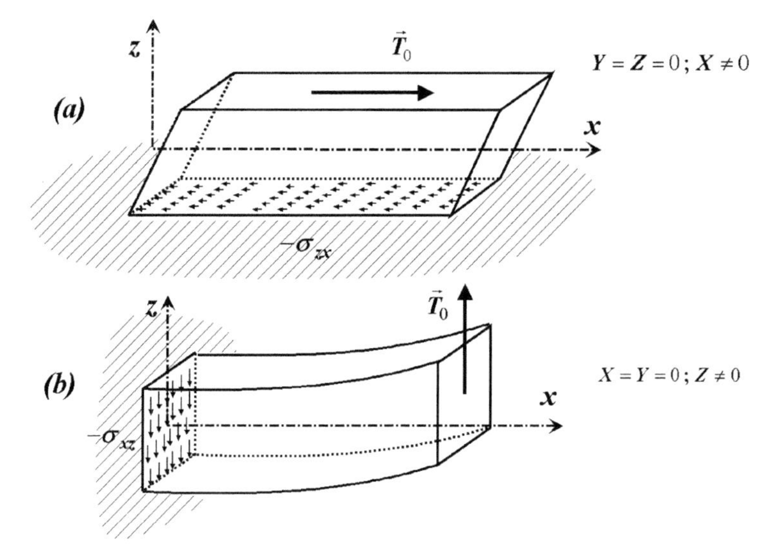

Figure 1.25. *Shear deformation of a slender parallelepiped*

As the stresses are again increasing functions of the strains, it follows that:

$$\frac{(\sigma_{zx})_{(a)}}{(\sigma_{xz})_{(b)}} = \frac{L_z}{L_x} \ll 1 \Rightarrow \frac{\partial X}{\partial z} \ll \frac{\partial Z}{\partial x} \qquad [1.150]$$

From similar calculations related to the other faces of the parallelepiped, it can also be stated that:

$$L_x \gg L_y, L_z \Rightarrow \frac{\partial X}{\partial z}; \frac{\partial X}{\partial y}; \frac{\partial Y}{\partial z}; \frac{\partial Z}{\partial y} \ll \frac{\partial Z}{\partial x}; \frac{\partial Y}{\partial x} \qquad [1.151]$$

A major conclusion arising from such orders of magnitude is that if L_x is large enough in comparison with the other two dimensions and if the local effects due to the loads and the boundary conditions are negligible, the gradient of the displacement field is also negligible, if taken along the directions L_y, L_z. As an immediate corollary, the strain components in the planes parallel to *Oyz* can be discarded. In other words, the cross-sections of the parallelepiped parallel to *Oyz* can be modelled as rigid bodies.

1.5.2.3 Validity of the simplification for a dynamic loading

The relations [1.136] and [1.137] defining the natural frequencies and mode shapes of a cubical solid are convenient to discuss the validity of the previous simplifications in the dynamic domain. Since we assume here that $L_x \gg L_y, L_z$, the first natural frequencies associated to the plane modes – indexed by the numbers $(n, 0, 0)$ – are much lower than that of the other modes. Therefore, if the frequency range of the excitation spectrum is much smaller than the lowest frequency ($f_{1,1,0}$ or $f_{1,0,1}$) of the first non-plane mode, then the dynamic response can be obtained with sufficient accuracy by taking into account the plane modes only (cf. for instance [AXI 04], Chapters 7 and 9). In other words, the natural modes of vibration which are related to transversal strains (*Oy, Oz* directions) can be safely neglected if their frequency is much higher than the frequency range of the excitation. It may be also noted that, according to such an approximation, the frequency spectrum of the guided waves is restricted to the plane wave branches, which are the lowest branches of the complete 3D frequency spectrum.

1.5.2.4 Structural elements in engineering

From the previous considerations, it is concluded that to make tractable the analysis of engineering structures, the first simplifying assumption is to replace the real three-dimensional continuous medium by an equivalent continuous medium of smaller dimension. In this way, structural elements can be defined, as *cables or beams* which are one-dimensional structures and as *plates and shells*, which are two-dimensional structures, see Figure 1.26. It is also worth emphasizing that if a dimension is changed, all the formulation of the equilibrium equations is deeply modified too, like the boundary conditions and space distribution of the loading. On the other hand, the advantages gained by using such simplified models in structural analysis are very important. The following chapters are devoted to work out such simplified models and to studying their properties both in the static and in the dynamic domains. However, the present book is not exhaustive by far and has been restricted, on purpose, to the structural elements which are the most commonly encountered in engineering. For mathematical convenience, we start by studying

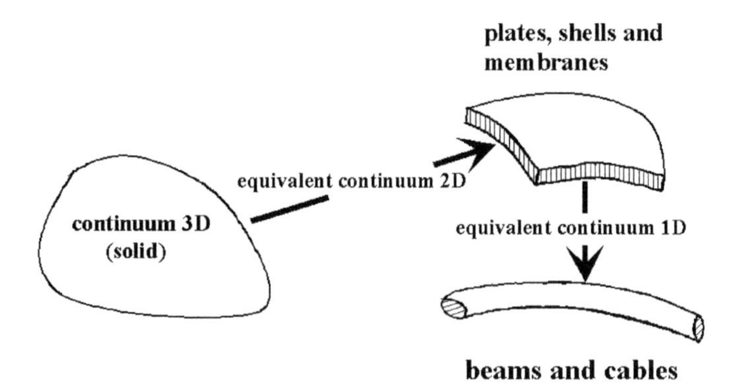

Figure 1.26. *From solids to structural elements*

the straight beams, then the plates and finally, the curved elements such as the arches and the shells. The present order of presentation is appropriate to take into account progressively the increasing mathematical difficulties encountered in modelling such structures. However, from a logical point of view, an exactly inverse order of presentation should be more concise, as, starting from a 3D solid body, it is possible to deduce all the other models as particular cases, as indicated in Figure 1.26.

Chapter 2

Straight beam models: Newtonian approach

Beams are slender structural elements which are employed to support transverse as well as axial loads. They are of very common use as columns, masts, lintels, joists etc, or as constitutive parts of supporting frames of buildings, cars, airplanes etc. They may be considered as the simplest structural element because of the relative simplicity of the equilibrium equations arising from a suitable condensation of the space variables into a single one, namely the abscissa along the beam length. However, the approximations used to represent in an appropriate manner the 3D solid as an equivalent 1D solid involve several subtleties, and several degrees of approximation can be proposed to refine the beam model when necessary. The object of this first of four chapters devoted to beams is to introduce the basic ideas of the elementary theory of beams. For the sake of simplicity at least, it is found convenient to adopt here a Newtonian instead of a variational approach. The latter will be adopted in the next chapters to establish more refined models than those presented here.

2.1. Simplified representation of a 3D continuous medium by an equivalent 1D model

2.1.1 *Beam geometry*

Let the dimensions of a structure be defined by three scale factors denoted L, D_1, D_2.

If one dimension L, called the length, is much larger than the two others, called transverse dimensions (see Figure 2.1 in which $D_1 \approx D_2 \approx D$) it is said to fall within the class of '*beams*'. The ratio L/D is called the *slenderness ratio*. As indicated in Chapter 1, subsection 1.5.2, the purpose of the present chapter is to model a beam structure as a one-dimensional equivalent medium. Having this purpose in mind, a first step is to simplify the 3D expression of the strain and displacement fields in a suitable way.

2.1.2 *Global and local displacements*

Let us cut a beam by a nearly transversal plane. The centre of the area of this section – which is supposed homogeneous and isotropic – is called the *centroid* denoted C. The *line of centroids*, or *central axis* is defined as the line that passes through the centroids along the beam. The cross-sections are then defined as the beam sections which are perpendicular to the central axis. For the sake of simplicity,

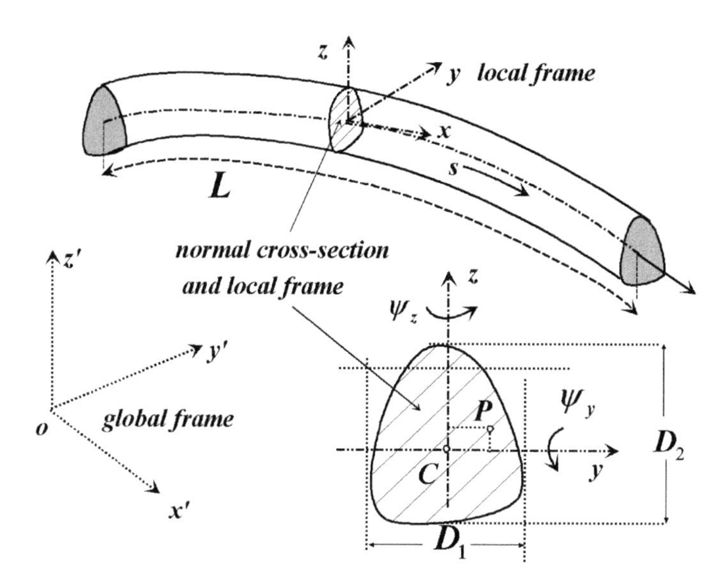

Figure 2.1. *Geometric representation of a beam*

the material is assumed in what follows to be isotropic and homogeneous in the same cross-section, but can vary along the beam central axis. The result obtained in subsection 1.5.2 of Chapter 1 in the particular case of a right parallelepiped is generalized as follows: if the slenderness ratio is high enough, the longitudinal strains are much larger than the transverse ones, for a given external loading. Therefore, it is assumed that:

The cross-sections remain rigid during any possible motion. Then, to describe the beam movement, six variables of displacement are needed which comprise three components of translation: the linear displacements of the centroid and three angular components: the small rotations of the cross-section about the centroid. These functions are dependent upon one space variable only, which is defined as the curvilinear coordinate along the line of centroids.

As shown in Figure 2.1, two Cartesian coordinate systems are used for writing the equilibrium equations. The global frame $Ox'y'z'$ serves to describe the structure as a whole, and the local frame $Cxyz$ serves to write the equilibrium equations of a beam element of infinitesimal length between the abscissa s and $s + ds$. $C(s)$ is the centroid of the cross-section located at s, Cx is tangent to the line of centroids, Cy and Cz are two principal axes of inertia of the cross-section. However, in this and in the next three chapters, consideration is restricted to straight beams, i.e. beams which have a straight central axis in the non deformed state and in which all the cross-sections have the same principal axes of inertia. Then the global frame $Oxyz$ is defined by the direction Ox merged with the central axis. Accordingly, the curvilinear abscissa s is replaced by the Cartesian coordinate x. On the other hand, the two transverse axes Oy and Oz are parallel to an orthogonal pair of principal axes of inertia of the cross-sections. As a further simplification, the cross-sections will be assumed to be symmetric with respect to the Cy and Cz axes, as is the case for instance in Figure 2.2. A lack of central symmetry of the cross-sections induces some specific complications which will be discussed separately, in subsection 2.2.6.

The displacement of the centroid is defined by the translation vector $\vec{X}$ of Cartesian components $X(x), Y(x), Z(x)$. The rotation of the cross-section is defined by the vector $\vec{\psi}$. In accordance with the hypothesis of 'small' motions, the Cartesian components of $\vec{\psi}$, hereafter denoted $\psi_x(x), \psi_y(x), \psi_z(x)$, are the small angles of rotation with respect to the axes of the local coordinate system (Figures 2.1 and 2.2). These six quantities are the components of the *global displacement field* of the straight beam. They depend on the x abscissa and on the time t if the problem is dynamic.

The *local displacement field* $\vec{\xi}(\vec{r})$ of a particle P, whose position is defined by the vector $\vec{r}$ (coordinates y, z in the local frame) is written as:

$$\vec{\xi}(x, y, z) = \vec{X}(x) + \vec{\psi}(x) \times \vec{r} \qquad [2.1]$$

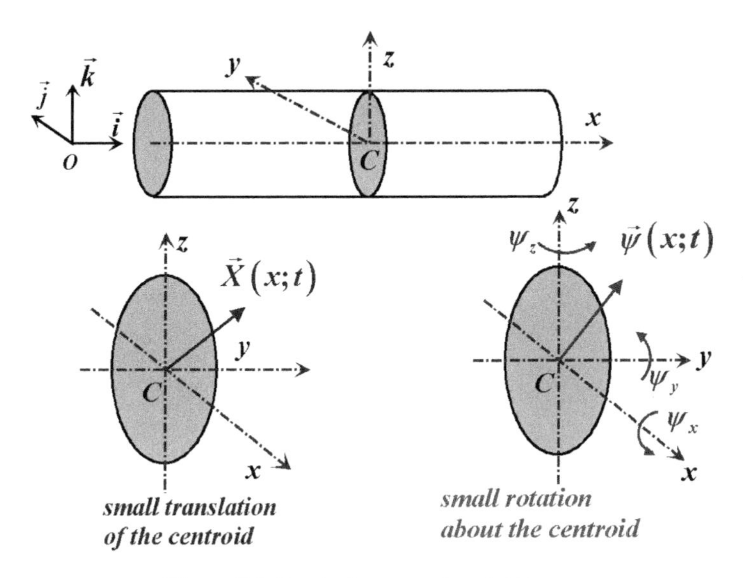

Figure 2.2. *Particular case of straight beams*

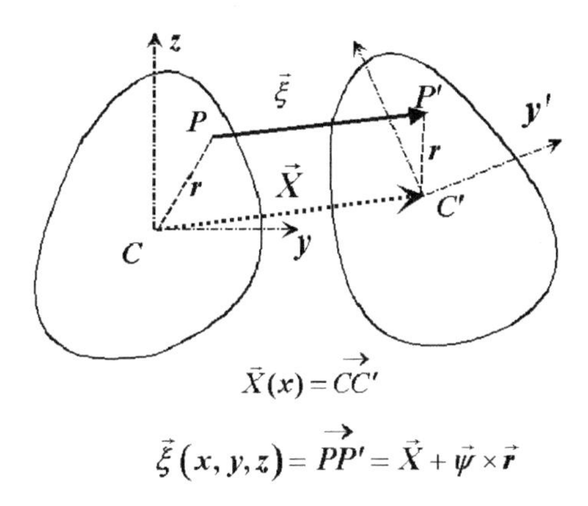

Figure 2.3. *Local displacement of the current point P of the cross-section*

This 3D field has the components:

$$\xi_x = X + z\psi_y - y\psi_z; \quad \xi_y = Y - z\psi_x; \quad \xi_z = Z + y\psi_x \qquad [2.2]$$

X is termed *axial* or *longitudinal* displacement, whereas Y and Z are termed *transverse* (or *lateral*) displacements. ψ_x is the *torsion* (or *twist*) angle, whereas ψ_y and ψ_z are the *flexure* (or *bending*) angles about the axes Oy and Oz respectively (Figure 2.3).

2.1.3 *Local and global strains*

The components of the *local strains* are obtained by substituting [2.2] into the expression [1.22] or [1.25] of the *local strain tensor*. The result is:

$$
\begin{aligned}
\varepsilon_{xx}(x,y,z) &= \frac{1}{2}\left(\frac{\partial \xi_x}{\partial x}+\frac{\partial \xi_x}{\partial x}\right)=\frac{\partial X}{\partial x}+z\frac{\partial \psi_y}{\partial x}-y\frac{\partial \psi_z}{\partial x}\\
\varepsilon_{xy}(x,y,z) &= \frac{1}{2}\left(\frac{\partial \xi_x}{\partial y}+\frac{\partial \xi_y}{\partial x}\right)=\frac{1}{2}\left(-\psi_z+\frac{\partial Y}{\partial x}-z\frac{\partial \psi_x}{\partial x}\right)\\
\varepsilon_{xz}(x,y,z) &= \frac{1}{2}\left(\frac{\partial \xi_x}{\partial z}+\frac{\partial \xi_z}{\partial x}\right)=\frac{1}{2}\left(+\psi_y+\frac{\partial Z}{\partial x}+y\frac{\partial \psi_x}{\partial x}\right)\\
\varepsilon_{yy}(x,y,z) &= \frac{\partial \xi_y}{\partial y}=0;\quad \varepsilon_{zz}(x,y,z)=\frac{\partial \xi_z}{\partial z}=0\\
\varepsilon_{yz}(x,y,z) &= \frac{1}{2}\left(\frac{\partial \xi_y}{\partial z}+\frac{\partial \xi_z}{\partial y}\right)=0
\end{aligned}
\qquad [2.3]
$$

Starting from the relations [2.3], the field of *global strains* is defined by the six following components:

$$\eta_{xx}(x)=\frac{\partial X}{\partial x};\quad \eta_{xy}(x)=\frac{\partial Y}{\partial x};\quad \eta_{xz}(x)=\frac{\partial Z}{\partial x} \qquad [2.4]$$

$$\chi_{xx}(x)=\frac{\partial \psi_x}{\partial x};\quad \chi_{yy}(x)=\frac{\partial \psi_y}{\partial x};\quad \chi_{zz}(x)=\frac{\partial \psi_z}{\partial x} \qquad [2.5]$$

These components are described individually in Figures 2.4, which illustrates their physical meaning:

1. η_{xx} measures the *longitudinal*(or *axial*) *strain* of the beam element, which is either *traction* (or *stretching*) or a *contraction* (or *compression*), whether it is positive or negative.
2. η_{xy}, η_{xz} are the *transverse shear strains*, in the planes Oxy or Oxz respectively.
3. χ_{xx} is the *torsion* (or *twist*) strain.
4. χ_{yy} and χ_{zz} are the *bending* (or *flexure*) *strains*, with respect to the Oy and Oz axes respectively.

In a straight beam, all these modes of global deformation are uncoupled from each other, since each of them is expressed in terms of only one component of the global displacement field, which differs from one mode of deformation to the other. In contrast to the global strains and angular displacements, the local strains are expressed as a superposition of distinct components of the global strains, as

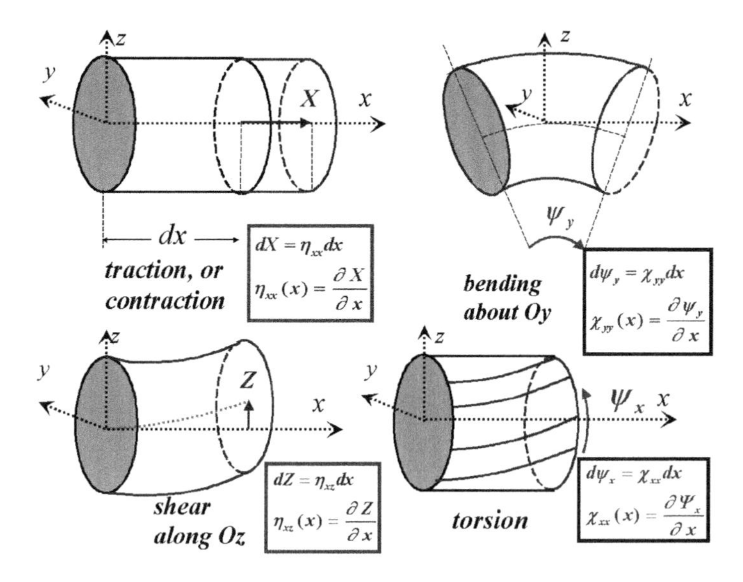

Figure 2.4. *Global deformations of a beam element*

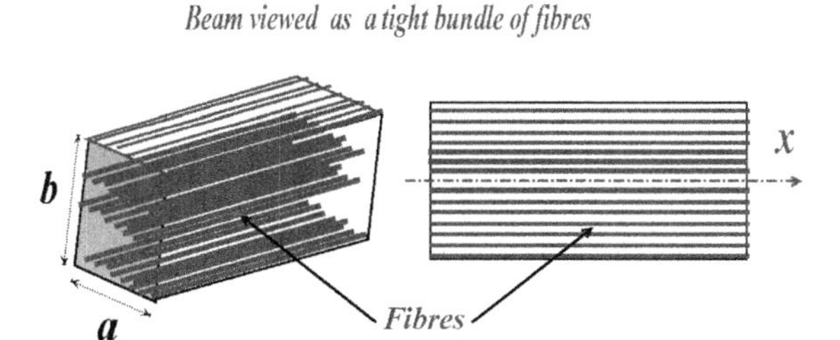

Figure 2.5. *Beam seen as a continuous set of fibres parallel to the beam axis*

evidenced in the relations [2.3]. To visualize the local strains, it is often found convenient to look at the beam as if it were made of a continuous set of material lines, or *fibres*, parallel to the Ox axis, see Figure 2.5. The change in length of a fibre passing through the point (x, y, z) depends on the bending and longitudinal global strains, see Figures 2.6a and 2.6b. Taking into account bending deformation only, the relation [2.2] reduces to $\xi_x = z\psi_y - y\psi_z$ so a layer of neutral lines whose length remains unchanged exists. If the beam is homogeneous, the neutral layer is defined by the condition $y = 0$, or $z = 0$, depending on the bending plane. So the central axis is often called the *neutral fibre*.

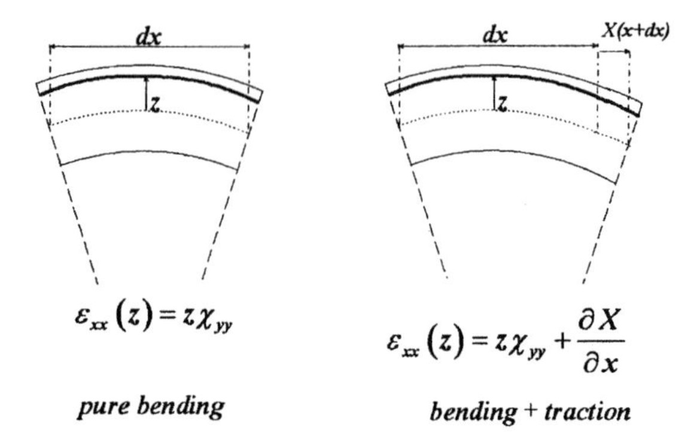

Figure 2.6a. *Local strains: superposition of bending and axial deformations*

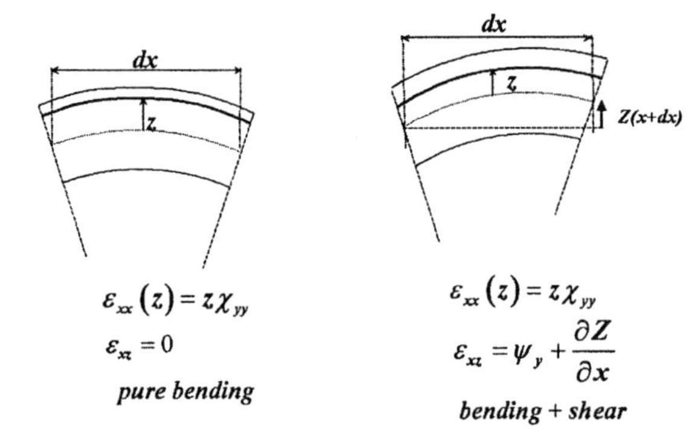

Figure 2.6b. *Local strains: superposition of bending and shear deformations*

2.1.4 *Local and global stresses*

The local stress tensor due to the strains acting on a cross-section has three components, which are associated with the stress vector:

$$\vec{t}_1 = \sigma_{xx}\vec{i} + \sigma_{xy}\vec{j} + \sigma_{xz}\vec{k} \qquad [2.6]$$

In agreement with the definition given in subsection 1.2.2, $\vec{t}_1$ is the force per unit cross-sectional area exerted by the right-hand part of the beam on the left-hand part, see Figure 2.7. Six variables of *global stresses* (also called *resultant stresses*) are associated with the global displacements and strains. They define a force vector $\vec{T}$ and a moment vector $\vec{\mathcal{M}}$:

$$\vec{T}(x) = \int_{(\mathcal{S})} \vec{t}_1 d\mathcal{S}; \quad \vec{\mathcal{M}} = \int_{(\mathcal{S})} (\vec{r} \times \vec{t}_1)\, d\mathcal{S} \qquad [2.7]$$

$\vec{r}$ is again the radius vector of a current point of the cross-section $(\mathcal{S})$.

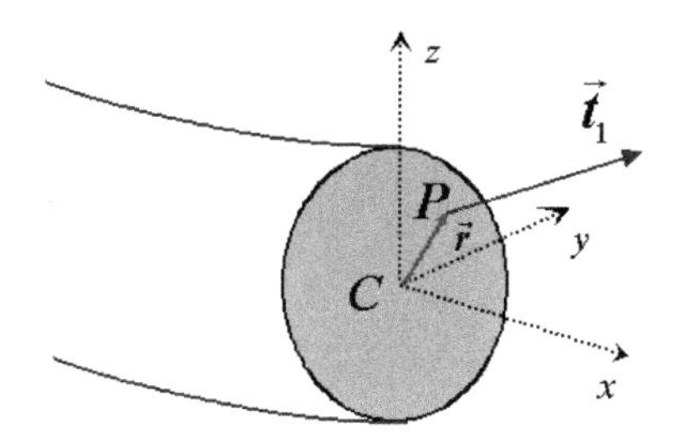

Figure 2.7. *Local stress vector on the beam cross-section*

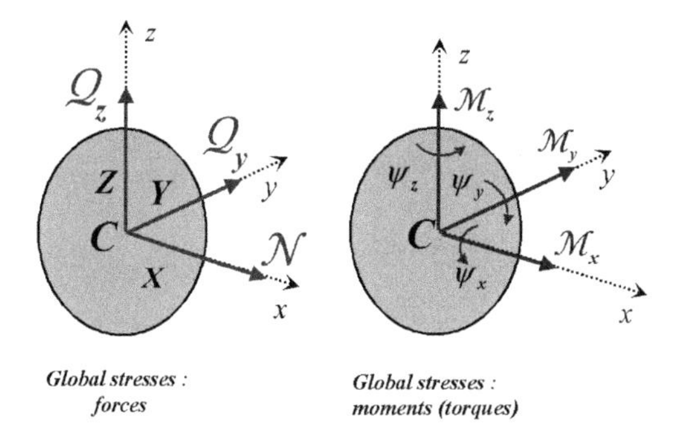

Figure 2.8. *Global stress components on a beam cross-section*

The components of $\vec{T}$ and $\vec{\mathcal{M}}$ are shown in Figure 2.8 and the relations [2.8], derived from [2.7], specify the correspondence between the components of global displacements and global stresses. Global displacements and global stresses are thus found to be related to each other as pairs of *conjugate quantities*. On the other hand, the sign of the bending moments is controlled by the usual convention of signs applied to the rotations in a direct Cartesian frame.

$$X \rightarrow \mathcal{N}(x) = \int_{(\mathcal{S})} \sigma_{xx} d\mathcal{S} : \quad \text{normal force in the } x \text{ direction (axial, or string force)}$$

$$Y \rightarrow \mathcal{Q}_y(x) = \int_{(\mathcal{S})} \sigma_{xy} d\mathcal{S} : \quad \text{transverse shear force in the } y \text{ direction} \qquad [2.8]$$

$$Z \rightarrow \mathcal{Q}_z(x) = \int_{(\mathcal{S})} \sigma_{xz} d\mathcal{S} : \quad \text{transverse shear force in the } z \text{ direction}$$

$$\psi_x \rightarrow \mathcal{M}_x(x) = \int_{(\mathcal{S})} (y\sigma_{xz} - z\sigma_{xy}) d\mathcal{S} : \quad \text{torsion moment}$$

$$\psi_y \rightarrow \mathcal{M}_y(x) = +\int_{(\mathcal{S})} z\sigma_{xx} d\mathcal{S} : \quad \text{bending moment about } Oy \qquad [2.9]$$

$$\psi_z \rightarrow \mathcal{M}_z(x) = -\int_{(\mathcal{S})} y\sigma_{xx} d\mathcal{S} : \quad \text{bending moment about } Oz$$

The term 'string force' used here as an equivalent to 'axial force' $\mathcal{N}(x)$ stems from common experience, since a string, or cable, can be seen as a beam whose cross-section is so small that resistance to a transverse load is provided almost exclusively by the tensile force $\mathcal{N}(x)$. A typical example is the strings of musical instruments. Study of the transverse equilibrium of beams subjected to tensile or compressive loads is postponed to Chapter 3, subsection 3.2.3 and Chapter 4, subsection 4.2.4.

2.1.5 *Elastic stresses*

The three-dimensional Hooke's law [1.37] is adapted here, taking into account the simplifications described just above. As cross-sections remain non deformed, the local stresses acting in the transverse directions are not related to the transverse strains but to the Lagrange multipliers related to the rigidity conditions. Accordingly, $\varepsilon_{yy}, \varepsilon_{zz}, \varepsilon_{yz}$ and the elastic stresses $\sigma_{yy}, \sigma_{zz}, \sigma_{yz}$ are assumed to be zero. From [1.38] we get:

$$\varepsilon_{xx} = \left(\frac{1+\nu}{E} - \frac{\nu}{E}\right)\sigma_{xx} = \frac{\sigma_{xx}}{E}; \quad \varepsilon_{xy} = \frac{1+\nu}{E}\sigma_{xy}; \quad \varepsilon_{xz} = \frac{1+\nu}{E}\sigma_{xz} \quad [2.10]$$

or,

$$\sigma_{xx} = E\frac{\partial \xi_x}{\partial x}; \quad \sigma_{xy} = G\left(\frac{\partial \xi_x}{\partial y} + \frac{\partial \xi_y}{\partial x}\right); \quad \sigma_{xz} = G\left(\frac{\partial \xi_z}{\partial x} + \frac{\partial \xi_x}{\partial z}\right)$$

In terms of global displacements [2.10] becomes:

$$\begin{gathered}\sigma_{xx} = E\left(\frac{\partial X}{\partial x} + z\frac{\partial \psi_y}{\partial x} - y\frac{\partial \psi_z}{\partial x}\right) \\ \sigma_{xy} = G\left(\frac{\partial Y}{\partial x} - z\frac{\partial \psi_x}{\partial x} - \psi_z\right); \quad \sigma_{xz} = G\left(\frac{\partial Z}{\partial x} + y\frac{\partial \psi_x}{\partial x} + \psi_y\right)\end{gathered} \quad [2.11]$$

By integrating [2.11] on the cross-sectional area $S(x)$, the following global stresses are found to be:

$$\mathcal{N} = ES\frac{\partial X}{\partial x}; \quad \mathcal{Q}_y = GS\left(\frac{\partial Y}{\partial x} - \psi_z\right); \quad \mathcal{Q}_z = GS\left(\frac{\partial Z}{\partial x} + \psi_y\right) \quad [2.12]$$

$$\mathcal{M}_x = GJ\frac{\partial \psi_x}{\partial x}; \quad \mathcal{M}_y = EI_y\frac{\partial \psi_y}{\partial x}; \quad \mathcal{M}_z = EI_z\frac{\partial \psi_z}{\partial x} \quad [2.13]$$

where I_y and I_z are the area inertia moments of the cross-section about the Oy and Oz axis respectively:

$$I_y = \int_{(S)} z^2\, dS; \quad I_z = \int_{(S)} y^2\, dS \quad [2.14]$$

The area polar moment of inertia is:

$$J = I_y + I_z \qquad [2.15]$$

2.1.6 *Equilibrium in terms of generalized stresses*

The equations of dynamical equilibrium are obtained by writing down the force and moment balance, including all the strain, inertia, and external terms, which act at the boundaries and within a beam element of infinitesimal length, limited by the cross-sections $\mathcal{S}(x)$ and $\mathcal{S}(x + dL)$. The actual loads exerted within and at the boundary of the 3D body are averaged in a similar way to the local stress vector in relation [2.7], to produce an equivalent one-dimensional loading comprising:

1. A force density (force per unit length) $\vec{F}^{(e)}(x,t)$ (Newton/meter in S.I. units) applied to the centroid.
2. A moment density (moment per unit length) about the centroid $\vec{\mathfrak{M}}^{(e)}(x,t)$ (Newton).

Equation [1.32] still holds and can be expressed either by using a vector or a simplified indicial notation, according to which the first index of the stress term is replaced by 'x', because $(\mathcal{S})$ is normal to Ox. In terms of local quantities it reduces to:

$$\begin{aligned} \rho\frac{\partial^2\vec{\xi}}{\partial t^2} - \frac{\partial\vec{t}_1}{\partial x} &= \vec{f}^{(e)}(x,y,z;t) \\ \rho\frac{\partial^2\xi_j}{\partial t^2} - \frac{\partial\sigma_{xj}}{\partial x} &= f_j^{(e)}(x,y,z;t); \quad (j = 1,2,3) \end{aligned} \qquad [2.16]$$

2.1.6.1 *Equilibrium of forces*

For the sake of simplicity, the force balance is established ignoring the contribution of the moment density $\vec{\mathfrak{M}}^{(e)}(x,t)$, which will be included later in subsection 2.2.4. Integration of the local force densities on a beam element of infinitesimal length dL gives:

$$\int_{x-dL/2}^{x+dL/2} dx \int_{(\mathcal{S})} \left(\rho\frac{\partial^2\vec{\xi}}{\partial t^2} - \frac{\partial\vec{t}_1}{\partial x}\right) d\mathcal{S} = \int_{x-dL/2}^{x+dL/2} dx \int_{(\mathcal{S})} \vec{f}^{(e)} d\mathcal{S} = \vec{F}^{(e)}(x;t)\, dL \qquad [2.17]$$

Using the relations [2.1], [2.6] and [2.8], the following one-dimensional equations are obtained:

$$\rho S \frac{\partial^2 X}{\partial t^2} - \frac{\partial \mathcal{N}}{\partial x} = F_x^{(e)}(x;t)$$

$$\rho S \frac{\partial^2 Y}{\partial t^2} - \frac{\partial \mathcal{Q}_y}{\partial x} = F_y^{(e)}(x;t) \qquad [2.18]$$

$$\rho S \frac{\partial^2 Z}{\partial t^2} - \frac{\partial \mathcal{Q}_z}{\partial x} = F_z^{(e)}(x;t)$$

A slightly different way to establish the equations [2.18] is to write directly the global force equilibrium as:

$$-\rho S \ddot{\vec{X}}\, dx + \vec{T}(x + dx) - \vec{T}(x) + \vec{F}^{(e)} dx = 0 \Rightarrow \rho S \ddot{\vec{X}} - \frac{\partial \vec{T}}{\partial x} = \vec{F}^{(e)} \qquad [2.19]$$

The axial and transverse force balance are visualized schematically in Figure 2.9.

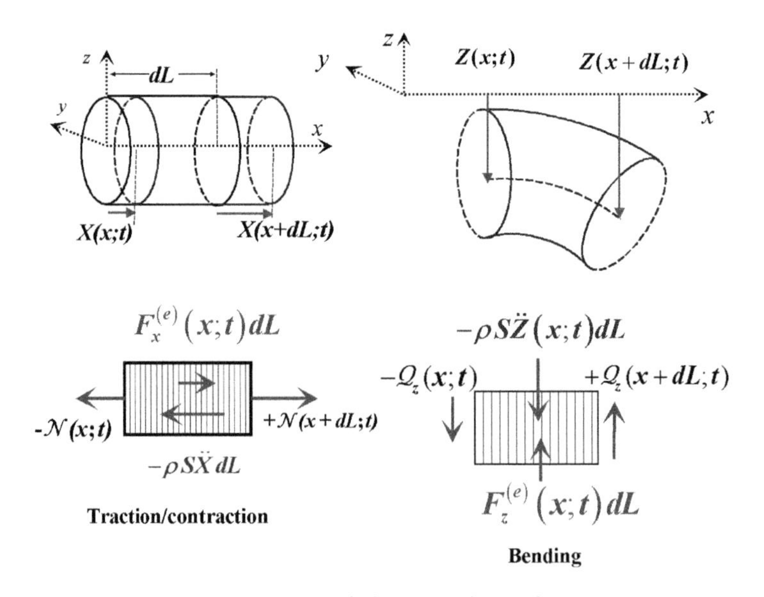

Figure 2.9. *Force balance in a beam element*

2.1.6.2 *Equilibrium of the moments*

The balance of moments is calculated with respect to the centre-of-mass of the beam element, located at abscissa x. It is written as:

$$\int_{x-dL/2}^{x+dL/2} \int_{(\mathcal{S})} \vec{r} \times \left(\rho \frac{\partial^2 \vec{\xi}}{\partial t^2} - \frac{\partial \vec{t}_1}{\partial x} \right) d\mathcal{S}\, dx \\ = \int_{x-dL/2}^{x+dL/2} \int_{(\mathcal{S})} (\vec{r} \times \vec{f}^{(e)})\, d\mathcal{S}\, dx + \vec{\mathcal{M}}^{(e)}(x;t)\, dL \qquad [2.20]$$

Using the relations [2.1], [2.8], and [2.9], the following one-dimensional equations are obtained:

$$\rho J \frac{\partial^2 \psi_x}{\partial t^2} - \frac{\partial \mathcal{M}_x}{\partial x} = \mathfrak{M}_x^{(e)}(x;t)$$

$$\rho I_y \frac{\partial^2 \psi_y}{\partial t^2} - \frac{\partial \mathcal{M}_y}{\partial x} + \mathcal{Q}_z = \mathfrak{M}_y^{(e)}(x;t) \qquad [2.21]$$

$$\rho I_z \frac{\partial^2 \psi_z}{\partial t^2} - \frac{\partial \mathcal{M}_z}{\partial x} - \mathcal{Q}_y = \mathfrak{M}_z^{(e)}(x;t)$$

The balances corresponding to torsion and to bending about Oy are sketched in Figure 2.10.

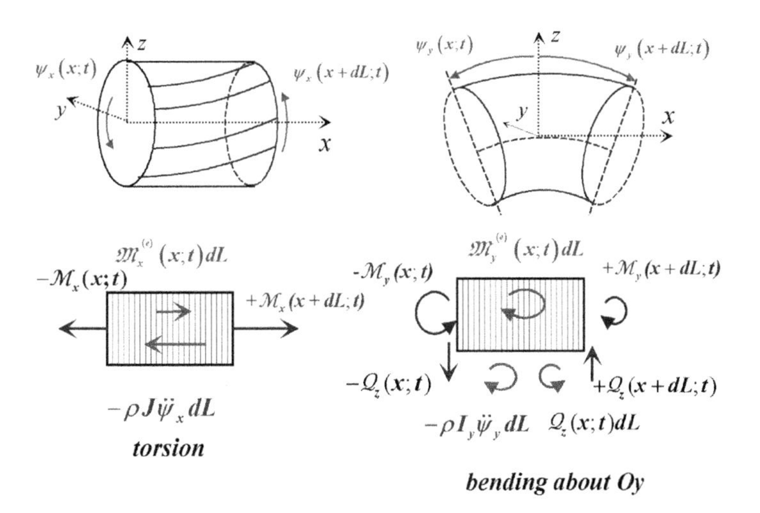

Figure 2.10. *Moment balance in a beam element for torsion and bending*

The equations [2.21] may also be obtained by balancing directly the global moments. The moment balance is conveniently written in matrix form:

$$-dx[\mathcal{I}][\ddot{\psi}] + [\mathcal{M}(x+dx)] - [\mathcal{M}(x)] + [\mathfrak{M}^{(e)}]dx - [\mathcal{Q}]dx = [0]$$

$$\Rightarrow \; [\mathcal{I}][\ddot{\psi}] - \frac{\partial[\mathcal{M}]}{\partial x} - [\mathcal{Q}] = [\mathfrak{M}^{(e)}]$$

$\bar{\bar{\mathcal{I}}}$ is the inertia tensor of the cross-section, written in matrix form as:

$$[\mathcal{I}] = \rho \begin{bmatrix} J & 0 & 0 \\ 0 & I_y & 0 \\ 0 & 0 & I_z \end{bmatrix}$$

The equilibrium equations [2.18] and [2.21] are clearly independent of the material law. In the following section, they are further developed in the case of linear elasticity.

2.2. Small elastic motion

As pointed out previously, the global deformations of straight beams may be categorized according to a few uncoupled modes, namely one longitudinal mode, one torsion mode, two transverse shear modes and two bending or flexure modes. So long as the study is restricted to the domain of small elastic motions, a specific vibration equation can be derived for each of these deformation modes, which is expressed in terms of the corresponding displacement variable.

2.2.1 *Longitudinal mode of deformation*

2.2.1.1 Local equilibrium

The equations [2.11] and [2.12] reduce to $\sigma_{xx} = E(\partial X/\partial x)$ and to $\mathcal{N} = ES(\partial X/\partial x)$ respectively. By substituting this elastic stress into the first force equation [2.18], we arrive at:

$$\rho S \frac{\partial^2 X}{\partial t^2} - \frac{\partial}{\partial x}\left(ES\frac{\partial X}{\partial x}\right) = F_x^{(e)}(x;t) \qquad [2.22]$$

Equation [2.22] was first established by Navier 1824, see [SOE 93] for a short historical survey of the vibration equations of structural elements. As detailed in Chapter 4, equation [2.22] governs the elastic vibrations in the axial direction, which can be identified with 1D dilatational, or longitudinal waves.

NOTE. – *Refinement of the beam model for longitudinal motion*

As a direct consequence of the assumption that the beam cross-sections are rigid, in the above equations they are implicitly assumed to remain undeformed when the beam is either lengthened or shortened in the axial direction. Actually, due to the Poisson effect, any axial strain induces a transverse one. Hence, by relaxing the condition of cross-sectional rigidity, the beam is found to either contract or expand laterally, in connection with any axial stretching or contraction respectively. As such transverse motions occur freely, they do not add any stress to the system, but they provide some additional inertia which can be taken into account, as further analysed in the subsection 3.2.1.2 of Chapter 3. However, it will be found that this effect is negligible, except for very high frequency vibrations which are of little interest in most engineering applications.

2.2.1.2 General solution of the static equilibrium without external loading

Here, the equation [2.22] is reduced to its static and homogeneous form:

$$-\frac{\partial}{\partial x}\left(ES\frac{\partial X}{\partial x}\right) = 0 \tag{2.23}$$

The solution is:

$$X(x) = a\int \frac{dx}{ES} + b \tag{2.24}$$

where a and b are two arbitrary constants.

If the beam is homogeneous and of constant cross-section, [2.24] gives:

$$X(x) = \frac{ax}{ES} + b \tag{2.25}$$

2.2.1.3 Elastic boundary conditions

The beam is supposed to be supported at the end $x = 0$ by a linear spring acting in the axial direction. The stiffness coefficient of the support is K_x. Therefore the axial movement is constrained by the support reaction $\mathcal{R} = -K_x X(0;t)$ as shown in Figure 2.11. At the cross-section (A), which stands for a boundary of the beam, the static equilibrium requires that the sum of the reaction $\mathcal{R}$ and of the axial stress be zero. As $\mathcal{R}$ acts on the left-hand side and $\mathcal{N}$ on the right-hand side of the beam cross-section, the condition is:

$$\mathcal{N}(0;t) + \mathcal{R} = 0 \Rightarrow ES\frac{\partial X}{\partial x}\bigg|_{x=0} - K_x X(0;t) = 0 \tag{2.26}$$

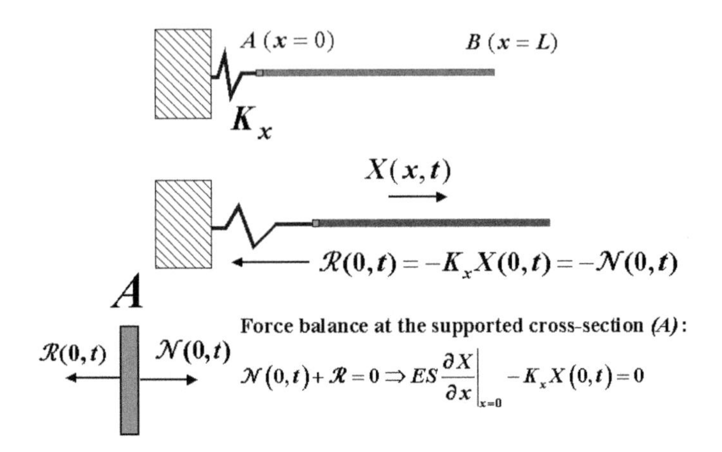

Figure 2.11. *Beam supported at the left end by an axial spring*

The two following limit cases are also of interest:

1. $K_x = 0$: *free end*
   The reaction due to the support is zero, then:

$$\mathcal{N}(0;t) = 0 \iff \left.\frac{\partial X}{\partial x}\right|_{x=0} = 0 \qquad [2.27]$$

2. $K_x = \infty$: *fixed end*
   Neither the reaction nor the stress can be infinite, so the appropriate condition is:

$$X(0,t) = 0 \qquad [2.28]$$

[2.28] is termed a *locking condition* which is applied to the left-hand end of the beam. The corresponding reaction can be determined by using a Lagrange multiplier.

The configuration in which the beam end $x = L$ is constrained by a spring (see Figure 2.12) is also studied as a short exercise about the sign convention of stresses. Obviously, the analysis is the same as above; however, the internal force becomes $-\mathcal{N}(L;t)$, which represents the effect of the beam part resting on the left-hand side of the cross-section at B. The spring acts on the right-hand side. The equilibrium balance is thus:

$$\begin{aligned} -\mathcal{N}(L;t) + \mathcal{R} = 0 \Rightarrow -ES\left.\frac{\partial X}{\partial x}\right|_{x=L} - K_x X(L;t) = 0 \\ \iff ES\left.\frac{\partial X}{\partial x}\right|_{x=L} + K_x X(L;t) = 0 \end{aligned} \qquad [2.29]$$

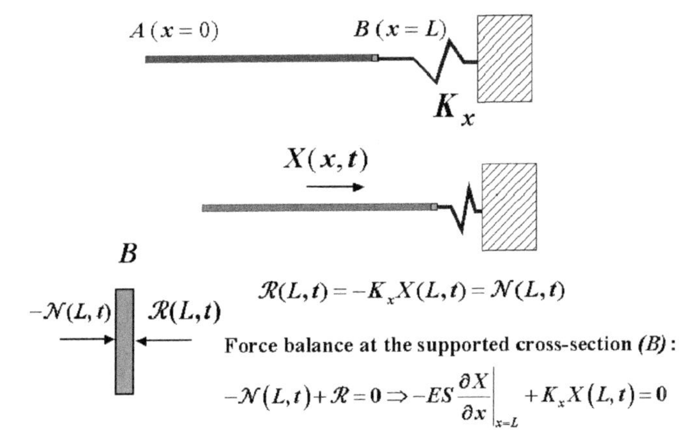

Figure 2.12. *Beam supported at its right end by an axial spring*

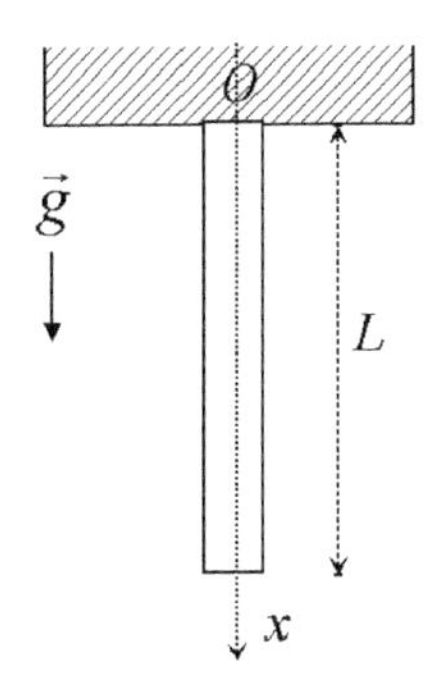

Figure 2.13. *Beam axially loaded by its own weight*

Of course, if X is positive, the left support pulls on the beam (positive stress) and conversely the right one pushes on it (negative stress).

Finally, it could be easily verified that the homogeneous problem has only the trivial solution $a = b = 0$ when the boundary conditions are accounted for, except if the beam is free at both ends, in which case $a = 0$ and b is arbitrary, appropriately since the beam is free to move axially without changing the potential of the system (state of indifferent equilibrium).

EXAMPLE. – *Beam loaded by its own weight*

The beam is assumed to hang vertically, as shown in Figure 2.13. To simplify the calculation, the beam is supposed homogeneous and with constant cross-section.

The problem is formulated as follows:

$$-ES\frac{d^2X}{dx^2} = \rho g S$$

$$X(0) = 0; \quad \left.\frac{dX}{dx}\right|_L = 0$$

The solution is found to be:

$$X(\xi) = \frac{\rho g L^2}{E}\left(1 - \frac{\xi}{2}\right)\xi \quad \text{where } \xi = \frac{x}{L}$$

As expected, the displacement is maximum at the free end; the stress is maximum at the fixed end and equal to the total weight.

2.2.1.4 *Concentrated loads*

An axial force acting on the beam cross-section at abscissa x_0 is considered. In Figure 2.14, two adjacent beam elements of infinitesimal length ε on each side of this cross-section are represented. $F_x^{(e)}(t)$ is the resultant of the axial external loading applied to the cross-section $\mathcal{S}(x_0) \cdot \mathcal{N}^- = -\mathcal{N}(x_0 - \varepsilon)$ is the normal stress acting through the cross-section $\mathcal{S}(x_0 - \varepsilon)$, induced by the beam part located on the left-hand of $\mathcal{S}(x_0 - \varepsilon)$. Conversely $\mathcal{N}^+ = \mathcal{N}(x_0 + \varepsilon)$ is the normal stress acting through $\mathcal{S}(x_0 + \varepsilon)$ due to the beam part on the right-hand side of $\mathcal{S}(x_0 + \varepsilon)$. The inertia force can be included, or already neglected at this step, because ε is an infinitesimal length. Hence, the equilibrium equation is written as:

$$\mathcal{N}^+ - \mathcal{N}^- + F_x^{(e)} = 0$$

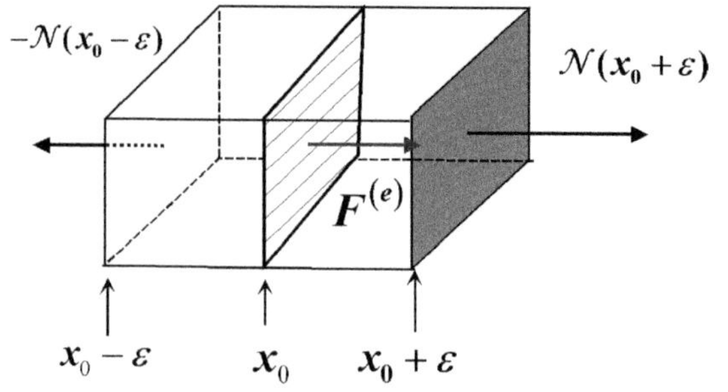

Figure 2.14. *Infinitesimal beam element loaded by a concentrated force*

Now, when ε tends to zero the above equation results in the finite discontinuity of the axial stress:

$$\mathcal{N}(x_{0+};t) - \mathcal{N}(x_{0-};t) = -F_x^{(e)}(t) \qquad [2.30]$$

For an elastic beam [2.30] becomes:

$$ES\left.\frac{\partial X}{\partial x}\right|_{x_{0+}} - ES\left.\frac{\partial X}{\partial x}\right|_{x_{0-}} = -F_x^{(e)}(t) \qquad [2.31]$$

This equation shows that a finite discontinuity of the displacement derivative takes place at the abscissa where the force is applied. Of course, the displacement itself remains a continuous function.

EXAMPLE. – *Beam fixed at $x = 0$, free at $x = L$ and loaded at x_0*

The boundary value problem of Figure 2.15 is governed by the following equations:

$$-ES\frac{d^2X}{dx^2} = 0; \quad X(0) = 0; \quad \left.\frac{dX}{dx}\right|_L = 0; \quad -ES\left.\frac{dX}{dx}\right|_{x_{0+}} + ES\left.\frac{dX}{dx}\right|_{x_{0-}} = F_0^{(e)}$$

In each of the two beam parts separated by the cross-section at x_0, the displacement X is a linear function of x. Because of the discontinuity, these functions differ in each part and the solution depends on four constants. They can be specified by applying the two boundary conditions plus the two conditions of finite stress discontinuity and of displacement continuity holding at $x = x_0$. So it is found that:

$$X(x) = \begin{cases} X_- = \dfrac{F_0^{(e)}x}{ES} & 0 \le x < x_0 \\[2ex] X_+ = \dfrac{F_0^{(e)}x_0}{ES} & x_0 \le x \le L \end{cases}$$

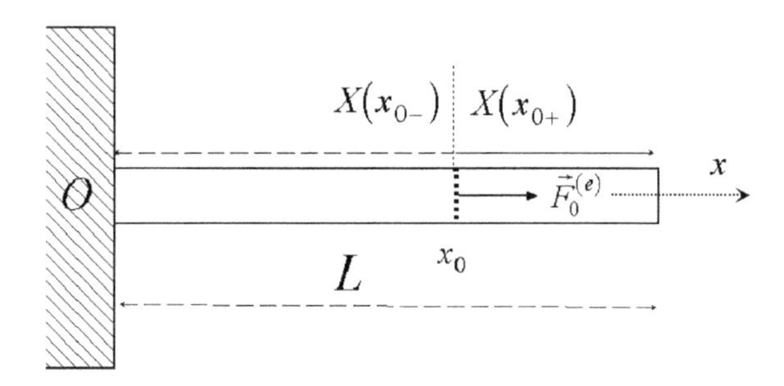

Figure 2.15. *Beam loaded by an axial force acting on the cross-section at x_0*

As expected, the displacement is constant and the stress is equal to zero on the right-hand side from the loaded cross-section, whereas, on the left-hand side, the displacement increases linearly (null at the fixed end) and the stress is constant. The displacement field may also be written as:

$$X(x) = \begin{cases} X_- = X_m \xi & 0 \le \xi \le \xi_0 \\ X_+ = X_m \xi_0 & \xi_0 \le \xi \le 1 \end{cases}$$

$$\text{where } K_\ell = \frac{ES}{L}; \quad X_m = \frac{F_0^{(e)}}{K_\ell}; \quad \xi = x/L$$

The interest in this reduced form is to bring out the relevance of the stiffness coefficient $K_\ell = ES/L$ for scaling the longitudinal stiffness of the beam.

2.2.1.5 *Intermediate supports*

Concentrated loads can be external or internal in nature. Internal loads arise as the reactions induced by local supports. In accordance with the result [2.31], a support at $x = x_0$ generates a finite discontinuity of the stress function at this point, which ensures the local equilibrium of the cross-section $\mathcal{S}(x_0)$. In particular, a support defined by a stiffness coefficient K_x implies the relation:

$$ES\left.\frac{\partial X}{\partial x}\right|_{x_{0+}} - ES\left.\frac{\partial X}{\partial x}\right|_{x_{0-}} - K_x X(x_0; t) = 0 \qquad [2.32]$$

It can be shown that the condition [2.32] agrees with the boundary conditions [2.26] and [2.29]. Indeed in the first boundary configuration (spring at the left-hand end) the point $x = 0_-$ lies out of the beam, the stress is then zero and:

$$ES\left.\frac{\partial X}{\partial x}\right|_{0+} - ES\left.\frac{\partial X}{\partial x}\right|_{0-} - K_x X(0; t) = 0 \Rightarrow ES\left.\frac{\partial X}{\partial x}\right|_{0} - K_x X(0; t) = 0$$

In the second boundary configuration (spring at the right-hand end) the point $x = L_+$ lies out of the beam and:

$$ES\left.\frac{\partial X}{\partial x}\right|_{L+} - ES\left.\frac{\partial X}{\partial x}\right|_{L_-} - K_x X(L; t) = 0 \Rightarrow -ES\left.\frac{\partial X}{\partial x}\right|_{L} - K_x X(L; t) = 0$$

EXAMPLE. – *Uniformly loaded beam supported by a spring at* x_0

Denoting by $F^{(e)}$ the external force per unit length applied to the beam in the axial direction, the boundary value problem is governed by the following equations:

$$-ES\frac{d^2X}{dx^2} = F^{(e)}$$

$$X(0) = X(L) = 0$$

$$ES\frac{\partial X}{\partial x}\bigg|_{x_{0+}} - ES\frac{\partial X}{\partial x}\bigg|_{x_{0-}} - K_x X(x_0;t) = 0$$

These equations are restated in a reduced form, by using the same scale factors as in the last example. This gives:

$$\frac{1}{X_0}\frac{d^2X}{d\xi^2} = -1$$

$$X(0) = X(1) = 0$$

$$\frac{\partial X}{\partial \xi}\bigg|_{\xi_{0+}} - \frac{\partial X}{\partial \xi}\bigg|_{\xi_{0-}} = \kappa X_+(\xi_0) = \kappa X_-(\xi_0)$$

where

$$X_0 = \frac{F_0}{K_\ell}; \quad F_0 = F^{(e)}L; \quad \kappa = \frac{K_x}{K_\ell}$$

Since the first derivative of X is discontinuous at the spring anchoring point, the solutions are of the type:

$$\frac{X_-}{X_0} = a\xi^2 + b\xi + c \text{ if } 0 \le \xi \le \xi_0 \quad \text{and} \quad \frac{X_+}{X_0} = a'\xi^2 + b'\xi + c' \text{ if } \xi_0 \le \xi \le 1$$

The beam is fixed at both ends, so we get:

$$\frac{X_-}{X_0} = a\xi^2 + b\xi \qquad \frac{X_+}{X_0} = a'(\xi^2 - 1) + b'(\xi - 1)$$

The load condition implies $a = a' = -1/2$ and the conditions at the spring provide the linear system:

$$\begin{bmatrix} (1+\kappa\xi_0) & -1 \\ -1 & 1+\kappa(1-\xi_0) \end{bmatrix} \begin{bmatrix} b \\ b' \end{bmatrix} = \frac{\kappa}{2}\begin{bmatrix} \xi_0^2 \\ 1-\xi_0^2 \end{bmatrix}$$

Then the solution is:

$$\frac{X_-}{X_0} = \frac{\left(1 + \kappa\xi_0\left(1 - \xi_0^2\right)\right)\xi}{2(1 + \kappa\xi_0(1 - \xi_0))} - \frac{\xi^2}{2}; \quad 0 \le \xi \le \xi_0$$

$$\frac{X_+}{X_0} = \frac{\left(1 + \kappa\xi_0\left(1 - \xi_0^2\right)\right)(\xi - 1)}{2(1 + \kappa\xi_0(1 - \xi_0))} + \frac{1 - \xi^2}{2}; \quad \xi_0 \le \xi \le 1$$

Limit cases:

$\kappa \to \infty$

$$\frac{X_-}{X_0} = \frac{\xi(\xi_0 - \xi)}{2}; \quad 0 \le \xi \le \xi_0; \qquad \frac{X_+}{X_0} = \frac{(\xi - \xi_0)(1 - \xi)}{2}; \quad \xi_0 \le \xi \le 1$$

$\kappa \to 0$

$$\frac{X_-}{X_0} = \frac{X_+}{X_0} = \frac{\xi(1 - \xi)}{2}$$

Several displacement functions for different values of κ are plotted in Figure 2.16 in dimensionless variables. As expected, decreasing values of displacement are associated with increasing κ values. On the other hand, the less the distance of the support to one end, the less is the displacement field in the corresponding beam part.

2.2.2 *Shear mode of deformation*

In most cases a transverse force applied to a beam induces both shear and bending deformations, though they are instances where bending remains negligible. In any case, the study of a pure shear model gives us the opportunity to single out the shear effects in a uniform and homogeneous beam. The transverse shear plus bending modes of deformation is studied in Chapter 3, subsection 3.2.2.2. The deformation of a beam element according to a shear mode in the Oz direction is shown in Figure 2.17.

2.2.2.1 Local equilibrium

The displacement field is reduced to the component $Z(x;t)$ (or $Y(x;t)$) and the stress field to the component Q_z (or Q_y). Using [2.12] and the third equation [2.18], the following equation of shear vibration is derived:

$$\rho S \frac{\partial^2 Z}{\partial t^2} - \frac{\partial}{\partial x}\left[GS\frac{\partial Z}{\partial x}\right] = F_z^{(e)}(x;t) \qquad [2.33]$$

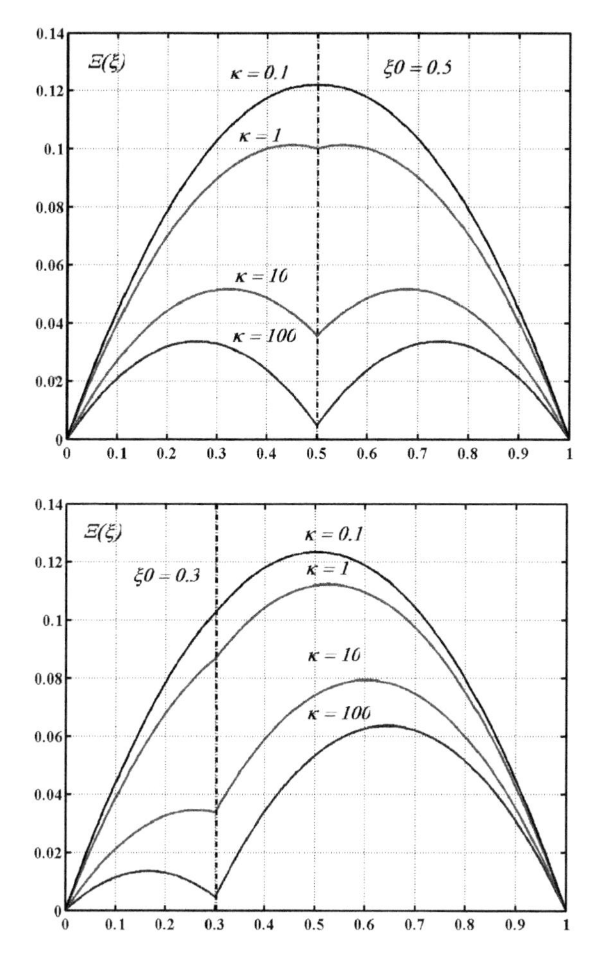

Figure 2.16. *Beam with an intermediate spring support: response to a concentrated axial force* ($\Xi = X/X_0$ and $\xi = x/L$)

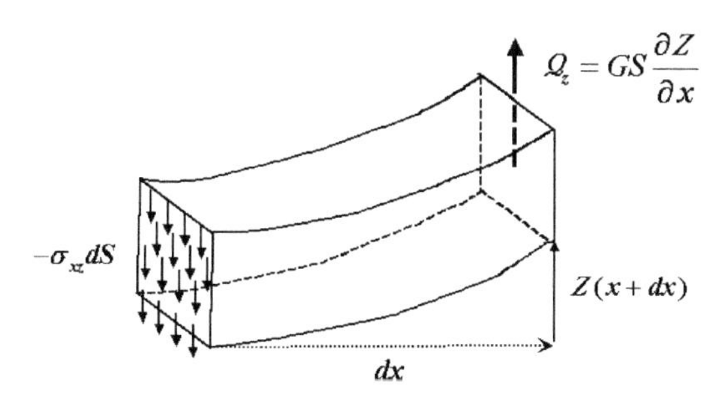

Figure 2.17. *Beam deformed according to a pure shear mode*

It has the same form as the equation which describes the longitudinal vibrations. Then, it is an easy task to adapt the analytical results already established in the case of the longitudinal mode of deformation to the shear case.

2.2.2.2 *General solution without external loading*

$$Z(x) = a \int_0^x \frac{dx}{GS} + b \qquad [2.34]$$

a and b are two arbitrary real constants.

2.2.2.3 *Elastic boundary conditions*

If the end $x = 0$ is supported by a spring of stiffness coefficient K_z, the displacement of the beam is constrained by the reaction $\mathcal{R} = -K_z Z(0;t)$. Then the condition at $x = 0$ is:

$$Q_z(0,t) + \mathcal{R} = 0 \Rightarrow GS \left.\frac{\partial Z}{\partial x}\right|_{x=0} - K_z Z(0,t) = 0 \qquad [2.35]$$

The two limit cases are:

$$\text{1. free end: } \left.\frac{\partial Z}{\partial x}\right|_{x=0} = 0 \qquad [2.36]$$

$$\text{2. fixed end: } Z(0,t) = 0 \qquad [2.37]$$

2.2.2.4 *Concentrated loads*

If a transverse force denoted $F_z^{(e)}(t)$ is applied on the cross-section located at x_0, the discontinuity condition is:

$$Q_z(x_{0+};t) - Q_z(x_{0-};t) = -F_z^{(e)}(t) \qquad [2.38]$$

For an elastic beam it is written as:

$$GS\left(\left.\frac{\partial Z}{\partial x}\right|_{x_{0+}} - \left.\frac{\partial Z}{\partial x}\right|_{x_{0-}}\right) = -F_z^{(e)}(t) \qquad [2.39]$$

2.2.2.5 Intermediate supports

An elastic support of stiffness K_z introduces the discontinuity condition:

$$GS\left(\left.\frac{\partial Z}{\partial x}\right|_{x_{0+}} - \left.\frac{\partial Z}{\partial x}\right|_{x_{0-}}\right) - K_z Z(x_0;t) = 0 \qquad [2.40]$$

2.2.3 *Torsion mode of deformation*

2.2.3.1 Torsion without warping

Figure 2.18 shows a beam element shaped as a circular cylinder, in which the twisting of a few superficial fibres due to torsion are sketched to help visualize the deformation.

2.2.3.2 Local equilibrium

The displacement field reduces to the axial rotation ψ_x and the stress field to the axial moment $\mathcal{M}_x$. The elastic expression of $\mathcal{M}_x$ given by the first equation [2.13] is substituted into the first equation [2.21], which gives the following vibration equation:

$$\rho J \frac{\partial^2 \psi_x}{\partial t^2} - \frac{\partial}{\partial x}\left[GJ\frac{\partial \psi_x}{\partial x}\right] = \mathfrak{M}_x^{(e)}(x;t) \qquad [2.41]$$

This equation has the same form as the two previous ones. It was first established by Poisson in 1827.

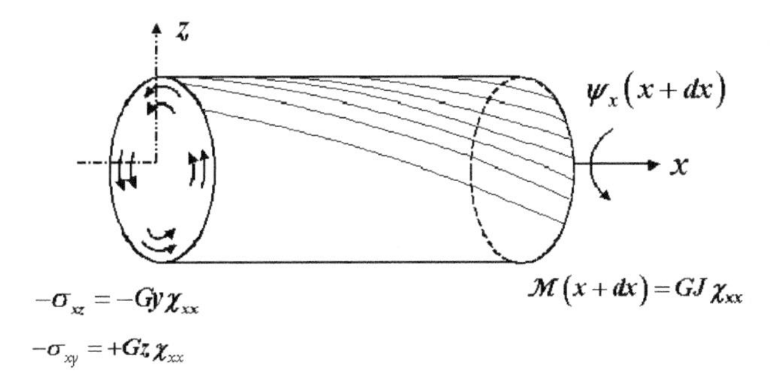

Figure 2.18. *Beam in torsion without warping*

2.2.3.3 General solution without loading

$$\psi_x(x) = a \int_0^x \frac{dx}{GJ} + b \qquad [2.42]$$

a and b are arbitrary real constants.

2.2.3.4 Elastic boundary conditions

If the end $x = 0$ is supported by a torsion spring of stiffness coefficient K_{ψ_x}, then:

$$\mathcal{M}_x(0,t) - K_{\psi_x}\psi_x(0,t) = 0 \Rightarrow GJ\frac{\partial \psi_x}{\partial x}\bigg|_{x=0} - K_{\psi_x}\psi_x(0,t) = 0 \qquad [2.43]$$

The limit cases are:

Free end: $\left.\dfrac{\partial \psi_x}{\partial x}\right|_{x=0} = 0$ [2.44]

Fixed end: $\psi_x(0,t) = 0$ [2.45]

2.2.3.5 Concentrated loads

If the axial moment $\mathcal{M}_x^{(e)}(t)$ is applied at x_0, the following discontinuity condition takes place:

$$\mathcal{M}_x(x_{0+};t) - \mathcal{M}_x(x_{0-};t) = -\mathcal{M}_x^{(e)}(t) \qquad [2.46]$$

If the beam is elastic this condition becomes:

$$GJ\left(\frac{\partial \psi_x}{\partial x}\bigg|_{x_{0+}} - \frac{\partial \psi_x}{\partial x}\bigg|_{x_{0-}}\right) = -\mathcal{M}_x^{(e)}(t) \qquad [2.47]$$

2.2.3.6 Intermediate supports

An elastic support of stiffness coefficient K_{ψ_x} leads to the discontinuity:

$$GJ\left(\frac{\partial \psi_x}{\partial x}\bigg|_{x_{0+}} - \frac{\partial \psi_x}{\partial x}\bigg|_{x_{0-}}\right) - K_{\psi_x} Z(x_0;t) = 0 \qquad [2.48]$$

2.2.3.7 Torsion with warping: Saint Venant's theory

We consider a beam which is fixed in axial rotation at one end and subject at the opposite end to a loading system defined by a null resultant force and a resultant axial moment $\mathcal{M}_x^{(e)}$, see Figure 2.19. The model established in the last subsection produces the solution:

$$\psi_x(x) = x\psi_0 \quad \text{where } \psi_0 = \frac{\psi_x(L)}{L} = \frac{\mathcal{M}_x^{(e)}}{GJ} \qquad [2.49]$$

and the components [2.2] of the local displacement reduce to:

$$\xi_x = 0; \quad \xi_y = -xz\psi_0; \quad \xi_z = xy\psi_0 \qquad [2.50]$$

However, experiment shows that some axial deformation of the cross-sections called *warping* takes place as soon as their shape is not circular. Warping is thus necessarily related to a non vanishing axial component ξ_x of the local displacement field. Accordingly, the local shear strains are written in agreement with [2.3] as:

$$\varepsilon_{xy} = \frac{1}{2}\left(\frac{\partial \xi_x}{\partial y} + \frac{\partial \xi_y}{\partial x}\right); \qquad \varepsilon_{xz} = \frac{1}{2}\left(\frac{\partial \xi_x}{\partial z} + \frac{\partial \xi_z}{\partial x}\right) \qquad [2.51]$$

where ξ_x depends at least on y and z; otherwise it could not be induced by an axial moment through the strain-stress elastic law. This supplementary deformation mode increases the flexibility of the beam. The effect depends on the cross-sectional geometry and is observed to be very large in the case of thin walled beams.

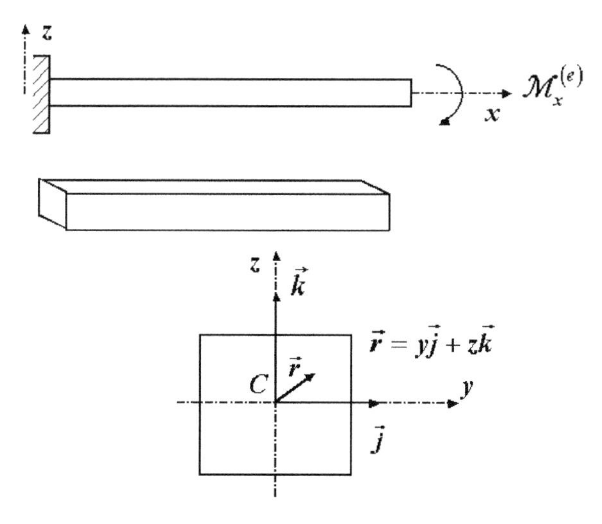

Figure 2.19. *Beam in torsion with warping*

To deal with the warping induced by torsion, Barré de Saint-Venant (1855) developed a simplified analytical model. It is instructive to outline the major features of it, in particular to emphasize that reducing a 3D continuum to an equivalent 1D continuum can be a rather involved and intellectually demanding task. For this purpose, it is sufficient to consider the particular problem shown in Figure 2.19. To account for warping, Saint-Venant assumed the following displacement field:

$$\xi_x = \psi_0 \Phi(z, y); \quad \xi_y = -xz\psi_0; \quad \xi_z = xy\psi_0 \qquad [2.52]$$

where the components [2.50] are completed by an axial component which represents a warping proportional to the maximum value of the twist angle. The unknown function $\Phi(y, z)$ is called the *warping function*; it has the physical dimension of an area. Φ is assumed to be independent of x for mathematical convenience, as this simplification allows one to determine it by solving a 2D elasticity problem, namely that of the equilibrium of a cross-section in plane stress conditions. According to such a model, all the beam sections warp in the same manner. Effects induced by restraining warping will be outlined briefly at the end of this subsection. The strains associated with the displacement field [2.52] are:

$$\begin{gathered}
\varepsilon_{xx} = \frac{\partial \xi_x}{\partial x} = 0; \quad \varepsilon_{yy} = \frac{\partial \xi_y}{\partial y} = 0; \quad \varepsilon_{zz} = \frac{\partial \xi_z}{\partial z} = 0 \\
\varepsilon_{xy} = \frac{1}{2}\left(\frac{\partial \xi_x}{\partial y} + \frac{\partial \xi_y}{\partial x}\right) = \frac{\psi_0}{2}\left(\frac{\partial \Phi}{\partial y} - z\right) \\
\varepsilon_{xz} = \frac{1}{2}\left(\frac{\partial \xi_x}{\partial z} + \frac{\partial \xi_z}{\partial x}\right) = \frac{\psi_0}{2}\left(\frac{\partial \Phi}{\partial z} + y\right) \\
\varepsilon_{yz} = \frac{1}{2}\left(\frac{\partial \xi_y}{\partial z} + \frac{\partial \xi_z}{\partial y}\right) = 0
\end{gathered} \qquad [2.53]$$

The local elastic stresses are:

$$\begin{gathered}
\sigma_{xx} = \sigma_{yy} = \sigma_{zz} = 0 \\
\sigma_{xy} = G\psi_0\left(\frac{\partial \Phi}{\partial y} - z\right); \quad \sigma_{xz} = G\psi_0\left(\frac{\partial \Phi}{\partial z} + y\right); \quad \sigma_{yz} = 0
\end{gathered} \qquad [2.54]$$

The local equilibrium of an unloaded cross-section is $\operatorname{div}\overline{\overline{\sigma}} = 0$, which reads here as:

$$\frac{\partial \sigma_{xy}}{\partial y} + \frac{\partial \sigma_{xz}}{\partial z} = 0; \quad \frac{\partial \sigma_{xy}}{\partial x} = 0; \quad \frac{\partial \sigma_{xz}}{\partial x} = 0 \qquad [2.55]$$

As ξ_x does not depend on x, the stress field [2.54] is found to fulfil automatically the two last equations [2.55]. The first can be written as:

$$\frac{\partial^2 \Phi}{\partial y^2} + \frac{\partial^2 \Phi}{\partial z^2} = 0 \qquad [2.56]$$

The problem is completed by several boundary conditions. First, at the end cross-sections of the beam ($x = 0$ and $x = L$), the following homogeneous conditions must hold:

$$\sigma_{xx}(0) = \sigma_{xx}(L) = 0$$
$$\int_{(\mathcal{S}(x=0))} \sigma_{xz}\, dy\, dz = \int_{(\mathcal{S}(x=0))} \sigma_{xy}\, dy\, dz = 0 \qquad [2.57]$$
$$\int_{(\mathcal{S}(x=L))} \sigma_{xz}\, dy\, dz = \int_{(\mathcal{S}(x=L))} \sigma_{xy}\, dy\, dz = 0$$

Then, the equilibrium of the loaded end $x = L$ gives:

$$\int_{(\mathcal{S}(x=L))} (y\sigma_{xz} - z\sigma_{xy})\, dy\, dz = \mathcal{M}_x^{(e)} \qquad [2.58]$$

Finally, the lateral boundary of the beam is free, so the stresses must comply with the condition:

$$\overline{\overline{\sigma}} \cdot \vec{n}(\vec{r}) = 0 \quad \forall \vec{r} \in (\mathcal{C}) \qquad [2.59]$$

where $\vec{n}(\vec{r})$ is the unit vector in the cross-sectional plane and normal to the cross-sectional contour $(\mathcal{C})$.

Using [2.54], [2.59] reduces to $\sigma_{yx}n_y + \sigma_{zx}n_z = \sigma_{xy}n_y + \sigma_{xz}n_z = 0$ and is finally written as:

$$\frac{\partial \Phi}{\partial y} n_y + \frac{\partial \Phi}{\partial z} n_z = \overrightarrow{\text{grad}\, \Phi} \cdot \vec{n} = z n_y - y n_z = (\vec{n} \times \vec{r}) \cdot \vec{i} \quad \forall \vec{r} \in (\mathcal{C}) \qquad [2.60]$$

where $\vec{i}$ is again the unit vector in the Ox direction.

Then, the warping function $\Phi(y, z)$ is found to be solution of the following boundary value problem:

$$\Delta\Phi(y, z) = 0$$
$$\overrightarrow{\text{grad}\, \Phi} \cdot \vec{n} = z n_y - y n_z = \vec{n} \times \vec{r} \cdot \vec{i} \quad \forall \vec{r} \in (\mathcal{C}) \qquad [2.61]$$

The displacement field [2.52] is thus fully determined, once the boundary value problem [2.61] is solved. However, to be acceptable such a solution has to comply also with the boundary conditions [2.57] and [2.58]. This is actually the case, as shown just below. Starting from the field [2.53], the conditions [2.57] are first transformed into:

$$\begin{aligned}\int_{(S)} \sigma_{xz}\,dy\,dz &= G\psi_0 \int_{(S)} \left(\frac{\partial \Phi}{\partial z} + y\right) dy\,dz = G\psi_0 \int_{(S)} \frac{\partial \Phi}{\partial z}\,dy\,dz \\ \int_{(S)} \sigma_{xy}\,dy\,dz &= G\psi_0 \int_{(S)} \left(\frac{\partial \Phi}{\partial y} - z\right) dy\,dz = G\psi_0 \int_{(S)} \frac{\partial \Phi}{\partial y}\,dy\,dz\end{aligned} \qquad [2.62]$$

where the y and z terms cancel out because of the assumed central symmetry of the cross-sections (cf. subsection 2.1.2).

Then, the mathematical trick is to transform the integrals [2.62] in such a manner as to let the right-hand side of the boundary condition of [2.61] appear. It is first noticed that, because Φ is a solution of [2.61], the following equalities hold:

$$\begin{aligned}\int_{(S)} \left(\frac{\partial \Phi}{\partial z}\right) dy\,dz &= \int_{(S)} \left\{\frac{\partial}{\partial y}\left(z\frac{\partial \Phi}{\partial y} - z^2\right) + \frac{\partial}{\partial z}\left(z\frac{\partial \Phi}{\partial z} + yz\right)\right\} dy\,dz \\ &= \int_{(C)} z\left\{\left(\frac{\partial \Phi}{\partial z} + y\right)n_z + \left(\frac{\partial \Phi}{\partial y} - z\right)n_y\right\} ds \\ &= \int_{(C)} z\{\overrightarrow{\text{grad}\,\Phi} \cdot \vec{n} + yn_z - zn_y\}\,ds = 0\end{aligned} \qquad [2.63]$$

where the surface integral is transformed into a contour integral by using the divergence theorem.

The same calculation applies to the second integral present in [2.62]. Thus, it is proved that not only the conditions [2.57] are satisfied but also that the transverse shear forces are nil, as it could be expected, based on the physics of the problem. Finally the relation [2.58] is written as:

$$G\psi_0 \int_{(S(x=L))} \left(y\left(\frac{\partial \Phi}{\partial z} + y\right) - z\left(\frac{\partial \Phi}{\partial y} - z\right)\right) dy\,dz = \mathcal{M}_x^{(e)} \qquad [2.64]$$

This relation is similar to that obtained without shear, provided the area polar inertia moment J of the cross-section is replaced by the *torsion constant* J_T defined by the integral:

$$\begin{aligned}\mathcal{M}_x^{(e)} &= GJ_T\psi_0 \\ J_T &= \int_{(S(x=L))} \left(y^2 + x^2 + y\frac{\partial \Phi}{\partial z} - z\frac{\partial \Phi}{\partial y}\right) dy\,dz\end{aligned} \qquad [2.65]$$

Then, to account for warping, it suffices to replace J by J_T in the elastic term of the vibration equation [2.41], whereas the inertia term remains unchanged. Indeed, the resultant inertia force related to warping motion is zero, since no axial global displacement is induced. The value of J_T obviously depends on the cross-sectional shape. The calculation is performed by solving first the system [2.61]. Analytical solutions are available for a few geometries, as illustrated by the two following examples.

EXAMPLE 1. – *Circular cross-section*

It is first of interest to check that no warping occurs in the case of circular cross-sections. Indeed, because of cylindrical symmetry, Φ must depend on the radial position r only and the radial vector $\vec{r}$ is parallel to the normal vector $\vec{n}$. Therefore (2.61) reduces to a homogeneous system which has only the trivial solution $\Phi \equiv 0$.

To prove it formally, it is however necessary to express the system [2.61] in polar coordinates. Anticipating the results presented later concerning the use of curvilinear coordinate systems, the result is:

$$\frac{1}{r}\frac{d}{dr}\left(r\frac{d\Phi}{dr}\right) = 0; \qquad \left.\frac{d\Phi}{dr}\right|_{r=R} = 0$$

The general solution of the differential equation is found to be $\Phi = a \ln r$ and the boundary condition implies that the constant a must be zero.

EXAMPLE 2. – *Warped rectangular cross-section*

Referring to the geometry shown in Figure 2.20, the system [2.61] is expressed in Cartesian coordinates as:

$$\frac{\partial^2 \Phi}{\partial y^2} + \frac{\partial^2 \Phi}{\partial z^2} = 0$$
$$\left.\frac{\partial \Phi}{\partial y}\right|_{y=\pm a/2} = z; \qquad \left.\frac{\partial \Phi}{\partial z}\right|_{z=\pm b/2} = -y$$

Before embarking in the task of solving this boundary value problem, it is of interest to note that due to the central symmetry of the cross-sections, $\Phi(y, z)$

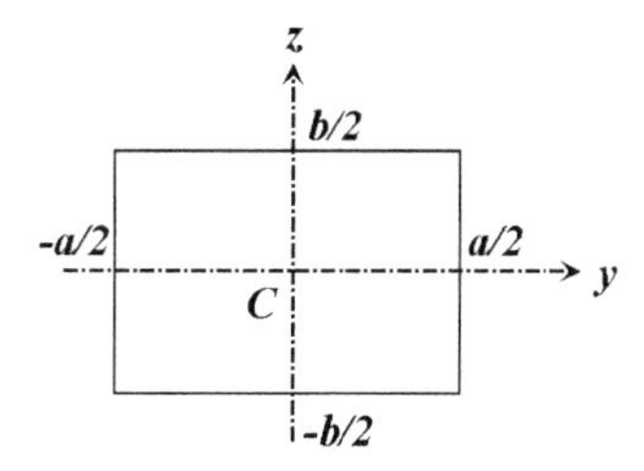

Figure 2.20. *Rectangular cross-section of a prismatic beam*

could be either an even or an odd function of the coordinates in such a way that $\Phi(y,z) = \Phi(-y,-z)$. However, only odd functions are appropriate due to the boundary conditions of the problem. That is, $\partial\Phi/\partial z$ is even with respect to z and odd with respect to y and $\partial\Phi/\partial y$ is even with respect to y and odd with respect to z. As a consequence, the conditions [2.57] are automatically verified since they read as:

$$\int_{(S)} \frac{\partial \Phi}{\partial z}\, dy\, dz = \int_{(S)} \frac{\partial \Phi}{\partial y}\, dy\, dz = 0$$

The general solution of the partial differential equation can be obtained by separating the variables, assuming $\Phi(y,z) = A(y)B(z)$. The calculation is similar to that already carried out in Chapter 1, subsection 1.4.7.2. We arrive at the following system of two ordinary differential equations:

$$\frac{d^2 A}{dy^2} + k^2 A = 0; \quad \frac{d^2 B}{dz^2} - k^2 B = 0$$
$$A'\left(\pm\frac{a}{2}\right) B(z) = z; \quad B'\left(\pm\frac{b}{2}\right) A(y) = -y$$

where the prime denotes a derivation.

The particular case $k = 0$ leads to $A(y) = ay$ and $B(z) = bz$, which produces the particular solution $\Phi_1 = \alpha yz$, where α is a constant. The solutions related to $k \neq 0$ are of the type:

$$A(y) = a_1 \sin ky + a_2 \cos ky$$
$$B(z) = b_1 \sinh kz + b_2 \cosh kz$$

It can be noted that interchanging the circular and the hyperbolic functions in the above expression is merely equivalent to interchanging the names of the variables; so it has no effect on the final result. As $A(y)$ and $B(z)$ are odd functions, the general solution of the homogeneous equation is written as:

$$\Phi(y,z) = \alpha yz + \beta \sin(ky)\sinh(kz)$$

Substitution of this form into the first boundary condition gives:

$$\left.\frac{\partial \Phi}{\partial y}\right|_{\pm a/2} = \alpha z + k\beta \cos\left(\frac{ka}{2}\right)\sinh(kz) = z$$

which is verified provided $\alpha = 1$ and $k = k_n = (2n+1)\pi/a,\ n = 0, 1, 2, \ldots$

The solution is thus expanded as the series:

$$\Phi(y,z) = yz + \sum_{n=1}^{\infty} \beta_n \sin\left(\frac{(2n+1)\pi y}{a}\right) \sinh\left(\frac{(2n+1)\pi z}{a}\right)$$

Substitution of the series into the second boundary condition gives:

$$\left.\frac{\partial \Phi}{\partial z}\right|_{\pm b/2} = \sum_{n=0}^{\infty} \beta_n \left(\frac{(2n+1)\pi}{a}\right) \cosh\left(\frac{(2n+1)\pi b}{2a}\right) \sin\left(\frac{(2n+1)\pi y}{2a}\right) = -2y$$

This relation can be interpreted as the Fourier series of a function which is equal to $-2y$ in the interval $[-a/2, +a/2]$. A suitable function is the so called *zig zag function* with periodicity $2a$ defined as:

$$f(y) = \begin{cases} y & -a/2 \leq y \leq +a/2 \\ a - y & +a/2 \leq y \leq +3a/2 \end{cases}$$

The Fourier series is found to be:

$$f(y) = \frac{4a}{\pi^2} \sum_{n=0}^{\infty} \frac{(-1)^n}{(2n+1)^2} \sin\left(\frac{(2n+1)\pi y}{a}\right)$$

Substituting the series expansion of $-2f(y)$ into the right-hand side of the boundary condition, and identifying term by term the two series, it is readily found that:

$$\beta_n = -\frac{8(-1)^n}{ak_n^3 \cosh(k_n b/2)}; \quad n = 0, 1, 2 \ldots; \ k_n = \frac{(2n+1)\pi}{a}$$

Finally the warping function is written as:

$$\Phi(y,z) = yz - \frac{8}{a} \sum_{n=0}^{\infty} \frac{(-1)^n}{k_n^3 \cosh(k_n b/2)} \sin(k_n y) \sinh(k_n z)$$

Convergence of the series is very fast. Figure 2.21 shows two warped rectangular cross-sections, referring to $a/b = 1$ and $a/b = 0.5$, respectively.

The calculation made just above gives an introductory view on the series expansion method to solve partial differential equations. A more systematic presentation will be given in the theoretical framework of modal analysis in Chapter 4. Application to the present problem will be described in Chapter 6, based on an analogy between the partial differential equation which governs the warping function and

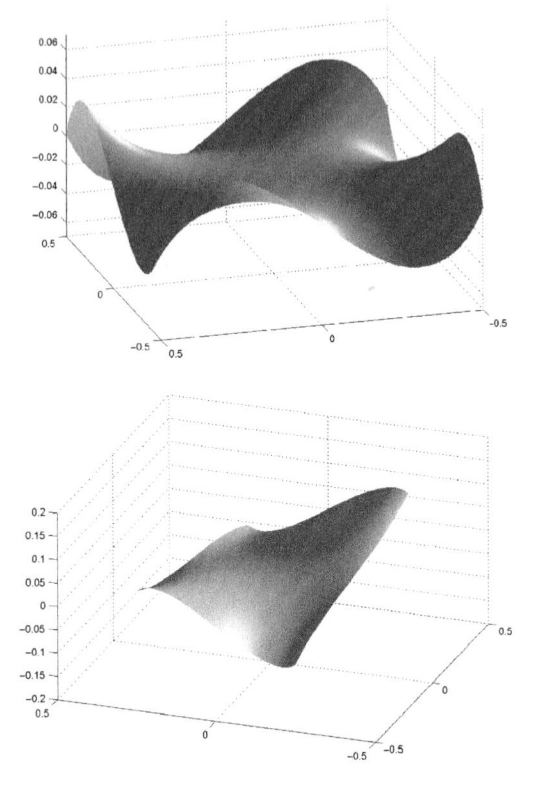

Figure 2.21. *Warping function of rectangular cross-sections*

that which governs the transverse displacement of a stretched membrane. More generally, the problem can be solved numerically by using the finite element method. As a general result, the following inequality holds:

$$J_T \leq J \qquad [2.66]$$

which means that the stiffness due to the elastic deformation of the cross-sections is less with warping than without. The equality holds for a circular cross-section only. The values of the torsion constant for various cross-sectional shapes may be found in several books, see for instance [BLE 79]. A few results are quoted below:

Ellipse: $J_T = \dfrac{\pi a^3 b^3}{a^2 + b^2}$ Square: $J_T = 0{,}1406\, a^4$

Rectangle: $J_T = \dfrac{\gamma a^3 b^3}{a^2 + b^2}$ with

a/b	2	4	8	∞
γ	0.286	0.299	0.312	1/3

Hollow rectangle (thickness: h_a and h_b): $J_T = \dfrac{2h_a h_b (a-h_a)^2(b-h_b)^2}{ah_a + bh_b - h_a^2 - h_b^2}$

Open thin walled cross-sections: $J_T = \mathcal{P}h^3/3$ ($\mathcal{P}$ length, h thickness)

Closed thin walled cross-sections: $J_T = \dfrac{4Sh}{\mathcal{P}}$

($\mathcal{P}$ perimeter, h thickness, S area of the closed part)

A few numerical applications of these formulae are useful to illustrate the importance of warping. Taking a square cross-section as a first example, the ratio J_T/J is found to be $0.1406 \times 6 \cong 0.84$. For a rectangular section of aspect ratio $a/b = 2$, $J = (5/6)b^4$ and $J_T = 0.2860 \times (5/6)b^4 \cong 0.46b^4$ or $J_T/J \cong 0.55$. Finally, if the aspect ratio of the rectangle is very large ($a/b \gg 1$) it is found that $J_T \simeq ab^3/3$, so $J_T/J \simeq 4(a/b)^2 \ll 1$.

To conclude on the subject, it may be added that if warping is prevented by clamping both ends, local axial stresses are generated, which are especially important in the case of open thin walled cross-sections. The reader interested in such aspects of the problem may be referred for instance to [TIM 51], [PIL 02].

2.2.4 *Pure bending mode of deformation*

As a preliminary we must emphasize that, in contrast to a frequently used definition according to which pure bending of a beam means that no shear forces arise, by *pure bending mode of deformation* we mean here that the beam is assumed to be flexed without any superposition of torsional or shear strains.

2.2.4.1 Simplifying hypotheses of the Bernoulli–Euler model

As a first approximation, transverse shear deformation is neglected. Such a simplification gives rise to the *Bernoulli–Euler model*, which is based on the three following simplifying assumptions:

1. As the beam is flexed, the cross-sections remain rigid and perpendicular to the centroid line in the deformed configuration.
2. The transverse shear forces are derived from the equilibrium equation written in terms of moments, and do not arise from elastic shear stresses.
3. The rotational inertia of the beam cross-sections is negligible in comparison with the translation inertia.

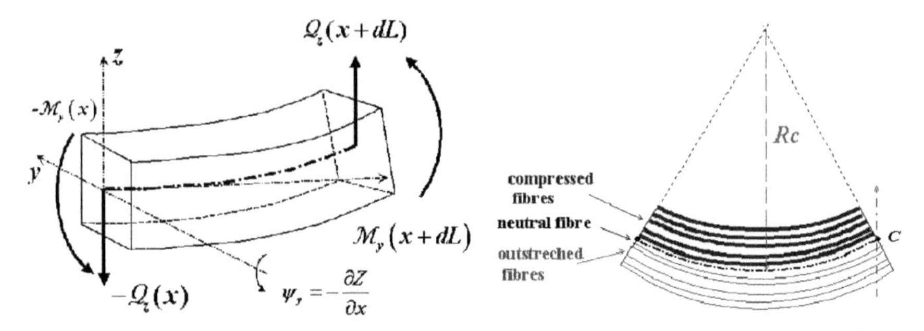

Figure 2.22. *Beam element: bending without shear*

Such assumptions, represented schematically in Figure 2.22, are adopted for mathematical convenience and their range of validity will be examined in Chapter 3 subsection 3.2.2.

The first assumption implies that the magnitude of the small rotation angle due to bending is equal to the slope of the deformed centroid line. In agreement with the sign convention for rotations in a direct reference frame, the following is obtained:

$$\psi_y = -\frac{\partial Z}{\partial x}; \quad \psi_z = +\frac{\partial Y}{\partial x} \Rightarrow \chi_{yy} = -\frac{\partial^2 Z}{\partial x^2}; \quad \chi_{zz} = \frac{\partial^2 Z}{\partial y^2} \qquad [2.67]$$

The global strains χ_{zz} and χ_{yy} may be interpreted as *small (linear) curvatures* of the deformed centroid line. According to the second assumption, the transverse shear forces are identified with the forces induced by the holonomic relations [2.67]. The third assumption implies the nullity of the inertia terms in the two last equations [2.21], which reduce thus to quasi-static equations. Validity of this simplification shall be discussed in Chapter 3, based on energy considerations. It will be shown that the kinetic energy due to the bending rotation of the cross-sections remains much smaller than the kinetic energy due to the transverse displacements provided the slenderness ratio of the beam is sufficiently large.

2.2.4.2 *Local equilibrium*

As shear is neglected, the relations [2.12] cannot be used to calculate the transverse internal forces which are required to ensure the equilibrium under transverse loading. This is the reason why $\mathcal{Q}_y$ and $\mathcal{Q}_z$ are defined from the static form of the two last equations [2.21]. It becomes:

$$\begin{aligned} \mathcal{Q}_y &= -\frac{\partial \mathcal{M}_z}{\partial x} - \mathfrak{M}_z^{(e)}(x;t) \\ \mathcal{Q}_z &= +\frac{\partial \mathcal{M}_y}{\partial x} + \mathfrak{M}_y^{(e)}(x;t) \end{aligned} \qquad [2.68]$$

Substitution of [2.67] into [2.13] leads to:

$$\mathcal{M}_y = -EI_y \frac{\partial^2 Z}{\partial x^2}; \quad \mathcal{M}_z = +EI_z \frac{\partial^2 Y}{\partial x^2} \qquad [2.69]$$

and with [2.68] transverse shear forces are given as:

$$\begin{aligned} \mathcal{Q}_y &= -\frac{\partial}{\partial x}\left(EI_z \frac{\partial^2 Y}{\partial x^2}\right) - \mathfrak{M}_z^{(e)}(x;t) \\ \mathcal{Q}_z &= -\frac{\partial}{\partial x}\left(EI_y \frac{\partial^2 Z}{\partial x^2}\right) + \mathfrak{M}_y^{(e)}(x;t) \end{aligned} \qquad [2.70]$$

The vibration equations are then obtained by substituting [2.70] into the transverse equilibrium equations [2.18]:

$$\begin{aligned} \frac{\partial^2}{\partial x^2}\left(EI_z \frac{\partial^2 Y}{\partial x^2}\right) + \rho S \frac{\partial^2 Y}{\partial t^2} &= F_y^{(e)}(x;t) - \frac{\partial \mathfrak{M}_z^{(e)}(x;t)}{\partial x} \\ \frac{\partial^2}{\partial x^2}\left(EI_y \frac{\partial^2 Z}{\partial x^2}\right) + \rho S \frac{\partial^2 Z}{\partial t^2} &= F_z^{(e)}(x;t) + \frac{\partial \mathfrak{M}_y^{(e)}(x;t)}{\partial x} \end{aligned} \qquad [2.71]$$

Equations [2.71] were established first by Daniel Bernoulli 1735 and solutions found by Euler 1744.

The external loading is expressed in terms of force and moment densities per unit length.

The sketches of Figure 2.23 are helpful to understand the equivalence between a non uniform bending moment and a transverse force field. Indeed, if $\vec{\mathfrak{M}}^{(e)}$ is depicted as torques distributed along the beam, it becomes clear that the axial variation $\mathrm{grad}(\vec{\mathfrak{M}}^{(e)})dx$ is equivalent to a transverse force density, which arises as the resultant of the torque components acting at abscissa x. The force balance is:

$$\begin{aligned} &-\rho S\ddot{Z}\,dx + (\mathcal{Q}_z(x+dx) - \mathcal{Q}_z(x)) + \left(\mathfrak{M}_y^{(e)}(x+dx) - \mathfrak{M}_y^{(e)}(x)\right) \\ &\qquad + F_z^{(e)}\,dx = 0 \Rightarrow \rho S\ddot{Z} - \frac{\partial \mathcal{Q}_z}{\partial x} = F_z^{(e)} + \frac{\partial \mathfrak{M}_y^{(e)}}{\partial x} \\ &-\rho S\ddot{Y}\,dx + (\mathcal{Q}_y(x+dx) - \mathcal{Q}_y(x)) - \left(\mathfrak{M}_z^{(e)}(x+dx) - \mathfrak{M}_z^{(e)}(x)\right) \\ &\qquad + F_y^{(e)}\,dx = 0 \Rightarrow \rho S\ddot{Y} - \frac{\partial \mathcal{Q}_y}{\partial x} = F_y^{(e)} - \frac{\partial \mathfrak{M}_z^{(e)}}{\partial x} \end{aligned}$$

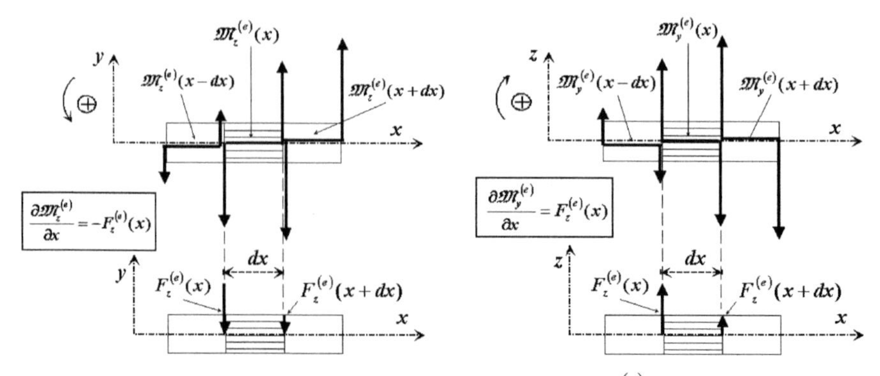

Figure 2.23. *Representation of the term* $\partial\vec{\mathfrak{M}}^{(e)}/\partial x$

2.2.4.3 Elastic boundary conditions

Let a beam be bent in the plane Oxz. The end $x = 0$ is supported by a spring of stiffness coefficient K_z acting in the Oz direction. The corresponding boundary condition is:

$$Q_z(0;t) - K_z Z(0;t) = 0 \Rightarrow EI_y \left.\frac{\partial^3 Z}{\partial x^3}\right|_{x=0} + K_z Z(0;t) = 0 \qquad [2.72]$$

If the supporting spring acts on the rotation with a stiffness coefficient K_{ψ_y}, the condition becomes:

$$\mathcal{M}_y(0;t) - K_{\psi_y}\psi_y(0;t) = 0 \Rightarrow -EI_y \left.\frac{\partial^2 Z}{\partial x^2}\right|_{x=0} + K_{\psi_y} \left.\frac{\partial Z}{\partial x}\right|_{x=0} = 0 \quad [2.73]$$

The two limit cases for which the spring stiffness tends to zero, or alternatively tends to infinity, provide us with the so called 'standard' boundary conditions which have specific names:

$$\begin{aligned}
&\text{free end: } \partial^2 Z/\partial x^2 = \partial^3 Z/\partial x^3 = 0 \\
&\text{pinned end: } Z = \partial^2 Z/\partial x^2 = 0 \\
&\text{sliding support: } \partial Z/\partial x = \partial^3 Z/\partial x^3 = 0 \\
&\text{clamped end: } Z = \partial Z/\partial x = 0
\end{aligned} \qquad [2.74]$$

A symbolic representation of these conditions is shown in Figure 2.24. In the three-dimensional case (bending in the Oxy and the Oxz plane) for a clamped end, all the six displacement components are zero and for a pinned end the three translations are zero and the two transverse rotations are free. The pinned support is represented here as a pair of triangles, which stand for two knife edge supports,

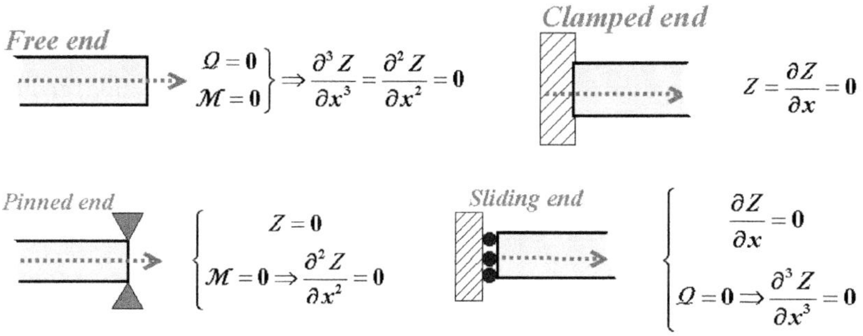

Figure 2.24. *Bending of a beam: symbolic representation of standard support conditions*

instead of only one as is most often accepted. This is to stress that the support condition is bilateral, i.e. transverse displacement is prohibited whatever its sign may be. Indeed, a unilateral support condition such as $Z \geq 0$ is nonlinear in nature, as further discussed in Chapter 4, in relation to problems of impacted beams.

2.2.4.4 *Intermediate supports*

Restricting the study to the case of elastic supports, the equations [2.73] and [2.74] become:

$$EI_y \left(\left. \frac{\partial^3 Z}{\partial x^3} \right|_{x_{0+}} - \left. \frac{\partial^3 Z}{\partial x^3} \right|_{x_{0-}} \right) + K_z Z(x_0;t) = 0$$
$$EI_y \left(\left. \frac{\partial^2 Z}{\partial x^2} \right|_{x_{0+}} - \left. \frac{\partial^2 Z}{\partial x^2} \right|_{x_{0-}} \right) - K_{\psi_y} \left. \frac{\partial Z}{\partial x} \right|_{x_0} = 0 \qquad [2.75]$$

2.2.4.5 *Concentrated loads*

There are two types of concentrated loads, namely transverse forces and bending moments, which induce the following finite discontinuities:

$$\mathcal{Q}_Z(x_{0+}) - \mathcal{Q}_Z(x_{0-}) + F_z^{(e)} = 0 \Rightarrow EI_y \left(\left. \frac{\partial^3 Z}{\partial x^3} \right|_{x_{0+}} - \left. \frac{\partial^3 Z}{\partial x^3} \right|_{x_{0-}} \right) = F_z^{(e)}(t)$$
$$\mathcal{M}_y(x_{0+}) - \mathcal{M}_y(x_{0-}) + \mathcal{M}_y^{(e)} = 0 \Rightarrow EI_y \left(\left. \frac{\partial^2 Z}{\partial x^2} \right|_{x_{0+}} - \left. \frac{\partial^2 Z}{\partial x^2} \right|_{x_{0-}} \right) = \mathcal{M}_y^{(e)}(t) \qquad [2.76]$$

2.2.4.6 General solution of the static and homogeneous equation

The beam is supposed to be homogeneous, of constant cross-section, and bent in the *Oxz* plane. In statics, if the right-hand side of the second equation [2.71] is zero, the solution reduces to:

$$Z(x) = ax^3 + bx^2 + cx + d \qquad [2.77]$$

where a, b, c, d are constants which are determined from the boundary conditions.

2.2.4.7 Application to some problems of practical interest

EXAMPLE 1. – *Cantilevered beam loaded by its own weight*

Let us consider a beam of constant cross-section set in a cantilevered configuration, i.e. clamped at one end and free at the other, see Figure 2.25. The beam lies horizontally in a uniform gravity field described by the acceleration vector $\vec{g}$. It is thus loaded uniformly by its own weight $\vec{F}_0 = \rho SL\vec{g}$. The loading function is the force density per unit length of beam $-F_0/L = f_0$.

The boundary value problem is written as:

$$EI\frac{d^4Z}{dx^4} = f_0$$
$$Z(0) = Z'(0) = 0$$
$$Z''(L) = Z'''(L) = 0$$

where the prime stands for a derivation with respect to x.

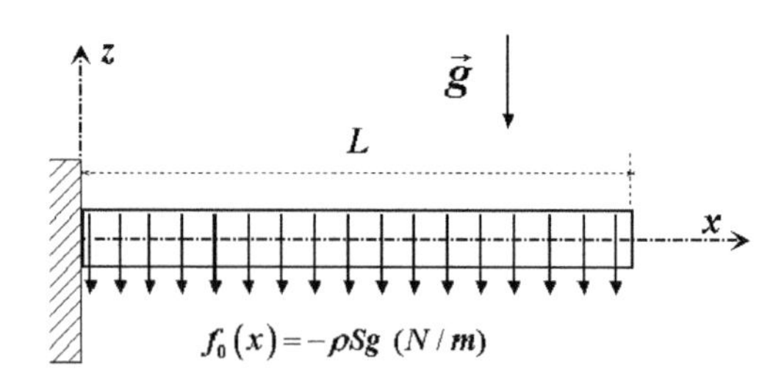

Figure 2.25. *Cantilevered beam loaded by its own weight*

The solution is of the type: $Z(x) = ax^4 + bx^3 + cx^2 + dx + e$; the constants are determined by the loading and the boundary conditions as follows.

Clamped end at $x = 0 \to d = e = 0 \to Z(x) = ax^4 + bx^3 + cx^2$

Loading: $24a = \dfrac{f_0}{EI} \to a = \dfrac{f_0}{24EI}$

Free end at $x = L$

$$Z'''(L) = 24ax + 6b = 0 \Rightarrow b = -\frac{f_0 L}{6EI}$$

$$Z''(L) = \frac{f_0 L^2}{2EI} - \frac{f_0 L^2}{EI} + 2c = 0 \to c = \frac{f_0 L^2}{4EI}$$

then:

$$Z(\xi) = Z_0(\xi^4 - 4\xi^3 + 6\xi^2) \quad \text{where } \xi = \frac{x}{L} \quad \text{and } Z_0 = \frac{F_0 L^3}{24EI}$$

The maximum deflection is $Z_{\text{max}} = Z(1) = 3Z_0 = F_0 L^3/(8EI)$

The reactions at the clamped end balance exactly the global internal stresses exerted on the corresponding cross-section. This result agrees both with the action-reaction principle and with the sign convention about stresses, according to which the stresses are the efforts (forces and moments) exerted by the right-hand part of the beam on the left-hand part. Vice versa, the efforts on the supports are the efforts exerted by the left-hand part on the right-hand part. So a clamped support acts as a force and as a moment given by:

$$\mathcal{R}_z = -\mathcal{Q}_z = EI \left.\frac{d^3 Z}{dx^3}\right|_{x=0} = 6bEI = -f_0 L = -F_0$$

$$\mathcal{M}_{Ry} = -\mathcal{M}_y = EI \left.\frac{d^2 Z}{dx^2}\right|_{x=0} = -2cEI = +\frac{F_0 L}{2}$$

As expected, the reactions are found to be independent of the material law. Further, as the global equilibrium of the beam is concerned, the above results indicate that the external loading is equivalent to the weight $\vec{F}_0$ concentrated at mid-span of the beam. This resultant is necessarily balanced by the support reaction $\mathcal{R}_z + F_0 = 0$. Then the support reaction is $\mathcal{R}_z = -\mathcal{Q}_z = -F_0$. In the same way, the external loading moment at $x = 0$ is $\mathcal{M}^{(e)} = -F_0 L/2$ and is exactly balanced by the reaction moment $\mathcal{M}_{Ry} = F_0 L/2$.

EXAMPLE 2. – *Loads applied to the free end*

Deflection of a cantilevered beam loaded by a static force concentrated at $x = L$ is found to be:

$$Z(\xi) = Z_0(3\xi^2 - \xi^3) \quad \text{where } \xi = \frac{x}{L} \text{ and } Z_0 = \frac{F_z^{(e)} L^3}{6EI}$$

If the end conditions and the load location are reversed the last result becomes:

$$Z(\xi) = Z_0(3(1-\xi)^2 - (1-\xi)^3) \quad = Z_0(\xi^3 - 3\xi + 2)$$

If an external moment $\mathcal{M}_y^{(e)}$ is exerted at $x = L$, instead of an external force, the deflection is:

$$Z(\xi) = -\frac{\mathcal{M}_y^{(e)} L^2}{2EI}\xi^2$$

If the end conditions and the load location are reversed the last result becomes:

$$Z(\xi) = \frac{\mathcal{M}_y^{(e)} L^2}{2EI}(\xi - 1)^2$$

The deflection in each of these cases is sketched in Figure 2.26.

EXAMPLE 3. – *Transverse force applied to a sliding support*

Figure 2.27 shows a beam which is supported at $x = 0$ by a sliding support and at $x = L$ by a pinned support. The external load is a transverse force applied at $x = 0$. It is recalled that, in the unloaded case, a sliding support is described by the conditions $Z'(0) = 0$ (no rotation) and $Z'''(0) = 0$ (no shear force). In the presence of the external load, these conditions have to be modified. To find out

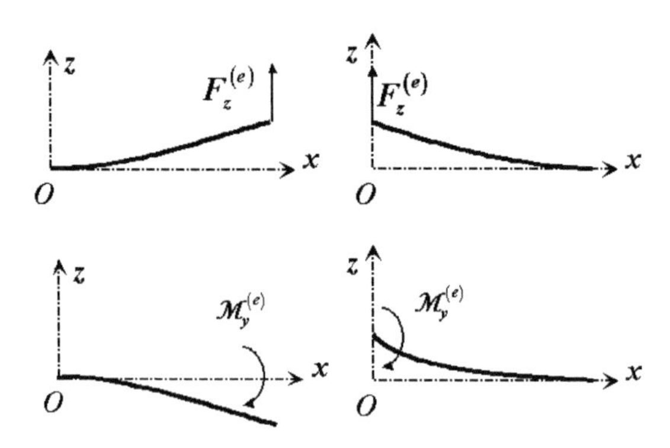

Figure 2.26. *Schematic view of deflected cantilevered beams subject to static end loads*

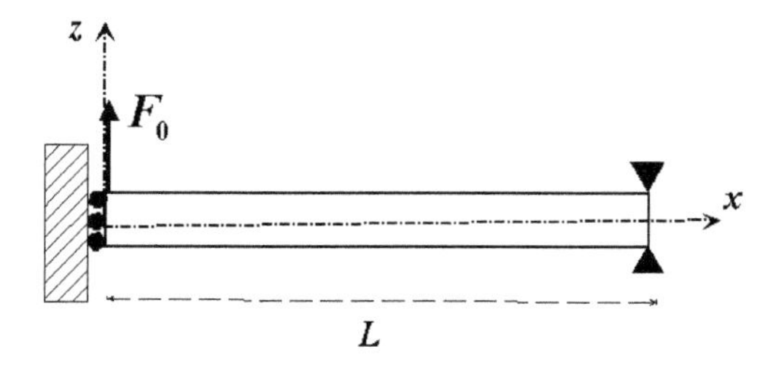

Figure 2.27. *Loading on a sliding support*

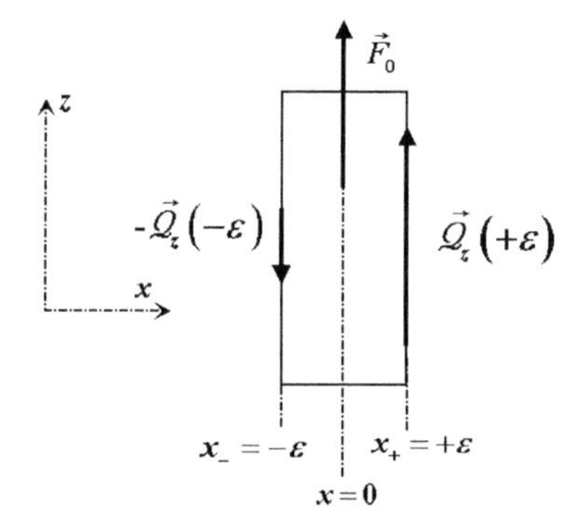

Figure 2.28. *Equilibrium of the loaded section*

how, the simplest way is to adapt the Figure 2.14 to the case of a transverse loading, as shown in Figure 2.28.

The balance of the transverse forces implies:

$$F_0 + Q_z(+\varepsilon) - Q_z(-\varepsilon) = 0;$$

but $Q_z(-\varepsilon) = 0$ since $-\varepsilon$ lies outside from the beam.

On the other hand, $\lim_{\varepsilon \to 0} Q_z(+\varepsilon) = -EI\partial^3 Z/\partial x^3$, which leads to the load condition $EI\partial^3 Z/\partial x^3 = F_0$.

Thus the boundary value problem to be solved is written as:

$$EI\frac{d^4 Z}{dx^4} = 0$$

$$Z'(0) = Z'''(0_-) = 0; \quad Z(L) = Z''(L) = 0$$

$$EIZ'''(0_+) = F_0$$

The solution is $Z(\xi) = F_0 L^3(\xi^3 - 3\xi^2 + 2)/(6EI)$ and the maximum deflection is $Z(0) = F_0 L^3/(3EI)$.

The support reactions are derived from the stresses at beam ends. The force exerted by the pinned support is $\mathcal{R}_z = -F_0$ and the moment applied by the sliding support is $\mathcal{M}_y = -LF_0$. Therefore it can be immediately seen that global equilibrium is satisfied as suitable.

EXAMPLE 4. – *Clamped-clamped beam loaded at mid-span by a transverse force*

Here, the beam is clamped at both ends (clamped-clamped configuration) and loaded at x_0 by a transverse force, see Figure 2.29. The boundary value problem is written as:

$$EI\frac{d^4Z}{dx^4} = 0; \quad Z(0) = Z'(0) = 0; \quad Z(L) = Z'(L) = 0$$
$$EIZ'''(x_{0+}) - EIZ'''(x_{0-}) = F_0$$

Since the third derivative of $Z(x)$ is discontinuous at the loaded cross-section, distinct solutions necessarily hold in the ranges $0 \le x < x_0$ and $x_0 \le x \le L$ respectively, i.e. on each side of the discontinuity. Then, using the dimensionless abscissa $\xi = x/L$, the solutions are of the following form:

$$\begin{cases} 0 \le \xi < \xi_0 & Z_1(\xi) = A_1\xi^3 + B_1\xi^2 + C_1\xi + D_1 \\ \xi_0 \le \xi \le 1 & Z_2(\xi) = A_2\xi^3 + B_2\xi^2 + C_2\xi + D_2 \end{cases}$$

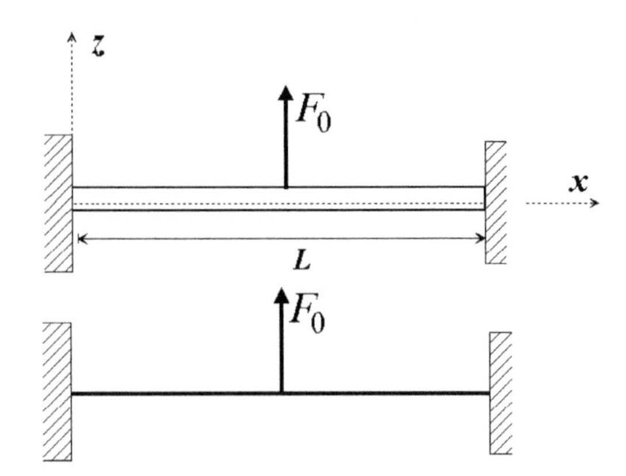

Figure 2.29. *Beam loaded at mid-span by a transversal force*

Eight constants have to be determined, which requires eight relations. Five are available from the boundary and the loading conditions and three are suitably provided by writing down that $Z(\xi)$ and its two first derivatives must be continuous at ξ_0. The following eight linear relations are thus obtained:

$$Z_1(0) = 0; \quad Z_1'(0) = 0; \quad Z_2(1) = 0; \quad Z_2'(1) = 0$$

$$EI(Z_2'''(\xi_0) - Z_1'''(\xi_0)) = F_0 L^3$$

$$Z_1(\xi_0) = Z_2(\xi_0); \quad Z_1'(\xi_0) = Z_2'(\xi_0); \quad Z_1''(\xi_0) = Z_2''(\xi_0)$$

Determination of the constants is straightforward, though rather tedious. If the load is concentrated at mid-span, calculation can be alleviated by using the symmetry conditions with respect to $\xi_0 = 1/2$. Incidentally, the present exercise gives us a good opportunity to emphasize that a mechanical continuous system can be said to be symmetric if, and only if, the same rules of symmetry hold for the geometry of the structure, the support conditions and the external loading. This is clearly the case here, because nothing is changed if the $\xi_0 > 0.5$ part of the system is exchanged with the $\xi_0 \leq 0.5$ part. As a consequence, the problem can be reduced to the study of half a beam, for instance the $\xi_0 \leq 0.5$ part, loaded by half the original force. Then the general form of the solution is:

$$0 \leq \xi \leq 0.5 \quad Z_1(\xi) = A_1\xi^3 + B_1\xi^2 + C_1\xi + D_1$$

The boundary conditions are:

$$Z_1(0) = 0; \quad Z_1'(0) = 0; \quad Z_1'(0.5) = 0; \quad -EIZ_1'''(0.5) = F_0 L^3/2$$

The third condition expresses the symmetry of the problem as evidenced by the following calculation:

$$Z_2(\xi) = Z_1(1-\xi) \Rightarrow \frac{dZ_2(\xi)}{d\xi} = -\frac{dZ_1(1-\xi)}{d\xi}$$

The continuity of the two first derivatives at mid-span gives:

$$Z_2'(0.5) = -Z_1'(0.5) \Rightarrow Z_2'(0.5) = Z_1'(0.5) = 0$$

Calculation of the four constants is quite straightforward:

$$Z_1(0) = Z_1'(0) = 0 \Rightarrow Z_1(\xi) = A_1\xi^3 + B_1\xi^2$$

$$Z_1'(0.5) = 0 \Rightarrow 3A_1(0.5)^2 + 2B_1(0.5) = 0 \Rightarrow B_1 = -\frac{3A_1}{4}$$

$$EIZ_1'''(0.5) = -F_0L^3/2 \Rightarrow 6EIA_1 = -F_0/2 \Rightarrow A_1 = -\frac{F_0L^3}{12EI}$$

Thus, the beam deflection is given by:

$$Z_1(\xi) = \frac{F_0 L^3}{12EI}\xi^2\left(\frac{3}{4} - \xi\right) \quad \xi \leq 0.5$$
$$Z_2(\xi) = \frac{F_0 L^3}{12EI}(1-\xi)^2\left(\xi - \frac{1}{4}\right) \quad \xi \geq 0.5$$

The maximum deflection is:

$$Z_{\max} = Z_1(0.5) = \frac{F_0 L^3}{192EI}$$

Turning to the non-symmetrical case $\xi_0 \neq 0.5$, the coefficients are found to be:

$$A_1 = \frac{F_0 L^3}{6EI}(3\xi_0^2 - 2\xi_0^3 - 1); \quad A_2 = \frac{F_0 L^3}{6EI}(3\xi_0^2 - 2\xi_0^3)$$
$$B_1 = \frac{F_0 L^3}{2EI}\xi_0(\xi_0 - 1)^2; \quad B_2 = \frac{F_0 L^3}{2EI}\xi_0^2(1 - \xi_0)$$

Figure 2.30 refers to a steel beam ($E = 2.110^{11}$ Pa) of length $L = 1$m and square cross-section ($a = 2$ cm). The transverse load of 1 kN is applied at $\xi_0 = 0.25$. The abscissa ξ_m of maximum deflection differs from that of the excitation point. The Figures 2.31 refer to the same beam and the same load and its point of application is varied. On the left-hand part of Figure 2.31a, ξ_m is plotted versus ξ_0 and on the right-hand part, maximum deflection $Z(\xi_m)$ is plotted versus ξ_0. It is noted that $Z(\xi_m)$ varies nonlinearly with ξ_0; as ξ_0 increases up to 0.5, ξ_m tends to ξ_0 and $Z(\xi_m)$ increases. On the other hand, it is also of interest to study the support reactions versus ξ_0. They are plotted in Figure 2.31b; the graph on the left-hand side refers to the transverse forces and that on the right-hand side to the moments. Though

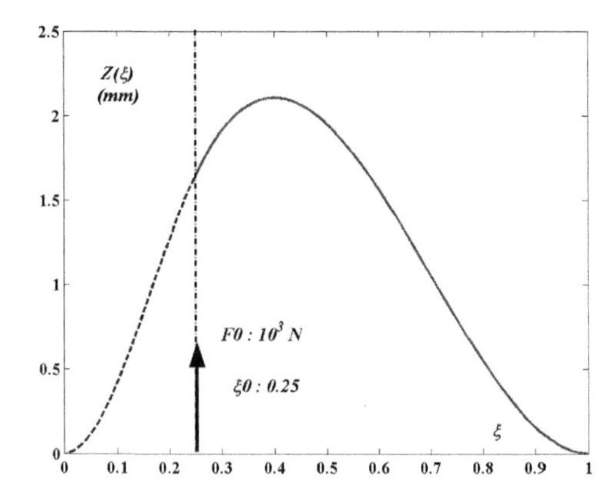

Figure 2.30. *Transverse deflection of the beam, non-symmetrical loading case*

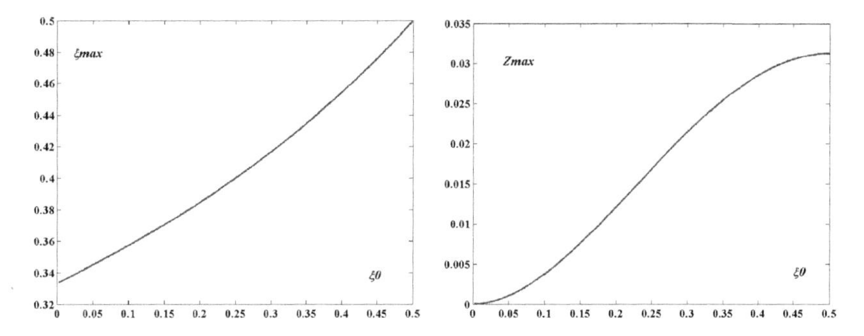

Figure 2.31a. *Position and magnitude of the maximum deflection*

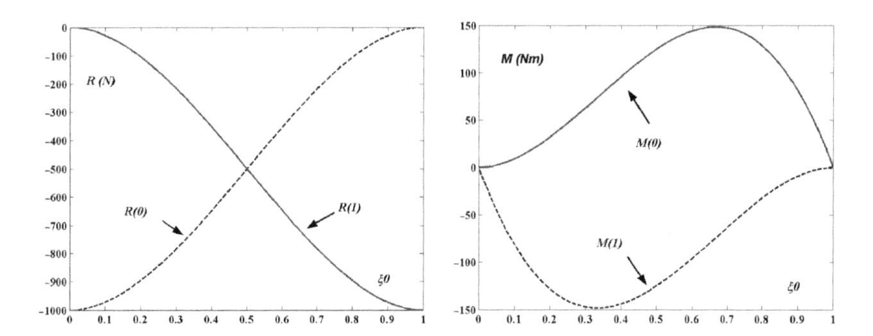

Figure 2.31b. *Support reactions: transverse forces and moments*

variation of these quantities with ξ_0 can be easily understood in a qualitative way, it would be however difficult to make a quantitative prediction without a detailed calculation. In this respect, it is suitable to emphasize that according to a rigid beam model – thus restricted to two degrees of freedom – it would be impossible to determine the reactions. Indeed, the clamping boundary conditions applied to the beam result in four holonomic conditions; therefore the discrete model turns out to be hyperstatic according to the definition given in [AXI 04], Chapter 4. This point can be further investigated by considering now the case of a beam provided with pinned supports at each end. A straightforward calculation adopting a flexible model gives:

$$Z_1(\xi) = A_1\xi^3 + C_1\xi \quad \xi \le \xi_0$$

$$Z_2(\xi) = A_2(\xi - 1)^3 + C_2(\xi - 1) \quad \xi_0 \le \xi \le 1$$

$$A_1 = \frac{F_0 L^3}{6EI}(\xi_0 - 1); \quad A_2 = \frac{F_0 L^3}{6EI}\xi_0$$

$$C_1 = \frac{F_0 L^3}{6EI}\xi_0(\xi_0 - 1)(\xi_0 - 2); \quad C_2 = \frac{F_0 L^3}{6EI}\xi_0(\xi_0^2 - 1)$$

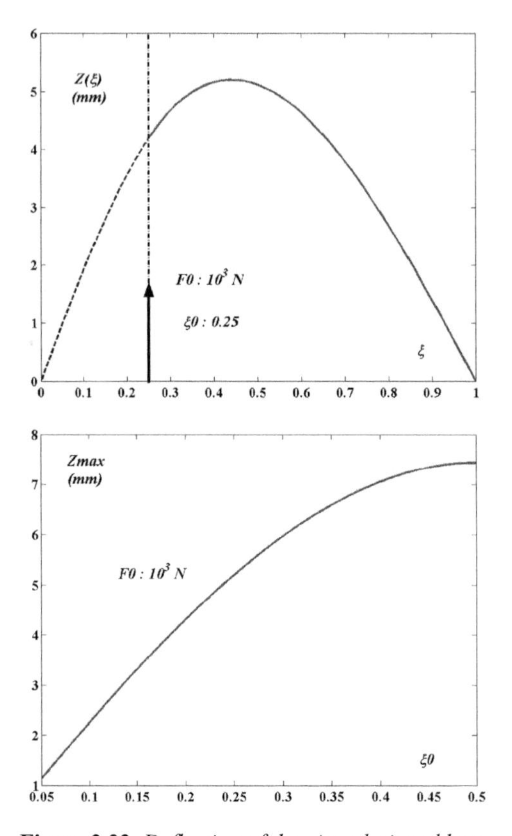

Figure 2.32. *Deflection of the pinned-pinned beam*

The first point to be noticed is that, for a given load, the maximum deflection is significantly larger for the pinned-pinned than for the clamped-clamped case, (compare Figures 2.32 and 2.30). Such a result can be easily understood since a clamped support is substantially stiffer than a pinned one. The second point is that in the pinned-pinned configuration, the support reactions are the same as those produced by using a rigid beam model, which turns out to be isostatic:

$$\mathcal{R}(0) = F_0(\xi - 1); \quad \mathcal{R}(1) = -F_0\xi.$$

NOTE. – *Support reactions and geometrical nonlinearities*

It is also of interest to discuss the support reactions which are actually observed when a beam configuration like that represented in Figure 2.29 is tested experimentally by mounting the supports on three-axial and stiff force transducers. The major point to be observed is that tensile forces occur which are not small in comparison with the shear forces, even if lateral deflection remains very small. This is in contradiction to the prediction of the linear model. More generally, the flexure of a beam provided with supports which prevent any axial displacement raises the same objection concerning the validity of a linear model. The reason is that, according to the latter, the length of the beam remains unchanged when deflected. Of course, this is not true and the actual length is:

$$L(Z_m) = \int_0^{L_0} \sqrt{1 + \left(\frac{dZ}{dx}\right)^2} dx$$

where L_0 is the beam length in the unloaded configuration and Z_m is the maximum deflection of the loaded beam. In so far as Z_m remains sufficiently small, the axial deformation can be defined as follows:

$$\eta_x(Z_m) = \frac{L(Z_m) - L_0}{L_0} \simeq \frac{1}{2} \int_0^{L_0} \left(\frac{dZ}{dx}\right)^2 dx$$

This can be interpreted as the axial deformation of a beam (or a string) subject to an axial tension. Hence, in the elastic domain the corresponding axial stress is:

$$\mathcal{N}(Z_m) = ES\eta_x(Z_m)$$

An 'order of magnitude' calculation is sufficient to point out that the axial stress induced by the geometrical nonlinearity is not at all negligible. As a gross approximation, let us adopt a simple analytical approximation of the lateral displacement field, for instance the parabolic shape:

$$Z(\xi) \simeq 4Z_m\xi(1 - \xi)$$

which is an appropriate form in the case of the symmetrical configuration depicted in Figure 2.29. It leads to the following axial strain:

$$\eta_x = 8\left(\frac{Z_m}{L_0}\right)^2 \int_0^1 (1 - 2\xi)^2 dx = \frac{8}{3}\left(\frac{Z_m}{L_0}\right)^2$$

The above expression is a quadratic form of the displacement amplitude. Hence, the axial deformation is a second order term, which can be safely neglected as the beam deflection is concerned. Nevertheless, as the load F_0 increases, the magnitude of the nonlinear tension is found to become rapidly larger than that of the transverse shear. Indeed, it is easy to verify that, according to the linear approximation, the maximum deflection is $Z_m = \mathcal{Q}_z L_0^3/(24EI)$ and consequently the axial global stress can be finally written as:

$$\frac{\mathcal{N}}{\mathcal{Q}_z} = \frac{2}{3} \frac{\mathcal{Q}_z}{Ea^2} \left(\frac{L}{a}\right)^4 \qquad [2.78]$$

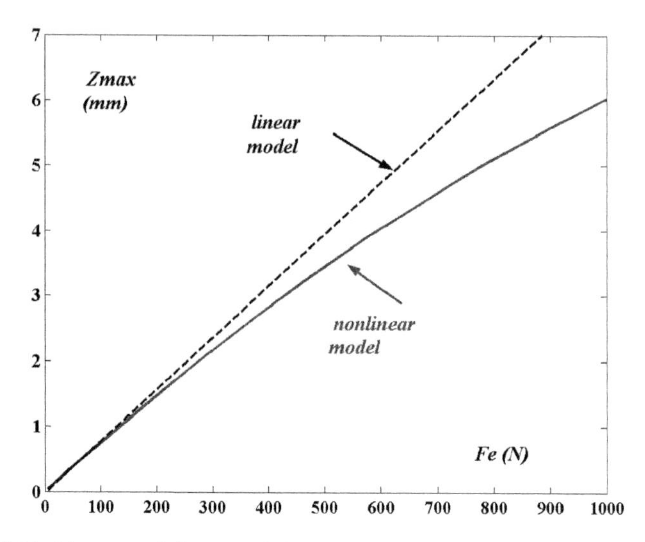

Figure 2.33. *Maximum deflection of the beam versus magnitude of the lateral loading*

Using the numerical data of the present example, it is found that $\mathcal{N}/\mathcal{Q}_z > 1$ as soon as $Q_z > 20N$. As clearly indicated by the approximate formula [2.78], this threshold decreases very rapidly with the slenderness ratio of the beam. Finally, it is also valuable to present here the results of a few numerical calculations performed in the nonlinear elastic domain, by using a finite element program. Figure 2.33 shows how the maximum lateral deflection varies with the magnitude of the external force, applied at mid-span of the clamped-clamped beam. It can be noted that as a consequence of the geometrical nonlinearity, the lateral deflection of the beam is smaller than that inferred from the linear approximation. Otherwise stated, in the present problem the geometrical nonlinearity has a stiffening effect. Here, the effect becomes well marked as soon as Z_m exceeds 2 mm. In Figure 2.34, the tensile stress is plotted versus the transverse shear force (full line); the dashed line corresponds to equality between the two components. This plot allows one to check the relevance of the order of magnitude calculation made just above. According to the nonlinear calculation, the axial force becomes larger than the transverse one as soon as their magnitude exceeds about 25 N, to be compared with the threshold of 20 N produced by the simplified analysis.

2.2.5 *Formulation of the boundary conditions*

2.2.5.1 Elastic impedances

Whatever the mode of beam deformation may be, the elastic boundary conditions are expressible analytically as linear homogeneous equations interrelating

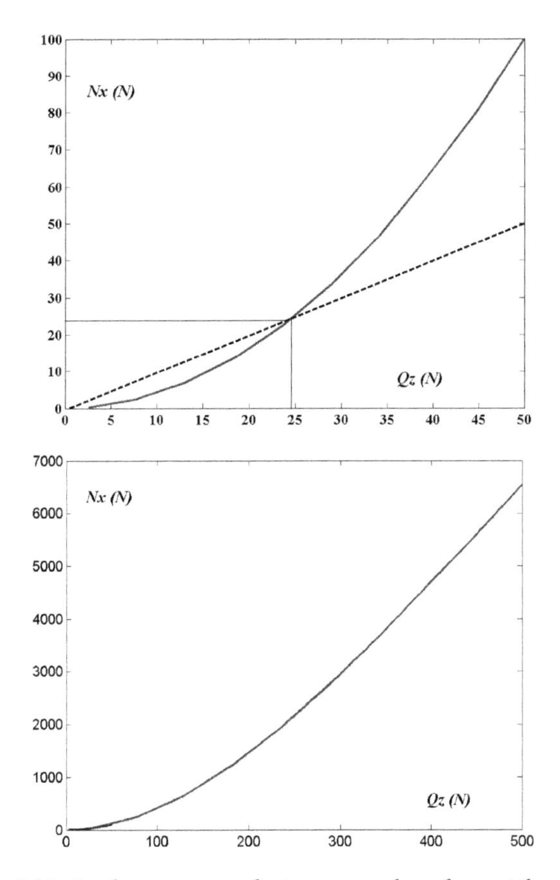

Figure 2.34. *Axial stress versus the transverse shear force at the support*

a conjugated pair of displacement and stress component. Such conditions may also be formulated by forming the ratio of the stress over the displacement related component:

$$\alpha q(x_0;t) + \beta Q(x_0;t) = 0 \iff -\frac{\alpha}{\beta} = \mathcal{Z}_e(x_0) = \left[\frac{Q(x_0;t)}{q(x_0;t)}\right]; \quad x_0 = 0, \text{ or } L \tag{2.79}$$

$q(x;t)$ is the component of the global displacement and $Q(x;t)$ the conjugate stress variable. By a slight abuse of language $\mathcal{Z}_e$ will be termed hereafter as an *elastic impedance*, which can vary from zero to infinity.

2.2.5.2 Generalized mechanical impedances

Besides the elastic supports, other kinds of linear boundary conditions can be useful, such as dissipative conditions of the type:

$$\alpha\dot{q}(x_0;t) + \beta\mathcal{Q}(x_0;t) = 0 \Longleftrightarrow Z_e(x_0) = \left[\frac{\mathcal{Q}(x_0;t)}{\dot{q}(x_0;t)}\right]; \quad x_0 = 0, \text{ or } L$$

or inertia conditions, which are of the type:

$$\alpha\ddot{q}(x_0;t) + \beta\mathcal{Q}(x_0;t) = 0 \Longleftrightarrow \mathcal{Z}_e(x_0) = \left[\frac{\mathcal{Q}(x_0;t)}{\ddot{q}(x_0;t)}\right]; \quad x_0 = 0, \text{ or } L$$

Actually, the term *mechanical impedance* in the strict sense of the word refers to the ratio between a force and a velocity and the concept is broadly used in connection with vibration studies, carried out in the frequency instead of the time domain. Accordingly, the name of the ratio which describes a given support condition is changed, depending on the physical nature of the conjugate variables used to form the ratio, see for instance [EWI 00]. However, the underlying concept remains basically the same in each particular case. Hence, the concept of *generalized mechanical impedance* is found appropriate to designate such ratios, independently of their physical nature, in the same way as the concept of generalized coordinates, or forces, is used to get free from the particular coordinate system used to formulate a mechanical problem. On the other hand, there is no difficulty in extending the concept of impedance to intermediate supports located at $0 \le x_0 \le L$, as it suffices to replace the force term by the corresponding force discontinuity at x_0.

2.2.5.3 Homogeneous and inhomogeneous conditions

To close this subsection, it is recalled that a boundary condition can be viewed as a discontinuity condition in which one stress is zero, because either the left-hand side ($x = 0_-$) or the right-hand side of the boundary ($x = L_+$) lies outside the beam. A homogeneous condition describes a support reaction while an inhomogeneous condition describes an external load applied to a boundary, cf. subsection 2.2.4.7, example 3. Nevertheless, an alternative and even more convenient formulation will be presented later in Chapter 3, via the unifying approach of local and continuous quantities offered by the Dirac distributions.

2.2.6 *More about transverse shear stresses and straight beam models*

The equations for a straight beam established in the preceding sections correspond to the basic engineering models, according to which axial, twist and bending deformations can be described separately. Nevertheless, in the presence of transverse shear stresses, the validity of such models must be revisited, based on the distribution of the local shear stresses which are induced by the deformations of the real 3D structure. The problem may be evidenced experimentally by subjecting cantilevered elastic beams of various shapes to a transverse load $\vec{F}$ concentrated at a point of the cross-section at the free end.

2.2.6.1 Asymmetrical cross-sections and shear (or twist) centre

Let us consider first a beam whose cross-sections are symmetrical about two principal axes of inertia taken as local axes Cy, Cz, where again C is the centroid of the cross-section. A cross-section subjected to a transverse force $\vec{F}$ moves rigidly as depicted schematically in Figure 2.35. If $\vec{F}$ is applied to C, and parallel to Cy, or Cz, the beam is deflected in the direction of the load and no torsion occurs. Then, if the load is tilted through a given angle, deflection can be obtained by superposing the individual responses to the load Cartesian components F_y, F_z and again no torsion arises. This is simply because of the central symmetry of the cross-section, the equivalent one-dimensional loading exerted on C reduces to $\vec{F}$ and no moment is generated. In contrast, if the load is applied to a point P distinct from C, the equivalent one-dimensional loading exerted on C comprises the force $\vec{F}$ and the moment of $\vec{F}$ about C, which now differs from zero. Hence, lateral deflexion of the beam is accompanied by torsion and, due to the central symmetry, the beam cross-section rotates about C.

Contrasting with this simple behaviour, if the cross-section is not symmetrical about the principal axes Cy, Cz, it is found that if $\vec{F}$ is applied to the centroid, both lateral deflection and torsion occur. However, there exists another point within the

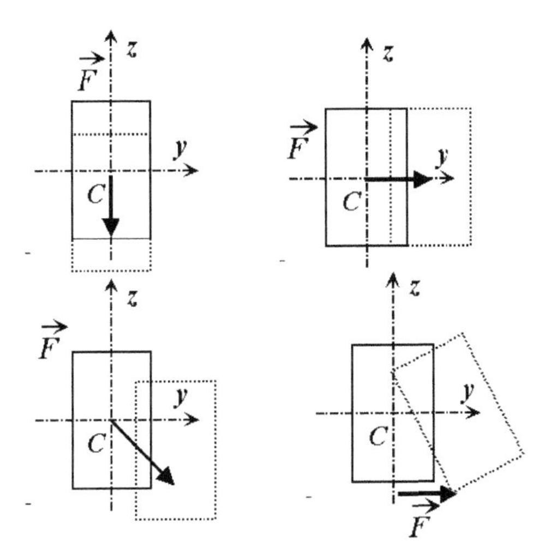

Figure 2.35. *Rigid displacement of a cross section with central symmetry subjected to a transverse force*

cross-section denoted S, at which the equivalent one-dimensional loading reduces to $\vec{F}$, so S can be defined as the point where the load can be applied to induce a lateral deflection without torsion.

Location of S is governed by the distribution over the cross-section of the local transverse shear-stresses which resist the external load and accordingly S is called the *shear-centre* of the cross-section. It cannot be defined starting from the basic model which assumes that the cross-sections are rigid. As in the case of warping related to torsion, the problem is analysed by using the 2D elasticity theory. The calculation process is however rather involved and will not be presented here. The interested reader is referred to more specialized literature like [TIM 51], [FUN 68], [PIL 02]. It turns out that various definitions of the shear centre can be made [FUN 68]. As a general result, the position of S depends upon the cross-sectional geometry and is distinct from C, in the case of asymmetrical cross-sections. Furthermore, if an axial moment is applied to S the cross-section rotates about S, which thus may also be defined as the *twist centre* of the cross-section. Accordingly, the line which passes through the shear (or twist) centres of the beam cross-sections are called the *bending axis* or *flexural axis*, which coincides with the centroidal axis in the particular case of central symmetry of the cross-sections.

2.2.6.2 *Slenderness ratio and lateral deflection*

As already stated, the lateral deflection of a beam can be induced either by a transverse shear mode or by a bending mode of deformation. In order to assess the relative importance of the two mechanisms in the global response of a beam to a transverse force, it is also found appropriate to relate the local transverse shear stresses induced by the load to the shear deformations of the cross-sections of the beam. The same 2D elastic calculation as evoked above can be used to define a global *shear deformation*, or *shear stiffness* coefficient denoted κ of the cross-section. Here again, there are various definitions of κ. As a general result, κ is used as a weighting factor leading to the one-dimensional model due to Timoshenko, which superposes the effects of bending and transverse shear deformations. This model will be described in Chapter 3, subsection 3.2.2.2, where it will be shown that the Bernoulli–Euler model remains of acceptable accuracy so long as the beam is slender enough.

2.3. Thermoelastic behaviour of a straight beam

2.3.1 *3D law of thermal expansion*

A temperature change $\Delta\theta = \theta_1 - \theta_0$ taking place in a homogeneous isotropic material produces a linear strain $\varepsilon_{ii} = \alpha\Delta\theta$ in each direction of the space; α is the *thermal dilatation coefficient*. Usually, α and the elastic coefficients E, ν may be assumed to be constant, in so far as $\Delta\theta$ is not too large. Here a square beam cross-section is considered (side length a, area S); the thermoelastic behaviour of the beam will be studied first in the longitudinal and then in the bending mode of deformation.

2.3.2 *Thermoelastic axial response*

In the initial unstressed state, the temperature is uniform and equal to θ_0. Then, the temperature is increased uniformly to a new value θ_1. The local strain ε_{xx} can be expressed as the sum of two components, namely an elastic strain related to the elastic stresses present in the system, and a thermal strain related to the thermal expansion law. Hence:

$$\varepsilon_{xx} = \frac{\sigma_{xx}}{E} + \alpha\Delta\theta \Rightarrow \sigma_{xx} = E\left(\frac{\partial X}{\partial x} - \alpha\Delta\theta\right) \qquad [2.80]$$

As the temperature is assumed uniform in the whole beam, the global stress is found to be:

$$\tilde{\mathcal{N}} = ES\left(\frac{\partial X}{\partial x} - \alpha\Delta\theta\right) \qquad [2.81]$$

The upper tilde is used to mark that the stress is thermoelastic.

Therefore the equilibrium equation turns out to be the same as in the isothermal elastic case:

$$-\frac{\partial\tilde{\mathcal{N}}}{\partial x} = -S\frac{\partial\sigma_{xx}}{\partial x} = 0 \Rightarrow \frac{\partial^2 X}{\partial x^2} = 0 \qquad [2.82]$$

The thermal effect is present via the boundary conditions only. If the beam is clamped at both ends, it cannot expand and thermal stresses are induced; conversely, if the beam is free at one end, it can expand freely and no thermal stresses are induced.

EXAMPLE 1. – *Clamped-clamped beam*

The general solution of [2.82] is $X(x) = ax + b$. The boundary conditions are $X(0) = X(L) = 0$, and then the solution is $X(x) = 0$, $\tilde{\mathcal{N}} = -E\alpha S\Delta\theta$. The axial stress is constant; if $\Delta\theta$ is positive, it represents a compressive force exerted on the left-hand side part of the beam by the part at the right-hand side (Figure 2.36).

EXAMPLE 2. – *Cantilevered beam*

The boundary conditions are $X(0) = 0; \tilde{\mathcal{N}}(L) = 0$, hence $X(x) = \alpha\Delta\theta x; \tilde{\mathcal{N}} = 0$. The displacement is maximum at the free end and the stress is zero. However, if

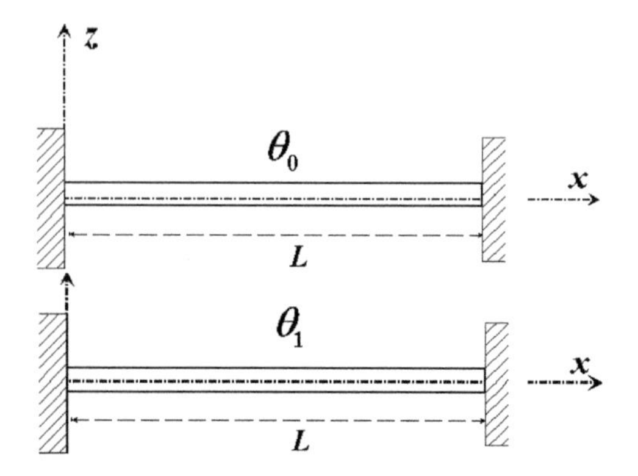

Figure 2.36. *Uniform temperature change in a clamped-clamped beam*

the temperature variation is not uniform, the thermoelastic equilibrium equation is changed; for instance, if the variation is linear it is found that:

$$\varepsilon_{xx} = \frac{\sigma_{xx}}{E} + \alpha\Delta\theta\frac{x}{L} \Rightarrow \tilde{\mathcal{N}} = ES\left(\frac{\partial X}{\partial x} - \alpha\Delta\theta\frac{x}{L}\right) \qquad [2.83]$$

and the local equilibrium is governed by the differential equation:

$$-\frac{\partial\tilde{\mathcal{N}}}{\partial x} = 0 \Rightarrow \frac{\partial^2 X}{\partial x^2} - \alpha\frac{\Delta\theta}{L} = 0 \qquad [2.84]$$

It is obvious that the thermal effect takes place here not only via the boundary conditions but also through the equilibrium equation itself.

EXAMPLE 3. – *Clamped-clamped beam*

The solution of [2.84] is: $X(x) = b + ax + \alpha\Delta\theta x^2/(2L)$. The boundary conditions $X(0) = 0$; $X(L) = 0$ giving $X(x) = \alpha\Delta\theta x(x - L)/(2L)$, which is parabolic and maximum at mid-span. The local stress is:

$$\sigma_{xx} = E\alpha\frac{\Delta\theta}{2}\left(\frac{2x}{L} - 1\right) - E\alpha\frac{\Delta\theta x}{L} = -E\alpha\frac{\Delta\theta}{2}$$

The global stress is $\tilde{\mathcal{N}} = -ES\alpha\Delta\theta/2$. This value is equal to the stress calculated above in the case of a uniform temperature equal to $\Delta\theta(x = L/2)$.

EXAMPLE 4. – *Cantilevered beam*

The boundary conditions are $X(0) = 0; \tilde{\mathcal{N}}(L) = 0$, so $X(x) = \alpha\Delta\theta x^2/(2L)$; $\tilde{\mathcal{N}} = 0$. As expected, X is maximum at the free end of the beam and the stress is zero.

2.3.3 *Thermoelastic bending of a beam*

In the unstressed initial state, the beam is at a uniform temperature θ_0. Then, the temperature at one lateral wall ($z = a/2 > 0$) is raised to θ_1. The temperature profile in the transverse direction z is denoted $\theta(z)$, see Figure 2.37. The local axial stress $\sigma_{xx}(z)$ related to the temperature field is:

$$\sigma_{xx}(z) = E\left(\frac{\partial X}{\partial x} - z\frac{\partial^2 X}{\partial x^2} - \alpha\Delta\theta(z)\right) \quad [2.85]$$

The global stresses $\tilde{\mathcal{N}}(x)$ and $\tilde{\mathcal{M}}_y(x)$ are then given by:

$$\tilde{\mathcal{N}}(x) = \int_{(S)} \sigma_{xx}(z)dS = ES\frac{\partial X}{\partial x} - \int_{-a/2}^{a/2} Eaz\frac{\partial^2 Z}{\partial x^2}dz - \alpha Ea\int_{-a/2}^{a/2} \Delta\theta(z)dz \quad [2.86]$$

$$\tilde{\mathcal{N}}(x) = ES\frac{\partial X}{\partial x} - \alpha Ea\int_{-a/2}^{a/2} \Delta\theta(z)\,dz$$

$$\tilde{\mathcal{M}}_y(x) = \int_{(S)} \sigma_{xx}(z)z\,dS = -EI\frac{\partial^2 Z}{\partial x^2} - \alpha Ea\int_{-a/2}^{a/2} \Delta\theta(z)z\,dz \quad \text{where } I = \frac{a^4}{12} \quad [2.87]$$

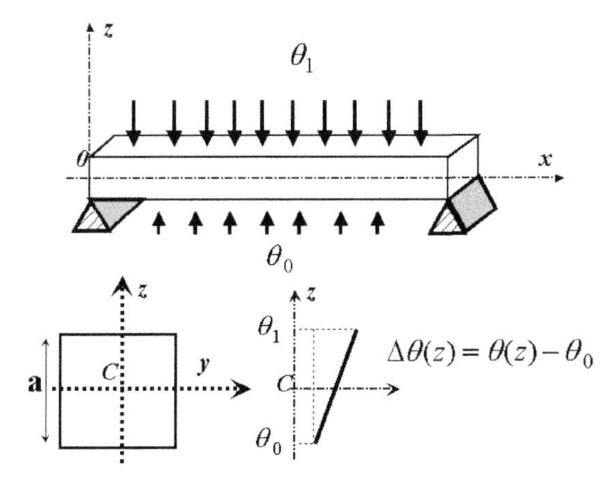

Figure 2.37. *Beam subjected to a transverse temperature gradient*

For the sake of simplicity, let us consider a linear variation of $\theta(z)$ across the beam sections. Introducing the quantities $\theta_m = (\theta_1 + \theta_0)/2$; $\Delta\theta = \theta_1 - \theta_0$,

$$\tilde{\mathcal{N}}(x) = ES\frac{\partial X}{\partial x} - \alpha Ea\int_{-a/2}^{a/2}\left(\theta_m + \frac{\Delta\theta}{a}z\right)dz = ES\frac{\partial X}{\partial x} - \alpha ES\theta_m = \alpha ES\frac{\Delta\theta}{2} \qquad [2.88]$$

$$\frac{\partial\tilde{\mathcal{N}}}{\partial x} = 0 \Rightarrow \frac{\partial}{\partial x}\left(\frac{\partial X}{\partial x} - \alpha\frac{\Delta\theta}{2}\right) = 0 \qquad [2.89]$$

As expected, if $\Delta\theta$ is independent of x, the axial balance is not modified. The bending stress field is:

$$\begin{aligned}\tilde{\mathcal{M}}_y(x) &= \int_{(S)}\sigma_{xx}(z)z\,dz = -EI\frac{\partial^2 Z}{\partial x^2} - \alpha Ea\int_{-a/2}^{a/2}\left(\frac{\Delta\theta}{a}\right)z^2\,dz\\ \tilde{\mathcal{M}}_y(x) &= -EI\left(\frac{\partial^2 Z}{\partial x^2} + \alpha\frac{\Delta\theta}{a}\right)\end{aligned} \qquad [2.90]$$

Then, the equilibrium equations are:

$$\begin{cases}\dfrac{\partial Q_z}{\partial x} = 0\\[2ex] -EI\dfrac{\partial}{\partial x}\left(\dfrac{\partial^2 Z}{\partial x^2} + \alpha\dfrac{\Delta\theta}{a}\right) - Q_z = 0\end{cases} \Rightarrow \begin{cases}\dfrac{\partial^4 Z}{\partial x^4} = 0\\[2ex] Q_z = -EI\dfrac{\partial^3 Z}{\partial x^3}\end{cases} \qquad [2.91]$$

Once more, thermal effects are introduced via the boundary conditions solely. If the beam is clamped at both ends, extension is prevented and thermal axial stresses arise. If one end is free, the beam expands freely and no axial stress is generated. It could be verified that in the case of a nonlinear temperature profile $\theta(z)$, a thermal term is present not only in the boundary conditions, but also in the local equilibrium equation.

EXAMPLE 1. – *Pinned-pinned beam*

The general solution of [2.91] is $Z(x) = Ax^3 + Bx^2 + Cx + D$. Application of the pinned boundary conditions gives:

$$\begin{aligned}Z(0) = 0 \Rightarrow D = 0;\quad Z(L) = 0 \Rightarrow C = -BL = \alpha\tfrac{\Delta\theta L}{2a}\\ \mathcal{M}_y(0) = 0 \Rightarrow B = -\alpha\Delta\theta/(2a);\quad \mathcal{M}_y(L) = 0 \Rightarrow A = 0\end{aligned}$$

So the solution is: $Z(x) = \alpha\Delta\theta(Lx - x^2)/2a$

EXAMPLE 2. – *Cantilevered beam*

$$Z(0) = Z'(0) = 0 \Rightarrow Z(x) = Ax^3 + Bx^2$$

$$\mathcal{M}_y(L) = 0 \Rightarrow Z''(L) = -\frac{\alpha\Delta\theta}{a} \Rightarrow 6AL + 2B = -\frac{\alpha\Delta\theta}{a}$$

$$\mathcal{Q}_z(L) = 0 \Rightarrow Z'''(L) = 0 \Rightarrow A = 0$$

$$Z(x) = -\frac{\alpha\Delta\theta x^2}{2a}$$

2.4. Elastic-plastic beam

A cantilevered beam with a uniform rectangular cross-section is considered (depth h, width a). A force $\vec{F}^{(e)}$ is applied to the free end, see Figure 2.38. The material is assumed to be elastic-plastic with negligible strain hardening (ideal plasticity). For the sake of simplicity, we will adopt the idealized 1D law of Figure 2.39, which is a plot of the local elastic-plastic stress versus strain. Starting from the initial unloaded and non deformed state, as the material is deformed in one direction, say for instance along the Ox axis, a stress component σ is induced which acts in the same direction. When $|\sigma| \leq \sigma_0$, where σ_0 stands for the yield stress, the material

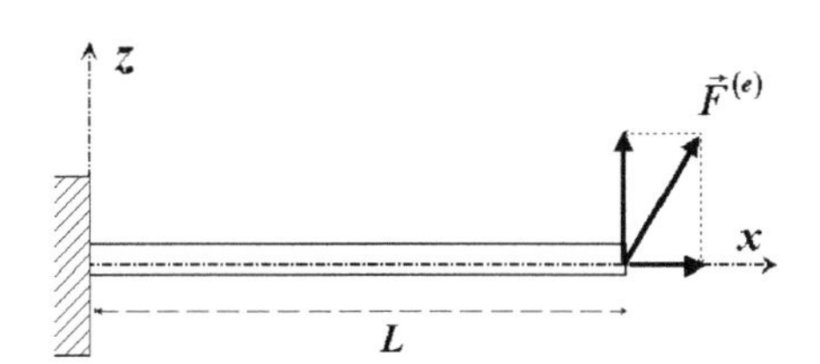

Figure 2.38. *Cantilevered beam loaded at the free end by a force*

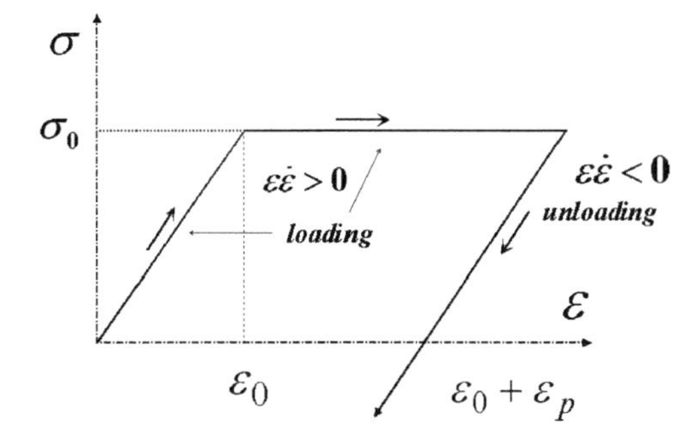

Figure 2.39. *Elastic-plastic behaviour without strain hardening (1D model)*

behaves elastically: $\sigma = E\varepsilon$. However, if deformation is further increased beyond $\varepsilon_0 = \sigma_0/E$, the stress remains equal to σ_0 and plastic deformation is initiated, which is irreversible. It is assumed that the total strain ε is the sum of an elastic and a plastic part $\varepsilon = \varepsilon_e + \varepsilon_p$, only ε_e is reversible. Starting now from a plastic state defined by the plastic strain $\varepsilon_p = \varepsilon - \varepsilon_0$, if ε is decreased the stress also decreases according to the elastic law $\sigma = E(\varepsilon - \varepsilon_p)$. Hence the material behaviour depends upon the sign of the strain rate and it is appropriate to distinguish between a loading step in which $\varepsilon\dot{\varepsilon} > 0$ and an unloading step, in which $\varepsilon\dot{\varepsilon} < 0$. In the static problems discussed below, we will be concerned with the loading part of the elastic-plastic law only.

2.4.1 *Elastic-plastic behaviour under uniform traction*

As is apparent in Figure 2.40, the equilibrium of a beam element reduces to $\mathcal{N} = F_x^{(e)}$ and the local stress is $\sigma_{xx}(x, y, z) = F_x^{(e)}/ah$. The limit load of elastic behaviour is $F_{x0} = \sigma_0 ah$. It is possible to define a global elastic-plastic law. In this elementary case, the global law is obtained from the local one simply by multiplying the local stress by the cross-sectional area. As strain hardening is not taken in account, the highest load the beam can sustain is F_{x0}. Indeed, if this threshold is exceeded, the material yields and plastic flow occurs, theoretically without any limit. This phenomenon is called *plastic failure* of the structure.

2.4.2 *Elastic-plastic behaviour under bending*

As shown in Figure 2.41, we consider now a cantilevered beam loaded by a transverse force $F_z^{(e)}$, applied at the free end.

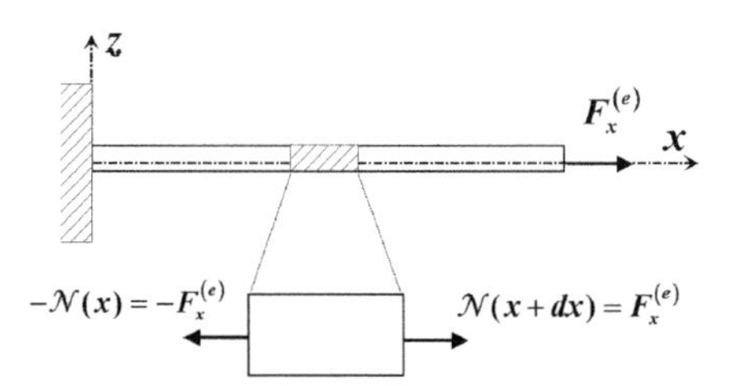

Figure 2.40. *Static equilibrium of a uniformly tensioned beam element*

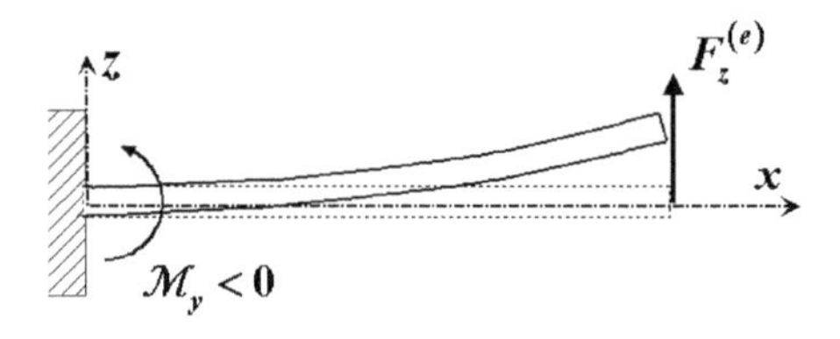

Figure 2.41. *Cantilevered beam loaded at the free end by a transverse force*

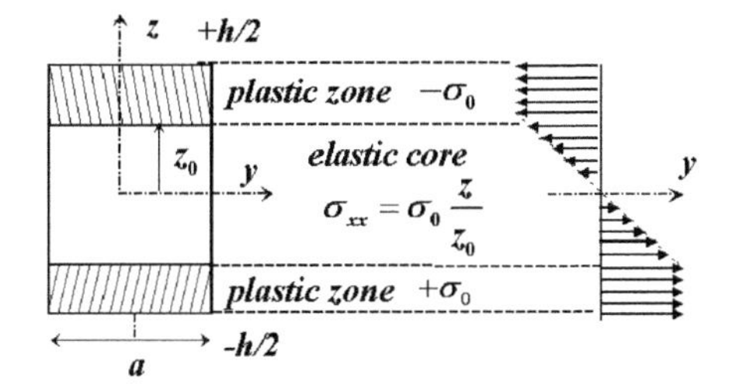

Figure 2.42. σ_{xx} *stress distribution in a beam cross-section*

2.4.2.1 Skin stress

The global equilibrium of a beam element implies a constant shear force $\mathcal{Q}_z = F_z^{(e)}$ and a bending moment which varies linearly along the beam $\mathcal{M}_y = -F_z(L-x)$. In the elastic domain, the moment can also be written as:

$$\mathcal{M}_y(x) = F_z^{(e)}(L-x) = -EI_y\frac{\partial^2 Z}{\partial x^2}; \quad I_y = \frac{ah^3}{12} \qquad [2.92]$$

Recalling that the local stress vanishes at the neutral fibres, in the elastic-plastic domain it is necessary to make the distinction between the parts of the cross-sections which are plastic and that which is still in the elastic domain, as shown in Figure 2.42.

Hence, one starts from the definition of the bending moment in terms of local stresses:

$$\mathcal{M}_y(x) = \int_{-a/2}^{a/2}\int_{-h/2}^{h/2} \sigma_{xx}(x,y,z)z\,dz\,dy \qquad [2.93]$$

The local elastic stress is:

$$\sigma_{xx}(x,z) = -zE\left.\frac{\partial^2 Z}{\partial x^2}\right|_x \qquad [2.94]$$

Its maximum magnitude is reached at the lateral boundaries of the beam ($z = \pm h/2$) so it is termed *skin stress*,

$$(\sigma_{xx}(x))_{skin} = \pm\left.\frac{Eh}{2}\frac{\partial^2 Z}{\partial x^2}\right|_x \qquad [2.95]$$

The curvature is given by:

$$\mathcal{M}_y = -F_z^{(e)}(L-x) = -EI_y\frac{\partial^2 Z}{\partial x^2} \Rightarrow \frac{\partial^2 Z}{\partial x^2} = \frac{\mathcal{M}_y}{EI_y} = \frac{12F_z^{(e)}(L-x)}{Eah^3} \qquad [2.96]$$

then,

$$(|\sigma_{xx}(x)|)_{skin} = \left| \frac{6F_z^{(e)}(L-x)}{ah^2} \right| \quad [2.97]$$

As could be expected, the local longitudinal stress is maximum at the clamped end. The load beyond which plasticity is initiated in the skin layer is given by:

$$\sigma_0 = \frac{6F_0 L}{ah^2} \Rightarrow F_0 = \frac{\sigma_0 a h^2}{6L} \quad [2.98]$$

This threshold increases in proportion to the cross-sectional area ah and with the reciprocal of the slenderness ratio.

2.4.2.2 *Moment-curvature law and failure load*

When the elastic limit is exceeded, the material yields first in the superficial layers, (the skin), in the vicinity of the clamped end, where magnitude of the elastic stresses is maximum. As the external load is further increased, the plastic zones spread in lateral and longitudinal directions, inward from the outer surface and along the beam axis. The stress distribution in a partially plastic cross-section is shown in Figure 2.42, where the material yields within the two intervals $z_0 \leq |z| \leq h/2$. However, so long as the load is not too large, an 'internal' elastic core subsists within $-z_0 < z < z_0$, which prevents the section collapsing by plastic failure. Therefore, the highest transverse load which can be sustained by the cantilevered beam is reached when $z_0 = 0$ at the clamped cross-section. The transverse shear force does not depend upon the elastic-plastic state of the beam, since according to the Bernoulli-Euler model it is independent of the material law. The bending moment becomes:

$$\mathcal{M}_y = -Ea\frac{\partial^2 Z}{\partial x^2}\int_{-z_0}^{z_0} z^2\, dz - 2\sigma_0 a \int_{z_0}^{h/2} z\, dz \Rightarrow$$
$$\mathcal{M}_y = -\left\{ \frac{2Eaz_0^3}{3}\frac{\partial^2 Z}{\partial x^2} + \sigma_0 a \left(\left(\frac{h}{2}\right)^2 - z_0^2 \right) \right\} \quad [2.99]$$

This moment balances the torque of the applied force $-F_z^{(e)}(L-x)$. Hence, the limit load for failure is given by:

$$F_{\text{fail}} = \sigma_0 a h \left(\frac{h}{4L}\right) \Rightarrow \frac{F_{\text{fail}}}{F_{z0}} = \frac{3}{2} \quad [2.100]$$

In contrast with the result obtained in the longitudinal mode, in bending, the load for plastic failure is found to be higher than the elastic limit. The difference is clearly due to the presence of an elastic core. Of course, the results, in particular the ratio between the two limit loads, depend on the shape of the cross-section.

2.4.2.3 Elastic-plastic bending: global constitutive law

In a partially plastic state, the maximum stress is $\sigma_0 = Ez_0\chi$ where χ is the beam curvature (rotation rate) at the cross-section. If this relation is substituted into the expression of the bending moment, we get:

$$\mathcal{M}_y = -\left\{\frac{2Eaz_0^3}{3}\chi + \sigma_0 a\left(\left(\frac{h}{2}\right)^2 - z_0^2\right)\right\}$$
$$= -\left\{\frac{2Eaz_0^3}{3}\chi + Eaz_0\chi\left(\left(\frac{h}{2}\right)^2 - z_0^2\right)\right\} = \left(\frac{Eah^2}{4}z_0 - \frac{Eaz_0^3}{3}\right)\chi \qquad [2.101]$$

It is convenient to introduce, as a pertinent scaling factor, the curvature χ_0 corresponding to the elastic limit beyond which plasticity is initiated somewhere in the beam. χ_0 is defined by the elastic condition $\sigma_0 = Eh/2\chi_0$. The boundary of the elastic core is thus written as:

$$z_0 = \frac{h}{2}\frac{\chi_0}{\chi} \qquad [2.102]$$

then,

$$\mathcal{M}_y = -\left\{\frac{Eah^3}{8}\chi_0 - \frac{Eah^3}{24}\frac{\chi_0^3}{\chi^2}\right\} \qquad [2.103]$$

The elastic limit of the bending moment is introduced by the elastic condition:

$$\mathcal{M}_0 = \frac{Eah^3}{12}\chi_0 \qquad [2.104]$$

The relation between the bending moment and the curvature is finally written as:

$$\text{sign}(\chi\dot{\chi}) > 0 \Rightarrow \frac{\mathcal{M}_y}{\mathcal{M}_0} = \frac{1}{2}\left(3 - \left(\frac{\chi_0}{\chi}\right)^2\right)$$
$$\text{sign}(\chi\dot{\chi}) < 0 \Rightarrow \frac{\mathcal{M}_y}{\mathcal{M}_0} = \frac{\chi - \chi_p}{\chi_0} \qquad [2.105]$$

[2.105] is known as a global law of elastic-plasticity in bending mode of deformation. The condition $\chi\dot{\chi} > 0$ means that the external load is increasing. As in the case of the local law, unloading is elastic. On the other hand, the global law [2.105] implies a positive strain hardening, even if the local law does not, as it is assumed to be the case here.

2.4.2.4 Superposition of several modes of deformation

In contrast with the local law, the global laws of plasticity depends not only of the material behaviour but also on the structural geometry and on the loading conditions. This feature restricts their range of application to a few basic problems. A simple example is sufficient to illustrate this remark. Turning back to the cantilevered beam of Figure 2.38, the case of an oblique external force is discussed on a qualitative basis. In the linear elastic domain, such a problem would be solved immediately by applying the principle of superposition. Accordingly, so long as the displacements remain small enough, F_x induces the uniform longitudinal stress $\mathcal{N} = F_x$ related to the uniform local stress field denoted $\sigma_a(x)$. The bending induced by F_z gives rise to a local stress field denoted $\sigma_b(x,z)$. The resultant field $\sigma_{xx}(x,z) = \sigma_a(x)+\sigma_b(x,z)$ is produced by superposition. Nevertheless, due to the nonlinear nature of plasticity, the superposition principle is invalidated and the actual profile of the local stresses is given by the nonlinear law:

$$\sigma_{xx}(x,z) = \begin{cases} \sigma_a(x) + \sigma_b(x,z) & \text{if } |\sigma_{xx}(x,z)| \leq \sigma_0 \\ \sigma_0 \operatorname{sign}(\sigma_a(x) + \sigma_b(x,z)) & \text{otherwise} \end{cases}$$

Such a profile referring to a partially plastic cross-section is shown in Figure 2.43. When such a 'compound profile' is used to derive the global laws, for instance moment-curvature law, the former result [2.105] is modified, becoming dependent upon the ratio F_z/F_x. The problem becomes even more intricate as the number of

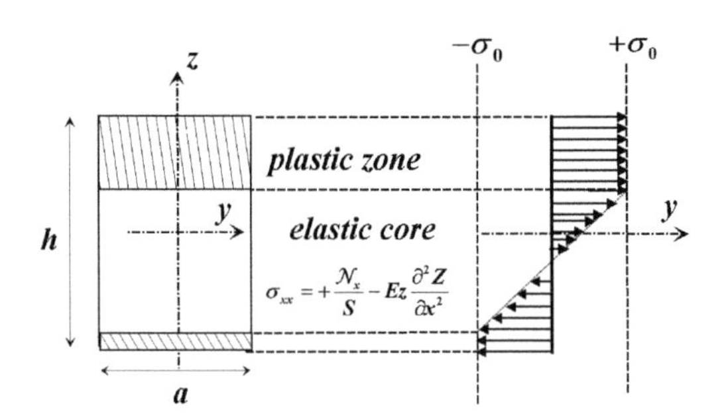

Figure 2.43. *Elastic-plastic profile of the normal stress $\sigma_{xx}(x,z)$*

modes of deformation is increased. In this case it is necessary to replace σ_0 by a yielding condition $f(\sigma_{ij})$ together with a loading and unloading criterion, which accounts for the relative importance of the several stress components induced by the loading. The reader interested in this question may consult specialized books, for instance [CHAK 87], [STR 93], [JON 89].

Chapter 3

Straight beam models: Hamilton's principle

The problem of modelling straight beams as an idealized 1D solid is further considered here with the aim of presenting a few distinct topics of theoretical and practical importance, namely:

1. Beam models with deformed cross-sections
2. Bending in the presence of axial loads and buckling
3. Concentrated quantities described in terms of singular distributions
4. Properties of symmetry of stiffness and mass operators
5. Finite element models.

It will be shown that Hamilton's principle provides a suitable theoretical framework to deal with these distinct subjects in a clear and unified manner.

3.1. Introduction

The variational approach introduced in Chapter 1 in the case of three-dimensional solids can be applied to beams without any difficulty. The global strains or stresses have to be substituted for local ones and the three-dimensional domain of integration has to be adapted to one-dimensional beam geometry. One major interest of the variational approach is its convenience for improving the basic models established in Chapter 2. Based on a global energy balance and making use of Hamilton's principle, corrective terms can be suitably defined, which take into account the local deformations of the beam cross-sections. With this aim in mind, it is found useful to split the transverse displacement field into two distinct components; that marked by the subscript 's' is related to transverse shear and the other, marked by the subscript 'b', is related to bending. Whatever the material law may be, by using the results given in subsections 2.1.3 and 2.1.4 of Chapter 2, the variation of strain energy $\mathcal{E}_s$ is written as:

$$\delta\mathcal{E}_s = \int_0^L \left\{\mathcal{N}\frac{\partial \delta X}{\partial x} + \mathcal{Q}_y\frac{\partial \delta Y_s}{\partial x} + \mathcal{Q}_z\frac{\partial \delta Z_s}{\partial x} +\right\} dx + \int_0^L \left\{\mathcal{M}_x\frac{\partial \delta \psi_x}{\partial x} + \mathcal{M}_y\frac{\partial \delta \psi_y}{\partial x} + \mathcal{M}_z\frac{\partial \delta \psi_z}{\partial x}\right\} dx \qquad [3.1]$$

where Y_s, Z_s are the fields of transverse displacement of the cross-sections specifically related to the transverse shear strains η_{xy}, η_{xz}. If they are neglected, the corresponding terms in [3.1] are set to zero. If the problem is restricted to linear and isotropic elasticity, using the results of Chapter 2 subsection 2.1.5, the elastic energy $\mathcal{E}_e$ is expressed as:

$$\mathcal{E}_e = \frac{1}{2}\int_0^L \left\{ES\left(\frac{\partial X}{\partial x}\right)^2 + GS\left(\left(\frac{\partial Y_s}{\partial x}\right)^2 + \left(\frac{\partial Z_s}{\partial x}\right)^2\right)\right\} dx + \frac{1}{2}\int_0^L \left\{GJ_T\left(\frac{\partial \psi_x}{\partial x}\right)^2 + \left(EI_z\left(\frac{\partial^2 Y_b}{\partial x^2}\right)^2\right) + \left(EI_y\left(\frac{\partial^2 Z_b}{\partial x^2}\right)^2\right)\right\} dx \qquad [3.2]$$

The kinetic energy $\mathcal{E}_\kappa$ is:

$$\mathcal{E}_\kappa = \frac{1}{2}\int_0^L \rho S\{\dot{X}^2 + (\dot{Y}_s + \dot{Y}_b)^2 + (\dot{Z}_s + \dot{Z}_b)^2\}\, dx + \frac{1}{2}\int_0^L \rho(J\dot{\psi}_x^2 + I_y\dot{\psi}_y^2 + I_z\dot{\psi}_z^2)\, dx \qquad [3.3]$$

In [3.3] the rotatory inertia of the cross-section with respect to the bending axes has been accounted for, though it is negligible if the beam is sufficiently slender.

In the next section, the variational calculus concerning first the longitudinal and then the bending modes of deformation are further detailed. In section 3.3, the weighted integral form of the equilibrium equations is introduced. This form generates various analytical and numerical methods used to find the exact, or approximate, solutions of the local equilibrium equations. Furthermore, it will be demonstrated that the weighted integral formulation is well suited to put in evidence the properties of symmetry which are verified by the mass and stiffness operators of the conservative mechanics. These properties extend those already evidenced in the study of discrete systems, cf. [AXI 04] to the continuous case. Finally, section 3.5 provides the reader with an introductory description of the finite element method, in short (F.E.M).

3.2. Variational formulation of the straight beam equations

3.2.1 *Longitudinal motion*

3.2.1.1 Model neglecting the Poisson effect

External loading comprises the density per unit length of an axial force denoted $F_x^{(e)}(x;t)$, which is assumed to vanish at the beam ends, plus axial forces denoted $T_x^{(e)}(0;t)$, $T_x^{(e)}(L;t)$ applied to the beam ends. Hamilton's principle is written as:

$$\delta[\mathcal{A}] = \int_{t_1}^{t_2} dt \left\{ \int_0^L \left\{ -\rho S\ddot{X}\delta X - \mathcal{N}\frac{\partial \delta X}{\partial x} + F_x^{(e)}\delta X \right\} dx \right\} + \int_{t_1}^{t_2} dt \left\{ \left[T_x^{(e)}\delta X \right]_L + \left[T_x^{(e)}\delta X \right]_0 \right\} = 0 \qquad [3.4]$$

Integration by parts of the second term of the first integral gives:

$$\delta[\mathcal{A}] = \int_{t_1}^{t_2} dt \left\{ \int_0^L \left\{ -\rho S\ddot{X} + \frac{\partial \mathcal{N}}{\partial x} + F_x^{(e)} \right\} \delta X\, dx \right\} + \int_{t_1}^{t_2} dt \left\{ \left[\left(T_x^{(e)} - \mathcal{N} \right) \delta X \right]_L + \left[\left(T_x^{(e)} + \mathcal{N} \right) \delta X \right]_0 \right\} = 0 \qquad [3.5]$$

From [3.5] the local equilibrium equation and the related boundary conditions can be immediately inferred as $\rho S\ddot{X} - \partial\mathcal{N}/\partial x = F_x^{(e)}(x;t)$ and $\mathcal{N}(L;t) = T_x^{(e)}(L;t); \mathcal{N}(0;t) = -T_x^{(e)}(0;t)$, which of course are identical to those already established in subsection 2.1.6 of Chapter 2.

3.2.1.2 Model including the Poisson effect (Love–Rayleigh model)

It is recalled that the basic idea for modelling a solid as a beam is to assume rigid cross-sections. Actually, when an axial load is applied to a beam, provided transverse motion is not prevented, any axial strain induces a transverse strain, which is proportional to Poisson's coefficient and to the axial strain. This can be easily shown by using the elastic strain-stress relationship [1.38] in the particular case of the uniaxial stress distribution $\sigma_{xx} \neq 0$. Indeed, from the conditions $\sigma_{yy} = \sigma_{zz} = 0$, it is immediately verified that:

$$\frac{\partial Y}{\partial y} = \frac{\partial Z}{\partial z} = -\nu\frac{\partial X}{\partial x} \Rightarrow Y = -\nu\frac{\partial X}{\partial x}y; \quad Z = -\nu\frac{\partial X}{\partial x}z \qquad [3.6]$$

The beam cross-sections are therefore contracted or dilated, depending on whether the beam is stretched out or compressed, in the axial direction. Poisson's effect produces an additional kinetic energy since the beam material is set in motion not only in the axial but also in the lateral directions. However, no additional elastic energy arises, since the only non-vanishing stress component is axial. Furthermore, the inertia effect related to the kinetic energy induced by the lateral motion can be easily accounted for in a 1D model by summing the local contributions over the whole cross-section area. Let us consider, for instance, the circular cylindrical rod of Figure 3.1. In cylindrical coordinates (see Appendix A.3, formulas [A.3.27,28]), the relations [3.6] become:

$$\frac{\partial U}{\partial r} = -\nu\frac{\partial X}{\partial x} \Rightarrow U = -\nu\frac{\partial X}{\partial x}r \qquad [3.7]$$

where U is the radial displacement and r the radial distance from the rod axis. The kinetic energy induced by the radial motion is:

$$\mathcal{E}_{Kr} = \int_0^L e_{Kr}\,dx \qquad [3.8]$$

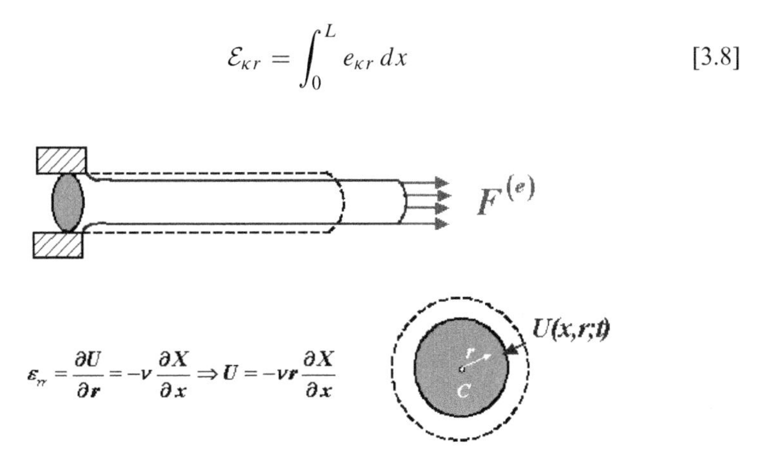

Figure 3.1. *Radial contraction of a rod related to its axial stretching*

where

$$e_{\kappa r} = \frac{\rho}{2}\left(\nu\frac{\partial^2 X}{\partial x \partial t}\right)^2 2\pi\int_0^R r^3\,dr = \rho\pi R^2\left(\frac{\nu R}{2}\frac{\partial^2 X}{\partial x \partial t}\right)^2$$

The variation of this term is:

$$\delta\mathcal{E}_{\kappa r} = 2\rho\pi R^2\left(\frac{\nu R}{2}\right)^2\int_{t_1}^{t_2}\int_0^L\left(\frac{\partial^2 X}{\partial x \partial t}\right)\left(\frac{\partial^2 \delta X}{\partial x \partial t}\right)dx\,dt \qquad [3.9]$$

Integrating [3.9] by parts, first with respect to t and secondly with respect to x, we obtain:

$$\delta\mathcal{E}_{\kappa r} = \rho\pi R^2\frac{(\nu R)^2}{2}\int_{t_1}^{t_2}\int_0^L\left(\frac{\partial^4 X}{\partial x^2 \partial t^2}\right)\delta X\,dx\,dt \qquad [3.10]$$

Indeed, any admissible virtual variation is assumed to vanish at t_1 and t_2; accordingly:

$$\int_0^L\left[\frac{\partial^2 X}{\partial t \partial x}\delta\left(\frac{\partial X}{\partial x}\right)\right]_{t_1}^{t_2}dx \equiv 0$$

On the other hand, according to the boundary conditions of the problem:

$$\int_{t_1}^{t_2}\left[\frac{\partial^3 X}{\partial t^2 \partial x}\delta X\right]_0^L dt \equiv 0$$

Based on these results, the so called Love–Rayleigh equation is suitably written as:

$$\rho S\left\{\frac{\partial^2 X}{\partial t^2} - (\nu\beta)^2\frac{\partial^4 X}{\partial x^2 \partial t^2}\right\} - ES\frac{\partial^2 X}{\partial x^2} = F_x^{(e)}(x;t) \quad S = \pi R^2;\ \beta = R/\sqrt{2} \qquad [3.11]$$

Note that the partial derivative equation which governs the longitudinal vibrations is of the fourth order, instead of two in the basic model, due to the additional inertia term induced by the Poisson effect.

In a similar way, in the case of a rectangular cross-section we would have $S = ab; \beta = (a^2 + b^2)/\sqrt{12}$.

3.2.2 *Bending and transverse shear motion*

3.2.2.1 Bending without shear: Bernoulli–Euler model

Here the beam is assumed to be deflected in the Oxz plane. The external loading comprises densities per unit length of a transverse force, denoted $F_z^{(e)}(x;t)$, and of a moment denoted $\mathfrak{M}_y^{(e)}(x;t)$, which vanish at the beam ends, plus transverse forces $T_z^{(e)}(0;t), T_z^{(e)}(L;t)$ and moments $\mathcal{M}_y^{(e)}(0;t), \mathcal{M}_y^{(e)}(L;t)$ applied to the beam ends. The variational form of the problem is:

$$\delta[\mathcal{A}] = \int_{t_1}^{t_2} dt \left\{ \int_0^L \{-\rho S\ddot{Z}\delta Z - \mathcal{M}_y\delta\chi_{yy} + F_z^{(e)}\delta Z + \mathfrak{M}_y^{(e)}\delta\psi_y\}\, dx \right\}$$
$$+ \int_{t_1}^{t_2} dt \left\{ + \left[T_z^e\delta Z\right]_L + \left[T_z^{(e)}\delta Z\right]_0 + \left[\mathcal{M}_y^{(e)}\delta\psi_y\right]_L + \left[\mathcal{M}_y^{(e)}\delta\psi_y\right]_0 \right\} = 0 \quad [3.12]$$

It is recalled that in the Bernoulli–Euler model $\psi_y = -\,\partial Z/\partial x$, $\chi_{yy} = -\,\partial^2 Z/\partial x^2$, then [3.12] takes the form:

$$\delta[\mathcal{A}] = \int_{t_1}^{t_2} dt \left\{ \int_0^L \left\{ -\rho S\ddot{Z}\delta Z + \mathcal{M}_y \frac{\partial^2 \delta Z}{\partial x^2} + F_z^{(e)}\delta Z - \mathfrak{M}_y^{(e)} \frac{\partial \delta Z}{\partial x} \right\} dx \right\}$$
$$+ \int_{t_1}^{t_2} dt \left\{ + \left[T_z^{(e)}\delta Z\right]_L + \left[T_z^{(e)}\delta Z\right]_0 - \left[\mathcal{M}_y^{(e)} \frac{\partial \delta Z}{\partial x}\right]_L - \left[\mathcal{M}_y^{(e)} \frac{\partial \delta Z}{\partial x}\right]_0 \right\}$$
$$= 0 \quad [3.13]$$

Integrating by parts and cancelling the boundary terms for each individual pair of conjugate displacement and stress components, we obtain:

$$\int_{t_1}^{t_2} dt \left\{ \int_0^L \left\{ -\rho S\ddot{Z} + \frac{\partial^2 \mathcal{M}_y}{\partial x^2} + F_z^{(e)} + \frac{\partial \mathfrak{M}_y^{(e)}}{\partial x} \right\} \delta Z dx \right\} = 0 \quad [3.14]$$

$$\int_{t_1}^{t_2} dt \left\{ \left[\left(T_z^{(e)} - \frac{\partial \mathcal{M}_y}{\partial x} \right) \delta Z \right]_L + \left[\left(T_z^{(e)} + \frac{\partial \mathcal{M}_y}{\partial x} \right) \delta Z \right]_0 \right\} = 0 \quad [3.15]$$

$$\int_{t_1}^{t_2} dt \left\{ \left[\left(\mathcal{M}_y - \mathcal{M}_y^{(e)} \right) \frac{\partial \delta Z}{\partial x} \right]_L + \left[\left(-\mathcal{M}_y - \mathcal{M}_y^{(e)} \right) \frac{\partial \delta Z}{\partial x} \right]_0 \right\} = 0 \quad [3.16]$$

Accordingly, the same local equilibrium equations and boundary conditions as those derived in Chapter 2, subsection 2.1.6 are found, written here as:

$$
\begin{aligned}
& -\rho S\ddot{Z} + \frac{\partial^2 \mathcal{M}_y}{\partial x^2} + F_z^{(e)} + \frac{\partial \mathfrak{M}_y^{(e)}}{\partial x} = 0 \\
& \left.\frac{\partial \mathcal{M}_y}{\partial x}\right|_L - \left.T_z^{(e)}\right|_L = 0; \quad \left.\frac{\partial \mathcal{M}_y}{\partial x}\right|_0 + \left.T_z^{(e)}\right|_0 = 0 \\
& \left.\mathcal{M}_y\right|_L - \left.\mathcal{M}_y^{(e)}\right|_L = 0; \quad \left.\mathcal{M}_y\right|_0 + \left.\mathcal{M}_y^{(e)}\right|_0 = 0
\end{aligned}
\qquad [3.17]
$$

3.2.2.2 Bending including transverse shear: the Timoshenko model in statics

As a first step, the equations for static equilibrium are solved for a cantilevered beam loaded by a transverse force F_0 applied to the free end $x = L$. It is recalled that the Bernoulli–Euler beam solution for this problem is (cf. Chapter 2, subsection 2.2.4.7, example 2):

$$
Z_b(x) = \frac{F_0 L^3}{6EI}\left(3\frac{x^2}{L^2} - \frac{x^3}{L^3}\right) \qquad [3.18]
$$

Here, the aim is to improve the Bernoulli–Euler model by taking into account shear elastic deformation. As shown in Figure 3.2, the total transverse deflection of the beam is assumed to be the sum of the bending component Z_b described by the Bernoulli–Euler model, and the shear component Z_s for which a suitable model is to be derived starting from the elastic energy produced by elastic transverse shear.

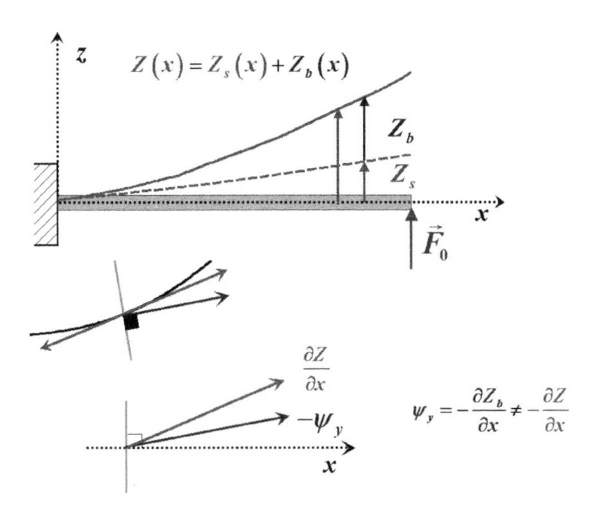

Figure 3.2. *Cantilevered beam: bending + shear*

By definition, the elastic energy related to shear deformation is:

$$\mathcal{E}_{es} = \frac{1}{2}\int_0^L G\,dx \int_{(S)} \varepsilon_{xz}^2\,dS \qquad [3.19]$$

As a gross approximation to be refined later, the shear strain $\varepsilon_{xz} = \partial Z_s/\partial x$ is assumed to be distributed uniformly over the beam cross-sections. Accordingly, the general expression [3.19] reduces to:

$$\mathcal{E}_{es} = \frac{1}{2}\int_0^L dx \int_{(S)} G\left(\frac{\partial Z_s}{\partial x}\right)^2 dS = \frac{1}{2}\int_0^L GS\left(\frac{\partial Z_s}{\partial x}\right)^2 dx \qquad [3.20]$$

which is identical to the corresponding term present in equation [3.2].

Nevertheless, the transverse shear strain and stress cannot be uniformly distributed over the cross-sections since in the absence of transverse loading, σ_{xz} must vanish at the boundary of the cross-sections and so the shear strain does too. Hence, knowledge of the 'local' function $\varepsilon_{xz}(x, y, z)$ is needed in order to calculate the surface integral involved in [3.19] more accurately than by adopting the simplified expression [3.20]. This calculation, based on 2D elasticity, has been already outlined in Chapter 2 subsection 2.2.6 for defining the transverse shear centre. As a result, a shear equivalent cross-section area κS is found, where κ stands for a global equivalent transverse shear deformation, or stress coefficient. A few values, taken from [BLE 79] and compiled in [COW 66], are given below:

$$\text{Full circular section } \kappa = \frac{6(1+\nu)}{7+4\nu}$$

$$\text{Hollow thin cylinder } \kappa = \frac{2(1+\nu)}{4+3\nu}$$

$$\text{Full rectangular section } \kappa = \frac{10(1+\nu)}{12+11\nu}$$

The shear strain energy is thus finally written as:

$$\mathcal{E}_{es} = \frac{1}{2}\int_0^L GS\kappa\left(\frac{\partial Z_s}{\partial x}\right)^2 dx \qquad [3.21]$$

The bending strain energy is still written in accordance with the Bernoulli–Euler model:

$$\mathcal{E}_{eb} = \frac{1}{2}\int_0^L EI\left(\frac{\partial^2 Z_b}{\partial x^2}\right)^2 dx \qquad [3.22]$$

The Lagrangian is:

$$\mathcal{L} = -\int_0^L (\mathcal{E}_{es} + \mathcal{E}_{eb})\,dx + F_0 Z(L) = \mathcal{L}_s + \mathcal{L}_b \qquad [3.23]$$

where:

$$\mathcal{L}_s = -\int_0^L (\mathcal{E}_{es})\,dx + F_0 Z_s(L); \quad \mathcal{L}_b = -\int_0^L (\mathcal{E}_{eb})\,dx + F_0 Z_b(L)$$

Since the Lagrangian can be split into the sum of a shear term $\mathcal{L}_s$ and a bending term $\mathcal{L}_b$, it is anticipated that shear and bending deflections are uncoupled from each other. Furthermore, bending is still governed by the Bernoulli–Euler model. Hence, the variational calculation is detailed here for the shear term only:

$$\delta\mathcal{L}_s = -\int_0^L \kappa GS \left(\frac{\partial Z_s}{\partial x}\right)\left(\frac{\partial(\delta Z_s)}{\partial x}\right) dx + F_0 \delta Z_s(L) = 0 \qquad [3.24]$$

Once more, expression [3.24] is integrated by parts and the cofactors of δZ_s are set to zero.

$$\delta\mathcal{L}_s = -\left[F_0 \delta Z_s - \kappa GS\left(\frac{\partial Z_s}{\partial x}\right)\delta Z_s\right]_0^L + \int_0^L \left(\kappa GS\left(\frac{\partial^2 Z_s}{\partial x^2}\right)\delta Z_s\right) dx = 0 \qquad [3.25]$$

The deflection due to elastic shear is thus governed by the equation:

$$\kappa GS\left(\frac{\partial^2 Z_s}{\partial x^2}\right) = 0 \qquad [3.26]$$

with the boundary conditions:

$$Z_s(0) = 0; \quad \kappa GS\left(\frac{\partial Z_s}{\partial x}\right)\bigg|_L = F_0 \qquad [3.27]$$

The system [3.26], [3.27] is consistent with equations [2.33] and [2.39], except the shear weighting factor κ.

The total deflection is:

$$Z(x) = Z_s(x) + Z_b(x) = F_0\left\{\frac{x}{\kappa GS} + \frac{(3x^2 L - x^3)}{6EI}\right\} \qquad [3.28]$$

This result may be used to discuss the relative importance of shear and bending deflections in relation to the slenderness ratio of the beam. Let us consider for

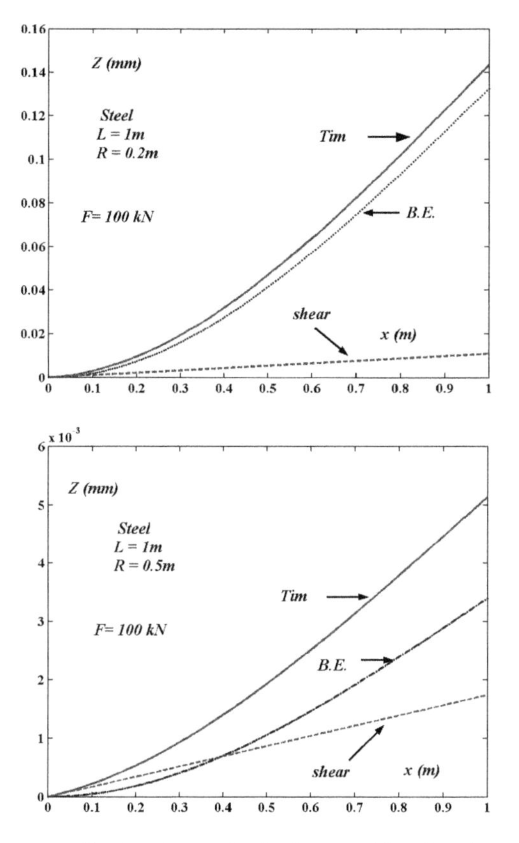

Figure 3.3. *Transverse deflection due to elastic shear and bending for two cantilever rods (upper plot L/R =10, lower L/R =2)*

instance a circular cylindrical rod of radius R and Poisson's ratio $\nu = 0.3$, $\kappa = 0.95 \cong 1$. We obtain:

$$Z(x) = \frac{F_0 L}{\pi R^2 E}\left\{2(1+\nu)\left(\frac{x}{L}\right) + \frac{2}{3}\left(\frac{L^2}{R^2}\right)\left(3\left(\frac{x}{L}\right)^2 - \left(\frac{x}{L}\right)^3\right)\right\} \qquad [3.29]$$

The result [3.29] is evidence that the relative contribution of bending to the total deflection increases as $(L/R)^2$, validating thus the Bernoulli–Euler model for slender beams, as illustrated in Figure 3.3.

3.2.2.3 The Rayleigh–Timoshenko dynamic model

The same problem as in the last subsection is studied now in dynamics. Focusing on the case of beams of small slenderness ratios, for which shear effects are important, the kinetic energy due to the rotation of the cross-sections with respect to the bending axis can no longer be neglected. The kinetic energy of a cross-section is thus written as:

$$\mathcal{E}_\kappa = \frac{1}{2}\int_0^L \rho S(\dot{Z}_s + \dot{Z}_b)^2 + \frac{1}{2}\rho I \left(\frac{\partial \dot{Z}_b}{\partial x}\right)^2 dx \qquad [3.30]$$

In contrast with the elastic energy, shear and bending deflections are found to be coupled due to the cross term arising in the translational kinetic energy. The variation of [3.30] is:

$$\int_{t_1}^{t_2} \delta[\mathcal{E}_\kappa]\, dt = \rho S \int_{t_1}^{t_2}\int_0^L -(\ddot{Z}_s + \ddot{Z}_b)(\delta Z_s + \delta Z_b) + \rho I \left\{\frac{\partial^2 \ddot{Z}_b}{\partial x^2}\right\} \delta Z_b\, dt\, dx \qquad [3.31]$$

Making use of the static results already derived in subsection 3.2.2.2, the following coupled dynamic equations are found:

$$\begin{aligned} \rho S \ddot{Z} - \rho I \frac{\partial^2 \ddot{Z}_b}{\partial x^2} + EI \frac{\partial^4 Z_b}{\partial x^4} &= 0 \\ \rho S \ddot{Z} - \kappa G S \frac{\partial^2 Z_s}{\partial x^2} &= 0 \\ Z &= Z_s + Z_b \end{aligned} \qquad [3.32]$$

Such equations can be further transformed to produce finally a single equation expressed in terms of the sole variable Z. At first, Z_b is eliminated from the bending equation to produce the intermediate result:

$$\rho S \ddot{Z} - \rho I \frac{\partial^2 \ddot{Z}}{\partial x^2} + \rho I \frac{\partial^2 \ddot{Z}_s}{\partial x^2} + EI \frac{\partial^4 Z}{\partial x^4} - EI \frac{\partial^4 Z_s}{\partial x^4} = 0 \qquad [3.33]$$

Then, from the second equation [3.32] we get:

$$\frac{\partial^2 Z_s}{\partial x^2} = \frac{\rho}{\kappa G}\ddot{Z} \Rightarrow \frac{\partial^4 Z_s}{\partial x^4} = \frac{\rho}{\kappa G}\frac{\partial^2 \ddot{Z}}{\partial x^2} \qquad [3.34]$$

Relations [3.34] can be used to eliminate Z_s from [3.33], providing thus the final result:

$$\rho S \frac{\partial^2 Z}{\partial t^2} - \rho I \left(1 + \frac{E}{\kappa G}\right) \frac{\partial^4 Z}{\partial x^2 \partial t^2} + \frac{\rho^2 I}{\kappa G}\frac{\partial^4 Z}{\partial t^4} + EI \frac{\partial^4 Z}{\partial x^4} = 0 \qquad [3.35]$$

Obviously, such an elimination of the variables Z_s and Z_b is possible only if the beam cross-sections are uniform.

3.2.3 *Bending of a beam prestressed by an axial force*

As is well known from common experience, the bending response of a beam can be significantly modified when an axial force of sufficient magnitude is applied to it. This effect is used in particular to adjust accurately the pitch of musical string instruments. More generally, in many instances one is interested in investigating bending resistant structures when their initial state of static equilibrium is stressed, due to the presence of an initial loading. The variational method is particularly well suited to modelling this kind of problem, because it brings out in a logical manner the central importance of nonlinear strains, even if the interest is restricted to the formulation of linear equations. Indeed, starting from an elastic medium whose initial state of static equilibrium is characterized by the local stress tensor $\overline{\overline{\sigma^{(0)}}}$, a further deformation $\overline{\overline{\varepsilon}}$ referred to this initial state induces an additional stress field denoted $\overline{\overline{\sigma^{(1)}}}$, and the actual stress tensor can be written as:

$$\overline{\overline{\sigma}} = \overline{\overline{\sigma^{(0)}}} + \overline{\overline{\sigma^{(1)}}} \qquad [3.36]$$

The central point of the problem is to realize that when such a prestressed system is set in motion, strain energy is varied due to the work done by the initial stress $\overline{\overline{\sigma^{(0)}}}$ as well as that done by $\overline{\overline{\sigma^{(1)}}}$. Both of them are related to the Green–Lagrange strain tensor [1.17] which comprises a linear plus a quadratic form of the displacement gradient. On the other hand, if the aim is restricted to derive linear equations of motion, it is suitable to approximate strain energy as a quadratic form of the displacement derivatives, as is already the case in the absence of $\overline{\overline{\sigma^{(0)}}}$. Because the elastic stress $\overline{\overline{\sigma^{(1)}}}$ is proportional to $\overline{\overline{\varepsilon}}$, whereas the initial stress $\overline{\overline{\sigma^{(0)}}}$ is independent of $\overline{\overline{\varepsilon}}$, it turns out that the elastic energy must be related to the linear component of the Green–Lagrange strain tensor, whereas the prestress energy must be related to the exact form of the Green–Lagrange strain tensor. This is further explained below taking the example of a cantilevered beam axially stressed by a static axial force $\vec{T}_0$ applied initially at its free end B, see Figure 3.4.

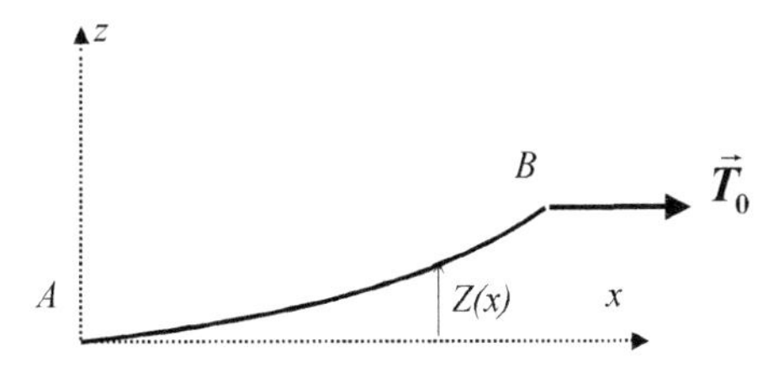

Figure 3.4. *Axially stressed cantilever*

3.2.3.1 Strain energy and Lagrangian

In so far as bending is described by the Bernoulli–Euler model, the stress tensor reduces to the axial component:

$$\sigma_{xx} = \sigma_{xx}^{(0)} + \sigma_{xx}^{(1)} = \frac{T_0}{S} + E\varepsilon_{xx} \qquad [3.37]$$

Making use of the relationship [1.47], the variation of the strain energy per unit volume is found to be:

$$\delta[e_s] = \sigma_{xx}\delta\varepsilon_{xx} = \left(\frac{T_0}{S} + E\varepsilon_{xx}\right)\delta\varepsilon_{xx} \qquad [3.38]$$

which implies:

$$e_s = \frac{T_0}{S}\varepsilon_{xx} + \frac{E\varepsilon_{xx}^2}{2} \qquad [3.39]$$

The longitudinal component of the Green–Lagrange tensor [1.17] is written as:

$$\varepsilon_{xx} = \frac{\partial \xi_x}{\partial x} + \frac{1}{2}\left(\left(\frac{\partial \xi_x}{\partial x}\right)^2 + \left(\frac{\partial \xi_z}{\partial x}\right)^2\right) \qquad [3.40]$$

where ξ_z is the local transverse displacement in the bending plane Oxz and ξ_x is the local axial displacement of a current point of the beam at a distance z from the bending axis. These components are related to the global displacement field defined in the prestressed state of equilibrium by:

$$\xi_z(x,z) = Z(x); \qquad \xi_x(x,z) = -z\frac{\partial Z}{\partial x} \qquad [3.41]$$

Substituting [3.41] into [3.40] we get:

$$\varepsilon_{xx} = -z\frac{\partial^2 Z}{\partial x^2} + \frac{1}{2}\left(\frac{\partial Z}{\partial x}\right)^2 + \frac{1}{2}\left(z\frac{\partial^2 Z}{\partial x^2}\right)^2 \qquad [3.42]$$

Since the aim is to derive linear equilibrium equations, it follows that strain energy density [3.39] must be expressed as a quadratic form in terms of the Z derivatives. Therefore, it is appropriate to use the exact form of [3.42] to formulate the energy involving the initial stress and to use its linear approximation to formulate the elastic energy, which is already quadratic in ε_{xx}^2. After making such manipulations and integrating e_s over the cross-sectional area, the following strain energy per unit beam length is obtained:

$$\int_{(\mathcal{S})} e_s \, d\mathcal{S} = T_0 \left\{ +\frac{1}{2}\left(\frac{\partial Z}{\partial x}\right)^2 + \frac{I}{2S}\left(\frac{\partial^2 Z}{\partial x^2}\right)^2 \right\} + \frac{EI}{2}\left(\frac{\partial^2 Z}{\partial x^2}\right)^2 \qquad [3.43]$$

Further, if the beam is sufficiently slender, only the first term proportional to the initial load may be retained, as the second one is less by a factor $(D/\mathcal{L})^2$, where D and $\mathcal{L}$ are the characteristic length scales in the transverse and in the axial directions respectively. The Lagrangian of the flexed beam is thus written as:

$$\mathcal{L} = \frac{1}{2}\int_0^L \left(\rho S \dot{Z}^2 - T_0 \left(\frac{\partial Z}{\partial x}\right)^2 - EI\left(\frac{\partial^2 Z}{\partial x^2}\right)^2 \right) dx \qquad [3.44]$$

3.2.3.2 Vibration equation and boundary conditions

Starting from the Lagrangian [3.44], Hamilton's principle is written as:

$$\delta[\mathcal{A}] = \int_{t_1}^{t_2} dt \int_0^L - \left\{ \rho S \ddot{Z} \delta Z + T_0 \left(\frac{\partial Z}{\partial x}\right)\left(\frac{\partial \delta Z}{\partial x}\right) + EI\left(\frac{\partial^2 Z}{\partial x^2}\right)\left(\frac{\partial^2 \delta Z}{\partial x^2}\right) \right\} dx = 0 \qquad [3.45]$$

leading to the vibration equation:

$$\rho S \ddot{Z} - T_0 \frac{\partial^2 Z}{\partial x^2} + EI \frac{\partial^4 Z}{\partial x^4} = 0 \qquad [3.46]$$

provided with the elastic boundary conditions:

$$\left[\left(T_0 \frac{\partial Z}{\partial x} - EI \frac{\partial^3 Z}{\partial x^3} \right) \delta Z \right]_0^L = 0; \quad \left[EI\left(\frac{\partial^2 Z}{\partial x^2}\right) \delta \left(\frac{\partial Z}{\partial x}\right) \right]_0^L = 0 \qquad [3.47]$$

In addition to the usual inertia and elastic stiffness terms, the vibration equation includes a prestress stiffness term proportional to the initial load T_0, which thus may be positive or negative, depending on the sign of T_0. On the other hand, the first boundary condition [3.47] related to the conjugate components of transverse shear stress and transverse displacement depends explicitly on the initial load T_0. In particular, if the loaded end is left free, as it is the case of the cantilevered beam, the boundary condition at $x = L$ is:

$$\left[EI \frac{\partial^3 Z}{\partial x^3} - T_0 \frac{\partial Z}{\partial x} \right]_L = 0 \qquad [3.48]$$

The second boundary condition, which is related to the conjugate components of bending moment and rotation of the cross-section, is independent of T_0.

NOTE. – *Newtonian approach*

Equations [3.46] provided with the suitable boundary conditions can also be derived by using the Newtonian approach. However, the balance of forces and moments must be written by referring to the deflected configuration of the beam, contrasting with the usual case in which the non-deflected configuration of static equilibrium is used as a reference. The effect of the preload (initial load) is shown in Figure 3.5. The preload vector is assumed to be constant; in particular it remains parallel to the beam axis direction of the non-deflected configuration, independently from the beam deflection. The forces $-T_0$ and T_0 act on the left- and right-hand side cross-sections bounding the infinitesimal beam element considered. As a first order approximation, used for projecting the forces, the small rotation of the cross-sections is assumed to be constant in the beam element and given by $-\partial Z/\partial x$. The projection of $\vec{T}_0$ onto a cross-sectional plane leads to the initial shear force:

$$Q_T \simeq T_s \simeq T_0\psi_y = -T_0\frac{\partial Z}{\partial x} \qquad [3.49]$$

The torque induced by the equilibrated axial forces $\pm\vec{T}_0$ acting at the ends of the beam element is:

$$Q_T dx = T_0\frac{\partial Z}{\partial x}dx \qquad [3.50]$$

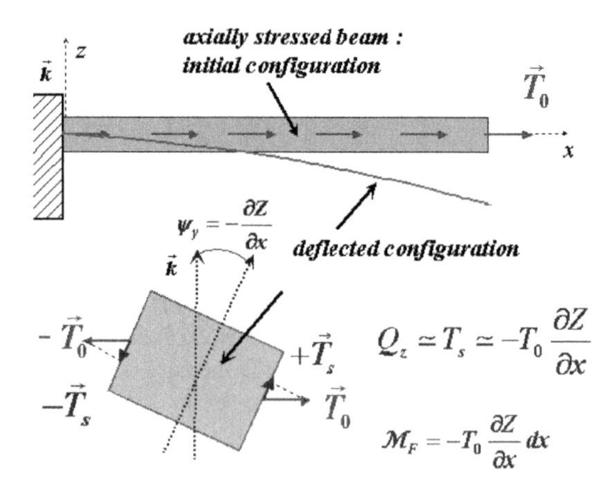

Figure 3.5. *Bent configuration of a beam stressed axially by an initial load*

In agreement with the Bernoulli–Euler model, the moment balance is written as:

$$\mathcal{M}_y(x+dx) - \mathcal{M}_y(x) + \mathcal{Q}_z dx + T_0 \frac{\partial Z}{\partial x} dx = 0 \qquad [3.51]$$

In so far as material behaves elastically, the resulting transverse shear force is:

$$\mathcal{Q}_z = -EI_y \frac{\partial^3 Z}{\partial x^3} + T_0 \frac{\partial Z}{\partial x} \qquad [3.52]$$

Substituting the expression [3.52] into the transverse equation [2.18], equation [3.46] is recovered. On the other hand, the boundary condition [3.48] is also recovered by cancelling the shear force $\mathcal{Q}_z$ at the free end of the cantilevered beam.

3.2.3.3 Static response to a transverse force and buckling instability

The physical effect of an axial preload on the bending response of the beam can be conveniently illustrated by solving the following static problem:

$$\begin{aligned} & EI \frac{d^4 Z}{dx^4} - T_0 \frac{d^2 Z}{dx^2} = 0 \\ & Z(0) = \left.\frac{dZ}{dx}\right|_0 = 0; \quad \left.\frac{d^2 Z}{dx^2}\right|_L = 0 \\ & EI \left.\frac{d^3 Z}{dx^3}\right|_L - T_0 \left.\frac{dZ}{dx}\right|_L = -F_0 \end{aligned} \qquad [3.53]$$

where the inhomogeneous boundary condition stands for a transverse force F_0 applied to the free end of the cantilevered beam.

The general solution of the homogeneous differential equation is found to be:

$$Z(x) = a e^{k_0 x} + b e^{-k_0 x} + cx + d \qquad [3.54]$$

where

$$k_0^2 = \frac{T_0}{EI}$$

Depending on the sign of the initial axial load T_0, k_0 is real or purely imaginary. Determination of the constants by using the boundary conditions of the problem presents no difficulty and the final result may be written as:

$$\begin{gathered}\text{Tensioned beam: } T_0 \geq 0 \\ Z(x) = \frac{F_0}{|T_0|}\left\{x - \frac{\sinh(k(x-L)) + \sinh(kL)}{k\cosh(kL)}\right\} \\ \text{Compressed beam: } T_0 \leq 0 \\ Z(x) = \frac{F_0}{|T_0|}\left\{-x + \frac{\sin(k(x-L)) + \sin(kL)}{k\cos(kL)}\right\}\end{gathered} \qquad [3.55]$$

where

$$k = |k_0| = \sqrt{\frac{|T_0|}{EI}}$$

In Figure 3.6 the deflection of the beam is represented in a reduced form for a few values of the tensile axial force, characterized by the dimensionless parameter $\Lambda = kL = L\sqrt{T_0/EI}$, where L is the beam length and $\xi = x/L$ is the dimensionless abscissa along the beam. $Z_m = Z(L; \Lambda = 0)$ is chosen as a relevant scaling factor to reduce the beam deflection. As is conspicuous in Figure 3.6, the magnitude of the deflection is significantly reduced as soon as $\Lambda > 0.1$, which means that initial tension enhances the effective stiffness of the deflected beam, as expected. Furthermore, the shape of the deflection curves is also modified being less curved as Λ increases, which is a mere consequence of the increasing importance of the tensioning term in comparison with the flexure term in the local equilibrium equation [3.53].

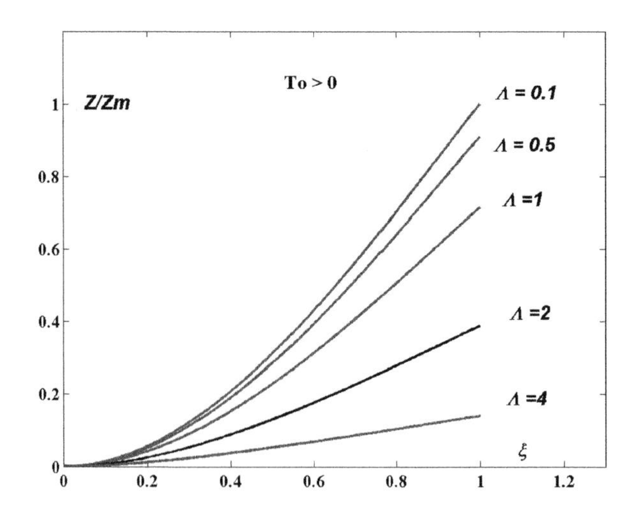

Figure 3.6. *Static response of a tensioned beam to a transverse force*

On the other hand, an initial compression decreases the effective stiffness, in such a way that the analytical response tends to infinity for the following infinite sequence of k values:

$$k_n = \frac{(2n+1)\pi}{2L}; \quad n = 0, 1, 2 \ldots \qquad [3.56]$$

The physical interpretation of such a result is as follows. When the magnitude of the compressive load is progressively increased starting from zero, the effective stiffness of the beam is progressively reduced, leading to a larger lateral deflection for the same transverse load F_0, see Figure 3.7. A critical compressive load T_c exists for which the effective stiffness is zero and the analytical deflection is infinite. Using [3.56], T_c is given by:

$$T_c = -EI\left(\frac{\pi}{2L}\right)^2 \qquad [3.57]$$

If the magnitude of the compressive load is further increased, according to the mathematical model [3.53] the effective stiffness becomes negative, as can be checked by looking at the sign of the analytical deflection [3.55], see the plots in dashed lines of Figure 3.7. As already indicated in [AXI 04], based on a few discrete systems such as articulated rigid bars, the axial compression of a beam can lead to static instability, commonly called '*buckling*' and validity of the linear model [3.53] to describe its physical response to F_0 is strictly restricted to the subcritical domain $T_0 > T_c$.

NOTE. – *Limit case of strings and cables*

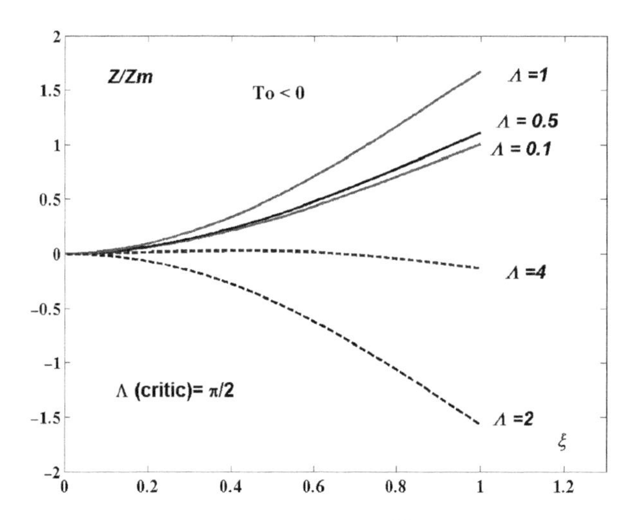

Figure 3.7. *Static response of a compressed beam to a transverse force*

It is easily shown that in an axially prestressed beam, the relative contribution of the flexural stiffness term to support a transverse load decreases continuously as the slenderness of the beam increases. A dimensional analysis of equation [3.53] gives immediately the dimensionless ratio:

$$\frac{\text{elastic term}}{\text{prestress term}} \simeq \frac{EI}{T_0\mathcal{L}^2} \propto \left(\frac{D}{\mathcal{L}}\right)^2 \qquad [3.58]$$

where again, the scale lengths D and $\mathcal{L}$ refer to the cross-sectional dimensions and to the axial variation of the deflection $Z(x)$.

Strings or *cables* correspond to the asymptotic and idealized case in which the slenderness ratio of the beam is so large that bending stiffness becomes practically zero. Accordingly, their buckling resistance is also negligible; so they can resist an external load by tensile stresses alone. In contrast with highly tensioned strings, as those used for instance in musical instruments, most cables sag significantly when loaded by their own weight and other in-service forces. As a consequence, geometrical nonlinearities must often be accounted for when analysing cable systems as illustrated in Appendix A.5 where an elementary catenary problem is solved.

3.2.3.4 *Follower loads*

By definition, a *follower force* retains the same orientation to the actual configuration of the structure in the course of motion, as illustrated in Figure 3.8 in the case of a cantilevered beam. A typical example is provided by a pressure field applied to the wall of a flexible structure. Starting from the results established just above, it is easy to see in which manner the preceding problem is modified if the tensile, or compressive, force $\vec{T}_0$ remains tangential to the deflected beam. The vibration equation [3.46] remains valid and the boundary condition [3.48] at a free end reduces to the classical expression, which holds in the absence of initial axial loading:

$$EI_y \frac{\partial^3 Z}{\partial x^3} = 0 \qquad [3.59]$$

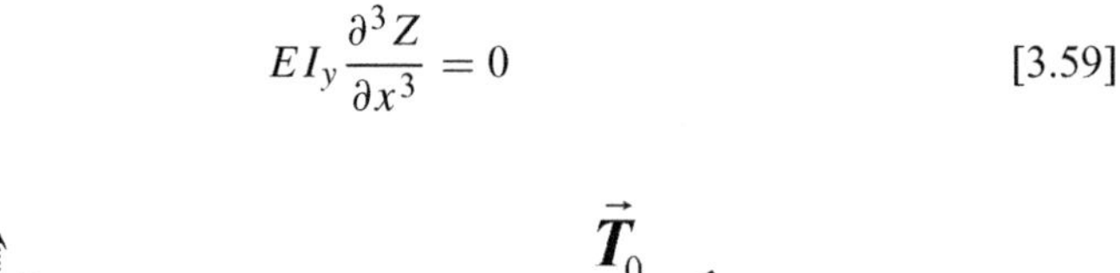

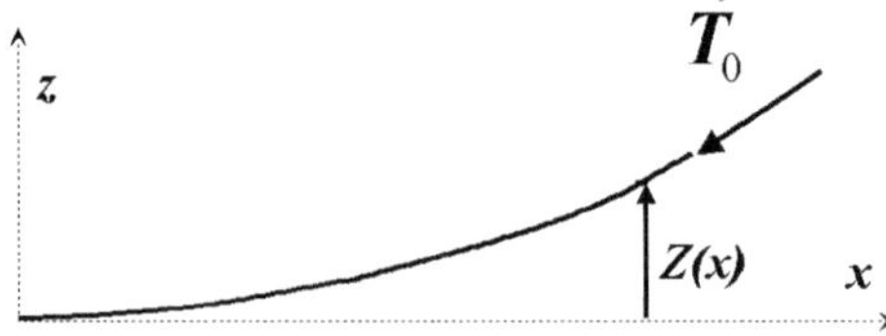

Figure 3.8. *Follower force applied to the free end of a cantilevered beam*

Hence, the static problem now takes the form:

$$
\begin{gathered}
EI\frac{d^4Z}{dx^4} - T_0\frac{d^2Z}{dx^2} = 0 \\
Z(0) = \left.\frac{dZ}{dx}\right|_0 = 0; \quad \left.\frac{d^2Z}{dx^2}\right|_L = 0 \\
EI\left.\frac{d^3Z}{dx^3}\right|_L = -F_0
\end{gathered}
\tag{3.60}
$$

Though at first sight the change when shifting from the system [3.53] to [3.60] may be thought as rather benign, the nature of the problem is actually profoundly modified. A first clue is provided by looking at the mathematical solution of [3.60]. Using the same calculation procedure as that used for solving the system [3.53] we obtain:

1. Tensioned beam: $T_0 \geq 0$

$$Z(x) = \frac{F_0}{|T_0|}\left\{x\cosh(kL) - \frac{\sinh(k(x-L)) + \sinh(kL)}{k}\right\} \tag{3.61}$$

2. Compressed beam: $T_0 \leq 0$

$$Z(x) = \frac{F_0}{|T_0|}\left\{-x\cos(kL) + \frac{\sin(k(x-L)) + \sin(kL)}{k}\right\} \tag{3.62}$$

Accordingly, it is found that the beam never losses stability! However, a closer inspection based on the concepts described in the next section shows that even in the absence of the external force F_0, the system [3.60] is non-conservative, in contrast with the system [3.53]. As a consequence its behaviour cannot be analysed based on a static model. Study of such dynamical problems is postponed to Volume 4. In the present example, which was investigated first by Beck [BEC 52], it is found that a compressive follower force induces a dynamical instability if its magnitude exceeds the critical value $T_c = 2.08\pi^2 EI/L^2$.

3.3. Weighted integral formulations

3.3.1 *Introduction*

So far, for analysing the mechanical response of flexible solids, the idea was to establish the differential equations which govern the local equilibrium of the body and the conditions to be satisfied at the boundaries. Restricting the study to the linear domain, a set of linear partial differential equations is obtained, which can be written according to the following canonical form:

$$K\left[\vec{X}\right]+C\left[\dot{\vec{X}}\right]+M\left[\ddot{\vec{X}}\right]=\vec{F}^{(e)}(\vec{r};t) \qquad [3.63]$$

where $K[\ ]$ is the *stiffness operator* (matrix if the mechanical system is discrete, differential if the system is continuous), $C[\ ]$ is the *viscous damping* operator and $M[\ ]$ is the *mass* operator. For example, for the Bernoulli–Euler beam model:

$$K[\,]=\frac{\partial^2}{\partial x^2}\left(EI\frac{\partial^2[\,]}{\partial x^2}\right); \quad M[\,]=\rho S[\,]$$

The operators depend on the coordinates of the Euclidean space used to describe the geometry of the structure and they operate in a functional vector space, namely the Hilbert space introduced in Chapter 1 subsection 1.3.2.2. On the other hand, even if the equations [3.63] are linear, they cannot be solved in an exact analytical way, except for a few simple configurations scarcely encountered in practice. In most cases, approximate analytical or numerical methods must be used to solve the problem. Amongst them, discretizing the system [3.63] by using either the finite elements method or the projection onto a modal basis are nowadays popular computational techniques in structural engineering. To introduce the basic theoretical aspects of these methods, a convenient way is to start by reformulating the system [3.63] in such a manner that the local aspect of the equations is removed. This is made possible by performing the functional scalar product of the local equation by a weighting functional vector $\vec{W}(\vec{r};t)$. The weighted integral formulation of continuous mechanics is introduced in subsection 3.3.2. It may be used as the starting point of various theoretical and numerical techniques. As a first theoretical application, subsection 3.3.3 shows the convenience of using singular Dirac distributions to describe local quantities as concentrated loads or support conditions. A second theoretical application, described in subsection 3.3.4 deals with the symmetry properties of the $K[\ ]$ and $M[\ ]$ operators in conservative systems. Such properties are a natural extension to the continuous case of those already evidenced for discrete systems in [AXI 04]. Finally, section 3.4 is devoted to an introductory description of the finite element method, emphasizing the most basic and salient features of the discretization procedure.

3.3.2 *Weighted equations of motion*

As indicated just above, the weighted integral formulation associated with the local system [3.63] is written as:

$$\left\langle \vec{W}, \left[K\left[\vec{X}\right] + C\left[\dot{\vec{X}}\right] + M\left[\ddot{\vec{X}}\right]\right]\right\rangle_{(\mathcal{V})} = \left\langle \vec{W}, \vec{F}^{(e)}\right\rangle_{(\mathcal{V})}$$
$$= \int_{(\mathcal{V})} \vec{W} \cdot \left[K\left[\vec{X}\right] + C\left[\dot{\vec{X}}\right] + M\left[\ddot{\vec{X}}\right]\right] d\mathcal{V} = \int_{(\mathcal{V})} \vec{W} \cdot \vec{F}^{(e)} d\mathcal{V} \qquad [3.64]$$

where use is made of a suitable weighting functional vector $\vec{W}(\vec{r};t)$.

Again, $(\mathcal{V})$ is the domain occupied by the structure. In the language of geometry, the weighted integral formulation stands for the projection of the local equilibrium equations [3.63] onto $\vec{W}(\vec{r};t)$; in agreement with the relationship [1.43], the projection is carried out both onto the Euclidean geometric space $(\mathcal{V})$ and onto the functional space of the Euclidean components of the vectors. If a displacement field is chosen for $\vec{W}(\vec{r};t)$, the local equations [3.63] have the physical meaning of an equilibrated balance of force densities and the weighted integral [3.64] is interpreted in terms of energy. As a particular case, if a virtual and admissible displacement (or velocity) field is selected for $\vec{W}$, [3.64] can be identified with the formulation of the principle of virtual work, or work rate. Moreover if $\vec{W}$ is equal to $\vec{X}e^{i\omega t}$, the time variable can be removed and the following *energy functionals* arise:

1. $\langle \vec{X}, K(\vec{r})\vec{X}\rangle_{(\mathcal{V})}$: internal (elastic and prestress) potential energy.
2. $\langle \vec{X}, C(\vec{r})\vec{X}\rangle_{(\mathcal{V})}$: viscous dissipation.
3. $\langle \vec{X}, M(\vec{r})\vec{X}\rangle_{(\mathcal{V})}$: kinetic energy.
4. $\langle \vec{X}, \vec{F}^{(e)}(\vec{r};t)\rangle_{(\mathcal{V})}$: external work.

3.3.3 *Concentrated loads expressed in terms of distributions*

In Chapter 2 it was shown that a load concentrated at the abscissa x_0 along a beam induces a finite discontinuity of the related global stress at x_0. By shifting to a weighted integral formulation of the problem, such a local effect can be described in terms of singular Dirac distributions. The Dirac delta distributions were already introduced in [AXI 04] Chapter 7, as a convenient analytical tool to model impulsive loading acting at a given time t_0 on discrete systems. Here, their use is extended to space variables, which allows one to unify the formulation of loads of any kinds, whatever their space and time repartition may be (concentrated or not, impulsive or not). At first, it is recalled that the Dirac distribution $\delta(x)$ is defined by the following integral:

$$\int_{x_1}^{x_2} f(x)\delta(x - x_0)\,dx = \begin{cases} f(x_0); & \text{if } x_0 \in [x_1, x_2] \\ 0; & \text{otherwise} \end{cases} \qquad [3.65]$$

$f(x)$ is any function complying with the sole condition that $f(x_0)$ exists. The integral [3.65] may be interpreted in terms of action, by saying that it specifies the action of the Dirac delta distribution on the ordinary function $f(x)$.

3.3.3.1 *External loads*

EXAMPLE 1. – *Beam loaded by a concentrated axial force*

Let us consider again the case of a beam loaded axially by an external force field distributed along the beam axis. It is recalled that the local equation of the problem is:

$$\rho S \frac{\partial^2 X}{\partial t^2} - \frac{\partial}{\partial x}\left(ES\frac{\partial X}{\partial x}\right) = F_x^{(e)}(x;t)$$

If the external load is concentrated at some position x_0, it may be expressed analytically as $F_x^{(e)}(t)\delta(x - x_0)$, provided that the equilibrium equation is formulated in terms of distributions instead of ordinary functions:

$$\rho S \frac{\partial^2 X}{\partial t^2} - \frac{\partial}{\partial x}\left(ES\frac{\partial X}{\partial x}\right) = F_x^{(e)}(t)\delta(x - x_0) \qquad [3.66]$$

To proceed into the solution of [3.66], the necessary step is to interpret the equation in terms of action, in a consistent way with the relation of definition [3.65]. Therefore, we shift from the local formulation [3.66] to the integral formulation:

$$\int_{x_1}^{x_2}\left(\rho S \frac{\partial^2 X}{\partial t^2} - \frac{\partial}{\partial x}\left(ES\frac{\partial X}{\partial x}\right)\right) dx = \int_{x_1}^{x_2} F_x^{(e)}(t)\delta(x - x_0)\,dx \qquad [3.67]$$

Here x_1 and x_2 may be chosen arbitrarily, except that they must comply with the condition of nonzero action of the loading, that is $x_0 \in [x_1, x_2]$. Accordingly, the integration of [3.67] leads to:

$$\int_{x_1}^{x_2}\left(\rho S \frac{\partial^2 X}{\partial t^2}\right) dx - \left[ES\frac{\partial X}{\partial x}\right]_{x_1}^{x_2} = F_x^{(e)}(t) \qquad [3.68]$$

Because the displacement field $X(x;t)$ is necessarily a continuous function, if x_1 and x_2 tend towards x_0, the action integral reduces to:

$$- ES\frac{\partial X}{\partial x}\bigg|_{x_0+} + ES\frac{\partial X}{\partial x}\bigg|_{x_0-} = +F_x^{(e)}(t) \qquad [3.69]$$

The finite jump of the axial stress at x_0 given by [3.69] is readily identified with that given by [2.31]. Here it arises as the action of the concentrated load on the beam. Hence the local equation [3.66] written in terms of distributions is found to be equivalent to the following system, written in terms of ordinary functions:

$$\begin{aligned} &\rho S\frac{\partial^2 X}{\partial t^2} - \frac{\partial}{\partial x}\left(ES\frac{\partial X}{\partial x}\right) = 0; \quad \forall x \in [0, L] \\ &ES\frac{\partial X}{\partial x}\bigg|_{x_0+} - ES\frac{\partial X}{\partial x}\bigg|_{x_0-} = -F_x^{(e)}(t) \end{aligned} \qquad [3.70]$$

Of course, for solving the problem, it is still necessary to use the system [3.70]. Nevertheless, the 'symbolic' form [3.66] is well suited to specify the discontinuity relations arising from concentrated loads by using the local equilibrium solely. This formalism can be generalized for other deformation modes as illustrated by the two following examples.

EXAMPLE 2. – *Beam loaded by a concentrated transverse force*

Again, we start from the classical case of a beam bent by a transverse external force field distributed along the beam axis. The local equation of the problem is (see equation [2.71]):

$$\frac{\partial^2}{\partial x^2}\left(EI_y\frac{\partial^2 Z}{\partial x^2}\right) + \rho S\frac{\partial^2 Z}{\partial t^2} = F_z^{(e)}(x;t)$$

If the external load is concentrated at x_0, the equation is written in terms of distributions as:

$$\frac{\partial^2}{\partial x^2}\left(EI_y\frac{\partial^2 Z}{\partial x^2}\right) + \rho S\frac{\partial^2 Z}{\partial t^2} = F_z^{(e)}(t)\delta(x - x_0) \qquad [3.71]$$

A similar calculation as that made just above shows immediately that [3.71] is equivalent to the system:

$$\begin{aligned} &\frac{\partial^2}{\partial x^2}\left(EI_y\frac{\partial^2 Z}{\partial x^2}\right)+\rho S\frac{\partial^2 Z}{\partial t^2}=0; \quad \forall x \in [0,L] \\ &\frac{\partial}{\partial x}\left(EI_y\frac{\partial^2 Z}{\partial x^2}\right)\bigg|_{x_0+}-\frac{\partial}{\partial x}\left(EI_y\frac{\partial^2 Z}{\partial x^2}\right)\bigg|_{x_0-}=F_z^{(e)}(t) \end{aligned} \qquad [3.72]$$

EXAMPLE 3. – *Beam loaded by a local transverse moment*

As indicated in the two last examples, it suffices to deal with the static version of the problem. Starting thus from the local equation:

$$\frac{\partial^2}{\partial x^2}\left(EI_y\frac{\partial^2 Z}{\partial x^2}\right)=\frac{\partial \mathfrak{M}_y^{(e)}(x)}{\partial x}$$

A moment applied at x_0 is written as the distribution $\mathcal{M}_y^{(e)}(t)\delta(x-x_0)$. When substituted into the right-hand side of the above equation we obtain:

$$\frac{\partial^2}{\partial x^2}\left(EI_y\frac{\partial^2 Z}{\partial x^2}\right)=\mathcal{M}_y^{(e)}\frac{d(\delta(x-x_0))}{dx}=\mathcal{M}_y^{(e)}\delta'(x-x_0) \qquad [3.73]$$

which is readily integrated to produce:

$$\frac{\partial}{\partial x}\left(EI_y\frac{\partial^2 Z}{\partial x^2}\right)=\mathcal{M}_y^{(e)}\delta(x-x_0)$$

where the constant of integration can be discarded.

Performing the action integral over the interval $x_0-\varepsilon, x_0+\varepsilon$ and letting ε tend to zero the expected discontinuity follows:

$$EI_y\frac{\partial^2 Z}{\partial x^2}\bigg|_{x_0+}-EI_y\frac{\partial^2 Z}{\partial x^2}\bigg|_{x_0-}=\mathcal{M}_y^{(e)} \qquad [3.74]$$

$\delta'(x)$ is called the Dirac dipole. Its action on a function $f(x)$ is obtained by integrating by parts the action of the Dirac delta on the derivative $f'(x)$ which is assumed to exist at x_0:

$$\int_{x_1}^{x_2} f'(x)\delta(x-x_0)\,dx=f'(x_0)=[f(x)\delta(x-x_0)]_{x_1}^{x_2}-\int_{x_1}^{x_2} f(x)\delta'(x-x_0)\,dx \qquad [3.75]$$

As $\delta(x)$ is zero everywhere except at $x = x_0$, it follows that the dipole action is found to be:

$$\int_{x_1}^{x_2} f(x)\delta'(x - x_0)\,dx = \begin{cases} -f'(x_0); & \text{if } x_0 \in [x_1, x_2] \\ 0; & \text{otherwise} \end{cases} \qquad [3.76]$$

NOTE. – *Informal differentiation of $\delta(x)$*

Let us assume that the derivative of $\delta(x)$ could be defined in the same manner as the derivative of an ordinary function. It would be thus given by the classical limiting process:

$$\delta'(x - x_0) = \lim_{dx \to 0} \left(\frac{\delta(x + dx/2 - x_0) - \delta(x - dx/2 - x_0)}{dx} \right)$$

In terms of concentrated forces, this expression can be interpreted physically as a torque of unit magnitude located at x_0, its sign depending on the plane which is considered, as indicated in Figure 3.9.

It can also be verified that the result arising from equations [3.74] is also consistent with those sketched in Figure 2.23 of Chapter 2. On this point, reference can also be made to the rigid rod maintained by two closely spaced knife-edged supports, as discussed in [AXI 04] Chapter 4.

3.3.3.2 Intermediate supports

Let us consider an elastic support located at x_0. It is characterized by a stiffness coefficient denoted K_q which acts on a generalized displacement $q(x;t)$. As a corollary of the results obtained in the last subsection, such a support may be described in terms of distributions, by defining the singular stiffness operator $K_q\delta(x - x_0)$ which produces the concentrated internal load $K_q q(x;t)\delta(x - x_0)$.

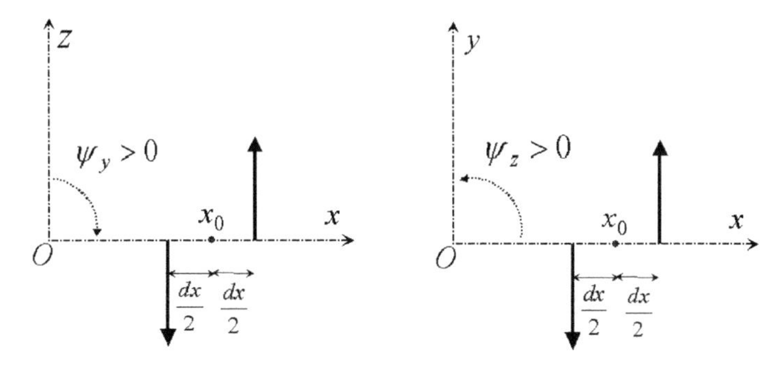

Figure 3.9. *Unitary torque equivalent to a Dirac dipole*

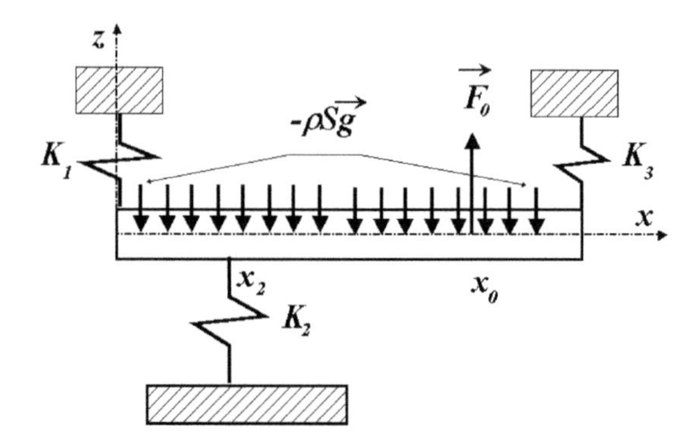

Figure 3.10. *Bending of beam with multiple supports and loads*

3.3.3.3 A comment on the use of distributions in mechanics

The use of the mathematical concept of distributions to describe the spatial distribution of physical quantities of any kind has the major advantage of unifying the formulation of the equilibrium conditions, in which the necessity to make a distinction between the local equilibrium equations and the boundary or discontinuity conditions vanishes, as illustrated by the example sketched in Figure 3.10. In terms of distributions, the problem can be entirely formulated by writing down the sole equation:

$$EI\frac{\partial^4 Z}{\partial x^4} + \{K_1\delta(x) + K_2\delta(x - x_2) + K_3\delta(x - L)\}Z + \rho S\ddot{Z}$$
$$= -\rho g S + F_0\delta(x - x_0)$$

All the local conditions are accounted for by the appropriate Dirac delta distributions. The use of the integral of action produces immediately the finite jump conditions verified by the ordinary functions which describe the displacement and the stress fields, and so one can proceed to the actual solution of the problem. Finally, it can be noted that the action integral can be thought of as a weighted integral in which the weighting function W is set to one if $0 \leq x \leq L$, where L is the beam length, and zero otherwise.

3.3.4 *Adjoint and self-adjoint operators*

As in the case of discrete systems, the dynamical behaviour of a continuous system depends upon the physical nature of the body and contact forces involved. In this respect, it is essential to draw a clear distinction between conservative and nonconservative forces. It is thus expected that some counterpart has to be found concerning the mathematical properties of the linear operators used to formulate these two kinds of forces. It is recalled (cf. for instance [AXI 04]) that the linear stiffness and mass operators of the discrete and conservative systems are symmetrical matrices denoted $[S]$, which obviously comply with the following condition of self-adjointness:

$$[W]^T[S][X] - [X]^T[S][W] = 0; \quad \forall [W], [X] \qquad [3.77]$$

Accordingly $[S]$ are also termed self-adjoint matrices. Furthermore, the mathematical properties concerning the eigenvalues and eigenvectors of symmetrical matrices were found to be of paramount importance for analysing the discrete mechanical systems. Extension of such properties to the continuous case can be made by starting from the weighted integral formulation [3.64], independently of the dimension of the geometrical space. For mathematical convenience, it is however appropriate to discuss the problem by referring to the one-dimensional case, which alleviates substantially the analytical notation. A linear differential operator in a one-dimensional space is of the general form:

$$\mathcal{L}[\;] = A_n(x)\frac{d^n}{dx^n} + \cdots + A_1(x)\frac{d}{dx} + A_0 = \sum_{k=0}^{n} A_k(x)\frac{d^k}{dx^k} \qquad [3.78]$$

where n is the order of the differential operator and where the coefficients $A_k(x)$ are assumed to be n-times differentiable functions.

In order to cope with finite discontinuities, we are interested in being able to define the quantity $\mathcal{L}[d]$ even if $d(x)$ is not a n-differentiable function. With this aim in mind, the weighted integral $\langle \mathcal{L}[d], \Psi \rangle_{(L)}$ is considered. By successive integrations by parts it is transformed into the adjoint form:

$$\langle \mathcal{L}[d], \Psi \rangle_{(L)} = \langle d, \mathcal{L}^{\#}[\Psi] \rangle_{(L)} \qquad [3.79]$$

$\Psi(x)$ is a test function which, by definition, can be differentiated at any order up to n, and is identically zero at the boundaries and out of the (L) domain, and so the derivatives. $\mathcal{L}^{\#}[\;]$ is called the *adjoint operator* of $\mathcal{L}[\;]$. It takes the form:

$$\mathcal{L}^{\#}[\Psi] = \sum_{k=0}^{n} (-1)^k \frac{d^k(A_k\Psi)}{dx^k} \qquad [3.80]$$

The important point in the relationship [3.79] is that even if $\mathcal{L}[d]$ cannot be defined directly because of the presence of singularities in d or its derivatives, the adjoint form $\mathcal{L}^{\#}[\Psi]$ holds, owing to the properties of the test functions. Hence, it is used to define the meaning of $\mathcal{L}[d]$, independently of the actual feasibility of the analytical operations involved. So, the relationship [3.79] holds even if $d(x)$ is a distribution, singular or not.

Let us consider now the forced equation, which holds also in terms of distributions:

$$\mathcal{L}[d] = f \qquad [3.81]$$

By definition $d(x)$ is a solution of [3.79] if for any test function ψ the following equivalence holds:

$$\langle \mathcal{L}[d], \Psi \rangle_{(L)} = \langle f, \Psi \rangle_{(L)} \iff \langle d, \mathcal{L}^{\#}[\Psi] \rangle_{(L)} = \langle f, \Psi \rangle_{(L)} \qquad [3.82]$$

The solutions can belong to one of the two following categories:

1. $d(x)$ can be differentiated a sufficient number of times to make the operation $\mathcal{L}[d]$ actually feasible, then the solution is termed 'classic'.
2. $d(x)$ is not differentiable up to the required order to make the operation $\mathcal{L}[d]$ feasible, then $d(x)$ is a solution in terms of distributions, the actual meaning of which is gained through the integral of action performed on the right-hand side of [3.82].

On the other hand, it can happen that $\mathcal{L}[\,] = \mathcal{L}^{\#}[\,]$. If this is the case, the operator is said to be *formally self-adjoint* and for any sufficiently regular pair of functions u, v which vanish identically at the vicinity of the boundaries of the domain (L) – and so the derivatives – the following equality is verified:

$$\langle u, \mathcal{L}[v] \rangle_{(L)} - \langle v, \mathcal{L}^{\#}[u] \rangle_{(L)} \qquad [3.83]$$

At this step, the property of self-adjointness is termed 'formal' because the relation of equivalence [3.82] does not enforce any boundary condition. However, if u and v, and/or their derivatives do not vanish at the boundaries, the integrations by parts produce the following result:

$$\langle u, \mathcal{L}[v] \rangle_{(L)} - \langle v, \mathcal{L}^{\#}[u] \rangle_{(L)} = [C\,(u, v)]_0^L \qquad [3.84]$$

where $C(u, v)$ is a bilinear form, called the *concomitant*, which depends upon the boundary conditions.

To be *self-adjoint* the operator has to comply with the conditions that it is formally self-adjoint, and that it is provided with appropriate boundary conditions, which make the concomitant disappear.

Abstract and arid as such theoretical concepts may be rightly felt, the attentive reader can still grasp the underlying physical meaning by putting them into the context of mechanics. Once more, if $\mathcal{L}[d]$ stands for a force density and u, v (or d) for a displacement field, the integrals considered above take the physical meaning of work, which suggests that self-adjointness is a specific property of conservative operators. The easiest way for evidencing this is to apply the mathematical formalism to a few mechanical systems.

EXAMPLE 1. – *Viscous damping operator*

Here the viscous damping operator is defined as a time differential operator:

$$A[\,] = A\frac{\partial}{\partial t} \quad \Rightarrow \quad A^{\#}[\,] = -A\frac{\partial}{\partial t}$$

It is readily found that $A[\,]$ is not self-adjoint and more generally any odd order operator is not self-adjoint for obvious reason of sign change included in the integration by parts.

EXAMPLE 2. – *Inertia operator*

Defining the inertia operator as a time differential operator,

$$M[\,] = M\frac{\partial^2}{\partial t^2} = M^{\#}[\,]$$

$M[\,]$ is formally self-adjoint as the integration by parts gives:

$$\langle u, M[v]\rangle = M\int_{-\infty}^{+\infty} u\frac{\partial^2 v}{\partial t^2}dt = M\,[u\dot{v} - v\dot{u}]_{-\infty}^{+\infty} + \langle v, M[u]\rangle$$

Going a little bit further, the concomitant vanishes, at least at infinity for any possible realistic motion which starts at some time and which does not last for ever. Thus the inertia operator provided with such realistic initial and final conditions is self-adjoint.

EXAMPLE 3. – *Beam on elastic supports*

Let us consider the bending operator of a beam, provided with elastic boundary conditions:

$$K[Z] = \frac{d^2}{dx^2}\left(EI(x)\frac{d^2 Z}{dx^2}\right) + \text{elastic supports conditions}$$

Elastic supports at the beam ends can be described either by linear relationships between the conjugate displacement and stress variables, as already seen in Chapter 2 subsection 2.2.5, or by concentrated stiffness forces as described in subsection 3.3.3.3. Adopting the first point of view, the elastic boundary conditions may be written as the following linear operators:

$$\mathcal{L}_1[Z(0)] = \left(K_1 Z + \frac{d}{dx}\left(EI\frac{d^2}{dx^2}\right)Z\right)_{x=0} = 0$$

$$\mathcal{L}_2[Z(L)] = \left(K_2 Z - \frac{d}{dx}\left(EI\frac{d^2}{dx^2}\right)Z\right)_{x=L} = 0$$

$$\mathcal{L}_3[Z(0)] = \left(K_3\frac{dZ}{dx} + EI\frac{d^2Z}{dx^2}\right)_{x=0} = 0$$

$$\mathcal{L}_4[Z(L)] = \left(K_4\frac{dZ}{dx} - EI\frac{d^2Z}{dx^2}\right)_{x=L} = 0$$

where K_1 to K_4 are the stiffness coefficients of the elastic supports.

The integration by parts of $\langle W, K[Z]\rangle_{(L)}$ gives:

$$\langle W, K[Z]\rangle_{(L)} - \langle Z, K[W]\rangle_{(L)}$$
$$= \left[W\frac{d}{dx}\left(EI\frac{d^2Z}{dx^2}\right) - Z\frac{d}{dx}\left(EI\frac{d^2W}{dx^2}\right) - \frac{dW}{dx}\left(EI\frac{d^2Z}{dx^2}\right) + \frac{dZ}{dx}\left(EI\frac{d^2W}{dx^2}\right)\right]_0^L$$

Then, the operator of bending stiffness is found to be formally self-adjoint. On the other hand, it is noted that the concomitant $C(W, Z)$ combines the conjugate displacement and stress variables of a flexed beam. Furthermore, the concomitant is found to vanish if W complies with the same elastic boundary conditions $\mathcal{L}_1$ to $\mathcal{L}_4$ as Z, since it takes the symmetrical form:

$$[WK_2Z - ZK_2W - W'K_4Z' + Z'K_4W']_{x=L}$$
$$+ [WK_1Z - ZK_1W - W'K_3Z' + Z'K_3W']_{x=0} = 0$$

where again the prime stands for a derivative with respect to x.

Then, the bending operator provided with elastic boundary conditions is self-adjoint. Adopting now the concentrated force point of view, the new operator arises:

$$\frac{d^2}{dx^2}\left(EI(x)\frac{d^2Z}{dx^2}\right) + K_1Z(0)\delta(x) + K_2Z(L)\delta(x-L)$$
$$+ K_3Z'(0)\delta'(x) + K_4Z'(L)\delta'(x-L)$$

Performing the action integral one obtains the same vanishing concomitant as above.

EXAMPLE 4. – *Cantilevered beam initially compressed by an axial force*

Let us consider the operator and boundary conditions of the system [3.53]:

$$K[Z] = EI\frac{d^4Z}{dx^4} - T_0\frac{d^2Z}{dx^2}$$

$$Z(0) = \left.\frac{dZ}{dx}\right|_0 = 0; \quad \left.\frac{d^2Z}{dx^2}\right|_L = 0; \quad EI\left.\frac{d^3Z}{dx^3}\right|_L - T_0\left.\frac{dZ}{dx}\right|_L = 0$$

It is easily shown that the initial stress operator is also formally self-adjoint and so is $K[Z]$. Furthermore, the concomitant [3.84] is found to be:

$$\left[\left(EI\frac{d^3Z}{dx^3} - T_0\frac{dZ}{dx}\right)W - \left(EI\frac{d^3W}{dx^3} - T_0\frac{dW}{dx}\right)Z\right]_0^L + \left[EI\frac{d^2Z}{dx^2}\frac{dW}{dx} - EI\frac{d^2W}{dx^2}\frac{dZ}{dx}\right]_0^L = 0$$

Accordingly, bending of an axially stressed beam is described by a stiffness operator which is self-adjoint.

EXAMPLE 5. – *Cantilevered beam initially compressed by a follower force*

Turning now to the system [3.60], the stiffness operator is the same as in the former case, but the boundary condition relative to the transverse shear force at the free end of the beam is changed, becoming:

$$EI\left.\frac{d^3Z}{dx^3}\right|_L = 0$$

as an immediate consequence the concomitant is not zero:

$$\begin{aligned}&\left[\left(EI\frac{d^3Z}{dx^3} - T_0\frac{dZ}{dx}\right)W - \left(EI\frac{d^3W}{dx^3} - T_0\frac{dW}{dx}\right)Z\right]_0^L + \left[EI\frac{d^2Z}{dx^2}\frac{dW}{dx} - EI\frac{d^2W}{dx^2}\frac{dZ}{dx}\right]_0^L \\ &= \left(T_0\left.\frac{dW}{dx}\right|_L\right)Z(L) - \left(T_0\left.\frac{dZ}{dx}\right|_L\right)W(L) \neq 0\end{aligned}$$

Accordingly, the bending of a beam initially stressed by a follower force is described by a stiffness operator which is not self-adjoint, but only 'formally' self-adjoint and the physics of the problem is drastically changed.

3.3.5 *Generic properties of conservative operators*

If the study is restricted to small elastic motions in the vicinity of a static stable (or indifferent, as a limit case) equilibrium position, the following inequalities must be verified:

$$0 \leq \left\langle \vec{X}, K\left[\vec{X}\right]\right\rangle_{(L)} < \infty; \quad 0 < \left\langle \dot{\vec{X}}, M\left[\dot{\vec{X}}\right]\right\rangle_{(L)} < \infty \quad \forall \vec{X} \qquad [3.85]$$

which means that the operators $K[\,]$, $M[\,]$ are respectively positive and positive definite.

Moreover $K[\,]$ and $M[\,]$ are self-adjoint as can be verified in equations [3.2], [3.3] which are quadratic symmetric forms; so the corresponding energy functionals comply with the conditions of self-adjointness:

$$\begin{aligned} \left\langle \vec{W}, M\left[\vec{X}\right]\right\rangle_{(L)} &= \left\langle \vec{X}, M\left[\vec{W}\right]\right\rangle_{(L)} = \left\langle M\left[\vec{W}\right], \vec{X}\right\rangle_{(L)} \\ \left\langle \vec{W}, K\left[\vec{X}\right]\right\rangle_{(L)} &= \left\langle \vec{X}, K\left[\vec{W}\right]\right\rangle_{(L)} = \left\langle K\left[\vec{W}\right], \vec{X}\right\rangle_{(L)} \end{aligned} \qquad [3.86]$$

Incidentally, it is worth noticing that the relations [3.86] extend the reciprocity theorem demonstrated by Betti which states that the virtual work of the stiffness forces $K[\vec{W}]$ for an admissible displacement field $\vec{X}$ is equal to the work done by the forces $K[\vec{X}]$ for the displacement field $\vec{W}$.

Self-adjointness is thus found to be a characteristic property of the linear operators of conservative mechanics and the following important results arise:

1. Any conservative problem is self-adjoint; this property characterizes not only the operator itself but also the boundary homogeneous conditions of the problem. The interpretation in terms of energy is very clear: the mechanical energy of a system is conserved if and only if the structure and the supports do not exchange mechanical energy with the surroundings. This explains, in particular, why the problem of the beam stressed by an axial force differs so much from that of a beam stressed by a follower force.
2. The mass operator is positive definite and if the static position is indifferent, or stable, the stiffness operator is positive, or positive definite. As a consequence, the amplitude of free oscillations in the vicinity of a stable position of static equilibrium must be constant because the mechanical energy is also constant.

3. These properties are a direct extension of those established in [AXI 04], in the case of conservative discrete mechanical systems. This indicates that it is possible to shift from the mathematical description of a discrete mechanical system to that of a continuous one, by letting the number of DOF of the system tend to infinity. However, it is generally found much more appropriate to do the reverse, i.e. discretizing the differential equations of continuous systems. This can be carried out using several methods and especially the finite element method, which nowadays is well implemented in computer science and mechanical engineering.

3.4. Finite element discretization

3.4.1 *Introduction*

The finite element method is a numerical procedure for obtaining an approximate solution of ordinary or partial differential equations. Restricting here drastically the field of applications to linear problems of structural mechanics, the main idea is to obtain approximate values of the kinetic and potential energy functionals introduced in subsection 3.3.2, by discretizing the geometrical domain into a set of N finite parts, which define the *mesh* of the finite element model, see Figures 3.11a,b,c. Then the functionals relative to the whole domain are approximated by summing the *element functionals* associated with every element of the discrete model:

$$\begin{aligned}\left\langle \vec{X}, K\left[\vec{X}\right]\right\rangle_{(\mathcal{V})} &\simeq \sum_{n=1}^{N}\left\langle \vec{X}, K\left[\vec{X}\right]\right\rangle_{(\mathcal{V}_n)} \\ \left\langle \vec{X}, M\left[\vec{X}\right]\right\rangle_{(\mathcal{V})} &\simeq \sum_{n=1}^{N}\left\langle \vec{X}, M\left[\vec{X}\right]\right\rangle_{(\mathcal{V}_n)}\end{aligned} \qquad [3.87]$$

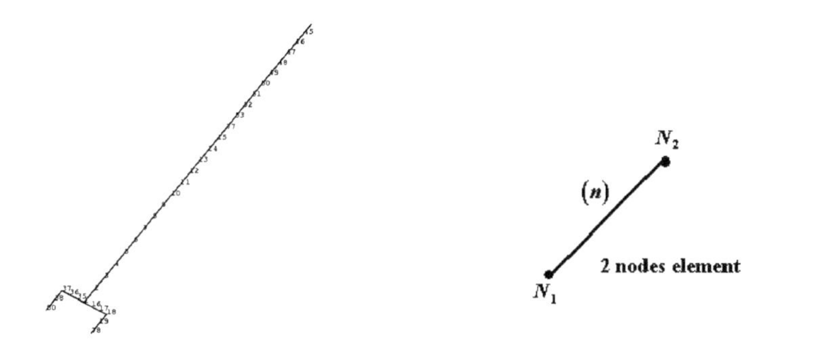

Figure 3.11a. *Mesh of a bolt spanner discretized by using straight beam elements*

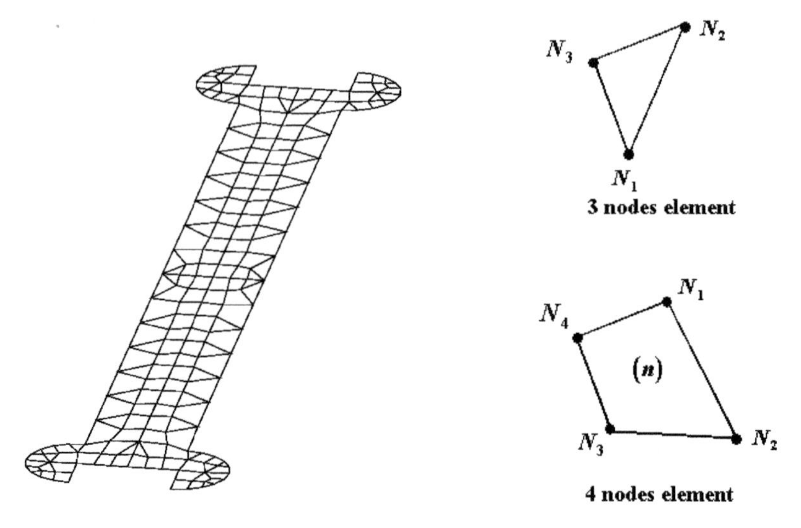

Figure 3.11b. *Mesh of a bolt spanner discretized by using plate elements*

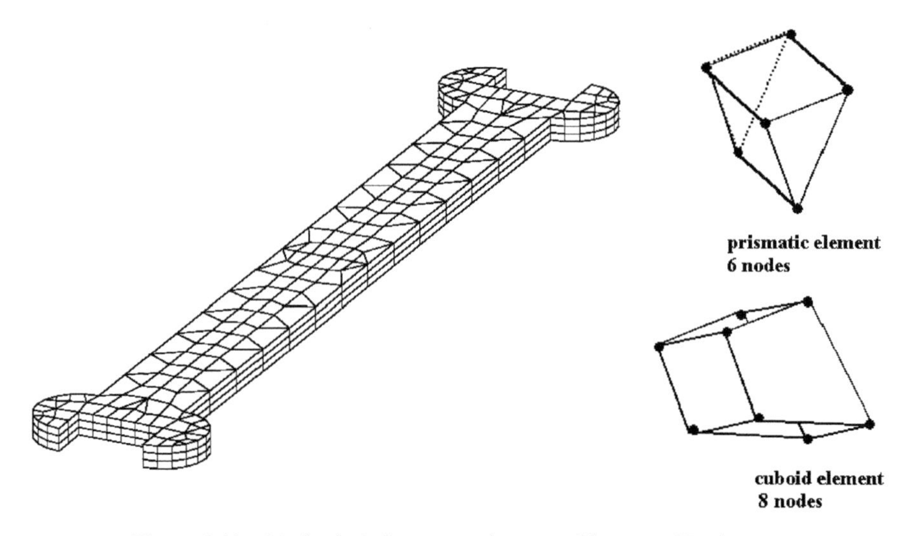

Figure 3.11c. *Mesh of a bolt spanner discretized by using 3D elements*

The symbol $\simeq$ is used here to emphasize that at this step an approximation is already likely to be made since the mesh of the finite element model does not map exactly the actual domain of the structure. Indeed, for convenience, the mesh is generated automatically by using geometrical supports of simple shapes such as straight segments, triangles, quadrilaterals, cuboids, prisms etc. They are defined by specifying a finite number of points which constitute the *nodes* of the mesh.

For instance, the geometric support of a straight beam element can be defined by using two nodes, that of a triangle by defining three nodes etc. The position of the nodes is specified by using a coordinate system termed *global* because it serves to describe the whole structure. Starting from the concept of nodes, one is led in a natural way to that of the *nodal displacement field* denoted $\vec{X}_N(t)$, where the subscript N specifies the node by its number. $\vec{X}_N(t)$ is the discrete version of the actual displacement field $\vec{X}(\vec{r};t)$ and the prime object of the method is to determine $\vec{X}_N(t)$ at each node of the mesh. Therefore overzealous mesh refinement is not advisable, at least in terms of computational cost. Actually, designing a suitable mesh to discretize a mechanical problem is not always a simple task, especially in the case of complex structures subjected to concentrated forces.

The next step is to calculate the functionals of every finite element of the model. By definition they are independent of time, so hereafter t can be dropped, or considered as fixed. As $\vec{X}$ is unknown, the problem is further simplified by defining in each element, numbered here by the integer n, a set of suitable analytical approximations for the components of $\vec{X}$, which results in an *interpolated displacement field* denoted $\vec{\Psi}_n(\vec{r})$ used to approximate the element functionals [3.87] as:

$$\left\langle \vec{\Psi}_n, K\left[\vec{\Psi}_n\right]\right\rangle_{(\mathcal{V}_n)}; \quad \left\langle \vec{\Psi}_n, M\left[\vec{\Psi}_n\right]\right\rangle_{(\mathcal{V}_n)} \qquad [3.88]$$

To perform the integrals involved in [3.88] it is particularly convenient to define the components of $\vec{\Psi}_n(\vec{r})$ by using "low degree" polynomials. They are defined within each individual element of the model and assumed to vanish outside of it. On the other hand, the degree p of the polynomials is chosen in such a way that it is possible to express all the monomial coefficients in terms of the nodal displacements of the element. The coefficients are determined by identifying the components of $\vec{\Psi}_n(\vec{r})$ with those of the corresponding $\vec{X}_n$ at each node of the element. As a result, $\vec{\Psi}_n(\vec{r})$ is found to be a linear form of the nodal displacements of the element, which may be written as:

$$\left[\vec{\Psi}_n(\vec{r})\right] = [N_n(\vec{r})]\,[X_n] \qquad [3.89]$$

NOTE. – *Degree of the interpolation polynomials and internal nodes*

To improve accuracy it may be desirable to increase the degree of the interpolation polynomials within a finite element. As p is governed by the number of conditions to be fulfilled at the nodes, this can be achieved by defining internal nodes. For instance, a beam finite element is often formulated by using either a two node element where the nodes are at the ends of the geometrical support (a straight segment) or a three node element, where the additional node is located judiciously within the segment. In a similar way, triangular elements can be provided with six nodes, cuboids with twenty nodes etc.

The dimension of $[X_n]$ is the number of generalized displacements per node times the number of nodes of the element, while the number of components of $\vec{\Psi}_n(\vec{r})$ is the number of generalized displacements per node. The functions $N_n(\vec{r})$ are the shape functions of the element; in [3.89] they are ordered as a line vector or as a rectangular matrix. The shape functions have unit values at a given node of the element and zero at all the other nodes. These conditions ensure the continuity of the interpolated displacement fields through the boundary between two elements, provided that:

1. A 'coherent finite element mesh' is used; that is a mesh in which adjacent elements have interpolation polynomials of the same degree.
2. The finite element displacement field is controlled by the nodal values of the boundary nodes only; that is a shape function associated with an interior node must be zero on the boundaries of the element.

The element stiffness $\left[K^{(n)}\right]$ or mass $\left[M^{(n)}\right]$ matrices of an element are then defined from [3.88] as:

$$\begin{aligned}\left\langle \vec{\Psi}_n(\vec{r}), K\left[\vec{\Psi}_n(\vec{r})\right]\right\rangle_{(\mathcal{V}_n)} &= [X_n]^T \left[\left[K^{(n)}\right]\right] [X_n] \\ \left\langle \vec{\Psi}_n(\vec{r}), M\left[\vec{\Psi}_n(\vec{r})\right]\right\rangle_{(\mathcal{V}_n)} &= [X_n]^T \left[\left[M^{(n)}\right]\right] [X_n]\end{aligned} \qquad [3.90]$$

Of course, the differential operators K and M depend on the mechanical properties of the element which have to be specified for each element as input data. In the same way the external loads are discretized by projecting the real force field $\vec{F}(\vec{r},t)$, at t fixed, on the displacement field within the element: $\langle \vec{F}(\vec{r},t), \vec{X}_n(\vec{r},t)\rangle_{(\mathcal{V}_n)}$. The resulting work is approximated in terms of nodal displacements and nodal forces by:

$$\langle \vec{F}^{(e)}(\vec{r},t), \vec{X}(\vec{r},t)\rangle_{(\mathcal{V}_n)} = \left[F_n^{(e)}\right]^T [X_n] \qquad [3.91]$$

where again, a polynomial interpolation is used to define the suitable nodal force vector $[F^{(n)}]$. The appropriate polynomials must be consistent with those already used for interpolating the displacement field. Computation of the element functionals [3.90] results in the formulation of the n-th finite element. It requires knowledge of the following information:

1. the element geometry,
2. the element shape functions (polynomials),
3. the material law of the element.

Calculation can be conveniently performed by using a local system of coordinates relevant to the specific element considered.

The following step is intended to collect in a suitable way the whole set of the element functionals to build the so called assembled matrices and assembled vectors describing the whole structure and the field of external load applied to it. As already stated in [3.87], this is done by summing all the element contributions. At this step, in most cases, it is necessary to transform the element functionals from the local frames to the global coordinate system, already used to describe the mesh of the whole structure. This point will be further detailed in subsection 3.4.3. In such a global system, the Lagrangian of the finite element model takes the quadratic form:

$$\mathcal{L} = \frac{1}{2}\left\{[\dot{X}]^T[M][\dot{X}] - [X]^T[K][X] + [F^{(e)}]^T[X]\right\} \qquad [3.92]$$

Accordingly, the equilibrium equations are immediately written as:

$$[K][X] + [M][\ddot{X}] = [F^{(e)}(t)] \qquad [3.93]$$

where the number *ND* of DOF is equal to the number of nodes of the whole mesh weighted by the number of nodal displacements.

The support conditions can be introduced into the system [3. 93], either by using Lagrange's multipliers or by defining specific elements which model the material law of the support, as outlined in subsection 3.4.2.6.

On the other hand, when the finite element method is applied to dynamic problems to be solved step by step in time, discretization of space must be followed by discretization of time. This can be achieved by using a finite difference algorithm as already described in [AXI 04]. Finally, the method can also be extended to nonlinear problems by using incremental steps in which the unbalance forces are reported on the right-hand side of [3.93]. Many algorithms exist but their descriptions are clearly out of the scope of this book. The reader interested in the nonlinear analysis of structures by the finite element method can consult many books, in particular [CRIS 86,96], where the subject is treated in an extensive and comprehensive manner.

3.4.2 *Beam in traction-compression*

In Figure 3.12, the straight beam considered has a non-uniform cross-section $S(x)$; we are interested in determining the elastodynamic response to an axial load distributed along the beam axis. The analytical formulation of the problem is:

$$-E\frac{\partial}{\partial x}\left(S(x)\frac{\partial X}{\partial x}\right) + \rho S(x)\frac{\partial^2 X}{\partial t^2} = F^{(e)}(x;t); \quad X(0) = 0; \quad \frac{\partial X}{\partial x|_L} = 0$$

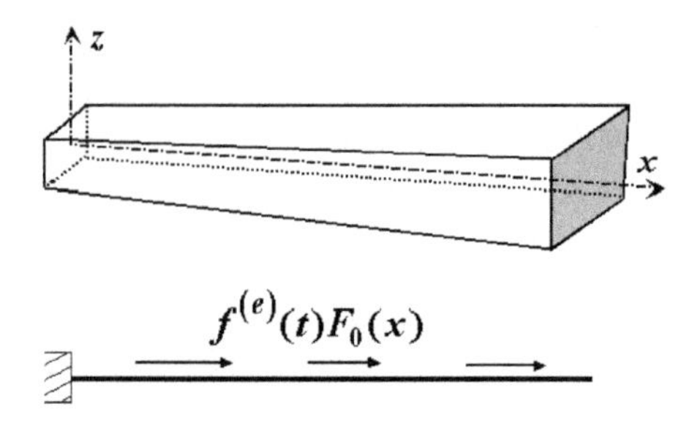

Figure 3.12. *Beam fixed at one end and free at the other, subjected to an axial load*

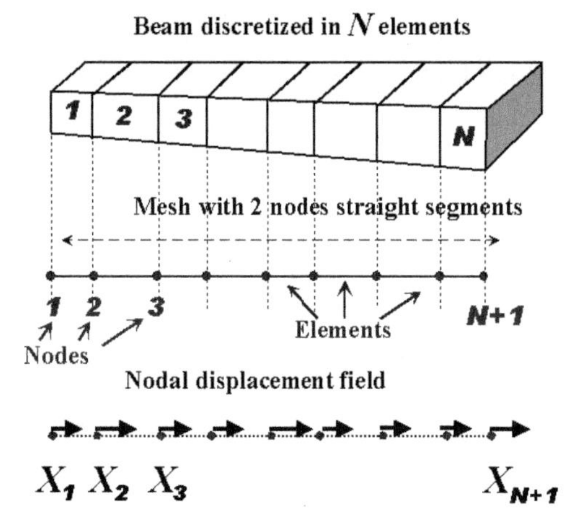

Figure 3.13. *Mesh and nodal displacement field of the beam*

For the sake of simplicity, the external load is assumed to be of the usual separated form:

$$F^{(e)}(x;t) = f^{(e)}(t)F_0(x); \quad \text{with the norm condition } \int_0^L F_0(x)\,dx = 1$$

Tractability of an analytical solution depends highly upon the analytical form of the variable coefficient $S(x)$ and of the external loading. Therefore, it is generally advisable to search for a numerical solution by using the finite element method, which is described below in the framework of this particular problem.

3.4.2.1 Mesh

As shown in Figure 3.13, the beam is divided in N adjacent elements. Then, the geometric supports of the finite elements are N straight segments of length $\ell_n(n = 1, N)$ and specified by $N + 1$ nodes.

3.4.2.2 *Shape functions*

Let the current finite element be specified by the nodes I and J and the index n. For the sake of simplicity, the corresponding beam cross-sectional area is approximated by a constant value S_n within the element. This is indeed equivalent to adopting a polynomial of zero degree for interpolating the function $S(x)$ within the element. A reasonable choice for S_n is the averaged nodal values $S_n = (S(x_I) + S(x_J))/2$. As the interpolation function of the axial displacement is concerned, it is found suitable to adopt a linear polynomial $X_n(x) = ax + b$ within $0 \leq x \leq \ell_n$, where x refers to a local system of coordinates with origin at node I and the x-axis along the segment IJ. According to such a choice, the polynomial coefficients a, b can be expressed in terms of the nodal displacements:

$$X_n(x) = ((X_J - X_I)/\ell_n)x + X_I = [1 - x/\ell_n, x/\ell_n]\begin{bmatrix} X_I \\ X_J \end{bmatrix} \qquad [3.94]$$

Hence, referring to [3.89], it may be verified that $N_I(0) = 1; N_J(\ell_n) = 0$ and $N_I(\ell_n) = 0; N_J(\ell_n) = 1$, as required. It is recalled that X_I and X_J are the prime unknowns of the discrete problem, i.e. the unknowns which are the first to be calculated. On the other hand, it should be noticed that the degree of the interpolation polynomial cannot be chosen arbitrarily since the number of the coefficients must fit the number of conditions to be fulfilled per element. Here, for instance a parabolic polynomial would not be appropriate as there are only two nodal conditions to be fulfilled. A constant would also be clearly unsuitable to account for the continuity and the non-uniformity of the actual displacement field.

3.4.2.3 *Element mass and stiffness matrices*

The element functional of kinetic energy is calculated using [3.90], written here as:

$$\left\langle \vec{X}, M\left[\vec{X}\right]\right\rangle_{(\ell_n)} \cong \langle \vec{X}_n, M^{(n)}\left[\vec{X}_n\right]\rangle_{(\ell_n)} = \rho_n\, S_n \int_0^{\ell_n} (((X_j - X_k)/\ell_n)x + X_j)^2 dx$$

where ρ_n and S_n are the mean values of the mass per unit volume and the cross-sectional area within the finite element n.

It is quite straightforward to write the functional as a quadratic and symmetrical form, which defines the mass-matrix of the n-th element, denoted $[M^{(n)}]$:

$$\langle \vec{X}_n, M^{(n)}\left[\vec{X}_n\right]\rangle_{(\ell_n)} = [X_n]^T[M^{(n)}][X_n] = m_n[X_j, X_k]\begin{bmatrix} 1 & 1/2 \\ 1/2 & 1 \end{bmatrix}\begin{bmatrix} X_j \\ X_k \end{bmatrix} \qquad [3.95]$$

where $m_n = (\rho_n\, S_n\, \ell_n)/3$.

$[M^{(n)}]$ is positive definite, as expected. In the particular case $X_I = X_J = 1$, which represents a rigid body axial displacement of unit magnitude, the mass of the element is recovered as:

$$[1,1][M^{(n)}]\begin{bmatrix}1\\1\end{bmatrix} = \frac{\rho_n S_n \ell_n}{3}[1,1]\begin{bmatrix}1 & 1/2\\1/2 & 1\end{bmatrix}\begin{bmatrix}1\\1\end{bmatrix} = \rho_n S_n \ell_n \qquad [3.96]$$

which can provide a satisfactory approximation of the actual mass of the corresponding structural element, provided the element is small enough.

The element functional of elastic energy is calculated in the same manner:

$$\langle \vec{X}, K[\vec{X}]\rangle_{(\ell_n)} \approx \langle \vec{X}_n, [K^{(n)}][\vec{X}_n]\rangle_{(\ell_n)} = -E_n S_n \int_0^{\ell_n} X_n \frac{\partial^2 X_n}{\partial x^2}\, dx$$

However, the calculation of the integral cannot be performed directly in the form just given above, since the linear approximation of the displacement field is clearly unsuited to describing the second derivative with respect to x. It could be thought that a difficulty has arisen because the interpolation polynomial cannot be a parabola if the geometrical support of the finite element is provided with only two nodes. Notwithstanding, keeping in mind the characteristic properties of symmetry evidenced in the self-adjoint operators, an integration by parts is found to be appropriate in order to decrease the degree of the derivation and at the same time to express the integral in a symmetrical form:

$$-E_n S_n \int_0^{\ell_n} X_n \frac{\partial^2 X_n}{\partial x^2} dx = \left[-E_n S_n X_n \frac{\partial X_n}{\partial x}\right]_0^{\ell_n} + E_n S_n \int_0^{\ell_n} \left(\frac{\partial X_n}{\partial x}\right)^2 dx \qquad [3.97]$$

The quantity within the brackets is closely related to the concomitant of the adjoint operator and to the boundary conditions of the problem. Accordingly, such terms are discarded here and discussed later in relation to the self-adjoint boundary conditions prevailing at the ends of the whole beam. The second term of [3.97] produces the suitable quadratic and symmetrical form:

$$\langle X_n, [K^{(n)}][X_n]\rangle_{(\ell_n)} = k_n[X_I, X_J]\begin{bmatrix}1 & -1\\-1 & 1\end{bmatrix}\begin{bmatrix}X_I\\X_J\end{bmatrix} \quad \text{where } k_n = E_n S_n/\ell_n \qquad [3.98]$$

The element stiffness matrix $[K^{(n)}]$ is symmetric and positive, but not positive definite. Indeed, if $X_I = X_J = 1$, the elastic energy is nil, as expected since the element moves like a rigid body, so no strain is developed within the element. On the other hand, considering the nodal field $X_I = 1; X_J = 0$, the form [3.98] identifies with the physical stiffness of the actual beam element, fixed at one end and free at the other.

3.4.2.4 *Equivalent nodal external loading*

Projection of the external loading is written as:

$$\langle F^{(e)}(x;t), X_n\rangle_{(\ell_n)} = f^{(e)}(t)\int_0^{\ell_n} F_0(x)\left(\left(\frac{(X_J - X_I)}{\ell_n}\right)x + X_I\right)dx$$
$$= \left[F_I^{(e)}, F_J^{(e)}\right]\begin{bmatrix} X_I \\ X_J \end{bmatrix} \qquad [3.99]$$

Thus, the nodal values of the external generalized forces depend on the polynomial interpolation of the displacement field and on the space distribution $F_0(x)$ of the external loading within the element. For instance, if F_0 is distributed uniformly along the element, it is readily found that:

$$F_I^{(e)}(t) = F_J^{(e)}(t) = f^{(e)}(t)\,\ell_n/2$$

If the load is varying linearly, $F^{(e)}(x;t) = f^{(e)}(t)(L - x)/L$ and the nodal forces are:

$$F_I^{(e)}(t) = f^{(e)}(t)\ell_n\left[\frac{1}{2} - \frac{\ell_n}{6L}\right]; \qquad F_J^{(e)}(t) = f^{(e)}(t)\ell_n\left(-\frac{\ell_n}{3L}\right)$$

Finally, if the load is concentrated at node J, $F^{(e)}(x;t) = f^{(e)}(t)\,\delta(x - x_J)$ and the nodal forces are:

$$\begin{bmatrix} F_I^{(e)} \\ F_J^{(e)} \end{bmatrix} = \begin{bmatrix} 0 \\ f_0^{(e)}(t) \end{bmatrix}$$

3.4.2.5 *Assembling the finite element model*

By summing all the element functionals of the finite element model, we obtain quadratic and symmetric forms for the internal energy and a linear form for the work of the external loads. The diagonal stiffness coefficient at the current line indexed by I is the sum of the corresponding element coefficients pertaining to the two adjacent finite elements bounded by the node I. In Figure 3.14, the structure of the Lagrange equations derived from such functionals is depicted schematically. The rectangles stand for the element matrices. The partial overlapping accounts for the summation

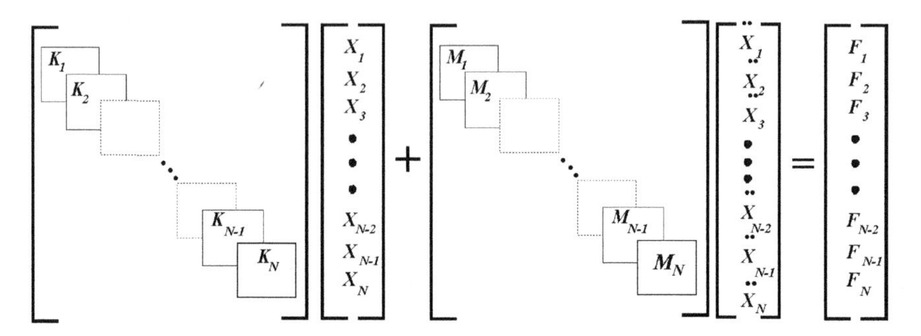

Figure 3.14. *Structure of the Lagrange equations of the finite elements model*

of diagonal terms of the contiguous elements. It is worth emphasizing that, from the Newtonian standpoint, this procedure of assembly ensures the equality of the nodal displacements from one element to the next and the exact force balancing at the nodes.

3.4.2.6 *Boundary conditions*

Turning back now to the element boundary terms arising in [3.97] and then disregarded for a while, at first sight they have no reason to vanish. Nevertheless, as they stand for internal forces, namely the generalized stresses acting at the boundaries of the elements, they are necessarily complying with the principle of action and reaction. Accordingly, in the assembling process, the element contributions of the boundary terms are found to cancel out exactly at each internal node of the model, i.e. a node bounding more than a single element. Indeed, shifting from one element to a contiguous one, the contact force exerted on the cross-section undergoes a sign change, as could be easily checked in a direct manner by calculating the functional [3.97] over two contiguous elements. Thus the summation of the element functional leads to zero, except for the two end nodes labelled 1 and $N + 1$ in Figure 3.13. Of course, at the boundaries of the structure there is no superposition of two opposite sign contributions of internal stresses. Going a little bit further in the reasoning, by assuming that the boundary term is zero at the end node $N + 1$, it is automatically stated that no stress is present at the boundary labelled $N + 1$. This remark leads us to discuss the way of completing the finite element model in a suitable way to include in it the boundary conditions at the ends of the beam. In the present problem as the first node is assumed to be fixed, the condition $X_1 = 0$ must be enforced. This may be achieved by two ways. The first method, described here, is to use a Lagrange multiplier, the second, outlined in the next subsection, is to use a penalty coefficient. A Lagrange multiplier λ_1 applied to the first node of the mesh may be used to model any prescribed motion, where the time-history of the prescribed displacement $D(t)$ is given. Of course, the condition which holds in the present problem is recovered as the particular case $D = 0$. The elastic functional of the first finite element, which is constrained by the condition $X_1 = D(t)$ is written as:

$$k_1[X_1, X_2]\begin{bmatrix} 1 & -1 \\ -1 & 1 \end{bmatrix}\begin{bmatrix} X_1 \\ X_2 \end{bmatrix} + \lambda_1(X_1 - D(t))$$

$$= [X_1, X_2, \lambda_1]\left[\begin{bmatrix} k_1 & -k_1 & 1 \\ -k_1 & k_1 & 0 \\ 0 & 0 & 0 \end{bmatrix}\begin{bmatrix} X_1 \\ X_2 \\ \lambda_1 \end{bmatrix} - \begin{bmatrix} 0 \\ 0 \\ D(t) \end{bmatrix}\right] \qquad [3.100]$$

It is recalled that in [3.100] the variables are the nodal displacements and the Lagrange multiplier which is used to determine the support reaction. Using the potential functional [3.100] and the kinetic energy functional, the finite element model of the first beam element, assumed to be loaded by the nodal forces F_1 and F_2 plus the prescribed displacement $X_1 = D(t)$ is given by the three corresponding Lagrange equations written in matrix form as:

$$\begin{bmatrix} k_1 & -k_1 & 1 \\ -k_1 & k_1 + k_2 & 0 \\ 1 & 0 & 0 \end{bmatrix}\begin{bmatrix} X_1 \\ X_2 \\ \lambda_1 \end{bmatrix} + \begin{bmatrix} m_1 & m_{12} & 0 \\ m_{12} & m_1 + m_2 & 0 \\ 0 & 0 & 0 \end{bmatrix}\begin{bmatrix} \ddot{X}_1 \\ \ddot{X}_2 \\ \ddot{\lambda}_1 \end{bmatrix} = \begin{bmatrix} F_1 \\ F_2 \\ D_1 \end{bmatrix} \qquad [3.101]$$

where k_2 and m_2 are the diagonal stiffness and mass coefficients arising from the adjacent element and where $D_1 = D(t)$. As appropriate, the last row of [3.101] identifies with the condition of prescribed displacement.

3.4.2.7 Elastic supports and penalty method

Let us consider an elastic support defined by a stiffness coefficient k_J which acts on the nodal displacement X_J. The corresponding elastic functional is $k_J X_J^2$. The contribution of this additional term into the assembled stiffness matrix is accounted for simply by adding k_J to the diagonal term K_{JJ}. The support reaction is given by:

$$\mathcal{R}_J = -(k_J + K_{JJ})X_J \qquad [3.102]$$

Since the reaction enters into the global force balance of the system, it cannot become infinite, so if k_J is large enough, the enforced condition is practically $X_J = 0$, which represents a rigid support. Thus the second method to lock a degree of freedom is to add an appropriate stiffness coefficient K_P to the diagonal term K_{JJ}. Used for such a numerical purpose, the method is broadly known as a *penalty method*, where the value of the penalty coefficient K_P is suitably chosen according to numerical criteria concerning the relative order of magnitude of the different coefficients of the assembled stiffness matrix. These conditions may be applied to any degree of freedom. Finally, in full agreement with the absence of stress at a free end, the condition of letting free a degree of freedom is automatically fulfilled by letting the related coefficients of $[K]$ remain unchanged, which corresponds precisely to $K_P = 0$.

3.4.3 *Assembling non-coaxial beams*

Let us consider the example of the portal frame depicted in Figure 3.15. It is made up of three steel straight beams of the same cross-section. The rooting points (A) and (D) are assumed to be clamped and a static load is applied to (P). The geometry of the structure is described by using a global Cartesian system *Oxyz*. For convenience it is assumed that *Oz* is vertical and *Ox* horizontal. Interest is focused first on the elastic stresses induced in the frame when subjected to a vertical force at P. Transverse shear stresses have to develop in the horizontal beam to oppose the load. Such forces are transmitted to the vertical legs of the frame as axial stresses. On the other hand, the shear stresses related to bending moments are also transmitted to the legs. As an immediate consequence, flexural and axial deformations of the frame are found to be coupled together. Accordingly, for solving the problem using the finite element method, it is necessary to build first a finite element of beam type, which describes both flexural and axial modes of deformation. Such modes are still uncoupled on the local scale of the finite elements, since the latter stand for straight beams.

3.4.3.1 The stiffness and mass matrices of a beam element for bending

To build the element matrices it is necessary to use a local system of coordinates, in which distinction is drawn between the axial and the transverse directions. In agreement with the analytical modelling described in Chapter 2, a suitable choice is to adopt a direct Cartesian frame $O'x'y'z'$ in which $O'x'$ is in the axial direction of the beam element, and the transverse axes $O'y'$, $O'z'$ are along the principal directions of inertia of the cross-section passing through the centroid O', see Figure 3.15. As in [3.97], the elastic functional of the bending element is written

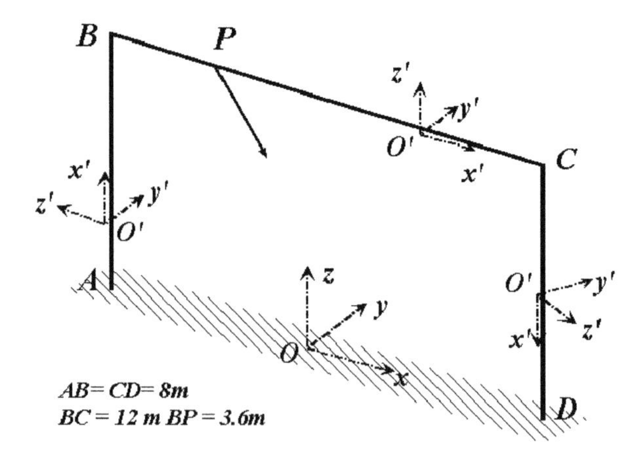

Figure 3.15. *Beam assembly used as a portal frame*

as the symmetrical form:

$$E_n I_n \int_0^{\ell_n} Z_n \frac{\partial^4 Z_n}{\partial x^4} dx = E_n I_n \left[Z_n \frac{\partial^3 Z_n}{\partial x^3} - \frac{\partial Z_n}{\partial x} - \frac{\partial^2 Z_n}{\partial x^2} \right]_0^{\ell_n} + E_n I_n \int_0^{\ell_n} \left(\frac{\partial^2 Z_n}{\partial x^2} \right)^2 dx \qquad [3.103]$$

where again the element boundary terms are disregarded for a while. Therefore, the integral is written in terms of the second derivative of the lateral displacement and the minimal degree of the shape polynomial is thus equal to two. However, it is not possible to determine the three coefficients in terms of the nodal displacements Z_I, Z_J. We are thus led in a natural way to enlarge the number of nodal displacements by stating that the rotation ψ_y of the cross-sections can also be treated as an independent variable. Accordingly, the nodal displacement field of a finite element now has four components which can be used to define the four coefficients of a cubic polynomial $Z(x) = ax^3 + bx^2 + cx + d$. Of course, if bending is described according to the Bernoulli–Euler model $\psi_y = Z'$, where the prime stands for the derivation with respect to x. Such a procedure is suitable in terms of accuracy, since it ensures the continuity of the transverse displacement and of its derivatives up to the second order. The nodal approximation is given by solving the linear system:

$$\begin{bmatrix} Z_I \\ Z'_I \\ Z_J \\ Z'_J \end{bmatrix} = \begin{bmatrix} 0 & 0 & 0 & 1 \\ 0 & 0 & 1 & 0 \\ \ell_n^3 & \ell_n^2 & \ell_n & 1 \\ 3\ell_n^2 & 2\ell_n & 1 & 0 \end{bmatrix} \begin{bmatrix} a \\ b \\ c \\ d \end{bmatrix} \Rightarrow \begin{aligned} a &= \frac{\ell_n (Z'_I + Z'_J) - 2(Z_J - Z_I)}{\ell_n^3} \\ b &= \frac{3(Z_J - Z_I) - \ell_n (2Z'_I + Z'_J)}{\ell_n^2} \\ c &= Z'_I \\ d &= Z_I \end{aligned} \qquad [3.104]$$

then,

$$\begin{aligned} Z_n(x) = &\left(1 - 3\left(\frac{x}{\ell_n}\right)^2 + 2\left(\frac{x}{\ell_n}\right)^3 \right) Z_I + \ell_n \left(\frac{x}{\ell_n} - 2\left(\frac{x}{\ell_n}\right)^2 + \left(\frac{x}{\ell_n}\right)^3 \right) Z'_I \\ &+ \left(3\left(\frac{x}{\ell_n}\right)^2 - 2\left(\frac{x}{\ell_n}\right)^3 \right) Z_J + \ell_n \left(\left(\frac{x}{\ell_n}\right)^3 - \left(\frac{x}{\ell_n}\right)^2 \right) Z'_J \\ Z''_n(x) = &\left(-\frac{6}{\ell_n^2} + \frac{12x}{\ell_n^3} \right) Z_I + \left(-\frac{4}{\ell_n} + \frac{6x}{\ell_n^2} \right) Z'_I + \left(\frac{6}{\ell_n^2} - \frac{12x}{\ell_n^3} \right) Z_J \\ &+ \left(\frac{6x}{\ell_n^2} - \frac{2}{\ell_n} \right) Z'_J \end{aligned} \qquad [3.105]$$

The strain energy is given by the symmetrical form of the integral [3.103]:

$$\int_0^{\ell_n} EI\left(\frac{\partial^2 Z}{\partial x^2}\right)^2 dx \cong E_n I_n \int_0^{\ell_n} \left(\frac{\partial^2 Z_n}{\partial x^2}\right)^2 dx$$

$$= \begin{bmatrix} Z_j \\ Z'_j \\ Z_k \\ Z'_k \end{bmatrix}^T \begin{bmatrix} K_{11} & K_{12} & K_{13} & K_{14} \\ K_{21} & K_{22} & K_{23} & K_{24} \\ K_{31} & K_{32} & K_{33} & K_{34} \\ K_{41} & K_{42} & K_{43} & K_{44} \end{bmatrix} \begin{bmatrix} Z_j \\ Z'_j \\ Z_k \\ Z'_k \end{bmatrix}$$

Using [3.105] the following coefficients are obtained:

$$\begin{gathered} K_{11} = \frac{12E_n I_n}{\ell_n^3} \quad K_{12} = \frac{6E_n I_n}{\ell_n^2} \quad K_{13} = -\frac{12E_n I_n}{\ell_n^3} \quad K_{14} = \frac{6E_n I_n}{\ell_n^2} \\ K_{22} = \frac{4E_n I_n}{\ell_n} \quad K_{23} = -\frac{6E_n I_n}{\ell_n^2} \quad K_{24} = \frac{2E_n I_n}{\ell_n} \quad K_{33} = \frac{12E_n I_n}{\ell_n^3} \\ K_{34} = -\frac{6E_n I_n}{\ell_n^2} \quad K_{44} = \frac{4E_n I_n}{\ell_n} \end{gathered} \qquad [3.106]$$

By using a similar computational procedure, the following mass matrix can be obtained:

$$\int_0^{\ell_n} \rho S Z^2 dx \cong \rho_n S_n \int_0^{\ell_n} Z^2 dx = \begin{bmatrix} Z_j \\ Z'_j \\ Z_k \\ Z'_k \end{bmatrix}^T \begin{bmatrix} M_{11} & M_{12} & M_{13} & M_{14} \\ M_{21} & M_{22} & M_{23} & M_{24} \\ M_{31} & M_{32} & M_{33} & M_{34} \\ M_{41} & M_{42} & M_{43} & M_{44} \end{bmatrix} \begin{bmatrix} Z_j \\ Z'_j \\ Z_k \\ Z'_k \end{bmatrix} \qquad [3.107]$$

As an exercise the reader can calculate the following coefficients:

$$\begin{gathered} M_{11} = M_{33} = 156 m_n \quad M_{12} = -M_{34} = 22 m_n \ell_n \\ M_{13} = 54 m_n \quad M_{14} = -M_{23} = -13 m_n \ell_n \\ M_{22} = M_{44} = 4 m_n \ell_n^2 \quad M_{24} = -3 m_n \ell_n^2 \end{gathered} \qquad [3.108]$$

where

$$m_n = \frac{\rho_n S_n \ell_n}{420}$$

3.4.3.2 Stiffness matrix combining bending and axial modes of deformation

It is clear that the individual modes of deformation of the beam can be superposed in the same way as the elastic functional. Combining the axial and the bending deformation we get:

$$\int_0^{\ell_n} Es\left(\frac{\partial X}{\partial x}\right)^2 dx + \int_0^{\ell_n} EI\left(\frac{\partial^2 Z}{\partial x^2}\right)^2 dx$$

$$\simeq E_n S_n \int_0^{\ell_n}\left(\frac{\partial X_n}{\partial x^2}\right)^2 dx + E_n I_n \int_0^{\ell_n}\left(\frac{\partial^2 Z_n}{\partial x^2}\right)^2 dx$$

$$= [X_I \;\; Z_I \;\; Z'_I \;\; X_J \;\; Z_k \;\; Z'_J]\begin{bmatrix} K_{11} & K_{12} & K_{13} & K_{14} & K_{15} & K_{16} \\ K_{21} & K_{22} & K_{23} & K_{24} & K_{25} & K_{26} \\ K_{31} & K_{32} & K_{33} & K_{34} & K_{35} & K_{36} \\ K_{41} & K_{42} & K_{43} & K_{44} & K_{45} & K_{46} \\ K_{51} & K_{52} & K_{53} & K_{54} & K_{55} & K_{56} \\ K_{61} & K_{62} & K_{63} & K_{64} & K_{65} & K_{66} \end{bmatrix}\begin{bmatrix} X_I \\ Z_I \\ Z'_I \\ X_J \\ Z_J \\ Z'_J \end{bmatrix}$$

where the element stiffness matrix $[K^{(n)}]$ is:

$$[K^{(n)}] = \begin{bmatrix} \frac{E_n S_n}{\ell_n} & 0 & 0 & -\frac{E_n S_n}{\ell_n} & 0 & 0 \\ 0 & \frac{12E_n I_n}{\ell_n^3} & \frac{6E_n I_n}{\ell_n^2} & 0 & -\frac{12E_n I_n}{\ell_n^3} & \frac{6E_n I_n}{\ell_n^2} \\ 0 & \frac{6EI_n}{\ell_n^2} & \frac{4EI_n}{\ell_n} & 0 & -\frac{6E_n I_n}{\ell_n^2} & \frac{2E_n I_n}{\ell_n^2} \\ -\frac{E_n S_n}{\ell_n} & 0 & 0 & \frac{E_n S_n}{\ell_n} & 0 & 0 \\ 0 & -\frac{12E_n I_n}{\ell_n^3} & -\frac{6E_n I_n}{\ell_n^2} & 0 & \frac{12E_n I_n}{\ell_n^3} & -\frac{6E_n I_n}{\ell_n^2} \\ 0 & \frac{6E_n I_n}{\ell_n^2} & \frac{2E_n I_n}{\ell_n^2} & 0 & -\frac{6E_n I_n}{\ell_n^2} & \frac{4E_n I_n}{\ell_n} \end{bmatrix} \quad [3.109]$$

This matrix is obtained by rearranging the coefficients already calculated, in accordance with the ordering chosen for the nodal displacements of the beam element. The zero terms appearing in this matrix are in agreement with the absence of coupling between bending and axial motions at the local scale of a finite element.

3.4.3.3 Assembling the finite element model of the whole structure

A coordinate transformation from the local reference frames $O'x'y'z'$ to the global reference frame *Oxyz* is necessary for assembling the $[K]$ matrix of the

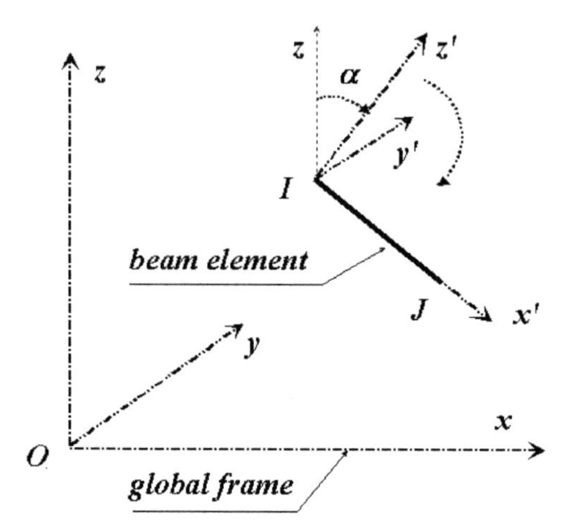

Figure 3.16. *Local and global coordinate systems*

structure since the local reference frames differ necessarily along the horizontal beam and along the legs. It is precisely this transformation which produces the coupling terms between bending and axial terms which arise in the $[K]$ matrix. Let us consider the beam element of Figure 3.16, which is assumed to lie in the Oxz plane. The local frame can be transformed into the global frame by the product of the translation $\overrightarrow{IO}$ and the plane rotation α. Neither the forces nor the displacements are modified by a translation, which thus can be disregarded. Obviously, the same is not true as far as the rotation is concerned. The rotation is described by the matrix $[R]$ which relates the displacements by $[q] = [R][q']$.

The transformation rule of the node labelled J is:

$$\begin{bmatrix} X_J \\ Y_J \\ Z_J \\ \psi_{xJ} \\ \psi_{yJ} \\ \psi_{zJ} \end{bmatrix} = \begin{bmatrix} \cos\alpha & 0 & \sin\alpha & 0 & 0 & 0 \\ 0 & 1 & 0 & 0 & 0 & 0 \\ -\sin\alpha & 0 & \cos\alpha & 0 & 0 & 0 \\ 0 & 0 & 0 & \cos\alpha & 0 & \sin\alpha \\ 0 & 0 & 0 & 0 & 1 & 0 \\ 0 & 0 & 0 & -\sin\alpha & 0 & \cos\alpha \end{bmatrix} \begin{bmatrix} X'_J \\ Y'_J \\ Z'_J \\ \psi'_{xJ} \\ \psi'_{yJ} \\ \psi'_{zJ} \end{bmatrix} \qquad [3.110]$$

More generally the rotation matrix takes the form:

$$[\mathfrak{R}] = \begin{bmatrix} [R] & [0] \\ [0] & [R] \end{bmatrix}; \quad \text{with } [R] = \begin{bmatrix} \ell_{x'} & m_{x'} & n_{x'} \\ \ell_{y'} & m_{y'} & n_{y'} \\ \ell_{z'} & m_{z'} & n_{z'} \end{bmatrix} \qquad [3.111]$$

where $\ell_{i'}, m_{i'}, n_{i'}$ are the director cosines of the local axes as expressed in the global system.

Since energy is invariant through a transformation of axes, the element matrices and force vectors are found to comply with the following conditions:

$$[q_I \quad q_J][K^{(n)}]\begin{bmatrix} q_I \\ q_J \end{bmatrix} = [q'_I \quad q'_J][K'^{(n)}]\begin{bmatrix} q'_I \\ q'_J \end{bmatrix} \quad \Rightarrow \quad [K^{(n)}] = [\mathfrak{R}]^T[K'^{(n)}][\mathfrak{R}]$$

$$[\dot{q}_I \quad \dot{q}_J][M^{(n)}]\begin{bmatrix} \dot{q}_I \\ \dot{q}_J \end{bmatrix} = [\dot{q}'_I \quad \dot{q}'_J][M'^{(n)}]\begin{bmatrix} \dot{q}'_I \\ \dot{q}'_J \end{bmatrix} \quad \Rightarrow \quad [M^{(n)}] = [\mathfrak{R}]^T[M'^{(n)}][\mathfrak{R}]$$

$$[q_I \quad q_J]\begin{bmatrix} Q_I \\ Q_J \end{bmatrix} = [q'_I \quad q'_J]\begin{bmatrix} Q'_I \\ Q'_J \end{bmatrix} \quad \Rightarrow \quad \begin{bmatrix} Q_I \\ Q_J \end{bmatrix} = [\mathfrak{R}]^T\begin{bmatrix} Q'_I \\ Q'_J \end{bmatrix}$$

[3.112]

Once the element matrices and vectors are suitably transformed into the global system of coordinates, the finite element model can be assembled according to the same rules as described in subsection 3.4.2.5.

Colour Plate 3 shows the elastic response of a portal frame to a concentrated force as computed by using the finite element software CASTEM 2000 [CAS 92]. The finite element model uses two nodes beam elements. The material is steel and the beams have uniform rectangular cross-sections 20 cm × 10 cm. The height of the vertical legs is 8 m and the length of the horizontal beam (lintel) is 12 m. The legs are assumed to be clamped at the lower ends. A concentrated force is applied at some point of the lintel, load magnitude is 2.5 kN. which is assumed successively to be vertical (case a), along the horizontal beam (case b), and finally in the transverse Oy direction (case c). For each loading case, two figures are shown. The figures on the left-hand side show the global reference frame in green, the finite element mesh in black, the external force represented as a red arrow and finally the support reactions as blue arrows for the forces and magenta arrows for the moments. The length of the arrows is determined by a magnifying factor 1000. The figures on the right-hand side show the undeformed and the deformed frame (the black and red lines respectively). Real displacements are magnified by a factor 100 to make deformations clearly noticeable.

In case (a), the lintel and the legs are deflected. As expected, maximum deflection occurs along the lintel. The reaction forces comprise a vertical component equal to the shear forces at the ends of the lintel, transmitted to the legs as compressive forces, and horizontal components in the plane of the frame, which are equal to the shear forces at the clamped ends of the legs. The moments are perpendicular to the plane of the frame. They are clearly related to the bending of the legs. As the loaded point is nearer to one leg than the other, the two legs are not flexed by the same amount, but in the ratio of the lengths of the lever arms defined by the loaded point and the extremities of the lintel.

In case (b), maximum deflection occurs in the legs, which are deflected by the same amount. The bending moments at the junction between the legs and the lintel induce in-plane bending of the latter. Deflection is found to be antisymmetric because the moments exerted at the ends of the lintel are the same. As a consequence, one leg is compressed and the other is stretched.

In case (c), the legs and the lintel are deflected out of the frame plane. However, due to the asymmetrical position of the loaded point with respect to the two legs, the portal frame is also twisted about the vertical direction. Accordingly, though the major component of the moment reactions is horizontal and, in the plane of the frame, equal to half the load times the length of the legs, they also have a vertical component which differs from one leg to the other and a horizontal component in the out-of-plane direction.

3.4.3.4 Transverse load resisted by string and bending stresses in a roof truss

The example of the portal frame was used in the last subsection to bring out that in an assembly of connected straight beams transverse shear stresses can be transmitted as longitudinal stresses to another part of the assembly due to geometrical effects. This leads also to a coupling of the basic modes of deformation of the individual beams. However, because of the large difference which usually exist between the coefficients of the longitudinal and flexure stiffness matrices [3.98] and [3.106], suitably scaled by the coefficients $K_\ell = ES/\ell$ and $K_b = EI/\ell^3$ respectively, coupling is barely detectable when looking at the deformed portal frame. Indeed, the axial displacements of the vertical legs are found to be quite negligible in comparison with the transverse deflection due to bending. To analyse this point further, it is instructive to study the assembly depicted in Figure 3.17 which stands

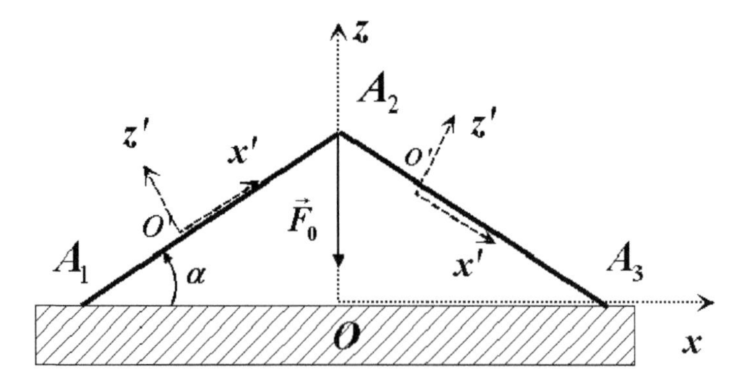

Figure 3.17. *Beam assembly used as roof truss*

for a roof truss made of two identical straight beams (A_1A_2) and (A_2A_3) tilted by the angle α with respect to the horizontal Ox direction. The beams are rigidly connected to each other at (A_2) and the rooting points (A_1) and (A_3) are assumed to be either pinned or clamped. A vertical force $\vec{F}_0$ is applied to (A_2). If $\alpha = 0$, the assembly responds to the load as a pinned-pinned, or clamped-clamped flexed beam and if $\alpha = 90°$, it responds as a compressed beam. It is thus of interest to analyse the response of the roof truss when α is varied from 0 to 90°. An analytical solution of this problem can be obtained conveniently by using a finite element model reduced to two beam elements and to the nodal displacements of interest.

Let us consider first the pinned-pinned configuration. In the local coordinate system of the first element (A_1A_2), the displacement variables are denoted U_1, U_2 (axial displacements), W_1, W_2 (transverse displacements in the Oxz plane), φ_1, φ_2 (small rotations about the Oy axis). These variables define the vector displacement $[U_1 \quad W_1 \quad \varphi_1 \quad U_2 \quad W_2 \quad \varphi_2]^T$. By using the matrix [3.109], the stiffness matrix of the first element is written as:

$$K_\ell \begin{bmatrix} 1 & 0 & 0 & -1 & 0 & 0 \\ 0 & 12\gamma & -6\gamma\ell & 0 & -12\gamma & -6\gamma\ell \\ 0 & -6\gamma\ell & 4\gamma\ell^2 & 0 & 6\gamma\ell & 2\gamma\ell^2 \\ -1 & 0 & 0 & 1 & 0 & 0 \\ 0 & -12\gamma & 6\gamma\ell & 0 & 12\gamma & 6\gamma\ell \\ 0 & -6\gamma\ell & 2\gamma\ell^2 & 0 & 6\gamma\ell & 4\gamma\ell^2 \end{bmatrix}$$

$$\text{where} \quad \gamma = \frac{K_b}{K_\ell} = \frac{I}{S\ell^2} \simeq \frac{1}{12\eta^2} \ll 1$$

η is the slenderness ratio of the element, assumed to be much larger than one.

In the same way, the vector displacement of the second element (A_2A_3) is written as:

$$\begin{bmatrix} U_2 & W_2 & \varphi_2 & U_3 & W_3 & \varphi_3 \end{bmatrix}^T$$

The stiffness matrix is the same as that of the first element.

However, due to the boundary conditions, $U_1 = U_3 = W_1 = W_3 = 0$ and the displacement vectors and stiffness matrices can be reduced as follows:

$$\begin{bmatrix} U_1 & W_1 & \varphi_1 & U_2 & W_2 & \varphi_2 \end{bmatrix}^T \Rightarrow \begin{bmatrix} \varphi_1 & U_2 & W_2 & \varphi_2 \end{bmatrix}^T$$

$$K_\ell \begin{bmatrix} 1 & 0 & 0 & -1 & 0 & 0 \\ 0 & 12\gamma & -6\gamma\ell & 0 & -12\gamma & -6\gamma\ell \\ 0 & -6\gamma\ell & 4\gamma\ell^2 & 0 & 6\gamma\ell & 2\gamma\ell^2 \\ -1 & 0 & 0 & 1 & 0 & 0 \\ 0 & -12\gamma & 6\gamma\ell & 0 & 12\gamma & 6\gamma\ell \\ 0 & -6\gamma\ell & 2\gamma\ell^2 & 0 & 6\gamma\ell & 4\gamma\ell^2 \end{bmatrix}$$

$$\Rightarrow K^{(1)} = K_\ell \begin{bmatrix} 4\gamma\ell^2 & 0 & 6\gamma\ell & 2\gamma\ell^2 \\ 0 & 1 & 0 & 0 \\ 6\gamma\ell & 0 & 12\gamma & 6\gamma\ell \\ 2\gamma\ell^2 & 0 & 6\gamma\ell & 4\gamma\ell^2 \end{bmatrix}$$

$$\begin{bmatrix} U_2 & W_2 & \varphi_2 & U_3 & W_3 & \varphi_3 \end{bmatrix}^T \Rightarrow \begin{bmatrix} U_2 & W_2 & \varphi_2 & \varphi_3 \end{bmatrix}^T$$

$$K_\ell \begin{bmatrix} 1 & 0 & 0 & -1 & 0 & 0 \\ 0 & 12\gamma & -6\gamma\ell & 0 & -12\gamma & -6\gamma\ell \\ 0 & -6\gamma\ell & 4\gamma\ell^2 & 0 & 6\gamma\ell & 2\gamma\ell^2 \\ -1 & 0 & 0 & 1 & 0 & 0 \\ 0 & -12\gamma & 6\gamma\ell & 0 & 12\gamma & 6\gamma\ell \\ 0 & -6\gamma\ell & 2\gamma\ell^2 & 0 & 6\gamma\ell & 4\gamma\ell^2 \end{bmatrix}$$

$$\Rightarrow K^{(2)} = K_\ell \begin{bmatrix} 1 & 0 & 0 & 0 \\ 0 & 12\gamma & -6\gamma\ell & -6\gamma\ell \\ 0 & -6\gamma\ell & 12\gamma & 2\gamma\ell^2 \\ 0 & -6\gamma\ell & 2\gamma\ell^2 & 4\gamma\ell^2 \end{bmatrix}$$

Then, $K^{(1)}$ is rotated by the angle α by using the coordinate transformation rule:

$$\begin{bmatrix} \psi_1 \\ X_2 \\ Z_2 \\ \psi_2 \end{bmatrix} = \begin{bmatrix} 1 & 0 & 0 & 0 \\ 0 & C & -S & 0 \\ 0 & S & C & 0 \\ 0 & 0 & 0 & 1 \end{bmatrix} \begin{bmatrix} \varphi_1 \\ U_2 \\ W_2 \\ \varphi_2 \end{bmatrix}; \quad \text{where } C = \cos\alpha \text{ and } S = \sin\alpha$$

The rotated stiffness matrix is found to be:

$$K_\alpha^{(1)} = K_\ell \begin{bmatrix} 4\gamma\ell^2 & 6\gamma\ell S & 6\gamma\ell C & 2\gamma\ell^2 \\ 6\gamma\ell S & C^2 + 12\gamma S^2 & CS(12\gamma - 1) & 6\gamma\ell S \\ 6\gamma\ell C & CS(12\gamma - 1) & S^2 + 12\gamma C^2 & 6\gamma\ell C \\ 2\gamma\ell^2 & 6\gamma\ell S & 6\gamma\ell C & 4\gamma\ell^2 \end{bmatrix}$$

$K^{(2)}$ is rotated by the angle $-\alpha$ by using the coordinate transformation rule:

$$\begin{bmatrix} X_2 \\ Z_2 \\ \psi_2 \\ \psi_3 \end{bmatrix} = \begin{bmatrix} C & +S & 0 & 0 \\ -S & C & 0 & 0 \\ 0 & 0 & 1 & 0 \\ 0 & 0 & 0 & 1 \end{bmatrix} \begin{bmatrix} U_2 \\ W_2 \\ \varphi_2 \\ \varphi_3 \end{bmatrix}$$

The rotated stiffness matrix is found to be:

$$K_\alpha^{(2)} = K_\ell \begin{bmatrix} C^2 + 12\gamma S^2 & -CS(12\gamma - 1) & 6\gamma\ell S & 6\gamma\ell S \\ -CS(12\gamma - 1) & S^2 + 12\gamma C^2 & -6\gamma\ell C & -6\gamma\ell C \\ 6\gamma\ell S & -6\gamma\ell C & 4\gamma\ell^2 & 2\gamma\ell^2 \\ 6\gamma\ell S & -6\gamma\ell C & 2\gamma\ell^2 & 4\gamma\ell^2 \end{bmatrix}$$

The finite element model of the problem is assembled as:

$$\begin{bmatrix} 4\gamma\ell^2 & 6\gamma\ell S & 6\gamma\ell C & 2\gamma\ell^2 & 0 \\ 6\gamma\ell S & 2(C^2 + 12\gamma S^2) & 0 & 12\gamma\ell S & 6\gamma\ell S \\ 6\gamma\ell C & 0 & 2(S^2 + 12\gamma C^2) & 0 & -6\gamma\ell C \\ 2\gamma\ell^2 & 12\gamma\ell S & 0 & 8\gamma\ell^2 & 2\gamma\ell^2 \\ 0 & 6\gamma\ell S & -6\gamma\ell C & 2\gamma\ell^2 & 4\gamma\ell^2 \end{bmatrix} \begin{bmatrix} \psi_1 \\ X_2 \\ Z_2 \\ \psi_2 \\ \psi_3 \end{bmatrix} = \begin{bmatrix} 0 \\ 0 \\ -F \\ 0 \\ 0 \end{bmatrix}$$

Finally, the solution can be conveniently obtained by noticing that because of the symmetry of the problem with respect to the Oz axis, it is necessary that $X_2 = 0$ and $\psi_3 = -\psi_1$. As a consequence:

$$Z_2(\alpha) = \frac{-F}{2(S^2 + 3\gamma C^2)K_\ell} = \frac{-F}{2(K_\ell(\sin\alpha)^2 + 3K_b(\cos\alpha)^2)} \qquad [3.113]$$

and

$$\psi_3 = -\psi_1 = \frac{3Z_2 \cos\alpha}{2\ell}$$

The solution [3.113] brings out that the structure resists the vertical force essentially through its longitudinal stiffness as soon as the tilt angle is larger than $1/\eta$. Such a result is of paramount importance for designing structures resistant

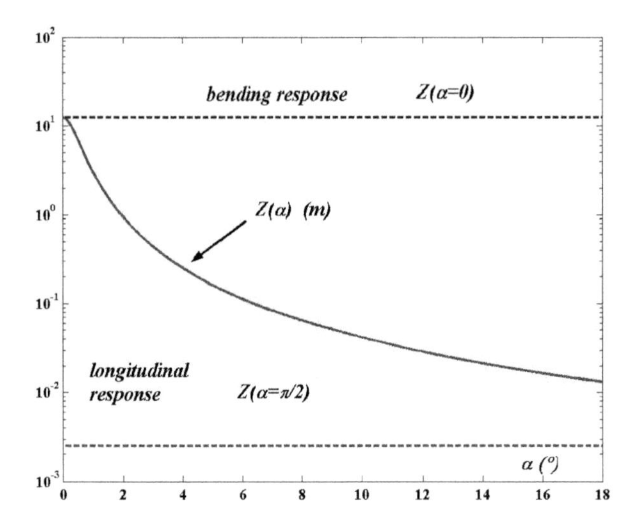

Figure 3.18. *Deflection versus the tilt angle of the roof truss*

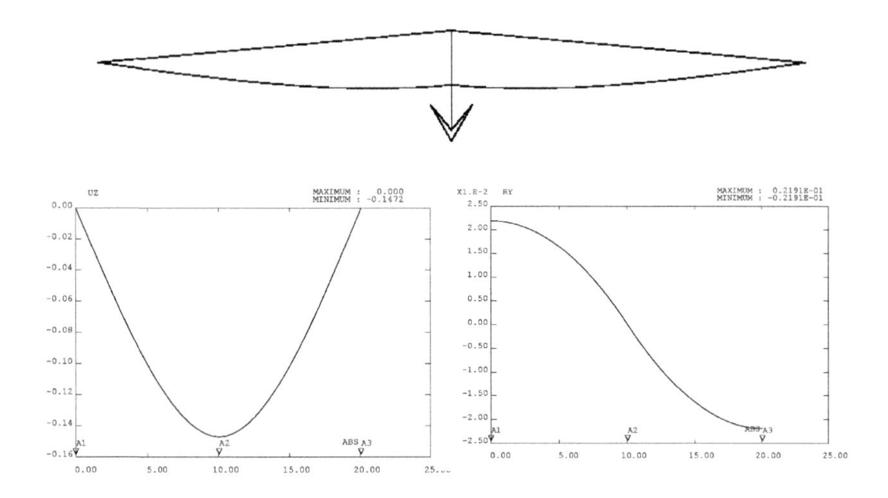

Figure 3.19. *Roof truss in pinned-pinned configuration, from the above to the bottom plots: unloaded and loaded structure, plots of the vertical displacement in meters and flexure angle in radians*

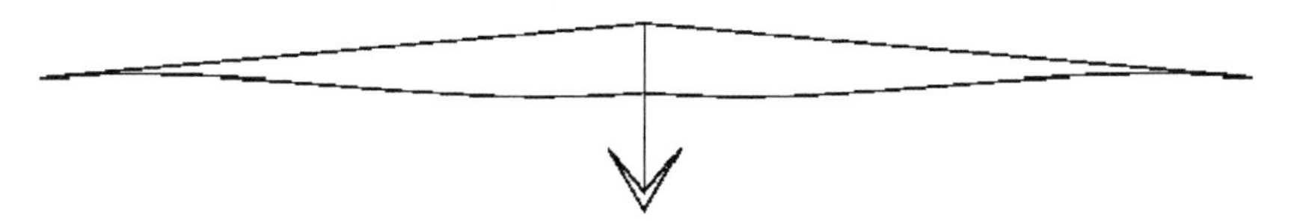

Figure 3.20. *Roof truss in the clamped-clamped configuration: unloaded and loaded structure*

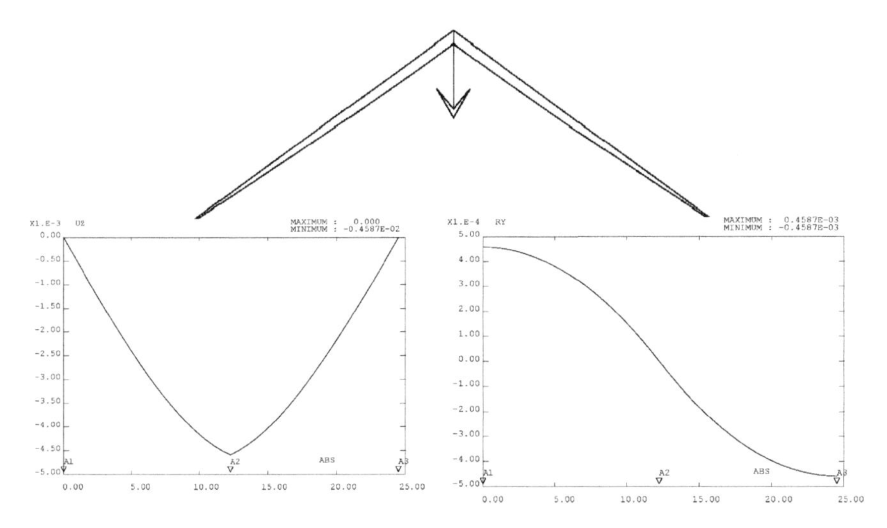

Figure 3.21. *Roof truss $\alpha = 45°$, in pinned-pinned configuration: unloaded and loaded structure, plots of the vertical displacement in meters and flexure angle in radians*

to transverse loads, as will be further detailed in Chapters 7 and 8 devoted to curved structures. Figure 3.18 refers to a roof truss made of wood beams ($E = 10^9$ Pa, $\ell = 10$ m, square cross-sections $a = 20$ cm) loaded by a vertical force $F = 10$ kN applied to (A_2). It plots the vertical displacement of (A_2) versus the tilt angle in degree. The dashed lines stand for the limit cases $\alpha = 0$ and $\alpha = 90°$. As soon as $\alpha > 5°$ magnitude of the displacement is less than 1% than its value at $\alpha = 0$.

Figure 3.19 refers to the pinned-pinned configuration. It shows the deflection of the whole structure for $\alpha = 5°$, as computed by using a finite element model comprising 40 two node beam elements. The load vector and the non-deformed structure are also represented. Vertical displacement and flexure angle are plotted along the beams.

Figure 3.20 refers to the clamped-clamped configuration. The magnitude of the maximum deflection is less than in the pinned-pinned configuration (10.9 cm instead of 14.7 cm) as the result of a stiffening effect in bending. A simple analytical calculation carried out in the same manner as above gives for the clamped-clamped configuration:

$$Z_2(\alpha) = \frac{-F}{2(K_\ell(\sin\alpha)^2 + 12K_b(\cos\alpha)^2)}$$

and

$$\psi_1 = \psi_2 = \psi_3 = 0$$

Finally, Figure 3.21 refers to a roof truss $\alpha = 45°$, in pinned-pinned configuration. The deflection is drastically reduced in comparison with the case $\alpha = 5°$.

3.4.4 *Saving DOF when modelling deformable solids*

The methodology followed to model deformable solids as equivalent beams is worth brief mention to emphasize the practical importance of minimizing judiciously the number of degrees of freedom when modelling continuous systems. This can be sketched graphically, as shown in Figure 3.22. Figure 3.22a sketches the 'real' structure as a 3D solid, supporting a load distributed as a 2D pressure field. In Figure 3.22b we take advantage of the slenderness of the body to reduce the size of the problem by shifting from a 3D medium to an equivalent 1D medium, via a beam model. Finally, in Figure 3.22c the beam is discretized in finite elements. In this way, the final model to be solved has a finite number of DOF

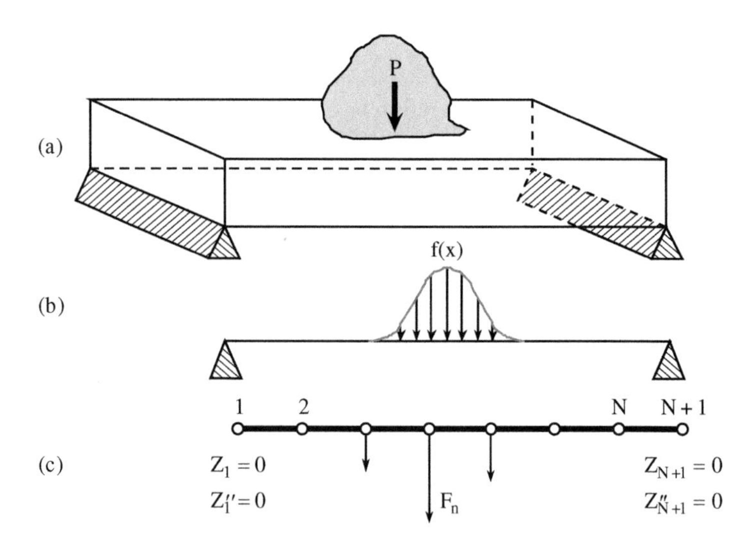

Figure 3.22. *From the deformable solid to the finite element model*

and can be solved by using a computer. The whole process justifies the interest in discretizing a continuum; one other specificity of this modelling procedure is its ability to identify the most pertinent DOF in order to minimize their number. This step is partially fulfilled by a good appreciation of the structure behaviour. In the following chapter another efficient method for suppressing superfluous degrees of freedom is described, based on the natural modes of vibration of the structures.

Chapter 4

Vibration modes of straight beams and modal analysis methods

As already outlined in Chapter 1, modes of vibration arise as a natural concept in the study of the free vibrations of mechanical systems, discrete or continuous. In the continuous case, they are defined mathematically as many solutions of an eigenvalue problem involving the stiffness and mass operators introduced in Chapter 3. Provided the system is self-adjoint and statically stable, the discrete and infinite sequence of eigenvalues are positive and the related eigenvectors are real. From a physical viewpoint, they describe standing waves in which the material of the structure vibrates harmonically about a static and stable position of equilibrium, at specific frequencies and according to specific space shapes. Amongst several interesting properties, the most important one certainly is that the mode shapes can be used as an orthonormal vector basis to transform the partial derivative equations of motion into a set of time differential equations described by modal stiffness and mass matrices which operate on the so-called natural, or modal, coordinates of the material system. This new discretization procedure gives rise to a so called modal model, in which the response properties of the structure are characterized by a set of modal oscillators, instead of a set of finite elements. Many linear and even nonlinear problems of continuous mechanics are much more efficiently solved starting from a modal model, instead of a finite element model.

4.1. Introduction

From the mathematical viewpoint, the natural modes of vibration of elastic structures arise as the solutions of an eigenvalue problem written in the canonical form as:

$$\left[K(\vec{r}) - \omega^2 M(\vec{r})\right]\vec{\varphi}(\vec{r}) = \vec{0}$$
$$+ \text{ self-adjoint homogeneous boundary conditions} \qquad [4.1]$$

where $K(\vec{r})$ and $M(\vec{r})$ are the linear stiffness and mass operators of the material system.

The differential system [4.1] is the continuous counterpart of the algebraic modal equation encountered in the analysis of discrete systems. It can be demonstrated that, when provided with a suitable set of self-adjoint boundary conditions, it has an infinite sequence of nontrivial solutions called *natural, or normal, modes of vibration*, which can be used to determine an orthonormal basis with respect to the stiffness and mass operators of the associated Hilbert space. These infinitely numerous solutions are enumerated by using one, two or three integer indices, in agreement with the dimension of the Euclidean space of the structural model. Considering here the 1D case of beams, the eigenvalues are denoted $\omega_n^2, n = 1, 2, \ldots$ and the related eigenvectors are denoted $\varphi_n(\vec{r})$. Furthermore, the ω_n are positive or eventually nil in the absence of supports, and the $\varphi_n(\vec{r})$ are real vectors. The relations of orthogonality of mode shapes are formulated as follows:

$$\left\langle \vec{\varphi}_i(\vec{r}), M[\vec{\varphi}_j(\vec{r})]\right\rangle_{(\mathcal{V})} = \begin{cases} M_n > 0; & \text{if } i = j = n \\ 0; & \text{if } i \neq j \end{cases}$$
$$\left\langle \vec{\varphi}_i(\vec{r}), K[\vec{\varphi}_j(\vec{r})]\right\rangle_{(\mathcal{V})} = \begin{cases} K_n \geq 0; & \text{if } i = j = n \\ 0; & \text{if } i \neq j \end{cases} \qquad [4.2]$$

These mathematical properties, of paramount importance from both the theoretical and the practical standpoints, extend to the continuous case those already established in the modal analysis of discrete systems (see [AXI 04], Chapters 5 and 6). The only new feature is that n extends to infinity. A formal proof of such results requires however rather advanced mathematical developments in the spectral theory of partial derivative operators which are beyond the scope of this book. The reader interested in the subject is referred to mathematically oriented textbooks for instance [COL 63], [STA 70].

The vibration equations of straight beams are simple enough to provide us with convenient structural examples in order to study the methods of modal analysis in the case of continuous systems. In particular, it will be shown that many problems can be efficiently solved by projecting first the local equations of motion on a suitable modal basis. Using this procedure, the problem is restated in terms of discrete, but still infinitely many unknowns, termed *modal displacements* or *normal coordinates*, which can be understood as a particularly convenient set of generalized displacements. At this step, the remaining problem is to truncate judiciously the modal series to a finite number of terms. It turns out that, in many applications, the dimension of the functional space can be efficiently reduced to a few DOF only, by truncating the modal series in accordance with the physical peculiarities of the specific problem to be solved.

Section 4.2 is concerned with the natural modes of vibration of straight beams. In section 4.3, the basic principles of mode shape expansion methods are introduced and criteria to truncate the modal series suitably are established. As could be anticipated, such criteria are based on the spatial and spectral features of the excitation signals in relation to those of the natural modes of vibration. Finally, the rate of convergence of modal series related to displacements and stresses are briefly discussed. All these notions are illustrated by means of a few examples.

In section 4.4 the so called *substructuring method* is introduced, according to which a complex structure may be partitioned into simpler substructures, each substructure being finally discretized by using a truncated modal basis. Finally, in section 4.5 a few impacting beam problems are worked out to demonstrate that the modal analysis method can be also recommended for dealing with nonlinear problems, especially when nonlinearities are concentrated in a few specific places in the physical system. In the following chapters other examples concerning arches, plates and shells will be given.

4.2. Natural modes of vibration of straight beams

4.2.1 *Travelling waves of simplified models*

The nature and the speed of the elastic waves propagating in a solid body depend on the kinematic assumptions made to model it. This arises as a consequence of the necessity to specify, at least implicitly, the boundary conditions which hold in the directions not considered in the simplified model, for instance at the lateral surface bounding a beam, at the faces of a plate, etc.

4.2.1.1 Longitudinal waves

For a longitudinal wave, the equilibrium equation of a straight uniform beam in which the Poisson effect is neglected is:

$$-E\frac{\partial^2 X}{\partial x^2} + \rho \ddot{X} = 0 \qquad [4.3]$$

From the calculation made in Chapter 1, subsection 1.4.3, the phase speed of the longitudinal wave is readily found to be:

$$c_\varphi = c_0 = \sqrt{E/\rho} \qquad [4.4]$$

As this velocity is independent of the pulsation, the model predicts non-dispersive waves which are the one-dimensional counterpart of the plane dilatation waves propagating in a three-dimensional elastic solid. The speed given by [4.4] is however less than that given by [1.82], in the ratio:

$$\frac{c_0}{c_L} = \sqrt{\frac{(1+\nu)(1-2\nu)}{(1-\nu)}} \qquad [4.5]$$

For instance, in the case of steel $\nu = 0.3$ and $c_0 \cong 0.862 c_L$. At first sight, such a result is somewhat surprising as it could be expected that the stiffness arising from the beam model would be higher than that arising from the corresponding 3D solid body; this because of the rigidifying assumption made in the beam model that the cross-sections are not deformed. Actually it turns out that, to the contrary, by neglecting the Poisson effect, the longitudinal stiffness of the beam is decreased. This is because no transverse stresses arise in the 1D model to oppose the longitudinal deformation, as is the case in the 3D solid. Indeed, according to the 3D elastic law [1.19] we get:

$$(\sigma_{xx})_{3D} = \frac{(1-\nu)E}{(1+\nu)(1-2\nu)}\frac{\partial X}{\partial x}; \quad (\sigma_{yy})_{3D} = (\sigma_{yy})_{3D} = \frac{\nu E}{(1+\nu)(1-2\nu)}\frac{\partial X}{\partial x}$$

whereas according to the 1D model:

$$(\sigma_{xx})_{1D} = E\frac{\partial X}{\partial x}; \quad (\sigma_{yy})_{1D} = (\sigma_{yy})_{1D} \equiv 0$$

So, in a rigorous analysis of the problem, it would be necessary to use a 3D elastic law and to prescribe vanishing local stresses at the lateral surface of the beam, in accordance with the kinematic conditions which prevail in an axially guided wave. As pointed out in Chapter 1 subsection 1.4.6, this difficult problem has been solved by Pochammer in the particular case of a circular cylindrical rod. The dispersion equation thus obtained defines multiple distinct branches $k_n(\omega)$ and only approximations to the first lower branches can be recovered by adopting simplified beam models. Fortunately, such approximations are found to be sufficiently accurate in most engineering applications.

On the other hand, it has been outlined in Chapter 3 subsection 3.2.1 that the basic beam model can be improved by taking into account the inertia related to the lateral deformations of the beam cross-sections due to the Poisson effect. This additional inertia becomes significant if the wavelength is not very large in comparison with the transverse dimensions of the beam. This leads to the Love–Rayleigh equation [3.11], written here in the case of a cylindrical rod as:

$$c_0^2 \frac{\partial^2 X}{\partial x^2} - \ddot{X} + \beta^2 \frac{\partial^2 \ddot{X}}{\partial x^2} = 0; \quad \text{where } \beta^2 = \frac{\nu^2 R^2}{2}$$

The phase speed of the Love–Rayleigh model, as expressed in terms of the wave number k, is found to be:

$$\frac{c}{c_0} = \sqrt{\frac{1}{1 + \frac{1}{2}(k\nu R)^2}}; \quad \text{where } k = \frac{2\pi}{\lambda} \qquad [4.6]$$

These waves are thus found to be dispersive and c decreases as kR increases, tending to $c_0 = \sqrt{E/\rho}$ when kR decreases, as shown in Figure 4.1, where λ/R is the reduced wavelength. As expected, the phase speed is less than that of the basic model because the Poisson ratio effect is accounted for in the inertia term only. Furthermore, as expressed in terms of the pulsation, the phase speed is found to be:

$$\frac{c}{c_0} = \sqrt{1 - \frac{\omega^2 \beta^2}{c_0^2}} \qquad [4.7]$$

According to [4.7], there is a cut-off frequency $f_c = c_0/2\pi\beta$ such that only the frequencies less than f_c can propagate.

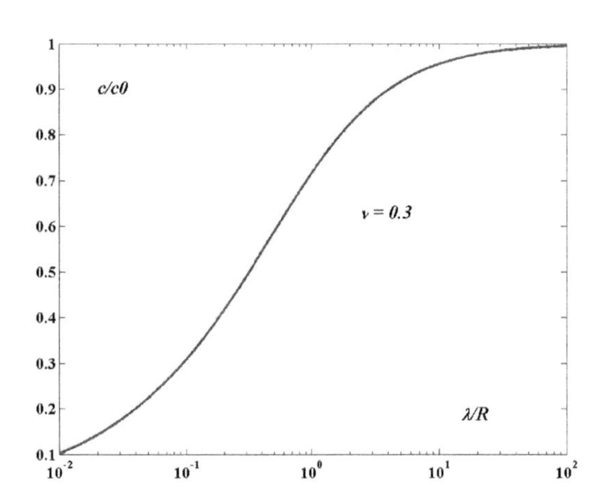

Figure 4.1. *Phase speed of the longitudinal waves according to the Love–Rayleigh model*

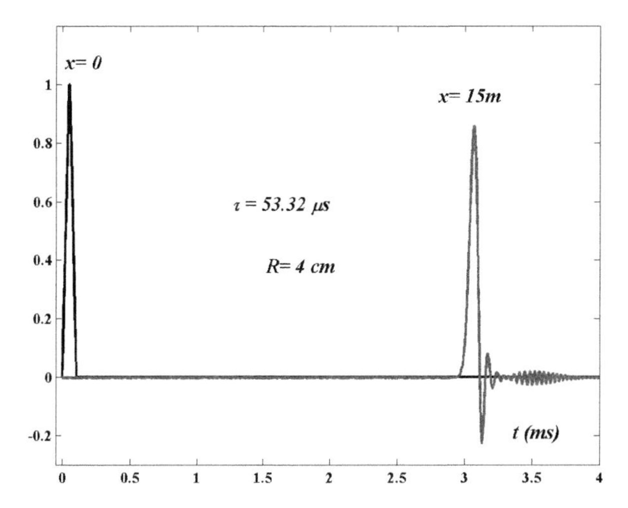

Figure 4.2. *Propagation of a longitudinal triangular wave according to the Love–Rayleigh model*

Figure 4.2 refers to the propagation of a longitudinal triangular wave of half-duration $\tau = 53.32\ \mu\text{s}$ along a cylindrical steel rod ($R = 4$ cm) according to the Love–Rayleigh model. The cut-off frequency is $f_c = 94$ kHz, whereas the maximum frequency contained in the triangular signal is $f_{\max} = 1/\tau = 19$ kHz. The travelling wave is computed by using the fast Fourier transform algorithm implemented in MATLAB. The wave at 15 m from the source is marked by a triangular peak followed by high frequency oscillations of much less magnitude.

4.2.1.2 Flexure waves

Starting from the Bernoulli–Euler model for a uniform beam:

$$EI \frac{\partial^4 Z}{\partial x^4} + \rho S \ddot{Z} = 0$$

The phase speed of the flexure waves is found to be:

$$c_\psi = k\sqrt{EI/\rho S} = (EI/\rho S)^{1/4}(\omega)^{1/2} \qquad [4.8]$$

These waves are highly dispersive. Furthermore, c_ψ is found to increase without limit with frequency. Such a behaviour is clearly unrealistic and simply indicates that the Bernoulli–Euler model is unsuitable at short wavelengths $\lambda/R \leq 1$, as already discussed in Chapter 3, section 3.2.2. When the wavelength is sufficiently small, contribution of shear deformation and rotatory inertia of the beam cross-sections must be accounted for. This is the object of the Rayleigh–Timoshenko model [3.35]:

$$\rho S \frac{\partial^2 Z}{\partial t^2} - \rho I \left(1 + \frac{E}{\kappa G}\right) \frac{\partial^4 Z}{\partial x^2 \partial t^2} + \frac{\rho^2 I}{\kappa G} \frac{\partial^4 Z}{\partial t^4} + EI \frac{\partial^4 Z}{\partial x^4} = 0$$

The related dispersion equation produces the two following branches for the phase velocity:

$$c_{sb} = c_s \sqrt{1 + \frac{3+2\nu}{4}(kR)^2 + \sqrt{\left(1 + \frac{3+2\nu}{4}(kR)^2\right)^2 - \frac{(1+\nu)(kR)^4}{2}}}$$

$$c_{bs} = c_s \sqrt{1 + \frac{3+2\nu}{4}(kR)^2 - \sqrt{\left(1 + \frac{3+2\nu}{4}(kR)^2\right)^2 - \frac{(1+\nu)(kR)^4}{2}}};$$

$$\text{where } c_s = \sqrt{\frac{G}{\rho}} \qquad [4.9]$$

In Figure 4.3, they are plotted versus the reduced wavelength λ/R. The lower branch identifies with the Bernoulli–Euler branch if λ/R is large enough, as shown in Figure 4.4. On the other hand, if λ/R is small enough the shear effect prevails, so the phase speed tends to the shear velocity c_s. As a consequence, the dispersive nature of the waves vanishes asymptotically in the range of short wavelengths.

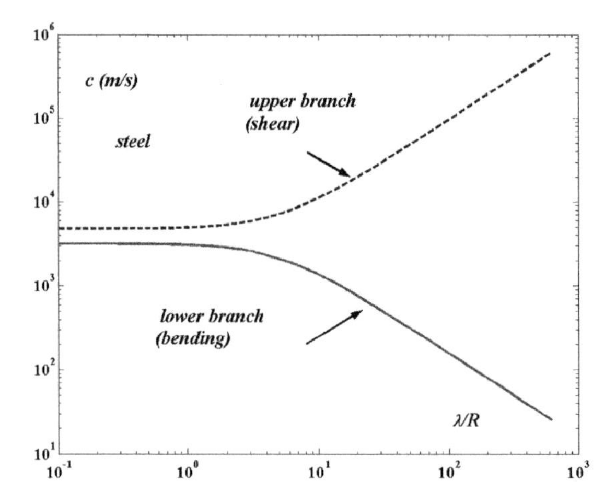

Figure 4.3. *Branches of the dispersion equation of the Rayleigh–Timoshenko model*

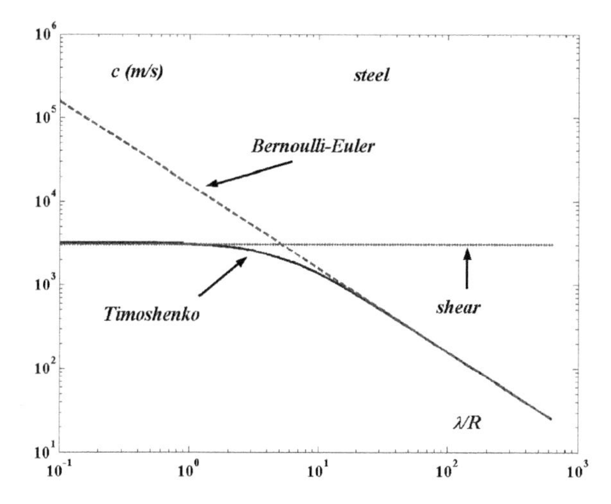

Figure 4.4. *Phase velocity of the bending-shear coupled waves*

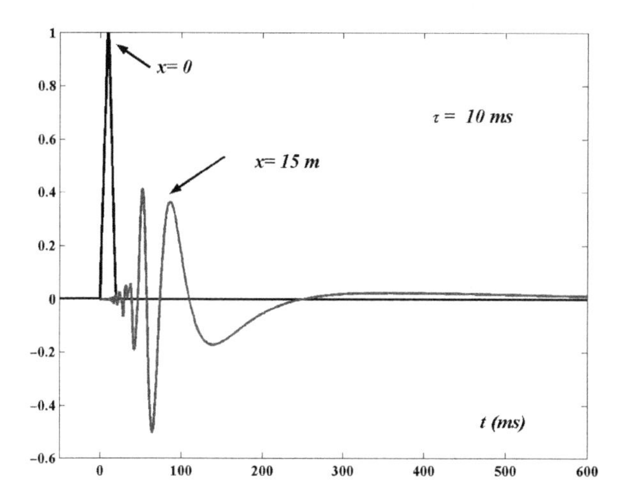

Figure 4.5. *Propagation of a transverse triangular wave according to the Rayleigh–Timoshenko model*

Figure 4.5 refers to the propagation of a transverse triangular wave of half-duration $\tau = 10$ ms along a cylindrical steel rod ($R = 4$ cm) according to the Rayleigh–Timoshenko model. The wave at 15 m from the source is shaped as a train of oscillations, the frequency of which decreases markedly as time elapses, in agreement with the dispersive law [4.9]. Transverse waves are clearly much more dispersive than the longitudinal waves.

4.2.2 *Standing waves, or natural modes of vibration*

The natural modes vibration of a structure are determined according to the general procedure already introduced in Chapter 1 subsection 1.4.7, which is briefly summarized here as follows:

1. Find the general solution of the modal equation.
2. Comply with the boundary conditions by adjusting the values of the integration constants.

A linear and conservative boundary condition is expressed as a linear and homogeneous relation between a pair of conjugated displacement and stress variables. The constant values are specified, except a multiplicative arbitrary factor which can be set according to a given norm. Thus infinitely numerous natural frequencies and related mode shapes are produced. However, the problem can be solved analytically in closed form only in a few cases, in particular when a uniform beam is provided with standard support conditions at both ends. In what follows, the presentation will be restricted essentially to those cases. As a general result, it will be noted that high order modes are found to be rather insensitive to boundary conditions; this indicates that the ratio of the elastic strain energy of the whole beam over that induced by the elastic supports increases with the wave number of the mode.

4.2.2.1 Longitudinal modes

According to the basic model, the modal equation is:

$$\frac{d^2X}{dx^2} + \left(\frac{\omega}{c_0}\right)^2 X = 0; \quad \text{with } c_0 = \sqrt{E/\rho} \qquad [4.10]$$

The natural circular frequencies and related mode shapes are of the general form:

$$\omega_n = \frac{\varpi_n c_0}{L}; \quad \varphi_n(\xi) = a\sin(\varpi_n \xi) + b\cos(\varpi_n \xi); \quad \text{where } \xi = x/L \qquad [4.11]$$

The modal stiffness and mass coefficients are given by:

$$M_n = \frac{\rho SL}{2} = \frac{M}{2}; \quad K_n = \frac{1}{2}\frac{ES}{L}\varpi_n^2 \qquad [4.12]$$

The modal coefficients ϖ_n, a, b, pertinent to standard boundary conditions, are reported in Table 4.1. It is worth emphasizing the remarkable result that the natural

Table 4.1. *Longitudinal modes of vibration of a straight beam*

B.C.	ϖ_n		a	b
free-free	$n\pi$	$n = 0, 1, 2, \ldots$	0	1
fixed-fixed	$n\pi$	$n = 1, 2, \ldots$	1	0
free-fixed	$(1+2n)\frac{\pi}{2}$	$n = 0, 1, 2, \ldots$	0	1
fixed-free	$(1+2n)\frac{\pi}{2}$	$n = 0, 1, 2, \ldots$	1	0

frequencies form an harmonic sequence and that the mode shapes identify with the functional vectors used as an orthonormal basis in the Fourier series. As detailed further in section 4.3, mode shape expansion methods can be rightly understood as a natural extension of the concept of Fourier series.

EXAMPLE. – *fixed-free beam.*

The detailed calculation leading to the results gathered in Table 4.1 and formulae [4.11] is illustrated by taking the case of a uniform beam provided with fixed-free end conditions. Starting from the equation of motion:

$$-ES\frac{\partial^2 X}{\partial x^2} + \rho S\frac{\partial^2 X}{\partial t^2} = 0; \quad X(0,t) = ES\frac{\partial X}{\partial x}\bigg|_L = 0$$

Its general solution of the harmonic type is written as:

$$X(x,t) = X_+ e^{i\omega(t - x/c_0)} + X_- e^{i\omega(t + x/c_0)}$$

At $x = 0$, the incident wave X_- is reflected and the sign is changed to comply with the fixed end condition. The free end condition at $x = L$, gives the transcendental equation:

$$\cos(\omega L/c_0) = 0 \Rightarrow \omega_n = \frac{\varpi_n c_0}{L}; \quad n = 0, 1, 2, \ldots.$$

where $\varpi_n = (1 + 2n)\pi/2$.

The mode shapes are obtained by superposing the incident and reflected waves, which gives the standing wave:

$$\varphi_n(x) = \frac{1}{2i}\left(e^{i\varpi_n x/L} - e^{-i\varpi_n x/L}\right) = \sin\left(\frac{\varpi_n x}{L}\right)$$

where the condition adopted to normalize the mode shapes is $\max\left(|\varphi_n(x)|\right) = 1$.

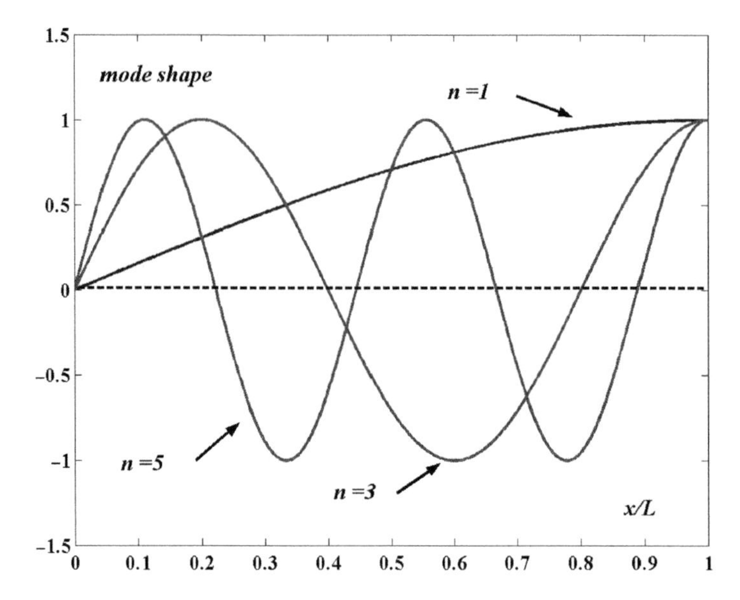

Figure 4.6. *Longitudinal mode shapes: fixed-free ends*

The modal wavelengths are equal to $\lambda_n = 4L/(1+2n)$.

Defining now the *vibration nodes* as the locations where the magnitude of displacement is zero and the *vibration antinodes* as the locations where it is maximum, it is readily verified that the first longitudinal mode of the fixed-free beam has a single node at $x = 0$ and a single antinode at $x = L$. The n-th mode has n nodes and n antinodes, as shown in Figure 4.6. More generally, the number of nodes and antinodes of the mode shapes is found to be proportional to n and the wavelength is inversely proportional to n. It is also easy to check that the modes are not affected in any way by providing the beam with additional supports located at the nodes. It is also interesting to note that by substituting ω_n into the boundary condition at $x = L$, it can be seen that the incident X_+ wave is reflected at the free end without change of sign:

$$\left.\frac{\partial X}{\partial x}\right|_0 = 0 \;\Rightarrow\; \frac{i\omega}{c_0}(X_+ - X_-) = 0$$

NOTE. – *Modal calculation using directly standing waves*

A slightly different manner of presenting the modal calculation follows even more closely the procedure described in [AXI 04]. Again, non-trivial harmonic solutions $\varphi(x)e^{i\omega t}$ of the equation [4.1] are looked for, which are governed by the modal equation:

$$\frac{d^2\varphi}{dx^2} + \left(\frac{\omega}{c_0}\right)^2 \varphi = 0; \quad \varphi(0) = \left.\frac{d\varphi}{\partial x}\right|_L = 0$$

However, instead of writing down the general solution as a linear superposition of travelling waves, it is directly expressed as a superposition of sine and cosine standing waves:

$$\varphi(x) = a \sin\left(\frac{\omega x}{c_0}\right) + b \cos\left(\frac{\omega x}{c_0}\right)$$

Finally, it is easy to check the orthogonality rule [4.2] of the mode shapes. Here it reduces to the ordinary orthogonality condition between functional vectors, since the stiffness and mass operators of a uniform beam are uniform:

$$\rho SL \int_0^1 \sin(\varpi_j \xi) \sin(\varpi_k \xi)\, d\xi = \begin{cases} \rho SL/2 = M/2; & \text{if } j = k \\ 0; & \text{otherwise} \end{cases}$$

$$ESL\varpi_j\varpi_k \int_0^1 \sin(\varpi_j \xi) \sin(\varpi_k \xi) \sin(\varpi_k \xi)\, d\xi = \begin{cases} \rho SL\varpi_j^2/2; & \text{if } j = k \\ 0; & \text{otherwise} \end{cases}$$

Turning now to the Rayleigh–Love model, the mode shapes are found to be the same as those related to the basic model, but the modal frequencies are lower, as could be expected. The modal system for a beam in the free-fixed support configuration is written in terms of dimensionless variables as follows:

$$\frac{d^2u}{d\xi^2} + \varpi^2\left(u - \gamma^2 \frac{d^2u}{d\xi^2}\right) = 0; \quad \left.\frac{du}{d\xi}\right|_0 = u(1) = 0$$

$$\text{where} \quad \varpi = \frac{\omega L}{c_0}; \quad \gamma^2 = \frac{1}{2}\left(\frac{\nu R}{L}\right)^2; \quad u = \frac{X}{L}; \quad \xi = \frac{x}{L}$$

$$u = Ae^{\mathfrak{k}\xi} + Be^{-\mathfrak{k}\xi} \quad \Rightarrow \quad \mathfrak{k}^2 + \varpi^2 - \mathfrak{k}^2\gamma^2\varpi^2 = 0 \quad \Rightarrow \quad \mathfrak{k} = \pm \frac{i\varpi}{\left(1 - \gamma^2\varpi^2\right)^{1/2}}$$

where $\mathfrak{k} = Lk$ is a dimensionless wave number.

The condition $\varphi(1) = 0$ leads to:

$$\mathfrak{k}_n = i(2n+1)\pi/2 \Rightarrow \varphi_n(\xi) = \cos(|\mathfrak{k}_n|\xi)\,; \; \varpi_n^2 = \frac{|\mathfrak{k}_n|^2}{1+|\mathfrak{k}_n|^2\gamma^2}$$

In accordance with the variation of the phase speed displayed in Figure 4.1, the modal frequencies increase nearly as an harmonic sequence only if the dimensionless number $|\mathfrak{k}_n\gamma|$ is small enough, which means if the modal wavelengths are much larger than the transverse dimensions of the beam.

4.2.2.2 *Torsion modes*

Torsion modes are discussed starting from equation [2.41], where the area polar moment of inertia J is replaced by the torsion constant J_T to account for the warping of the cross-sections. The corrective term for warping inertia is usually discarded. This equation has the same form as the longitudinal equation [4.10]. Then the results are similar, provided the stiffness and mass coefficients are suitably modified. The modal equation is:

$$\frac{d^2\psi_x}{dx^2} + \left(\frac{\omega}{c_T}\right)^2 \psi_x = 0; \quad \text{where } c_T = \left(\sqrt{G/\rho}\right)\left(\sqrt{J_T/J}\right) = c_S\sqrt{J_T/J} \tag{4.13}$$

The modal frequencies and mode shapes are immediately found to be:

$$\omega_n = \frac{\varpi_n c_T}{L}; \quad \varphi_n(\xi) = a\sin(\varpi_n\xi) + b\cos(\varpi_n\xi) \tag{4.14}$$

The modal coefficients a, b, ϖ_n are again those found in Table 4.1. The generalized stiffness and mass coefficients are:

$$M_n = \frac{\rho JL}{2}; \quad K_n = \frac{GJ_T\varpi_n^2}{2} \tag{4.15}$$

4.2.2.3 *Flexure (or bending) modes*

In so far as the wavelengths are sufficiently larger than the cross-sectional dimensions of the beam (roughly $\lambda_n/D \geq 10$), the Bernoulli–Euler model is satisfactory, as shown in Figure 4.4. The vibration equation for bending in the plane Oxz is :

$$EI\frac{d^4Z}{dx^4} - \omega^2\rho SZ = 0 \tag{4.16}$$

Table 4.2. *Flexure modes of a straight beam*

$\cos\varpi_n \cosh\varpi_n - 1 = 0$	$a_n = \dfrac{\cosh\varpi_n - \cos\varpi_n}{\sinh\varpi_n - \sin\varpi_n}$	ϖ_n
free-free : $\varepsilon_n = +1$		4.73; 7.85; 11.0
clamped-clamped $\varepsilon_n = -1$		14.1; 17.3; $(2n+1)\pi/2$
$\tan\varpi_n + \tanh\varpi_n = 0$		ϖ_n
free-sliding $\varepsilon_n = +1$	$a_n = \dfrac{\sinh\varpi_n - \sin\varpi_n}{\cosh\varpi_n + \cos\varpi_n}$	2.37; 5.50; 8.64
clamped-sliding $\varepsilon_n = -1$		11.8; 14.9; $(4n-1)\pi/4$
$\cos\varpi_n \cosh\varpi_n + 1 = 0$	$a_n = \dfrac{\sinh\varpi_n - \sin\varpi_n}{\cosh\varpi_n + \cos\varpi_n}$	ϖ_n
cantilever $\varepsilon_n = -1$		1.88; 4.69; 7.85
		11.0; 14.1; $(2n-1)\pi/2$
$\tan\varpi_n - \tanh\varpi_n = 0$		ϖ_n
pinned-free: $\varepsilon_n = +1$	$a_n = \dfrac{\cosh\varpi_n - \cos\varpi_n}{\sinh\varpi_n - \sin\varpi_n}$	(0); 3.93; 7.07; 10.2
pinned-clamped: $\varepsilon_n = -1$		13.4; 16.5; $(4n+1)\pi/4$

The corresponding natural frequencies and related mode shapes are:

$$\omega_n = \frac{\varpi_n^2 c_b}{L^2}; \quad \text{where } c_b = \sqrt{\frac{EI}{\rho S}} \text{ and } \xi = x/L \qquad [4.17]$$

$$\varphi_n(\xi) = \cos(\varpi_n\xi) + \varepsilon_n \cosh(\varpi_{n\xi}) - a_n \{\sin(\varpi_n\xi) + \sinh(\varpi_n\xi)\}$$

The modal stiffness and mass coefficients are of the type:

$$M_n = \mu_n M; \quad K_n = \mu_n EI\varpi_n^4/L^3 \qquad [4.18]$$

where μ_n is a dimensionless coefficients which depends on the mode shapes and their norm.

The formulas giving the modal coefficients for various standard boundary conditions are gathered in Table 4.2. Whatever the boundary conditions are, if n becomes sufficiently large, $a_n$tends to 1 and ϖ_n to $n\pi$. It is worth noticing that care should be taken in numerical applications as the results are increasingly sensitive to small errors in λ_n as n increases. In Figures 4.7 to 4.12, the shapes of the first three modes are plotted for different boundary conditions. Most of them are normalized with respect to the mass, in such a way that the modal mass is constant and equal to the physical mass of the beam when dimensioned quantities are used. However, the numerical values reported in the figures are dimensionless.

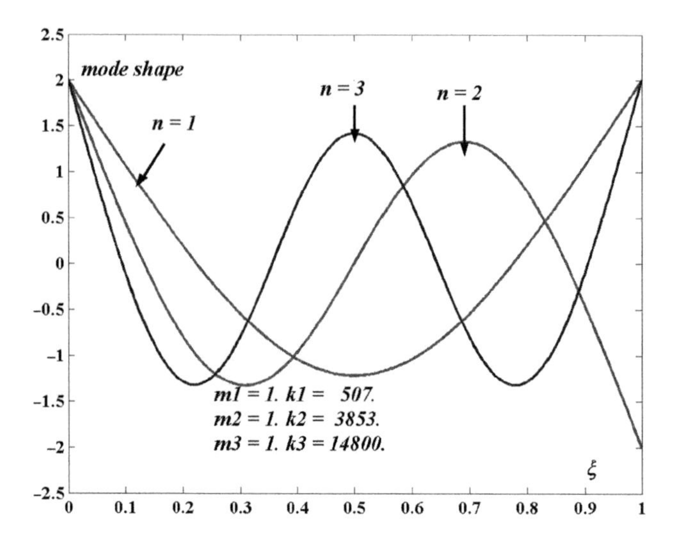

Figure 4.7. *Bending mode shapes: free-free ends*

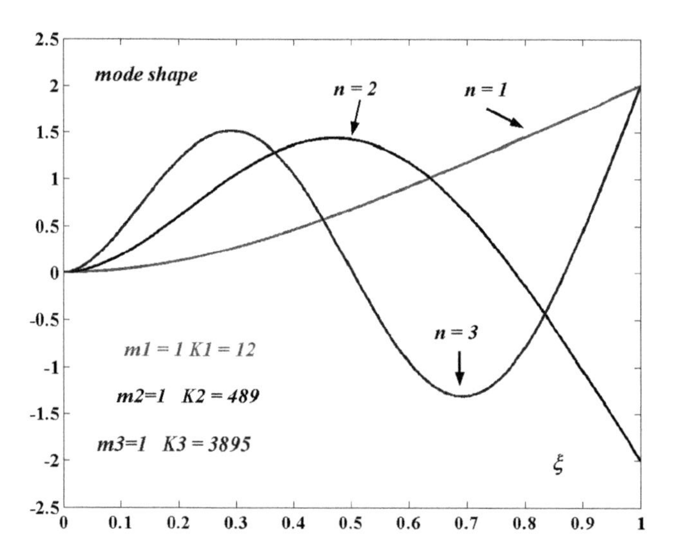

Figure 4.8. *Bending mode shapes: cantilevered beam*

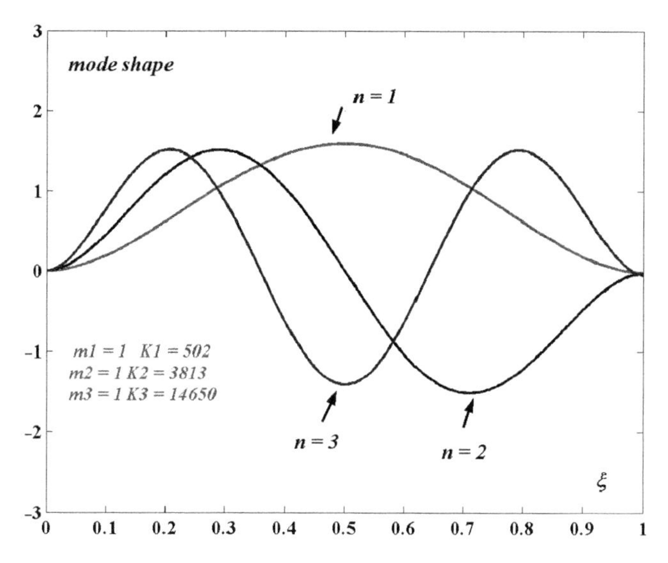

Figure 4.9. *Bending mode shapes: clamped-clamped ends*

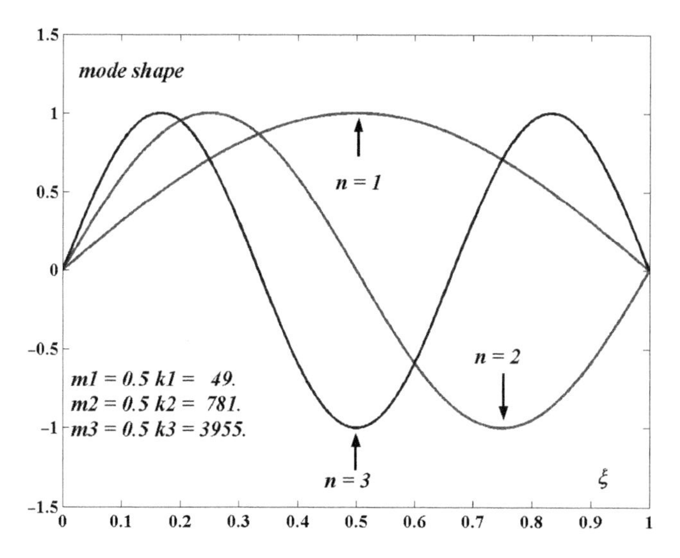

Figure 4.10. *Bending mode shapes: pinned-pinned ends*

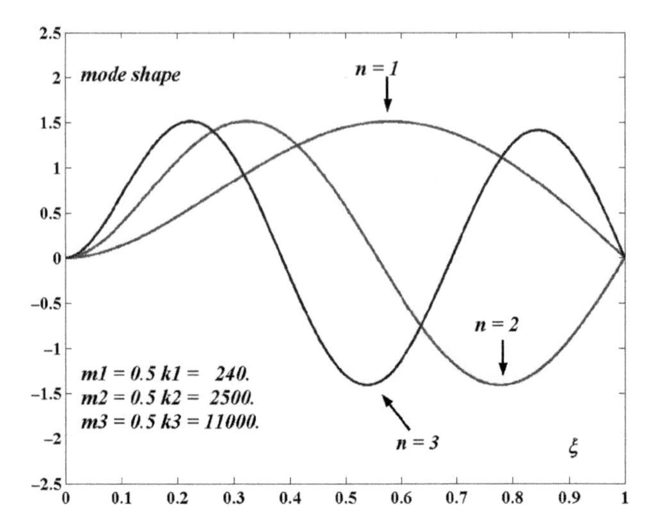

Figure 4.11. *Bending mode shapes: clamped-pinned ends*

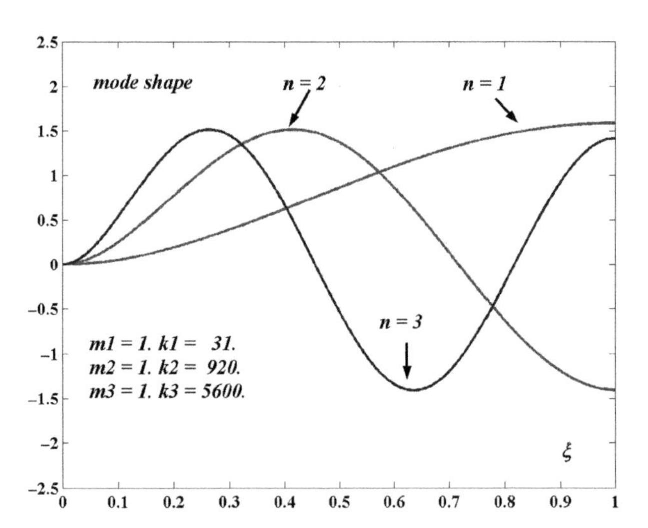

Figure 4.12. *Bending mode shapes: clamped-sliding ends*

4.2.2.4 Bending coupled with shear modes

The mode shapes of a Rayleigh–Timoshenko beam cannot be expressed analytically in closed form, except in the case of pinned-pinned end supports. Nevertheless, the study of this particular case suffices to bring out the relative importance of shear with respect to flexure in relation to the wavelength of the mode. For this reason the calculation is detailed below. From the equations [3.34] and [3.35] the following modal equations are derived:

$$\frac{\partial^2 Z_s}{\partial x^2} = -\omega^2 \frac{\rho}{\kappa G} Z; \quad \text{where } Z = Z_s + Z_b$$
$$\omega^2 \rho \left\{ \left(\omega^2 \frac{\rho I}{\kappa G} - S \right) Z + I \left(1 + \frac{E}{\kappa G} \right) \frac{\partial^2 Z}{\partial x^2} \right\} + EI \frac{\partial^4 Z}{\partial x^4} = 0 \qquad [4.19]$$

The pinned ends imply that:

$$Z(0) = Z(L) = 0; \quad \left. \frac{\partial^2 Z_b}{\partial x^2} \right|_{0,L} = 0 \qquad [4.20]$$

However, as a consequence of the flexure equation [4.19], [4.20] can be written as:

$$Z(0) = Z(L) = 0; \quad \left. \frac{\partial^2 Z}{\partial x^2} \right|_{0,L} = 0 \qquad [4.21]$$

It is easy to check that the mode shapes are not changed by the presence of shear:

$$\varphi_n(x) = \sin\left(\frac{n\pi x}{L}\right)$$

Natural frequencies are obtained by substituting $\varphi_n(x)$ into the second equation [4.19], which is suitably written in a dimensionless form by using the following reduced quantities:

$r = \sqrt{I/S}$: radius of gyration of the cross-sections about the bending axis, $\mathfrak{k}_n = n\pi r/L$: reduced wave number; $\gamma = E/\kappa G$; $\varpi = \omega_{BE}/\omega_{RT}$: frequency ratio, ω_{BE} refers to the Bernouilli–Euler model, ω_{RT} to the Rayleigh–Timoshenko model.

The result is:

$$\varpi^2 - \left(1 + (1+\gamma)\mathfrak{k}_n^2\right)\varpi + \gamma \mathfrak{k}_n^4 = 0 \qquad [4.22]$$

Only one of the two solutions of [4.22] is found to be positive, as appropriate to the present problem. Finally, the relative displacement due to bending compared

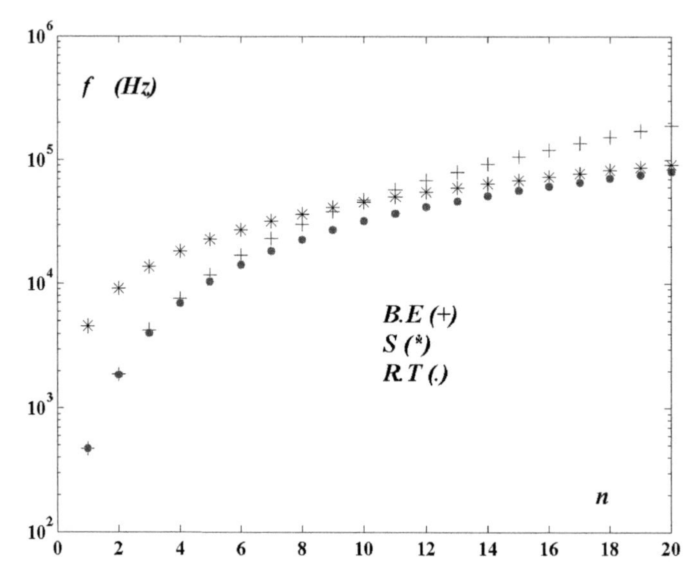

Figure 4.13. *Natural frequencies versus modal index n*

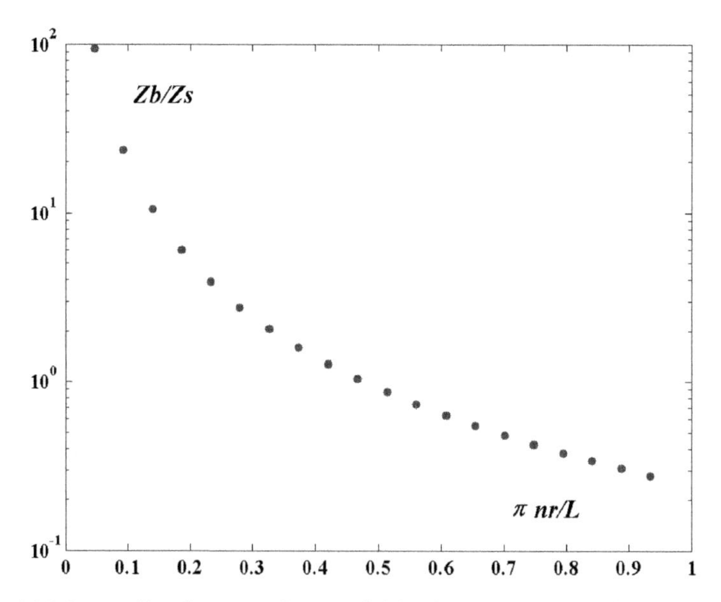

Figure 4.14. *Ratio of bending over shear modal displacement versus reduced wave number*

to shear can be obtained from the coupled equations [4.19]. The result can be written as:

$$\frac{Z_b}{Z_s} = \left(\frac{\varphi_b}{\varphi_s}\right)_n = \frac{1-\alpha_n}{\alpha_n};$$

$$\text{where} \quad \alpha_n = \frac{\omega_n^2 \rho}{\kappa G}\left(\frac{L}{n\pi}\right)^2 = \gamma\left(\frac{\omega_n r}{c\mathfrak{k}_n}\right)^2; \quad c = \sqrt{E/\rho} \qquad [4.23]$$

As expected, Z_b/Z_s is a decreasing function of the reduced wave number. Figures 4.13 and 4.14 refer to a thin circular cylinder made of steel, $L =$ 25 cm, $R = 5$ mm, $h = 0.5$ mm. The frequency plots of Figure 4.13 refer to the Bernoulli–Euler model (crosses), the pure shear model (stars) and finally the Rayleigh–Timoshenko model (heavy dots). Figure 4.14 is a plot of the modal ratio Z_b/Z_s.

4.2.3 *Rayleigh's quotient*

Many structural elements can be modelled as beams with nonuniform cross-sections and material properties. Generally, to determine their natural modes of vibration, it is necessary to use numerical techniques, amongst them the finite element method is by far the most popular. Nevertheless, in many instances it is still possible to obtain a satisfactory approximation of the first modal frequencies by assuming first some reasonable trial functions denoted $\vec{\psi}_n(\vec{r})$ to describe the actual mode shapes $\vec{\varphi}_n(\vec{r})$ and by calculating then the related natural frequencies through the Rayleigh quotient:

$$\omega_n^2 = \frac{K_n}{M_n} = \frac{\langle \vec{\varphi}_n(\vec{r}), K\,[\vec{\varphi}_n(\vec{r})]\rangle_{(\mathcal{V})}}{\langle \vec{\varphi}_n(\vec{r}), M\,[\vec{\varphi}_n(\vec{r})]\rangle_{(\mathcal{V})}} \simeq \frac{\langle \vec{\psi}_n(\vec{r}), K\big[\vec{\psi}_n(\vec{r})\big]\rangle_{(\mathcal{V})}}{\langle \vec{\psi}_n(\vec{r}), M\big[\vec{\psi}_n(\vec{r})\big]\rangle_{(\mathcal{V})}} \qquad [4.24]$$

The method is illustrated in the two following subsections.

4.2.3.1 Bending of a beam with an attached concentrated mass

A cantilevered beam of length L and mass M_b, is clamped at $x = 0$ and loaded at its free end by a concentrated mass M which is assumed to be much larger than M_b (see Figure 4.15). In such a system, most of the kinetic energy is related to the motion of the end mass and most of the potential energy is related to the bending of the beam. As a consequence, a satisfactory approximation of the first natural frequency can be produced by calculating the elastic functional related to the static deflection $Z(x)$ of the beam loaded by an end transverse force and by calculating the kinetic energy functional related to the motion of the end mass. However, it is an easy task to extend the kinetic energy functional to the whole system, which becomes necessary if M_b is not negligible. So far as the same deflection function $Z(x)$ remains valid, some additional elastic or kinetic energies can be included

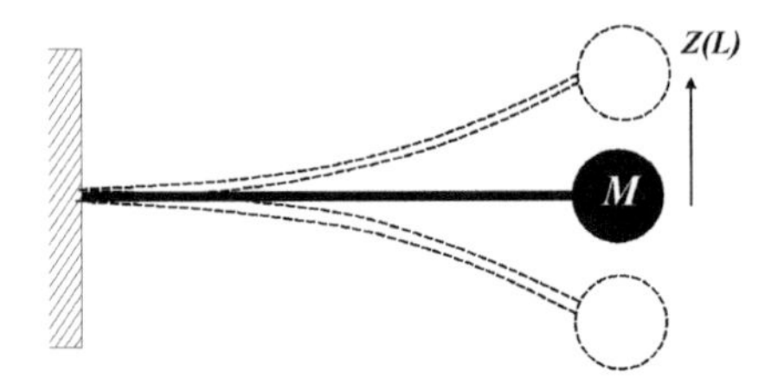

Figure 4.15. *Cantilevered beam loaded by a mass concentrated at the free end*

in the calculation, depending on the possible occurrence of some other perturbing nonuniformities in the system. The functionals are rewritten as:

$$\langle \mathcal{E}_K \rangle_{(L)} = MZ_M^2 + \rho S \int_0^L (Z(x))^2 dx; \quad \langle \mathcal{E}_e \rangle_{(L)} = EI \int_0^L \left(\frac{\partial^2 Z}{\partial x^2} \right)^2 dx$$

The admissible deflection $Z(x)$ is the solution of the static boundary value problem:

$$EI \frac{d^4 Z}{dx^4} = 0; \quad Z(0) = \left. \frac{dZ}{dx} \right|_{x=0} = 0; \quad \left. \frac{d^2 Z}{dx^2} \right|_{x=L} = 0; \quad EI \left. \frac{d^3 Z}{dx^3} \right|_{x=L} = -F_0$$

where the load magnitude F_0 can be chosen arbitrarily since we are interested in the ratio of the functionals only. The result is:

$$Z(x) = \frac{F_0 L^3}{6EI} \left(3(x/L)^2 - (x/L)^3 \right) \Rightarrow Z_M = Z(L) = \frac{F_0 L^3}{3EI}$$

Then the elastic functional, which can also be interpreted as a generalized stiffness is:

$$\langle \mathcal{E}_e \rangle_{(L)} = K_g = \frac{F_0^2 L}{3EI} = F_0 Z_M$$

and the kinetic functional, which can also be interpreted as a generalized mass is:

$$\langle \mathcal{E}_K \rangle = M_g = (M + 0.24 M_b) Z_M^2; \quad \text{where} \int_0^L Z^2(x)\, dx = 0.24 L Z_M^2$$

Finally, the first natural frequency of the system is given by the Rayleigh quotient:

$$\omega_1^2 = \frac{K_g}{M_g} = \frac{F_0}{(M + 0.24 M_b) Z_M} = \frac{3EI}{(M + 0.24 M_b) L^3}$$

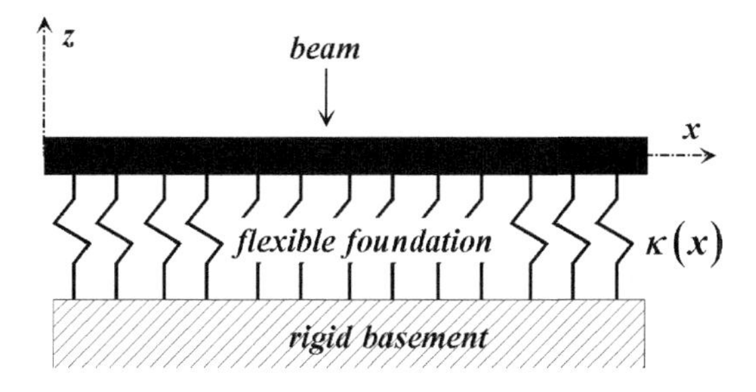

Figure 4.16. *Beam on elastic foundation*

4.2.3.2 Beam on elastic foundation

In some instances, one has to deal with beams which are lying on a continuum medium – solid, or even fluid, used as an elastic foundation. This kind of support can be modelled as an equivalent continuous stiffness density $\kappa(x)$ per unit length, as sketched in Figure 4.16. In so far as the foundation stiffness does not modify significantly the mode shapes of the non-supported beam, its effect can be accounted for in a simple manner, by adding the elastic functional of the beam to that of the foundation. Then the Rayleigh's quotient gives the perturbed natural frequencies:

$$f'_n = f_n \left(1 + \frac{\kappa_n L^3}{\varpi_n^4 EI}\right)^{1/2} \quad ; \quad \text{where } \kappa_n = \int_0^L \kappa(x)(\varphi_n(x))^2 \, dx \qquad [4.25]$$

f'_n and f_n are the natural frequencies with and without foundation respectively. κ_n is the generalized foundation stiffness coefficient. Of course, as this corrective term becomes larger, the perturbation calculus becomes less valid. Then, it can be necessary to solve the following equation:

$$EI\frac{d^4 Z}{dx^4} + \kappa(x)Z - \omega^2 \rho S Z = 0 \qquad [4.26]$$

with the appropriate boundary conditions.

However, it is noticed that if the foundation stiffness coefficient is constant, the mode shapes are not affected, because the spatial distribution of the elastic forces induced by the foundation is identical to that of the inertia forces. So the mass and the stiffness foundation operators differ from each other by a constant only, and the mode shapes are the same with and without foundation. In this particular case, the formula 4.25 which gives the shifted natural frequencies turns out to be exact.

4.2.4 *Finite element approximation*

It is of interest to check the errors which are likely to occur when a structural component is discretized by using the finite element method to compute the natural modes of vibration. A comprehensive analysis of this important subject is far beyond the scope of the present book. Nevertheless, it is worth emphasizing that large errors can be expected as soon as the size of the finite element is not small enough in comparison with the mode wavelength. According to the results presented so far, the mode wavelength of a beam can be roughly scaled by taking the ratio of the beam length to the mode index n.

4.2.4.1 Longitudinal modes

The plots in Figure 4.17 refer to the natural frequencies of the longitudinal modes of a non-supported beam, which is assumed to be uniform and homogeneous. The finite element model is made up of 50 two nodes elements of equal lengths. The dimensionless frequencies are shown in the left-hand side plot versus the mode order n. The dotted straight line refers to the theoretical values $(n-1)/2, n = 1, 2, \ldots$. The stars refer to a computation performed with the finite element mass matrix as derived in Chapter 3, subsection 3.4.2. The computed frequencies are found to be larger than the theoretical values and the plot on the right-hand side shows the relative error, which is found to be larger than 1 per cent if $n > 7$ and larger than 10 per cent if $n > 30$. Inspection of the computed modal mass and stiffness coefficients, plotted from $n = 2$ to $n = 50$ in Figure 4.18, clearly indicate that the finite element method produces underestimated values of both the mass and the stiffness coefficients. Simulations performed with other support conditions produce similar results. Finally, turning back to the left-hand side plot of Figure 4.17, the heavy dots refer to the so called *lumped mass matrix* model in which the diagonal terms of the mass matrix are taken as the half of the physical mass of the finite

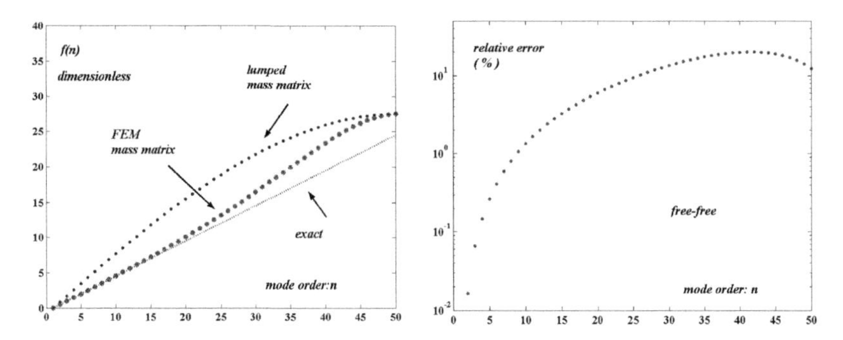

Figure 4.17. *Longitudinal modes: finite element approximation of the natural frequencies of a non-supported beam*

element and the nondiagonal terms are assumed to be zero. Use of lumped mass matrix is rather popular for computational convenience. However, caution should be taken when using it as it produces even worse results than the consistent FEM matrix model.

4.2.4.2 Bending modes

Errors made in the FEM computation of bending modes are also rather important, though significantly smaller than in the case of longitudinal modes, as shown in Figure 4.19, which refers to a pinned-pinned beam. Here modal mass and stiffness coefficients were computed by using the matrices of the finite element model and the exact mode shapes, in an attempt to concentrate on the errors related to the element shape functions solely. Accordingly, the natural frequencies were derived using the Rayleigh's quotient. The improvement in accuracy with respect to the

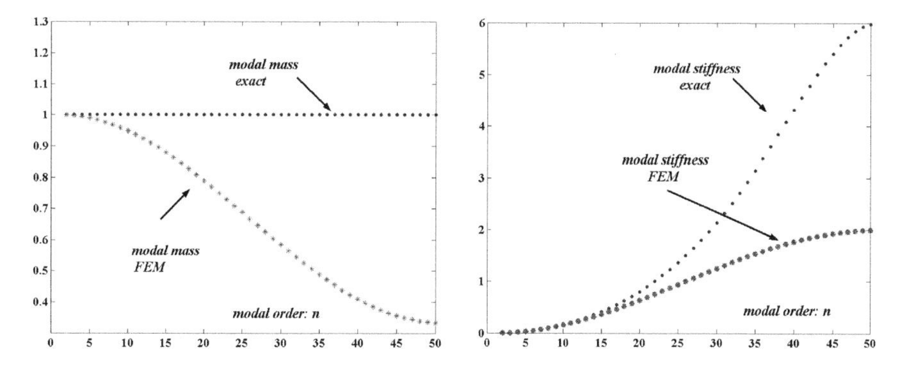

Figure 4.18. *Modal mass and stiffness coefficients*

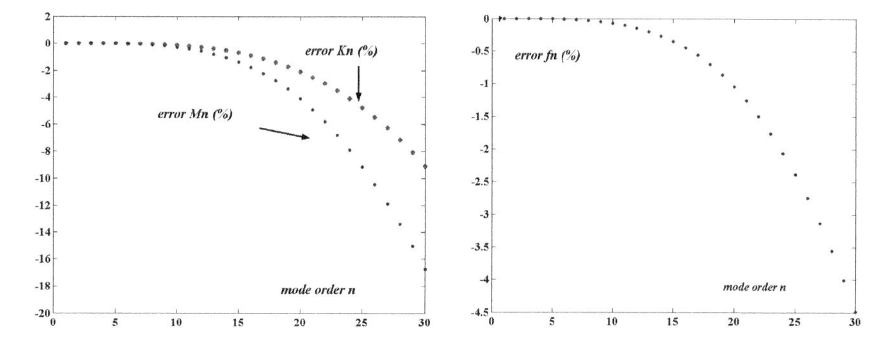

Figure 4.19. *Bending modes: finite element approximation of the natural frequencies of a pinned-pinned beam*

case of longitudinal modes is clearly due to the fact that the cubic shape function used in the bending element provides a better approximation to the actual mode shapes than the linear shape function used in the traction element.

More generally, when dealing with beam assemblies such as the portal frame of Chapter 3 subsection 3.4.3, the natural modes of vibration are mixing, bending, stretching and torsion of the beam components. As shown in Figure 4.20,

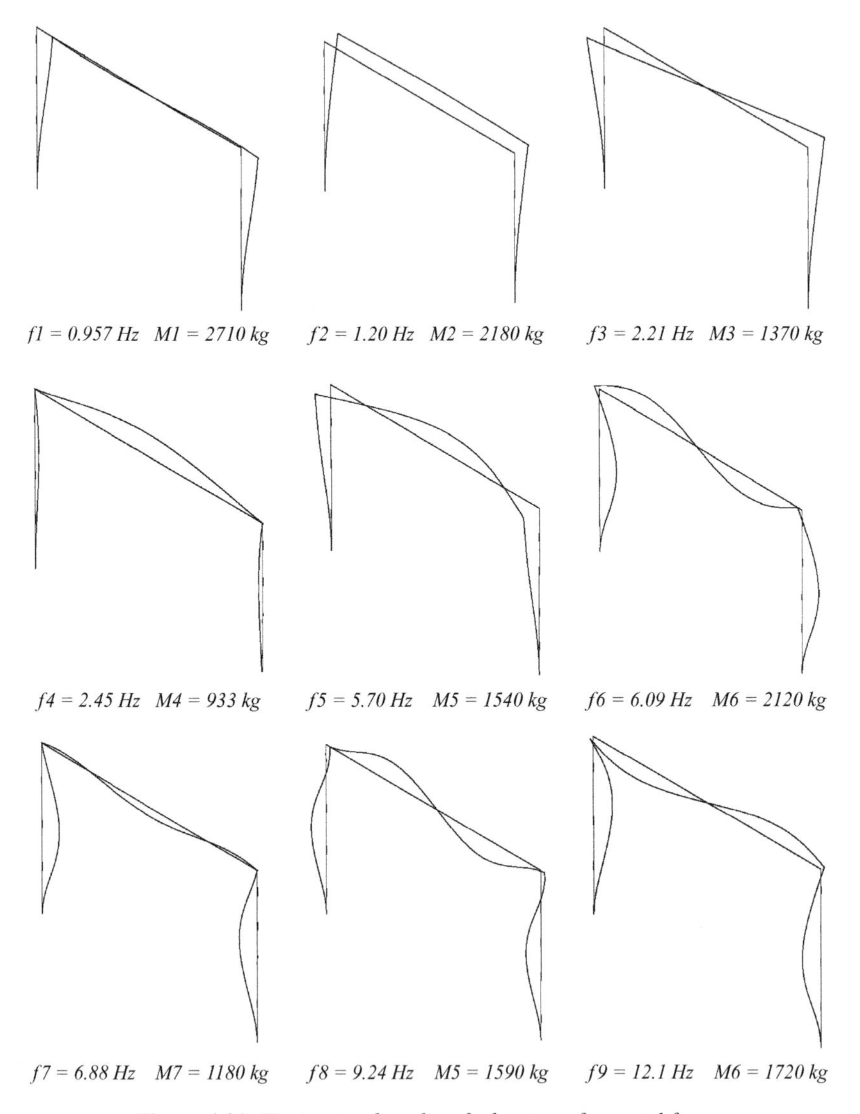

Figure 4.20. *First natural modes of vibration of a portal frame*

Colour plates

The following colour plates refer to topics covered in the text. They result from finite element computations performed with the software CASTEM2000 [CAS 92]. A symbolic colour scale is used as a particularly convenient manner to visualise the space distribution of elastic stress and support reaction fields induced on structural elements by static loads. Depending on the case, the stress isovalues are shown either on the non deformed or the deformed structure. Salient qualitative features of the results are briefly discussed here to illustrate interesting points discussed in a more general and abstract way in the text.

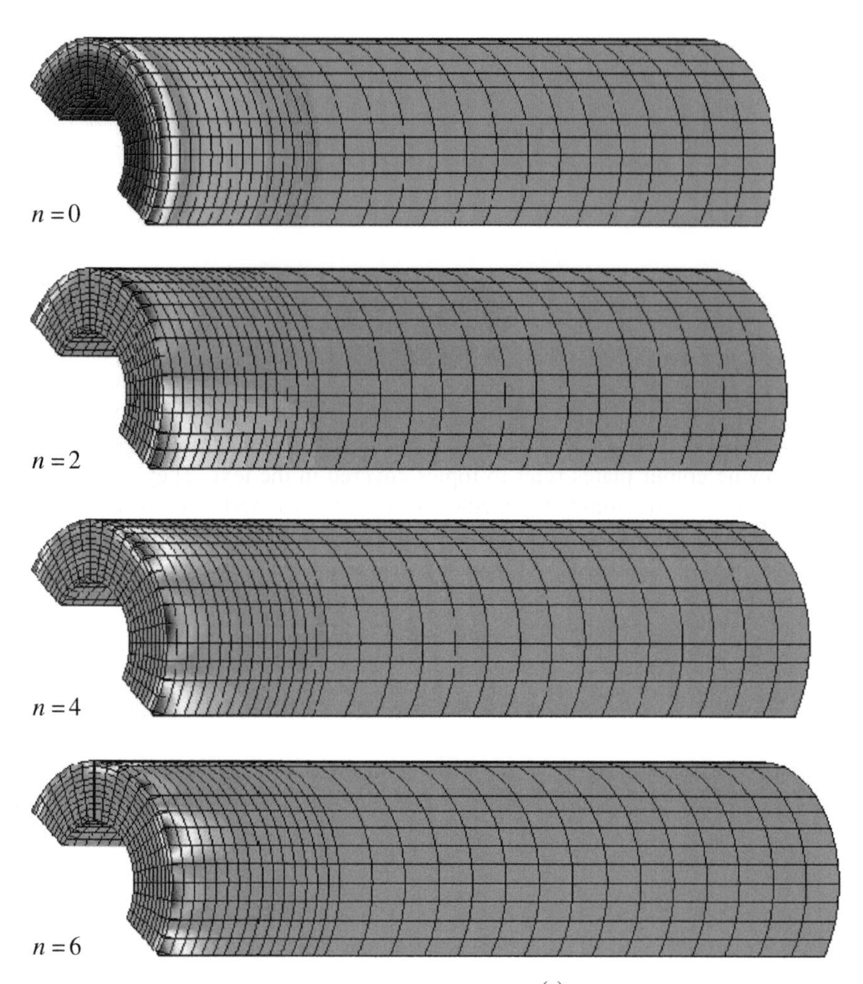

Plate 1. *Axial stress field induced by an axial force* $f_x^{(e)} = F_0 \sin(n\theta)$ *applied to the external contour of one base of a thick cylindrical shell of revolution*

The thickness ratio of the shell is $\varepsilon = 0.5$ and aspect ratio is $\eta = 5$. The shell is rigidly fixed at the right-hand side base and left free at the other. The material is steel. The external contour of the free base is loaded by an axial force density, distributed according to Fourier even harmonics. The finite element model uses massive cubical elements with eight nodes. In the case $n = 0$, a concentration of stresses is clearly evidenced by the red coloured zone in the immediate vicinity of the loaded contour. The blue colour indicates a stress level less than the peak value by a factor of at least ten. For $n = 2$ to $n = 6$, the stresses represented in blue and in red are of the same magnitude but of opposite sign. The green colour indicates a stress level less than the peak values by a factor of at least ten. Hence, it is found again that the zone of stress concentration is confined to the vicinity of the loaded contour, in agreement with the Saint-Venant principle.

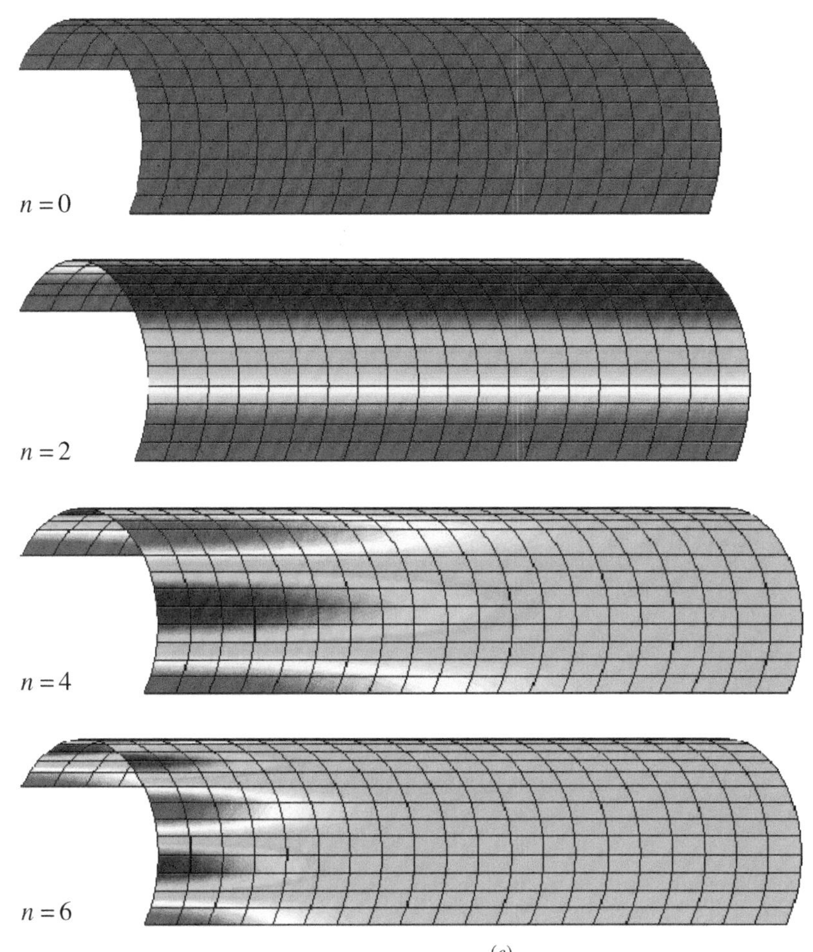

Plate 2. *Axial stress field induced by an axial force* $f_x^{(e)} = F_0 \sin(n\theta)$ *applied to one base of a thin cylindrical shell of revolution*

The same computation as in Plate 1 is made in the case of a thin shell ($\varepsilon = 0.005$). The finite element model uses harmonic shell elements with two nodes. When $n = 0$, the axial stress is found to be uniformly distributed over the whole shell; that is, no specific local response can be evidenced. Apparently, this remains true so far as the $n = 2$ loading is concerned. In the case of higher harmonics, an axial variation in the shell response can be identified starting from the loaded base up to some axial distance, the value of which decreases rather quickly as n increases. A same computation performed on a shell of higher aspect ratio would show that a similar feature holds for $n = 2$, extending to a characteristic length of about twenty shell radius. Such results are clearly in contradiction to the Saint-Venant principle.

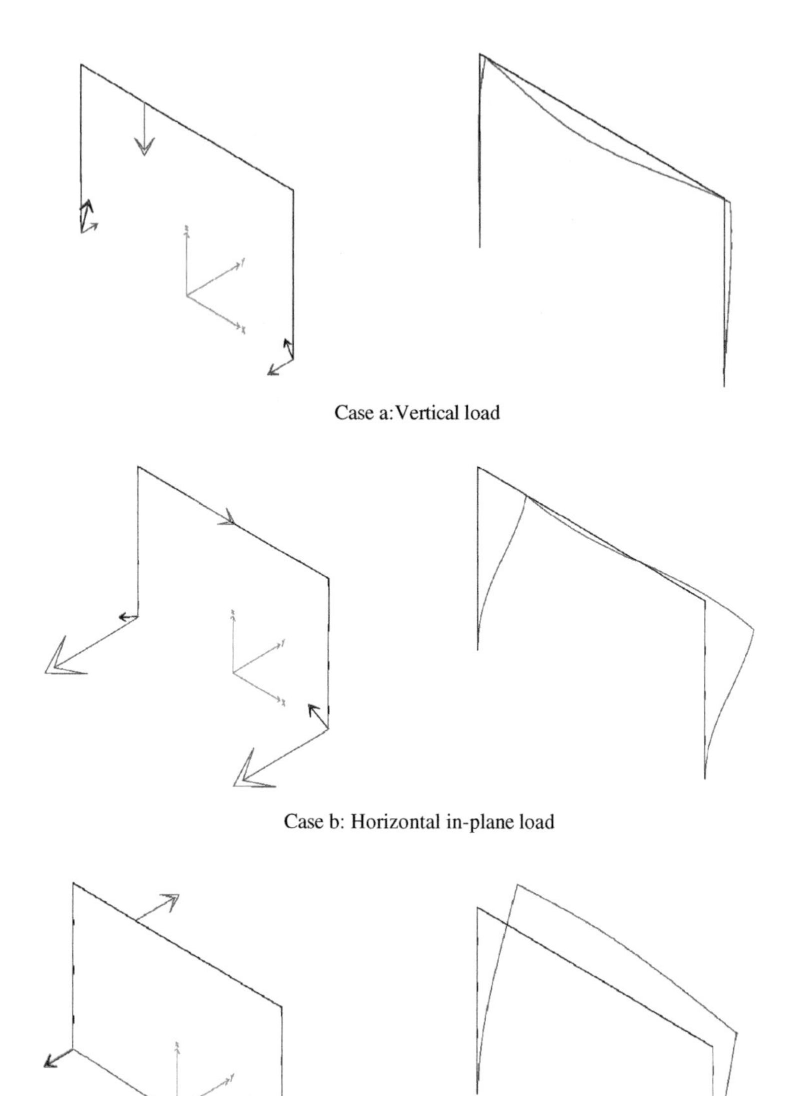

Case a: Vertical load

Case b: Horizontal in-plane load

Case c: Horizontal out-of-plane load

Plate 3. *Portal frame subjected to a concentrated force; for comments see Chapter 3, subsection 3.4.3.3. Global frame (-), mesh (-), deformed frame (-), load (→), reaction forces (→), reaction moments (→)*

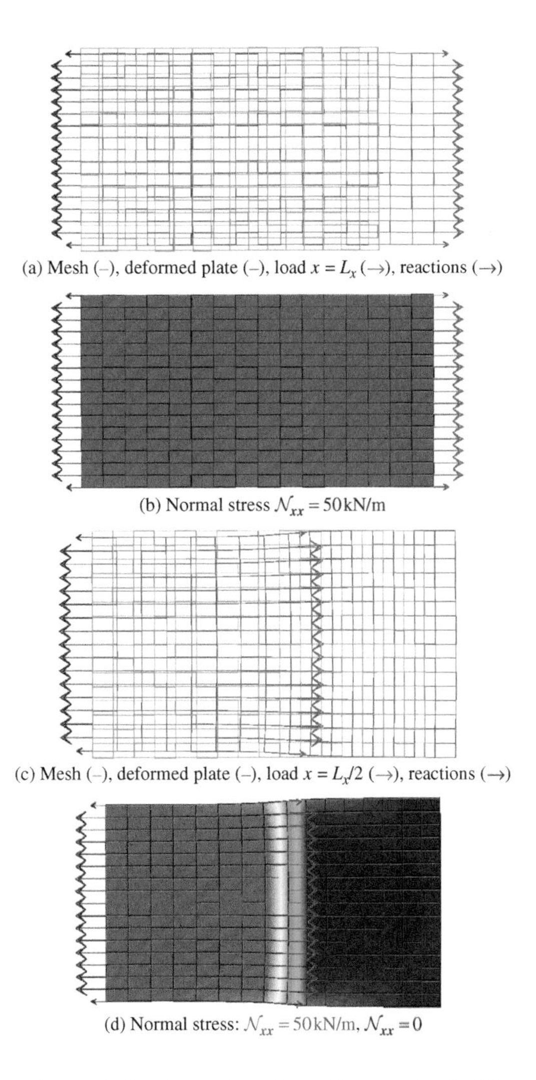

(a) Mesh (–), deformed plate (–), load $x = L_x$ (→), reactions (→)

(b) Normal stress $\mathcal{N}_{xx} = 50\,\text{kN/m}$

(c) Mesh (–), deformed plate (–), load $x = L_x/2$ (→), reactions (→)

(d) Normal stress: $\mathcal{N}_{xx} = 50\,\text{kN/m}$, $\mathcal{N}_{xx} = 0$

Plate 4. *Rectangular steel plate $L_x = 3\,m$; $L_y = 2\,m$ provided with a sliding support along the edge $x = 0$, and loaded by a tensile force field $f_x^{(e)} = 50\,kN/m$ uniformly distributed along $x = L_x$: (a), (b) and $x = L_x/2$: (c), (d)*

The isovalues of the global stresses are shown and values are given in the legend in accordance with the colour scale of computation. Please note that the theoretical finite stress jump across the loaded line $x = L_x/2$ is replaced by a continuous transition due to the smoothing effect of the finite element method. The characteristic scale of the transition zone is equal to the length of a finite element. The magnitude of the nodal forces (load and reactions) is two times less at the corners than at the other nodes, due to the discretization process, cf. Chapter 3, subsection 3.4.2.4.

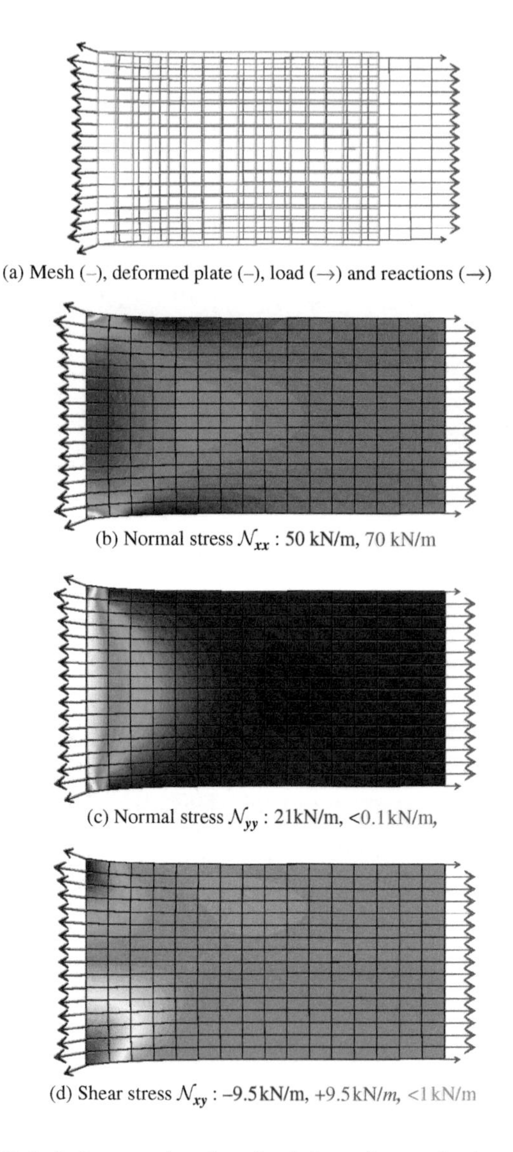

(a) Mesh (–), deformed plate (–), load (→) and reactions (→)

(b) Normal stress $\mathcal{N}_{xx}$: 50 kN/m, 70 kN/m

(c) Normal stress $\mathcal{N}_{yy}$: 21kN/m, <0.1kN/m,

(d) Shear stress $\mathcal{N}_{xy}$: –9.5kN/m, +9.5kN/*m*, <1 kN/m

Plate 5. *Rectangular plate fixed along the* $x = 0$ *edge*

Lateral contraction is prevented along the fixed edge, whereas it can develop freely everywhere else. As a consequence, normal stresses in the lateral direction and in-plane shear stresses are induced in the vicinity of the fixed edge. Note that the normal stress in the longitudinal direction presents peak values closely concentrated at the corners corresponding to the fixed edges. The non uniformity of the distribution of the reaction forces is merely a consequence of that of the stress-field.

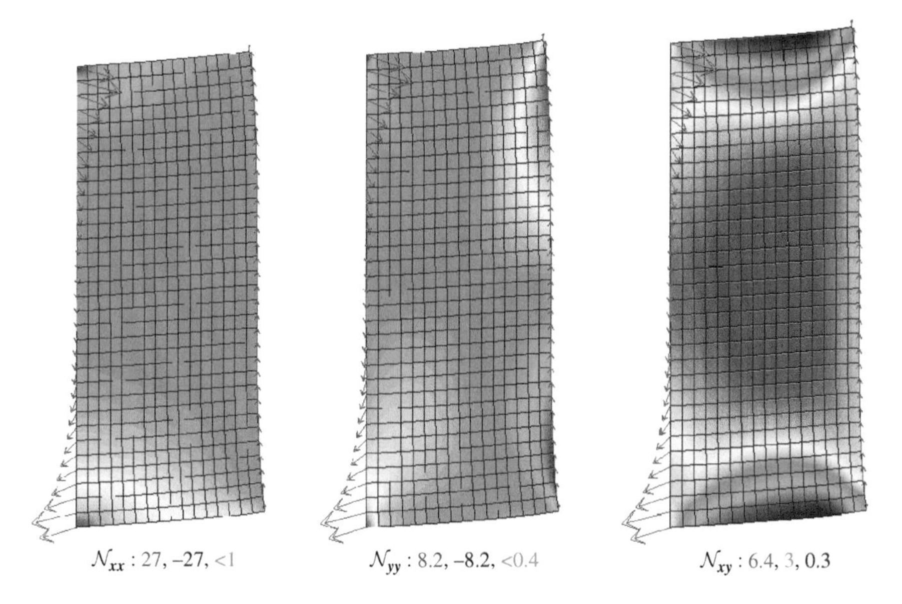

load(→), reaction forces (→), stress levels are given in *kN/m*

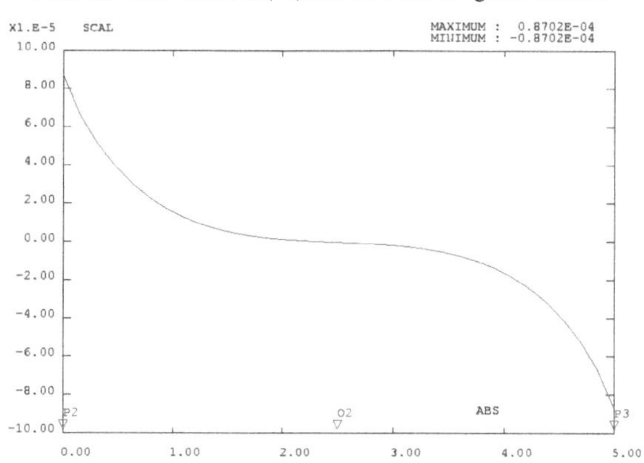

Plate 6. *Axial displacement of the loaded edge demonstrating the warping effect* $(X_{\max} = 87\ \mu m)$

The plate parameters are : $L_x = 2\,\text{m}$, $L_y = 5\,\text{m}$, $h = 5\,\text{cm}$, $E = 5 10^{10}\,\text{Pa}$, $\upsilon = 0.3$. It is fixed along the edge $x = 0$ and loaded by a lateral force of 25 kN, uniformly distributed along the opposite edge $x = L_x$. Due to the small aspect ratio ($\eta = 0.4$), shear deformations are important, leading to warping and a rather complicated distribution of in-plane stresses.

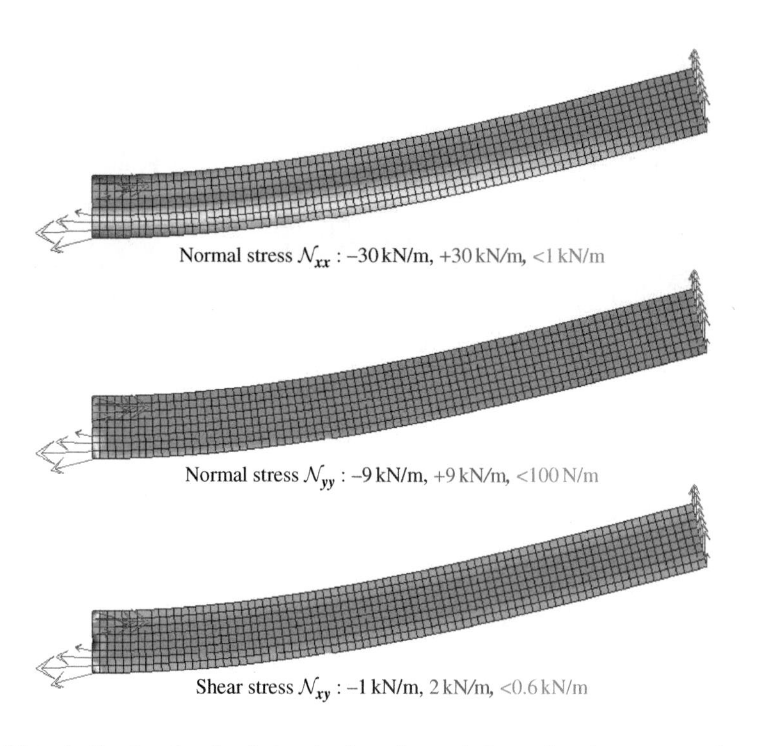

Plate 7. *Slender plate loaded at the free edge and clamped at the opposite edge*

The same calculation as in Plate 6, but related to a plate of much higher aspect ratio ($\eta = 10$), shows that shear deformations and warping are much less important than in the former case. The displacement field reduces essentially to a lateral deflection (maximum *0.7* mm) and a rotation of the lateral cross-sections. The distribution of the longitudinal stresses agrees with that expected in a flexed beam. However, plate effects are still noticeable locally in the vicinity of the clamped and the longitudinal edges. As a consequence, the reaction force field (magenta arrows) departs to some extent from the idealized distribution of the Bernoulli–Euler model. Nevertheless, for most current applications, the structure could be satisfactorily modelled as an equivalent beam.

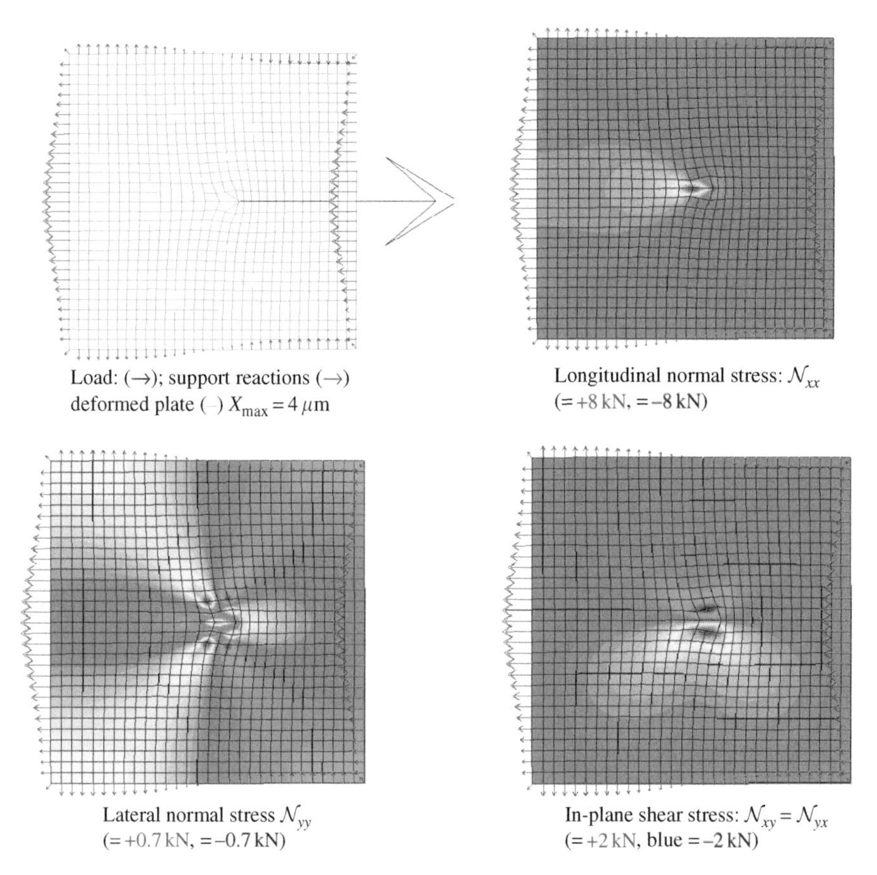

Plate 8. *Square steel plate $L = 4\,m$, $h = 5\,mm$ with sliding edges, loaded by a longitudinal force $F_x^{(e)} = 100\,kN$ applied to its centre*

The response to the concentrated load is so strongly marked by local effects that the distribution of the reaction forces is of great help in understanding its main features. They indicate that the plate is stretched in the left-hand side part and compressed in the right-hand side. Lateral displacement due to Poisson's effect is prevented at the longitudinal edges, which explains the presence of lateral reactions distributed antisymetrically with respect to the middle line $x = L/2$. As the plate elements are subjected to a non uniform normal stress field, they are also sheared.

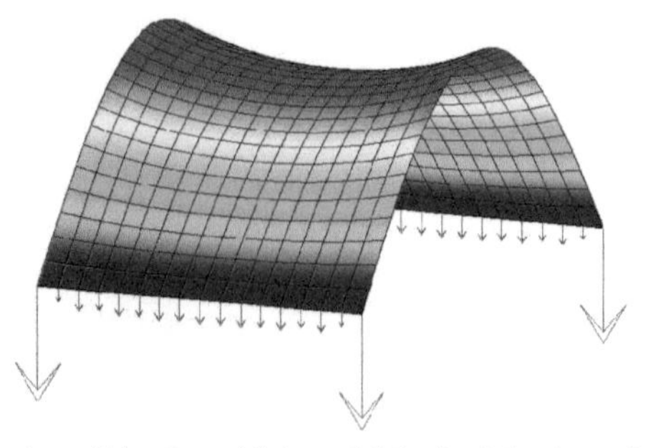

Deformed plate (magnifying factor 15) (central deflection 3.4 cm), reaction forces ($\rightarrow$), $\mathcal{M}_{xx}$ (kN): 3, 1.8, 0.2

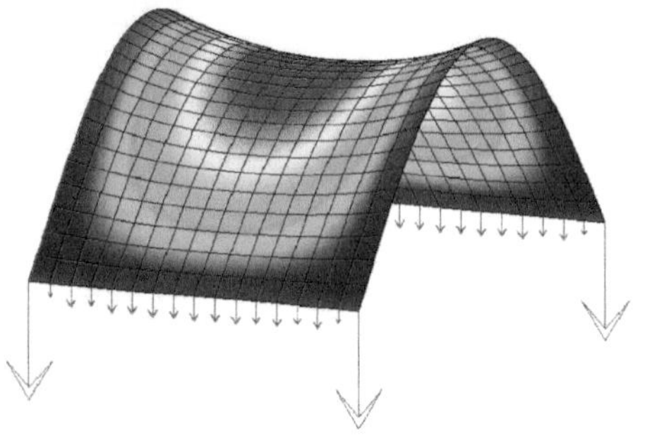

$\mathcal{M}_{yy}$ (kN): 0.6, 0.36, 0.03

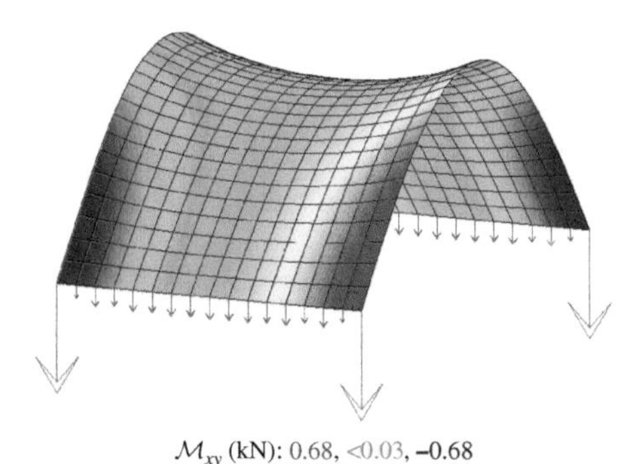

$\mathcal{M}_{xy}$ (kN): 0.68, <0.03, -0.68

Plate 9. *Rectangular plate hinged at two opposite edges and loaded by a uniform pressure field $P = 10$ kPa*

The plate parameters are: $L_x = 1.5\,\text{m}$, $L_y = 1\,\text{m}$, $h = 1\,\text{cm}$, $E = 210^{11}\,\text{Pa}$, $\upsilon = 0.5$. As the aspect ratio of the plate is rather small ($\eta = 1.5$) and as υ is large, the anticlastic deformation of the plate is clearly indicated in the figures. As expected, bending moments are maximum in the central portion of the plate. Note also the presence of large corner forces. They are related to the torsion moment field, the magnitude of which is maximum at the corners.

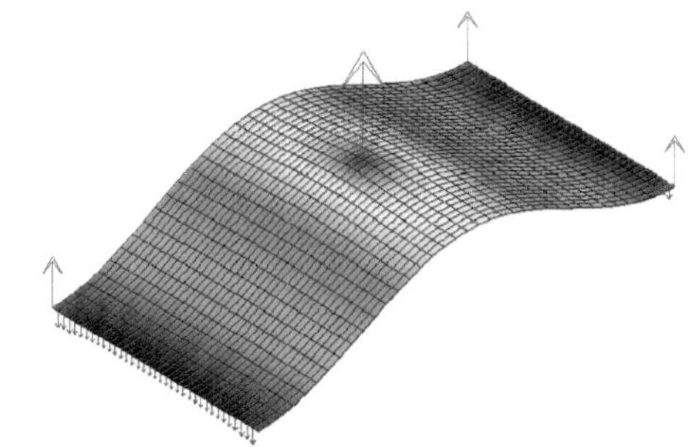

Load (→), deformed plate (magnifying factor 30) and reaction forces (→)

$\mathcal{M}_{xx}$ (N): 120, 30, –73

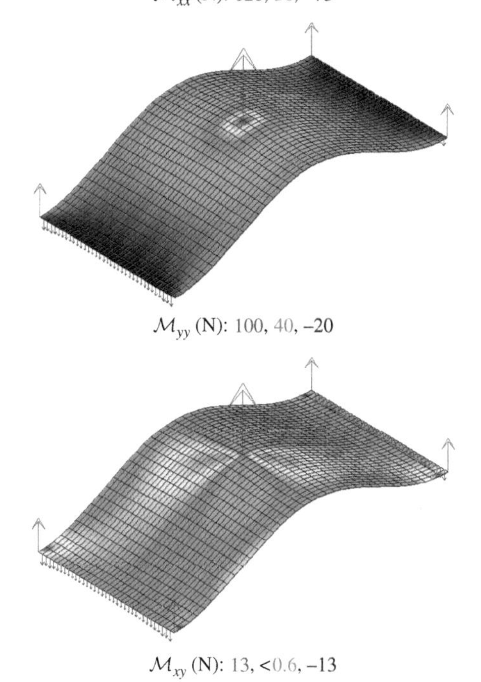

$\mathcal{M}_{yy}$ (N): 100, 40, –20

$\mathcal{M}_{xy}$ (N): 13, <0.6, –13

Plate 10. *Rectangular plate clamped along two opposite edges and loaded at its centre by a normal force $F = 300$ N*

The plate parameters are: $L_x = 4\,\text{m}$, $L_y = 2\,\text{m}$, $h = 5\,\text{mm}$, $E = 210^{11}$ Pa, $\upsilon = 0.3$. Plate deflection is marked by a global bending about the direction of the clamped edges and by a rather broad bump centred at the loaded point, which is well evidenced on the isovalues of bending and torsion stresses. This local component opposes the anticlastic deflection which would be observed for a uniform loading (see Plate 9). Torsion moments are important in the bumped zone and at the plate corners. Note also the striking differences in the distribution of the edge shear forces and of corner forces in comparison with Plate 9.

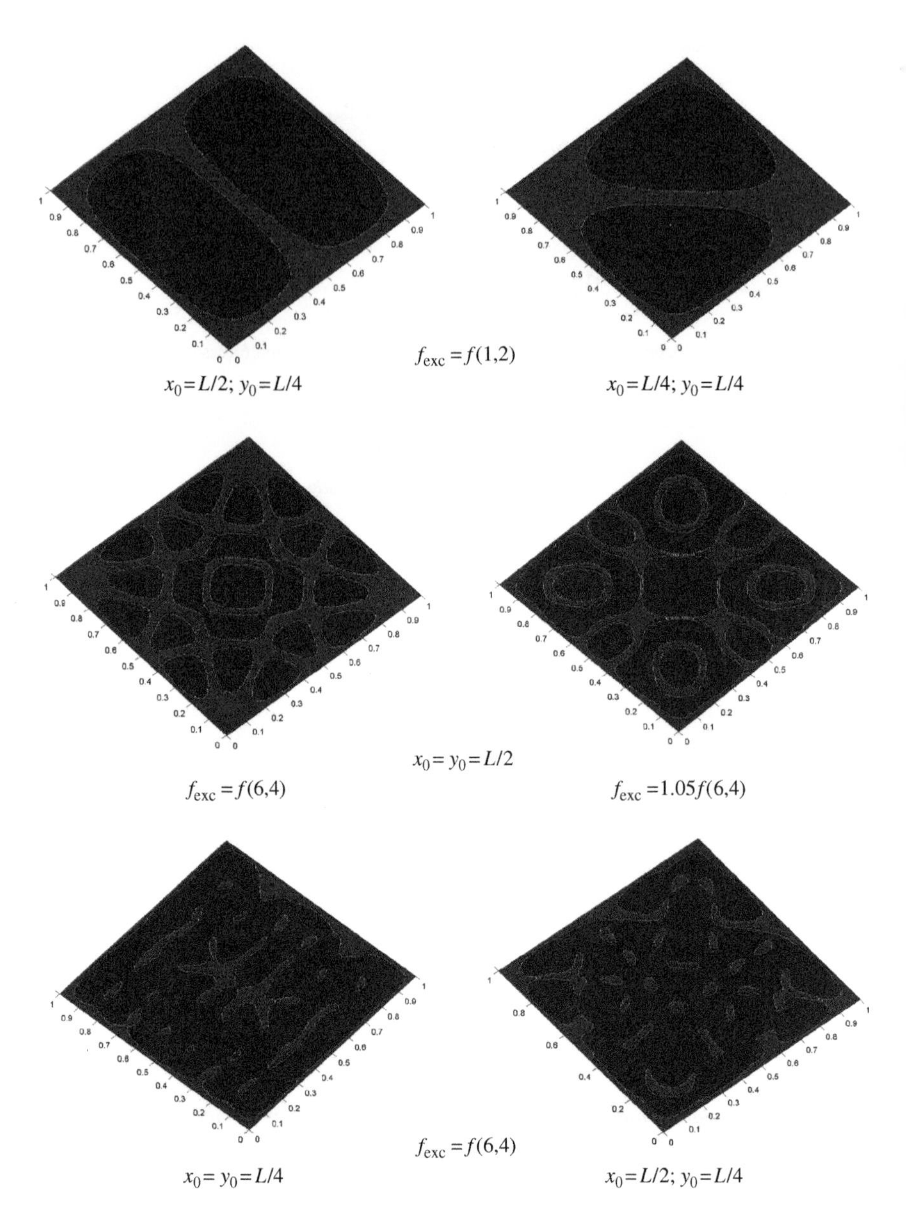

Plate 11. *Numerical simulation of Chladni figures for a square plate*

f_{exc} denotes the frequency of the excitation and (x_0, y_0) the location of the loaded point. The natural frequencies are denoted $f(n, m)$. Zones coloured in red correspond to vibration minima (less than 10% of the maximum amplitude of transverse displacement).

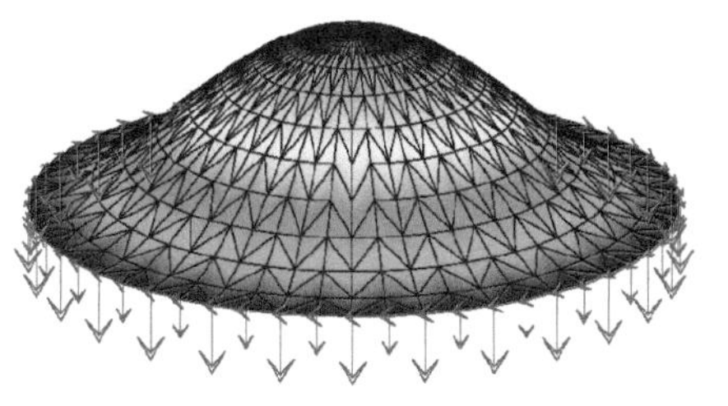

Deformed plate (magnifying factor 100), central deflection 5.5 mm, reaction moments (→), reaction forces (→)
$\mathcal{M}_{\theta\theta}$ (kN) : 9.5, 3, –2.7

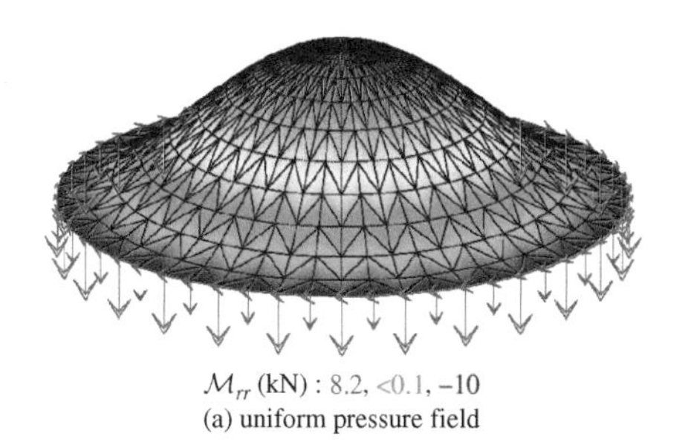

$\mathcal{M}_{rr}$ (kN) : 8.2, <0.1, –10
(a) uniform pressure field

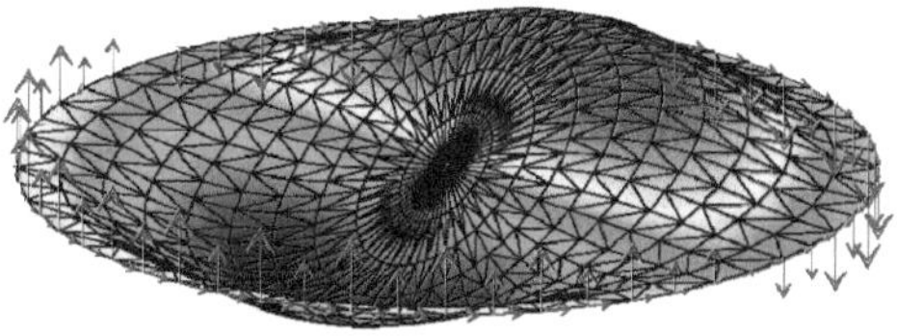

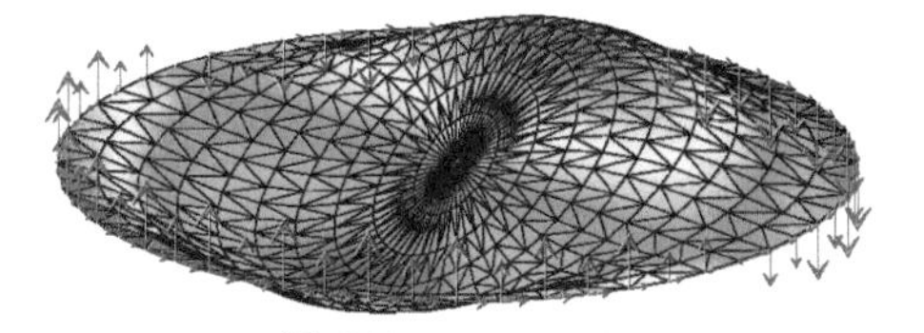

Deformed plate (magnifying factor 100), maximum deflection 2.4 mm, reaction moments (→), reaction forces (→)
$\mathcal{M}_{\theta\theta}$ (kN) : 2.9, <0.1, –2.9

M_{rr} (kN) : 6.1, <0.2, –6.1
(b) antisymmetrical pressure field

Plate 12. *Circular plate clamped at the edge contour and loaded by a pressure field* $P = 10^5\,Pa$

The plate parameters are: $R = 1\,\mathrm{m}$, $h = 2\,\mathrm{cm}$, $E = 2 10^{11}$ Pa, $\upsilon = 0.3$.

Note that the reaction forces vectors are not all of the same length. This is merely a consequence of the finite element discretization process, cf. Chapter 3, subsection 3.1.2.4.

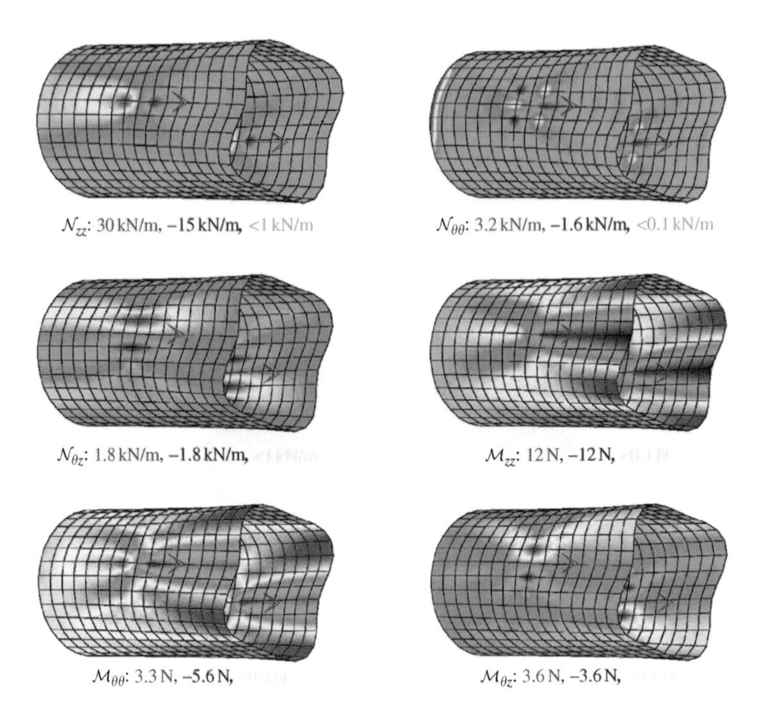

Plate 13. *Circular cylinder in clamped-free configuration,* $R = 3\,m,\ H = 12\,m,\ h = 1\,cm;$ $E = 2.10^{11}\ Pa,\ v = 0.3$

Load: two axial forces of same magnitude $F = 30\,\text{kN}$, applied to two diametrically opposed points, see red arrows.

In the figure, the deformed shape of the shell is presented with a magnifying factor of 1000. The isovalues of the global stresses are shown and approximate values are written in the legend in accordance with the colour scale of the computation.

The deformation of the free base is essentially the superposition of the harmonics $n = 2$ and $n = 4$, giving rise to a radial deflection which is greater along the loaded meridians than along those at 90°. Such an asymmetry is also conspicuous on the stress components. Axially, the shell is compressed on the right-hand side by the load and stretched on the left hand-side. Differences between the compressive and tensile stress levels are easily understood as the base on the left of the shell is clamped, whereas the base on the right is free. Hoop stress is maximum at the clamped base, three zones of compression also extend from the vicinity of the loaded points. Two shear stress zones of opposite sign extend near the loaded point. They are shaped as two elongated spots parallel to the loaded meridian. Distribution of the axial and tangential moments is in qualitative agreement with a $n = 4$ mode of deformation perturbed by the contribution of the $n = 2$ mode. Distribution of the torsion moment presents some similarities to that of the shear force. They are shaped as two crescents of opposite sign.

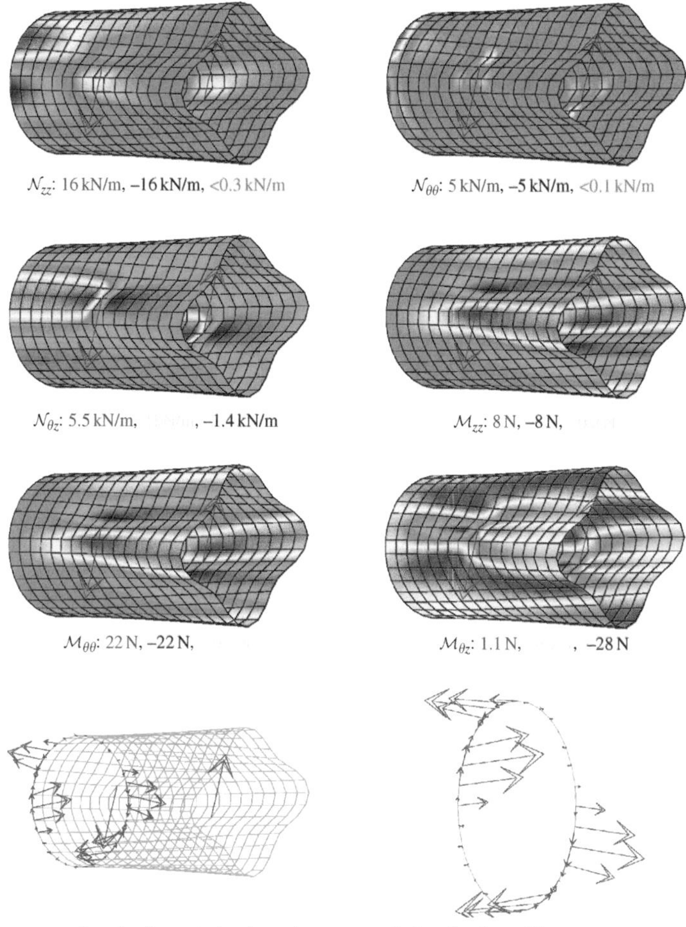

Reaction forces (→) and reaction moments (→), at the clamped base

Plate 14. *Circular cylinder in clamped-free configuration,* $R = 3\,m$, $H = 12\,m$, $h = 1\,cm$; $E = 2.10^{11}\,Pa$, $\upsilon = 0.3$

Load: two tangential forces of same magnitude $F = 10\,kN$, applied to two diametrically opposed points, see red arrows.

In the figure, the deformed shape of the shell is presented with a magnifying factor of 1000. Again, the deformation of the free base is essentially the superposition of the harmonics $n = 2$ and $n = 4$, and is similar to that of Plate 13. However, the distribution of the stresses is entirely different. As expected, shear and torsional components are much more important than in the former case. Nevertheless, it would be difficult to anticipate the distribution of the other stresses, in particular the normal forces which are found to be more intense in the axial than in the tangential direction. This explains the direction of the reaction forces at the clamped basis. On the other hand, the moment vectors are very small and essentially tangential.

the first three modes of vibration may be described as an in-plane bending mode, a symmetrical and an antisymmetrical out-of plane bending mode respectively, though some stretching is also present in the second mode and torsion in the third one. So in the case of complex structures, the accuracy of the FEM results is not easily quantified and it is advisable to check their sensitivity to mesh refinement.

4.2.5 *Bending modes of an axially preloaded beam*

4.2.5.1 Natural modes of vibration

Let us consider again a straight and uniform beam preloaded by the axial force T_0 in the absence of any external fluctuating excitation (Figure 4.21). Its transverse motion is governed by the equilibrium equation [3.46], established in Chapter 3 and rewritten here as:

$$\rho S\ddot{Z} - T_0\frac{\partial^2 Z}{\partial x^2} + EI\frac{\partial^4 Z}{\partial x^4} = 0 \qquad [4.27]$$

provided with the boundary conditions [3.47] written in the variational form:

$$\left[\left(T_0\frac{\partial Z}{\partial x} - EI\frac{\partial^3 Z}{\partial x^3}\right)\delta Z\right]_0^L = 0; \quad \left[EI\left(\frac{\partial^2 Z}{\partial x^2}\right)\delta\psi_Y\right]_0^L = 0 \qquad [4.28]$$

In most cases, solving analytically the modal system [4.27], [4.28] is not straightforward because the mode shapes are found to depend on T_0 and cannot be expressed in a closed analytical form. As in the case of the Rayleigh–Timoshenko beam, the difficulty is due to the presence of two distinct stiffness operators of distinct differential order. So, a mode shape holding for $T_0 = 0$ is not likely to comply with the orthogonality properties with respect to the prestress stiffness operator. In subsection 4.4.1.3, we will describe a semi-analytical method to produce approximate solutions of the problem. However, before embarking on such a subject, it is in order to start by analysing the particular case of a pinned-pinned beam, where the mode shapes are independent of T_0. This is because the functions

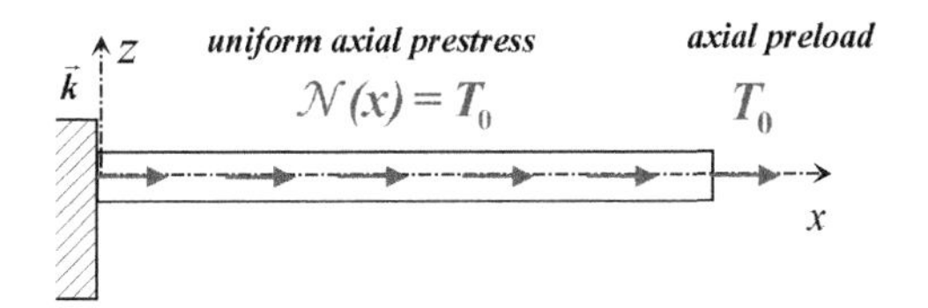

Figure 4.21. *Axially preloaded beam*

$\varphi_n(x) = \sin(n\pi x/L)$ are orthogonal with respect to both the elastic and the prestress stiffness operators and the boundary conditions do not depend upon T_0, since $\delta Z(0) = \delta Z(L) = 0$. Substituting these mode shapes into [4.27], [4.28] the following natural pulsations are obtained:

$$\omega_n = \sqrt{\frac{1}{\rho S}\left\{EI\left(\frac{n\pi}{L}\right)^4 + T_0\left(\frac{n\pi}{L}\right)^2\right\}} \qquad [4.29]$$

If the preload is positive (stretching), the natural frequencies increase with T_0. Moreover, if the tensile force is sufficiently large, such that $T_0 \gg EI(n/L)^2$, the ω_n form an harmonic sequence, which is appropriate for musical instruments. Accordingly, string instruments make use of very small diameter and highly tensioned threads or wires, and for lower pitch strings – where significant mass is unavoidable – a special inner structure of the string is devised to minimize the bending effect, see for instance [FLE 98]. In contrast, if the preload is negative, the natural frequencies steadily decrease with $|T_0|$ and there exist modal critical values of compressive preload given by:

$$T_n^{(c)} = -EI\left(\frac{n\pi}{L}\right)^2 \qquad [4.30]$$

at which modal stiffness and natural frequency vanish. Considering first the n-th mode, when T_0 is swept through $T_n^{(c)}$, the corresponding modal stiffness changes from positive to negative. Accordingly, the natural frequency bifurcates from a single positive value to a pair of pure imaginary conjugated values and the beam becomes unstable. Such an instability, already described in [AXI 04] in the case of discrete systems, is termed *buckling* in the engineering context, and *divergence* in the more theoretically oriented literature. Turning back now to the beam behaviour, it is rather obvious that the threshold for buckling is given by the smallest modal critical load and the beam buckles according to the corresponding mode, which turns out to be the first one ($n = 1$). The motion of the buckled beam is marked by large displacements since the equilibrium state taken as a reference is unstable. Therefore, the post-buckling behaviour of the beam is beyond the scope of linear mechanics and will not be further discussed in this book.

4.2.5.2 Static buckling analysis

The critical axial compressive load required to buckle a beam can also be derived by analysing the nonzero solutions of the following static equation:

$$\Lambda^2 \frac{d^2\psi}{dx^2} + \frac{d^4\psi}{dx^4} = 0; \quad \text{where } \Lambda^2 = \frac{T_0}{EI} \qquad [4.31]$$

In the same way as in the case of discrete systems, the problem defined by [4.31] is similar to a modal problem in which the load factor Λ is substituted by ω^2 and the preload operator by the mass operator. The nonzero solutions are termed *buckling modes*. The differential equation [4.31] may be rewritten as:

$$\frac{d^2}{dx^2}\left(\frac{d^2\psi}{dx^2} - \Lambda^2\psi\right) = 0 \qquad [4.32]$$

leading immediately to the general solution:

$$\psi(x) = A\sinh(\Lambda x) + B\cosh(\Lambda x) + Cx + D \qquad [4.33]$$

If the load is compressive $T_0 < 0 \Rightarrow \Lambda = \pm i|T_0|/EI$ and the buckling mode shape is of the type:

$$\psi(x) = A\sin(\Lambda x) + B\cos(\Lambda x) + Cx + D \qquad [4.34]$$

In [4.34] the constants are again determined by the boundary conditions, as detailed below for a few support conditions. The mode shapes are normalized by the condition of a unit maximum deflection.

Pinned-pinned beam. The results are the same as those obtained in dynamics:

$$T_{cn} = -\left(\frac{n\pi}{L}\right)^2 EI; \quad T_c = \min(T_{cn}) = -\left(\frac{\pi}{L}\right)^2 EI; \quad \psi_n(x) = \sin\left(\frac{n\pi x}{L}\right)$$

Cantilevered beam. The boundary conditions are:

$$\psi(0) = \psi'(0) = 0; \quad \psi''(L) = 0;$$
$$EI\psi'''(L) + |T_0|\psi'(L) = 0 \;\Rightarrow\; \psi'''(L) + \Lambda^2\psi'(L) = 0$$

where again the prime symbol denotes a space derivation.

The result is:

$$T_{cn} = -\left(\frac{(2n+1)\pi}{2L}\right)^2 EI; \quad T_c = -\left(\frac{\pi}{2L}\right)^2 EI;$$

$$\psi_n(x) = \frac{1}{2}\left(1 - \cos\left(\frac{(2n+1)\pi x}{2L}\right)\right)$$

Clamped-sliding beam. The boundary conditions are:

$$\psi(0) = \psi'(0) = 0; \quad \psi'(L) = 0;$$
$$\psi'''(L) + \Lambda^2\psi'(L) = 0$$

The result is:

$$T_{cn} = -\left(\frac{n\pi}{L}\right)^2 EI; \quad T_c = -\left(\frac{\pi}{L}\right)^2 EI$$

The buckling modal loads are the same as in the pinned-pinned case but the buckling mode shapes are different:

$$\psi_n(x) = \frac{1}{2}\left(1 - \cos\left(\frac{n\pi x}{L}\right)\right)$$

Pinned-sliding beam. The boundary conditions are:

$$\psi(0) = \psi'(0) = 0; \quad \psi'(L) = 0$$
$$\psi'''(L) + \Lambda^2\psi'(L) = 0$$

The modal buckling loads are the same as for the cantilevered beam and the buckling mode shapes are:

$$\psi_n(x) = \sin\left(\frac{n\pi x}{2L}\right)$$

Returning very briefly to the follower load problem introduced in Chapter 3 subsection 3.2.3.4, the reader can easily verify that the buckling eigenvalue problem has only trivial solutions and no buckling modes occur in such a system.

Finally, it may be also of interest to complete the theoretical results concerning buckling by two numerical examples to illustrate the practical importance of such problems. Let us consider first a cantilevered steel beam of length 1 m and square cross-sectional area 1 cm^2 which is erected vertically and loaded axially at its free end by a mass M much larger than the beam mass ($m_b = 0.8$ kg). The critical load is found to be nearly equal to 411 N that is, $M = 40$ kg in the terrestrial gravity field. In the second example, the same beam is provided with clamped-clamped end supports and its temperature is T_0. The critical load for buckling is found to be four times the value for a clamped-sliding beam and is equal here to 6900 N. This result can be derived from the preceding one by noticing that the deflection of the clamped-clamped beam can be deduced from that of the clamped-sliding beam through a symmetry about the sliding cross-sectional plane. On the other hand, by using the results established in Chapter 2, section 2.3, it is seen that the critical buckling load can be reached by increasing the initial beam temperature by 27°C, provided a thermal dilatation coefficient of $12^{-6}/°C$ is assumed. Going a step further, it can be noted that thermal buckling does not necessarily lead to catastrophic failure, because bending induces an axial tension – via the geometrical nonlinearities shortly discussed in Chapter 2 subsection 2.2.4.7, example 4, which tends to alleviate the thermal compressive load. The amount of axial strain which is necessary to relax entirely the thermal stress turns out to be very small $\varepsilon = \alpha \Delta \theta_c \cong 3.310^{-4}$; so if the temperature increase is still insufficient to induce plastic deformation, the beam recovers its original shape when the temperature decreases again to its initial value.

4.3. Modal projection methods

As in the case of discrete systems, a fruitful approach to deal with a large variety of mechanical problems is to start by transforming the equilibrium equations – initially written in terms of the so called *physical coordinates* – into an equivalent system of equations written in the so called *modal*, or *normal coordinates*. In geometrical language, such a transformation can be interpreted as a projection of the equations in their initial form onto a modal basis. A priori, in the case of continuous systems, the dimension of the modal basis is infinite, which is rather inconvenient since the analytical results must be expressed as series comprising infinitely many terms. Fortunately, it will be shown that, in many applications, the modal basis can be efficiently truncated by retaining a relatively few number of modes only. Furthermore, the suitable criteria to truncate the modal series to a finite number of terms can be derived based on simple physical considerations concerning the spatial and spectral properties of the external excitation and the excited system. The methods described below can be applied whatever the Euclidean dimension of the problem may be. However, for mathematical convenience the examples used to illustrate the general formalism refer here to straight beam systems. A few other examples about plates and shells will be described in the following chapters.

4.3.1 *Equations of motion projected onto a modal basis*

Let us consider the forced dynamical problem governed by the linear partial differential equations of the general type:

$$\left.\begin{array}{l} K[\vec{X}] + C[\dot{\vec{X}}] + M[\ddot{\vec{X}}] = \vec{F}^{(e)}(\vec{r};t) \\ + I.C. \text{ and } B.C. \end{array}\right\} \quad [4.35]$$

$I.C.$ stands for the initial conditions and $B.C.$ for the conservative boundary conditions. The stiffness, damping and mass operators $K[\;], C[\;], M[\;]$ can depend on the vector position $\vec{r}$ in the Euclidean space. To the forced problem governed by [4.35], we associate the modal problem [4.36], which satisfies the same boundary conditions:

$$\left.\begin{array}{l} [K(\vec{r}) - \omega^2 M(\vec{r})][\vec{\varphi}] = 0 \\ + B.C. \end{array}\right\} \quad [4.36]$$

solutions of which are defining the following modal quantities:

$$\{\vec{\varphi}_n\}; \quad \{\omega_n\}; \quad \{K_n\}; \quad \{M_n\}; \quad n = 1, 2, 3, \ldots \quad [4.37]$$

As in the case of discrete systems, the mode shapes can be used to determine an orthonormal basis with respect to the stiffness and mass operators, in which the solution $\vec{X}(\vec{r}, t)$ of [4.35] can be expanded as the modal series:

$$\vec{X}(\vec{r}, t) = \sum_{n=1}^{\infty} q_n(t)\vec{\varphi}_n(\vec{r}) \quad [4.38]$$

where the time functions $q_n(t)$ $n = 1, 2, \ldots$, termed modal displacements, are the components of the displacement field $\vec{X}(\vec{r}; t)$ in the modal coordinate system.

Formal proof of such a statement is not straightforward and is omitted as already mentioned in the introduction. As the dimension of the functional vector space is infinite, a delicate problem of convergence and space completeness arises. Substitution of [4.38] into the equation of motion [4.35], leads to:

$$\left.\begin{array}{l} \sum_{n=1}^{\infty} \{K[\varphi_n(\vec{r})]q_n(t) + C[\varphi_n(\vec{r})]\dot{q}_n(t) + M[\varphi_n(\vec{r})]\ddot{q}_n(t)\} = \vec{F}^{(e)}(\vec{r}; t) \\ + I.C. \end{array}\right\} \quad [4.39]$$

The system of equations [4.39] is projected on the k-th mode shape $\vec{\varphi}_k$ by using the scalar product [1.43] and the orthogonality properties [4.2]. The transformation results into the following ordinary differential equation:

$$K_k q_k + \sum_{n=1}^{\infty} C_{kn}\dot{q}_n + M_k \ddot{q}_k = Q_k^{(e)}(t) \qquad [4.40]$$

$$\text{where} \quad C_{kn} = \langle \vec{\varphi}_k, C[\vec{\varphi}_n] \rangle_{(\mathcal{V})} \quad \text{and} \quad Q_k^{(e)}(t) = \langle \vec{\varphi}_k, \vec{F}^{(e)} \rangle_{(\mathcal{V})}$$

The subscript $(\mathcal{V})$ accounts for the space integration domain involved in the scalar product. $Q_k^{(e)}(t)$ $k = 1, 2, \ldots,$ termed modal forces, are the components of the external force field $\vec{F}^{(e)}(\vec{r}; t)$, as expressed in the modal coordinate system. As already discussed in [AXI 04], Chapter 7, the result [4.40] can be drastically simplified if coupling through the damping operator is negligible. If it is the case, the mode shapes of the conservative system can be assumed to be orthogonal with respect to the C damping operator and [4.40] reduces to a single equation which governs the forced vibration of an harmonic oscillator, called *modal oscillator*:

$$M_k(\omega_k^2 q_k + 2\omega_k \varsigma_k \dot{q}_k + \ddot{q}_k) = Q_k^{(e)}(t)$$

$$\text{where} \quad \langle \vec{\varphi}_k, C[\vec{\varphi}_n] \rangle_{(\mathcal{V})} = \begin{cases} c_k = 2\omega_k M_k \varsigma_k > 0; & \text{if } n = k \\ 0; & \text{otherwise} \end{cases} \qquad [4.41]$$

By using the Laplace transform, the image of the modal displacement is obtained as:

$$\tilde{q}_k(s) = \frac{\tilde{Q}_k(s)}{M_k\left(\omega_k^2 + 2\omega_k \varsigma_k s + s^2\right)}$$

$$\text{where} \quad \tilde{Q}_k(s) = \tilde{Q}_k^{(e)}(s) + \dot{q}_k(0) + (2\omega_k \varsigma_k + s) q_k(0)$$

The Laplace transform of the displacement field in the physical coordinates system is finally obtained by using the modal series [4.38]:

$$\tilde{\vec{X}}(\vec{r}, s) = \sum_{k=1}^{\infty} \frac{\tilde{Q}_k(s) \vec{\varphi}_k(\vec{r})}{M_k\left(\omega_k^2 + 2\omega_k \varsigma_k s + s^2\right)} \qquad [4.42]$$

Here, the number of terms of the series is infinite, which was obviously not the case for N-DOF systems. Thus, in practice, to compute modal expansions like [4.42], the series must be truncated to a finite number of terms. Truncation criteria which lead to acceptable errors are discussed in the next subsections.

4.3.2 *Deterministic excitations*

To study the time response of continuous systems it is necessary to know both the spatial distribution of the excitations and their variation with time. The deterministic external forcing functions are described by a vector $\vec{F}^{(e)}(\vec{r};t)$ which may stand either for an external force field which fluctuates with time, or for a prescribed motion assigned to some degrees of freedom of the mechanical system. Firstly, it is useful to make the distinction between forcing functions in which space and time variables are separated and those where they are not. It is convenient to start with the former case which is more common.

4.3.2.1 Separable space and time excitation

The general form of this kind of excitation can be written as:

$$\vec{F}^{(e)}(\vec{r},t) = F_0 \Psi(\vec{r}) \vec{u}(\vec{r}) f(t) \qquad [4.43]$$

$\Psi(\vec{r})$ is a function, or a distribution, which specifies the space distribution of the loading and which complies with the norm condition:

$$\int_{(\mathcal{V})} |\Psi(\vec{r})| \, d\mathcal{V} = 1 \qquad [4.44]$$

$(\mathcal{V})$ is the domain in which Ψ is defined. F_0 is the scale factor of the load magnitude. The unit vector $\vec{u}(\vec{r})$ specifies the load direction in the Euclidean space and finally $f(t)$ describes the time evolution, or time-history of the loading. Ψ and $f(t)$ are assumed to be both integrable and square integrable. The resultant of the loading is given by:

$$\vec{R}(t) = \int_{(\mathcal{V})} \vec{F}^{(e)}(\vec{r},t) \, d\mathcal{V} = f(t) \int_{(\mathcal{V})} \Psi(\vec{r}) \vec{u}(\vec{r}) \, d\mathcal{V} \qquad [4.45]$$

The point $\vec{r}_a$ at which the resultant is applied is time independent. It is given by the barycenter of the function Ψ:

$$\vec{r}_a = \frac{\displaystyle\int_{(\mathcal{V})} \vec{r} \Psi \, d\mathcal{V}}{\displaystyle\int_{(\mathcal{V})} \Psi \, d\mathcal{V}} \qquad [4.46]$$

The modal forces related to this type of excitation are written as:

$$Q_n^{(e)}(t) = \left\langle \vec{\varphi}_n, \vec{F}^{(e)}(\vec{r};t) \right\rangle_{(\mathcal{V})} = F_0 f(t) \Psi_n; \quad \text{with } \Psi_n = \langle \vec{\varphi}_n, \Psi \vec{u} \rangle_{(\mathcal{V})} \qquad [4.47]$$

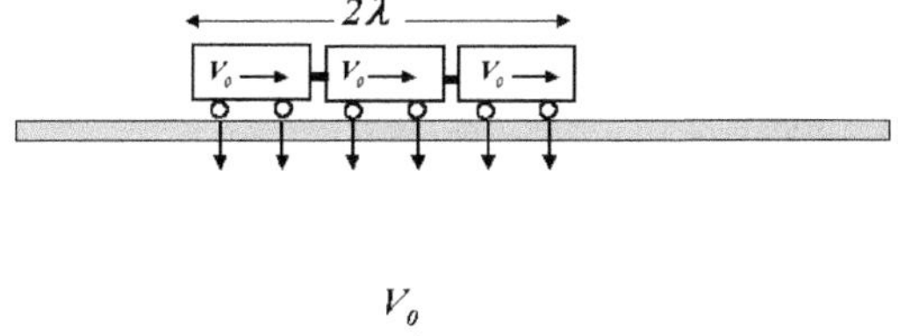

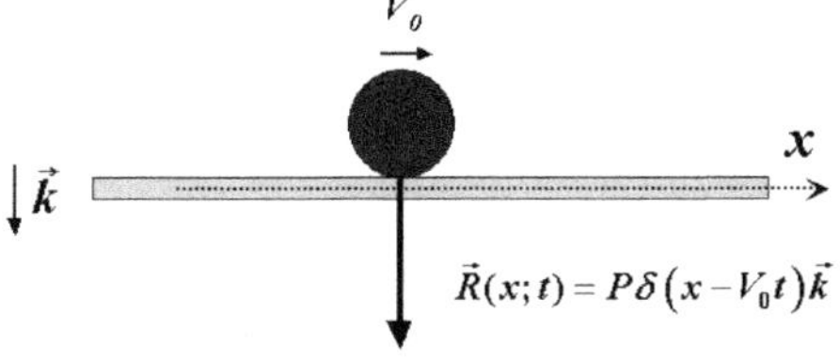

Figure 4.22. *Example of a travelling load*

4.3.2.2 Non-separable space and time excitation

When the position vector used to describe the spatial distribution of the load is time dependent it becomes impossible to separate the time and space variables. This is typically the case of the so called *travelling loads*, which are of practical importance in many applications. Let us consider, for instance, a train running on a flexible bridge at a constant cruising speed V_0, see Figure 4.22. One is interested in analysing the response of the bridge loaded by the weight of the running train. Let 2λ designates the length of the train. As a first approximation made here for the sake of simplicity, the total weight $\vec{P} = -M\vec{g}$ is assumed to be distributed uniformly along the train. So, the forcing function is written as:

$$\vec{F}^{(e)}(x,t) = \frac{\vec{P}}{2\lambda}[\mathcal{U}(x - V_0 t + \lambda) - \mathcal{U}(x - V_0 T - \lambda)]$$

$\mathcal{U}(x)$ is the Heaviside step function and the running abscissa $V_0 t$ is taken at the middle of the train. The resultant and the abscissa $x_a(t)$ at which it is applied are obtained by using the relationships [4.45] and [4.46]:

$$\vec{R} = \frac{\vec{P}}{2\lambda}\int_{-\infty}^{+\infty} [\mathcal{U}(x - V_0 t + \lambda) - \mathcal{U}(x - V_0 t - \lambda)]\,dx$$

$$= \frac{\vec{P}}{2\lambda}\left(\int_{V_{t-\lambda}}^{+\infty} dx - \int_{V_{t+\lambda}}^{+\infty} dx\right) = \vec{P}$$

$$x_a(t) = \frac{1}{2\lambda}\int_{-\infty}^{+\infty} x[\mathcal{U}(x - V_0 t + \lambda) - \mathcal{U}(x - V_0 t - \lambda)]\,dx$$

$$= \int_{Vt-\lambda}^{Vt+\lambda} x\,dx = V_0 t$$

Now, if the length of the train remains much smaller than the bridge span denoted L, the load can be reasonably modelled as a concentrated load which leads to the modal forces:

$$\mathcal{Q}_n^{(e)} = \langle \vec{\varphi}_n, \vec{P}\delta(x - V_0 t)\rangle = \vec{P} \cdot \vec{\varphi}_n(Vt) \qquad [4.48]$$

According to the result [4.48], the whole series of modes is excited as time elapses. If the actual length of the train is accounted for, the modal forces are found to be:

$$\begin{aligned}\mathcal{Q}_n^{(e)} &= \frac{1}{2\lambda}\int_0^L (\vec{P} \cdot \vec{\varphi}_n(x))[\mathcal{U}(x - V_0 t + \lambda) - \mathcal{U}(x - V_0 T - \lambda)]\,dx \\ &= \frac{1}{2\lambda}\int_{VT-\lambda}^{VT+\lambda} (\vec{P} \cdot \vec{\varphi}_n(x))\,dx \end{aligned} \qquad [4.49]$$

So, the modal excitation is found to depend on the ratio of the modal wavelength on the train length. Assuming for instance that the bridge deck can be modelled as a pinned-pinned beam, the modal force is expressed as:

$$\mathcal{Q}_n^{(e)} = \frac{PL}{2n\pi\lambda}\left(\cos\left(\frac{n\pi(V_0T - \lambda)}{L}\right) - \cos\left(\frac{n\pi(V_0T + \lambda)}{L}\right)\right)$$

which is found to vanish for all the modes such that $\lambda = mL/n$ where m is an integer less than n. So, such modes do not contribute to the bridge motion and can be removed from the modal basis.

4.3.3 *Truncation of the modal basis*

4.3.3.1 Criterion based on the mode shapes

The last result of the preceding subsection can be restated as a general rule, according to which all the mode shapes which are orthogonal to the spatial distribution of the excitation can be discarded from the modal model. In other words, the only modes which contribute to the response series [4. 42] are those modes which are not orthogonal to the spatial distribution of the excitation:

$$\langle \vec{\varphi}_n, \Psi(\vec{r})\vec{u}(\vec{r})\rangle_{(\mathcal{V})} = \Psi_n \neq 0 \qquad [4.50]$$

where for convenience the criterion [4.50] is formulated in the case of a separated variables forcing function.

It may be noted that a modal truncation based on the mode shape criterion is similar to the elimination of some components of the physical displacements in a solid body, based on the orthogonality with the loading vector field.

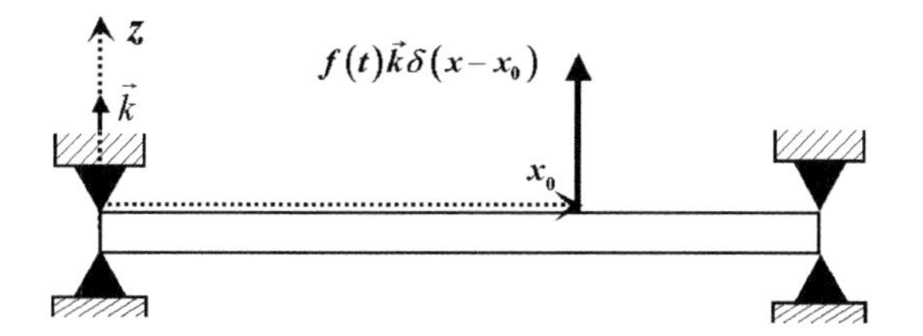

Figure 4.23. *Pinned-pinned beam loaded by a transverse concentrated force*

EXAMPLE. – *Beam loaded by a concentrated transverse force*

The problem is sketched in Figure 4.23. The equilibrium equation is written as:

$$EI\frac{\partial^4 Z}{\partial x^4} + C\dot{Z} + \rho S\ddot{Z} = f(t)\delta(x - x_0); \quad Z(0) = Z(L) = 0;$$

$$\left.\frac{\partial^2 Z}{\partial x^2}\right|_0 = \left.\frac{\partial^2 Z}{\partial x^2}\right|_L = 0$$

The beam being provided with pinned support conditions at both ends, the modal quantities relevant to the problem are:

$$\varphi_n(\xi) = \sin(n\pi\xi); \quad M_n = \rho SL/2 = M_b/2; \quad K_n = (n\pi)^2\frac{EI}{2L^3};$$

$$\omega_n^2 = \frac{(n\pi)^4}{L^4}c^2, \quad \text{where } \xi = x/L \text{ and } c^2 = EI/\rho S$$

The Laplace transform of the response is:

$$Z(\xi, \xi_0; s) = \frac{2\tilde{f}(s)}{M_b}\sum_{n=1}^{\infty}\frac{\sin(n\pi\xi)\sin(n\pi\xi_0)}{(c^2(n\pi/L)^4 + 2sc\varsigma_n(n\pi/L)^2)}$$

If $\xi_0 = 0.5$, the non-vanishing terms of the series are related to the odd modes $n = 2k + 1,\ k = 0, 1, 2, \ldots$ only:

$$Z(0.5, \xi; s) = \frac{2\tilde{f}(s)}{M_b}\sum_{k=0}^{\infty}\frac{(-1)^k\sin((2k+1)\pi\xi)}{c^2((2k+1)\pi/L)^4 + 2sc\varsigma_n^2((2k+1)\pi/L)^2}$$

EXAMPLE. – *Beam symmetrically or skew symmetrically loaded*

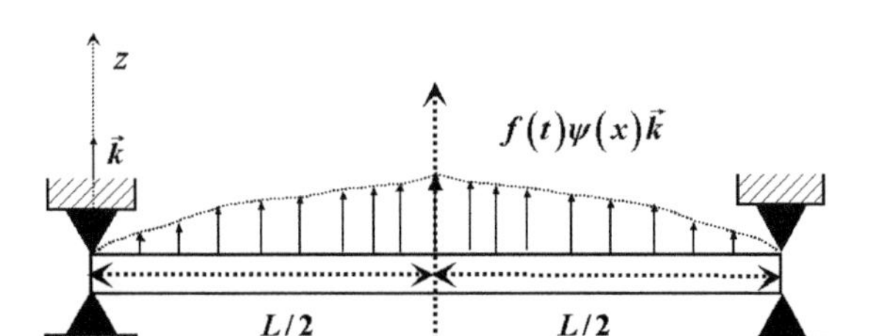

Figure 4.24. *Symmetric loading of a beam provided with symmetric supports*

If the load is symmetrically, or skew symmetrically, distributed with respect to the cross-section at mid-span of the beam and if the boundary conditions are symmetric, the only modes which contribute to the response must verify the same conditions of symmetry as the loading function, see Figure 4.24. Although this rule is correct from the mathematical standpoint, it is still necessary to be careful when using it, because real structures present inevitably material and geometrical defects which spoil the symmetry of the ideal model. One striking example of the importance of such 'small' defects will be outlined in subsection 4.4.3.3.

4.3.3.2 Spectral criterion

The spectral considerations made in [AXI 04] Chapter 9, concerning the dynamical response of forced N-DOF systems, can be extended to continuous structures to produce a very useful criterion for restricting the modal basis to a finite number of modes. Figure 4.25 is a plot of the power density spectrum of some excitation signal, which in practice extends over a finite bandwidth, limited by a lower cut-off frequency f_{c1} and by a upper cut-off frequency f_{c2}; that is, outside the interval f_{c1}, f_{c2} the excitation power density becomes negligible. The dots on the frequency axis mark the sequence of the natural frequencies of the excited structure, which of course extends to infinity.

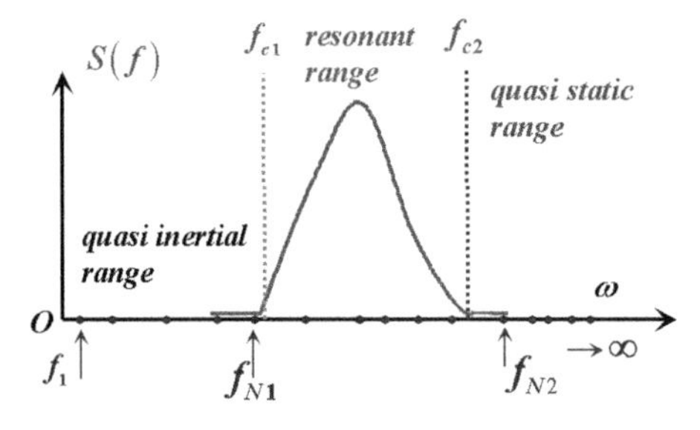

Figure 4.25. *Spectral domains of excitation versus structure response properties*

Let us consider first the response of a single mode f_n to the excitation signal. The response is found to be quasi-inertial if $f_n/f_{c1} \ll 1$, in the resonant range if $f_{c1} < f_n < f_{c2}$, and quasi-static if $f_n/f_{c2} \gg 1$. Therefore, a finite number of low frequency modes can lie in the quasi-inertial range, and infinitely many other modes lie in the quasi-static range. The contribution to the total response of the modes lying in the quasi-inertial range can be accounted for by neglecting the stiffness and damping terms of the modal oscillators and only a finite number of such modal contributions are to be determined. On the other hand, the contribution to the total response of the modes lying in the quasi-static range can be accounted for by neglecting the damping and inertial terms of the modal oscillators; however there are still infinitely many modal contributions to be accounted for. The method for avoiding the actual calculation of such an infinite series is best described starting from a specific example.

Let us consider again a vehicle of mass M, travelling at speed V_0 on a flexible bridge. Assuming the bridge deck is modelled as an equivalent straight beam provided with pinned supports at both ends and damping is neglected, the equations of the problem are written as:

$$EI\frac{\partial^4 Z}{\partial x^4} + \rho S\ddot{Z} = -Mg\delta(x - V_0 t)$$

$$Z(0) = Z(L) = 0; \qquad \left.\frac{\partial^2 Z}{\partial x^2}\right|_0 = \left.\frac{\partial^2 Z}{\partial x^2}\right|_L = 0$$

By projecting this system on the pinned-pinned modal basis, we get the system of uncoupled ordinary differential equations, comprising an infinite number of rows of the type:

$$\omega_n^2 q_n + \ddot{q}_n = -g\frac{2M}{M_b}\sin\left(\frac{n\pi V_0 t}{L}\right); \quad \text{where } \omega_n^2 = \left(\frac{n\pi}{L}\right)^4 \frac{EI}{\rho S}$$

It immediately appears that the dynamic response strongly depends upon the cruising speed V_0 of the load. In particular, the resonant response of the n-th mode occurs if the following condition is fulfilled:

$$\frac{n\pi V_n}{L} = \left(\frac{n\pi}{L}\right)^2 \sqrt{\frac{EI}{\rho S}} \Rightarrow V_n = \frac{n\pi}{L}\sqrt{\frac{EI}{\rho S}}$$

Shifting to the spectral domain, the Fourier transform of the beam deflection is found to be:

$$\widehat{Z}(x,\omega) = -g\frac{M}{M_b}\sum_{n=1}^{\infty}\frac{(\delta(\Omega_n - \omega) - \delta(\Omega_n + \omega))}{i(\omega_n^2 - \omega^2)}\sin\left(\frac{n\pi x}{L}\right)$$

$$\text{where } \Omega_n = \frac{n\pi V_0}{L}.$$

which becomes infinite at the undamped resonances $\omega_n = \Omega_n$.

Nevertheless, it may be also realized that in most cases of practical interest even the smallest cruising speed V_1 needed to excite the first resonance of the beam is likely to be far beyond the realistic speed range of the vehicle. This is because a bridge is designed to withstand large static transverse loads. As a consequence, the dynamical response of the loaded bridge can be determined entirely by using the quasi-static approximation:

$$\left(\frac{n\pi}{L}\right)^4 \frac{EI}{\rho S} q_n = -g\frac{2M}{M_b}\sin\left(\frac{n\pi V_0 t}{L}\right) \Rightarrow q_n = -\frac{gM}{K_n}\sin\left(\frac{n\pi V_0 t}{L}\right);$$

$$\text{where } K_n = \frac{M_b}{2}\left(\frac{n\pi}{L}\right)^4 \frac{EI}{\rho S}$$

and the bridge deflection is found to be:

$$Z(x;t) = -gM\sum_{n=1}^{\infty}\frac{1}{K_n}\sin\left(\frac{n\pi V_0 t}{L}\right)\sin\left(\frac{n\pi x}{L}\right)$$

To restate the conclusions of this example as a general rule, one-dimensional problems are considered for mathematical convenience. Further extension to the case of two or three-dimensional problems is straightforward as it suffices to deal in the same way with each of the Euclidean dimensions of the problem. As outlined above, it is relevant to discuss the relative importance of the modal expansion terms of the response in relation to the spectral content of the excitation. The Fourier transform of the response is written as:

$$\widehat{Z}(x;\omega) = F_0\widehat{f}(\omega)\sum_{n=1}^{\infty}\frac{\varphi_n(x)\Psi_n}{M_n\left(\omega_n^2 + 2i\omega_n\omega\varsigma_n - \omega^2\right)} \quad [4.51]$$

The excitation spectrum $S_{ff}(\omega) = 2\widehat{f}(\omega)\widehat{f}^*(\omega)$ is again assumed to be negligible outside the finite interval ω_{c1}, ω_{c2}. If ω_n designates the natural frequencies of the structure, ordered as an increasing sequence, the three following distinct cases have to be discussed:

1. *Resonant response range:* $\omega_n \in [\omega_{c1}, \omega_{c2}]$
   It is clearly necessary to retain all the modes whose frequencies lie within the spectral range of the excitation, except if they are orthogonal to the spatial distribution of the excitation, that is if ψ_n vanishes. The contribution of such modes to the response is given by the full expression [4.51]. The spectrum

of the modal response can be conveniently related to the excitation spectrum (cf. Volume 1 Chapter 9) as:

$$S_{ZZ}^{(n)}(x;\omega) = \left(\frac{\varphi_n(x)\Psi_n}{K_n}\right)^2 \left(\frac{1}{(1-(\omega/\omega_n)^2)^2 + 4(\omega\varsigma_n/\omega_n)^2}\right) S_{ff}(\omega)$$ [4.52]

Spectral relationships like [4.52] are especially useful to characterize the magnitude of the vibration without entering into the detailed time-history of the response. It is recalled that the mean square value of the modal response can be inferred from the modal response spectrum $S_{ZZ}^{(n)}(x;\omega)$ by integration with respect to frequency as detailed in [AXI 04], Chapter 8.

2. *Quasi-static responses*: $\omega_{c2} \ll \omega_N$

In the same way as in the travelling load example, the series [4.51] can be reduced to the quasi-static form:

$$\widehat{Z}(x;\omega,N) = F_0 \widehat{f}(\omega) \sum_{n=N}^{\infty} \frac{\varphi_n(x)\Psi_n}{K_n}$$ [4.53]

The displacement is synchronous with the time evolution of the excitation. Furthermore, the series [4.53] is found to converge, as appropriate from the physical standpoint. Here, convergence can be immediately checked as the modal stiffness coefficients K_n are proportional to n^4 for bending modes and to n^2 in the case of torsion and longitudinal modes. Furthermore, the series [4.53] calculated from $n = 1$ instead of $n = N$, must converge to the static solution Z_s of the forced problem related to the system [4.35]:

$$\left.\begin{aligned} K(x)Z_s &= F_0\Psi(x) \\ &+ B.C. \end{aligned}\right\}$$ [4.54]

This is because the Hilbert space of the solutions is complete by definition. So, the modal expansion of Z_s is found to be:

$$Z_s(x) = F_0 \sum_{n=1}^{\infty} \frac{\varphi_n(x)\psi_n}{K_n}$$ [4.55]

As a consequence, the quasi-static part of the dynamical response [4.51] may be conveniently written by using a finite number of terms only:

$$\widehat{Z}(x;\omega,N) = F_0 \widehat{f}(\omega) R_N(x)$$ [4.56]

where

$$R_N(x) = \frac{Z_s(x)}{F_0} - \sum_{n=1}^{N} \frac{\varphi_n(x)\psi_n}{K_n}$$

$R_N(x)$ is termed the *quasi-static mode* or *pseudo-mode*. It gives the resultant of the individual quasi-static contributions to the response of the infinitely many natural modes of vibration which lie in the quasi-static range, based on the spectral criterion $\omega_{c2} \ll \omega_N$. Provided the solution of the static problem [4.54] may be made available from a direct analytical or numerical calculation, the pseudo-mode can be determined by using [4.56]. Then, the Fourier transform of the solution to the dynamical problem [4.35] is expanded as:

$$\widehat{Z}(x;\omega) = F_0 \widehat{f}(\omega) \left\{ \sum_{n=1}^{N} \frac{\varphi_n(x)\Psi_n}{M_n\left(\omega_n^2 + 2i\omega_n\omega\varsigma_n - \omega^2\right)} + R_N(x) \right\} \quad [4.57]$$

The quasi-static correction term $R_N(x)$ may also be interpreted in a slightly distinct way, by making use of the concept of equivalent stiffness, defined here as the force to displacement ratio:

$$K_{eq}(x,\Psi) = \frac{F_0}{Z_s(x)}; \quad \text{where } \frac{1}{K_{eq}(x,\Psi)} = \sum_{n=1}^{\infty} \frac{\varphi_n(x)\Psi_n}{K_n} \quad [4.58]$$

$K_{eq}(x,\Psi)$ depends on the position along the beam and on the spatial distribution $\Psi(x)$ of the load through the static deflection $Z_s(x)$, as evidenced in equation [4.55]. In agreement with [4.57], the equivalent stiffness coefficient which characterizes the truncation of the modal basis to the order N, is defined as:

$$\frac{1}{K_N(x,\Psi)} = \sum_{n=N+1}^{\infty} \frac{\varphi_n(x)\Psi_n}{K_n} = \frac{1}{K_{eq}(x,\Psi)} - \sum_{n=1}^{N} \frac{\varphi_n(x)\Psi_n}{K_n} \quad [4.59]$$

This way of introducing the quasi-static corrective term for the truncated model presents the physical interest to put clearly in evidence that by dropping out the modes $n > N$, the truncated model is artificially stiffened, as expected since the number of degrees of freedom is decreased. The effect of adding the pseudo-mode contribution to the solution is precisely to provide the appropriate correction by adding a spring 'mounted in series' with the truncated model. Stated briefly, the inverse of the truncation stiffness accounts for the flexibility of the neglected modes.

3. *Quasi-inertial response*: $\omega_p \ll \omega_{c1}$

   By analogy with the quasi-static approximation, it is also possible to simplify the contribution of the modal responses related to the modes whose frequencies are within the quasi-inertial range. The corrective term is defined by:

$$F_0 \widehat{f}(\omega) R_p(x;\omega); \quad \text{where } R_p(x;\omega) = -\sum_{n=1}^{p} \frac{\varphi_n(x)\Psi_n}{\omega^2 M_n} \qquad [4.60]$$

Following the same procedure as in the quasi-static case, it is possible to define an equivalent mass of the system and an equivalent truncation mass coefficient. Accordingly, the expansion [4.57] can be further simplified as:

$$\widehat{Z}(x;\omega) = F_0 \widehat{f}(\omega) \left\{ R_p(x,\omega) + \sum_{n=p+1}^{N} \frac{\varphi_n(x)\psi_n}{M_n\left(\omega_n^2 + 2i\omega_n\omega\varsigma_n - \omega^2\right)} + R_N(x) \right\} \qquad [4.61]$$

Modal truncation in the quasi-inertial range is of practical interest when a large number of modes lie in the low frequency range $\omega_p \ll \omega_{c1}$.

4.3.4 *Stresses and convergence rate of modal series*

The modal series of the type [4.38] are found to converge for any realistic excitation, as anticipated based on physical reasoning. This can be checked mathematically by noting that the infinite number of terms involved in the quasi-static term $R_N(x)$ form a sequence decreasing at least as $1/n^2$. The convergence of the stress and strain expansions is also granted almost everywhere along the beam. However, convergence rate of stress or strain series is significantly slower than that of the displacement series because they are obtained through a space derivation of the latter. It is worth illustrating this important point by taking an example.

EXAMPLE. – *Stretching of a straight beam*

The static response of a cantilevered beam loaded by an axial force is analysed first by solving directly the boundary value problem (local formulation) and then by using the modal expansion method (Figure 4.26).

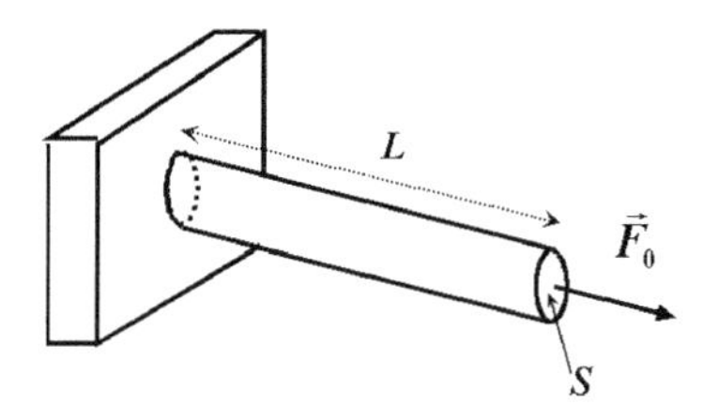

Figure 4.26. *Cantilevered beam loaded by an axial force*

1. *Local formulation*

   The solution of the static problem is immediate. The local equilibrium equation and boundary conditions are:

$$\left.\begin{array}{l} -ES\dfrac{d^2X}{dx^2} = F_0\delta(x-L) \\ X(0)=0 \end{array}\right\} \Leftrightarrow \quad \begin{array}{r} -ES\dfrac{d^2X}{dx^2} = 0 \\ X(0)=0;\, ES\left.\dfrac{dX}{dx}\right|_L = F_0 \end{array}$$

   The solution is:

$$X_s(\xi) = \frac{LF_0}{ES}\xi; \quad \text{where } \xi = \frac{x}{L}$$

   The equivalent stiffness is $K(\xi,1) = ES/(L\xi)$ and the axial stress is $\mathcal{N}(\xi) = (ES/L)(dX/d\xi) = F_0$.

2. *Modal formulation*

   The modal basis complying with the boundary conditions of the problem are:

$$\varphi_n(\xi) = \sin\varpi_n\xi; \quad \varpi_n = \pi(1+2n)/2; \quad \omega_n = \frac{c_0\varpi_n}{L}; \text{ where } c_0^2 = E/\rho$$

$$M_n = \rho SL/2; \quad K_n\frac{\pi^2 ES}{8L}(1+2n)^2$$

   The modal (or generalized) forces are:

$$Q_n^{(e)} = \langle\varphi_n, F_0\delta(\xi-1)\rangle_{(L)} = F_0\sin\varpi_n = (-1)^n F_0, \quad n = 0,1,2\ldots$$

   The modal expansion of the type [4.55] gives the displacement field as the Fourier series:

$$X(\xi,1) = \frac{F_0L}{ES}\frac{8}{\pi^2}\sum_{n=0}^{\infty}\frac{(-1)^n\sin[\pi(1/2+n)\xi]}{(1+2n)^2} \qquad [4.62]$$

   The series is found to converge rather quickly to the correct result as shown in Figure 4.27, in which the displacement is plotted versus the modal cut-off index N at $\xi = 0.75$ and $\xi = 0.1$. Incidentally, this kind of calculation is a convenient method to determine the value of a fairly large number of series. For instance, here it is found that:

$$\sum_{n=0}^{\infty}\frac{1}{(1+2n)^2} = \frac{\pi^2}{8}$$

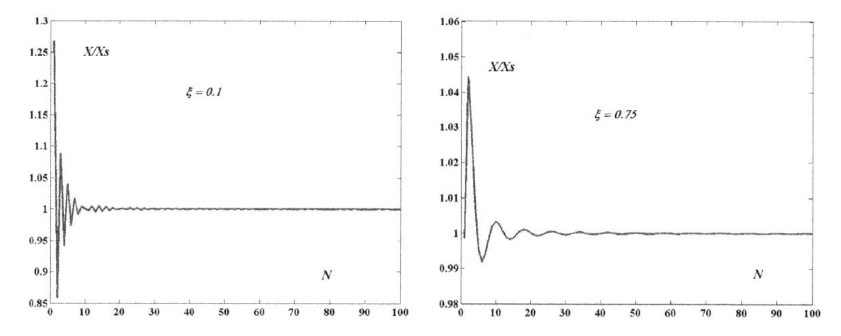

Figure 4.27. *Convergence of modal series of displacement*

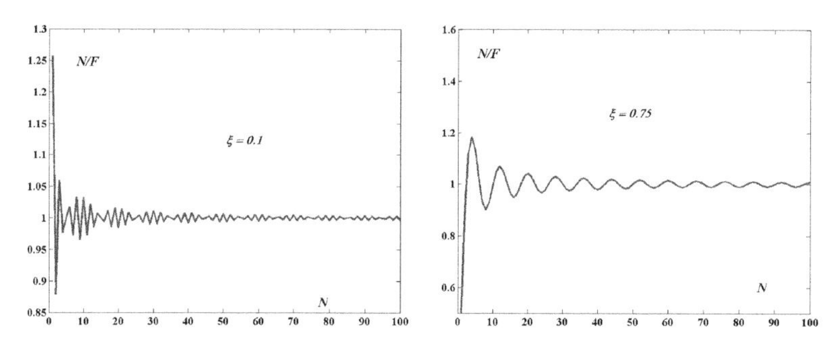

Figure 4.28. *Convergence of modal series of stress*

To determine the stress field, the displacement series is differentiated term by term, which results in:

$$\mathcal{N}_n(\xi, 1) = -F_0 \frac{4}{\pi} \sum_{n=0}^{\infty} \frac{(-1)^n \cos[\pi(1/2+n)\xi]}{(1+2n)} \qquad [4.63]$$

It may be immediately noticed that convergence of the series is not uniform, as all the terms vanish at $\xi = 1$. This is a direct consequence of the mode shapes used for the expansion, which refer to a free end condition at $\xi = 1$. Anywhere else the series converges in an alternative way to the correct value as shown in Figure 4.28.

4.4. Substructuring method

4.4.1 *Additional stiffnesses*

Let us consider first the case of a single structure described by the dynamic equation:

$$K(\vec{r})\vec{X}(\vec{r};t) + M(\vec{r})\ddot{\vec{X}}(\vec{r};t) = 0 + \text{elastic } B.C. \qquad [4.64]$$

$\{\vec{\varphi}_n(\vec{r})\}$ is the related modal basis. If other elastic supports are added to the structure, the equation becomes:

$$K(\vec{r})\vec{X}(\vec{r};t) + K_a(\vec{r})\vec{X}(\vec{r};t) + M(\vec{r})\ddot{\vec{X}}(\vec{r};t) = 0 + \text{elastic } B.C. \qquad [4.65]$$

where $K_a(\vec{r})$ stands for the stiffness operator of the additional supports, which can be either concentrated at some discrete locations, or continuously distributed at the boundary of the structure. It is possible to approximate the new modal basis $\{\vec{\phi}_n(\vec{r})\}$ by projecting the system [4.65] on $\{\vec{\varphi}_n(\vec{r})\}$; which gives a truncated modal expansion of the type:

$$\vec{\phi}_n(\vec{r}) = \sum_{j=1}^{N} q_j \vec{\varphi}_j(\vec{r}) \qquad [4.66]$$

Substituting [4.66] into [4.65] and projecting the result onto the mode $\vec{\varphi}_n(\vec{r})$, the current row of the coupled modal system is obtained as:

$$\left(K_n + K_n^{(a)} - \omega^2 M_n\right) q_n + \sum_{j \neq n}^{N} K_j^{(a)} q_j = 0 \qquad [4.67]$$

The coupling terms result from the lack of orthogonality of the $\vec{\varphi}_n(\vec{r})$ with respect to the operator $K_a(\vec{r})$. The solution of [4.67] produces N new mode shapes $\vec{\phi}_n(\vec{r})$ as expressed in the $\{\vec{\varphi}_n(\vec{r})\}$ basis. They can be written in terms of the physical coordinates $\vec{r}$ by using [4.66]. The method is illustrated by some examples in the next subsections. As could be anticipated, it will be shown that the higher the truncation order and the less the order of the calculated modes, the more accurate is the method.

4.4.1.1 Beam in traction-compression with an end spring

In this first example, the beam is fixed at one end and supported at the other end by a linear spring, as shown in Figure 4.29. The spring is successively modelled as an elastic impedance, i.e. a boundary condition, or as a connecting force, that is an additional loading term in the dynamic equation. In both cases, the object is to derive the longitudinal modes of the modified system.

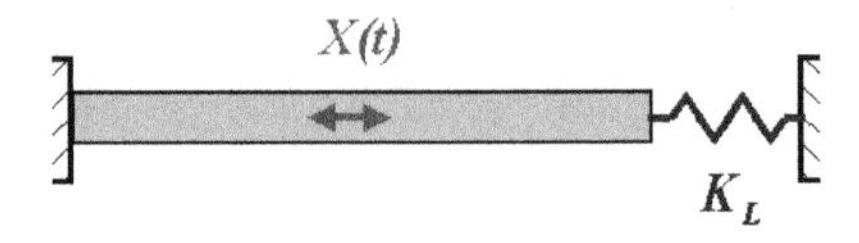

Figure 4.29. *Longitudinal modes of a beam fixed at one end supported by a spring at the other end*

1. *Spring modelled as an elastic impedance*:
   The modal system is written as:

$$-ES\frac{d^2X}{dx^2} - \omega^2\rho SX = 0; \; X(0) = 0; \qquad ES\frac{d^2}{dx^2}\bigg|_L = K_L X(L) \qquad [4.68]$$

To discuss the effect of the additional support it is found convenient to use dimensionless variables based on a few pertinent scaling factors, as follows:

$$\xi = x/L; \quad \bar{X} = X/L; \quad \varpi = \omega/\omega_0; \quad \gamma_{\mathrm{L}} = K_L L/ES; \quad \omega_0 = \frac{1}{L}\sqrt{\frac{E}{\rho}} \qquad [4.69]$$

So the modal system is written as:

$$\bar{X}'' + \varpi^2\bar{X} = 0; \quad \text{with } \bar{X}(0) = 0; \bar{X}'(1) = \gamma_L\bar{X}(1)$$

Again, the general solution of the differential equation is $\overline{X}(\xi) = a\sin(\varpi\xi) + b\cos(\varpi\xi)$ and the boundary conditions give the characteristic equation:

$$b = 0; \quad \varpi_n\cot(\varpi_n) = \gamma_L \qquad [4.70]$$

It is of interest to examine the following specific cases for further discussion:

(a) Spring stiffness coefficient is zero: $\gamma_L = 0$

$$\varpi_n = (1+2n)\frac{\pi}{2}; \; \varphi_n(\xi) = \sin(\varpi_n\xi); \quad n = 0, 1, 2\ldots$$

(b) Spring stiffness coefficient is infinite: $1/\gamma_L = 0$

$$\omega_n = n\pi; \quad \varphi_n(\xi) = \sin[\omega_n\xi]$$

(c) Spring stiffness coefficient has a nonzero finite value, for instance: $\gamma_L = \pi/4$

The first roots of the transcendental equation $x\cot(x) - \pi/4$ are found to be:

$$\varpi_1 = 1.9531; \quad \varpi_2 = 4.8722; \quad \varpi_3 = 7.9524; \quad \varpi_4 = 11.067$$

2. *Modal projection of the constrained system*
   The equilibrium equation of the constrained system is:

$$-ES\frac{d^2X}{dx^2} - \omega^2\rho SX = -K_L X\delta(x-L); \quad \text{with } X(0) = 0 \qquad [4.71]$$

Using the dimensionless quantities already defined above, the modal system is written as:

$$-\bar{X}'' - \varpi^2 \bar{X} = -\gamma_L \bar{X} \delta(\xi - 1); \quad \bar{X}(0) = 0 \qquad [4.72]$$

The mode shapes ϕ_n are expanded as a linear combination of the fixed-free beam modes:

$$\phi_n(\xi) = \sum_j^{\infty} q_j \varphi_j(\xi)$$

The projection of [4.72] onto φ_n results in:

$$\left[\varpi_n^2 - \varpi^2\right] q_n = -2\gamma_L (-1)^n \sum_j q_j \qquad [4.73]$$

Then the solution is obtained by solving a linear homogeneous coupled system. The modal frequencies of [4.73] are expected to lie within the range:

$$\varpi_n(\gamma_L = 0) < \varpi_n(\gamma_L = \pi/4) < \varpi_n(\gamma_L \to \infty) \Rightarrow \frac{(2n-1)\pi}{2} < \varpi_n < n\pi;$$
$$n = 1, 2, \ldots$$

So, if the system [4.73] is truncated to the order N, accuracy is expected to degrade progressively as n increases up to N. Taking for instance $N = 4$, [4.73] is written as follows:

$$\begin{bmatrix} 2\gamma_L + \varpi_1^2 - \varpi^2 & -2\gamma_L & 2\gamma_L & -2\gamma_L \\ -2\gamma_L & 2\gamma_L + \varpi_2^2 - \varpi^2 & -2\gamma_L & 2\gamma_L \\ 2\gamma_L & -2\gamma_L & 2\gamma_L + \varpi_3^2 - \varpi^2 & -2\gamma_L \\ -2\gamma_L & 2\gamma_L & -2\gamma_L & 2\gamma_L + \varpi_4^2 - \varpi^2 \end{bmatrix} \begin{bmatrix} q_1 \\ q_2 \\ q_3 \\ q_4 \end{bmatrix} = \begin{bmatrix} 0 \\ 0 \\ 0 \\ 0 \end{bmatrix}$$

by solving this system, the following values of the modified natural frequencies are obtained numerically:

$$\varpi_1 = 1.9652; \quad \varpi_2 = 4.8788; \quad \varpi_3 = 7.9571; \quad \varpi_4 = 11.071$$

If the projection is restricted to the first two modes, the values become:

$$\varpi_1 = 1.9784; \quad \varpi_2 = 4.8889$$

As a consequence of the stiffening effect due to modal truncation, these values are found to be higher (though by a small amount only) than the exact values. On the other hand, it is also interesting to check the convergence of the method when γ_L becomes so large as to enforce practically a fixed boundary condition; the numerical results presented below refer to a spring stiffness coefficient hundred times larger than the first modal stiffness. Six modes are

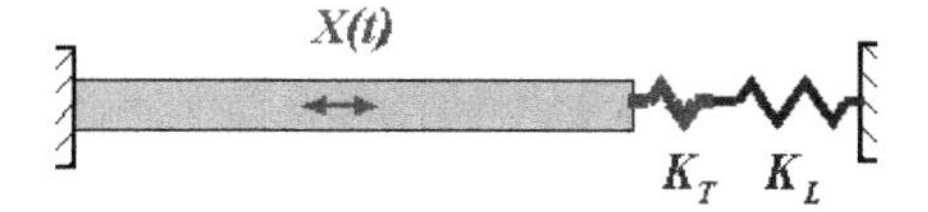

Figure 4.30. *Added truncation spring*

retained in the basis. The four first pulsations are found to be:

$$\varpi_1 = 3.1426; \quad \varpi_2 = 6.2915; \quad \varpi_3 = 9.4551; \quad \varpi_4 = 12.665$$

The relative error with respect to the exact values is an excess of about 3 percent. Further, if the number of selected modes is increased, to sixteen for instance, it is found that the relative error is reduced to about 1.3 percent. However, as indicated below, to improve accuracy it is more efficient to make use of the quasi-static corrective term [4.59] than to increase the size of the modal basis.

3. *Correction of the modal truncation*

   It is possible to improve the model presented above by adding to it a stiffness coefficient K_T which accounts suitably for the truncation of the modal basis. This numerical spring is mounted in series with the physical spring, as sketched in Figure 4.30, in such a way that the equivalent stiffness of the support is decreased, as given by:

$$K_e = \frac{K_T K_L}{K_T + K_L} \Rightarrow \frac{1}{\gamma_e} = \frac{1}{\gamma_T} + \frac{1}{\gamma_L} \qquad [4.74]$$

   Using the relation [4.59], the dimensionless flexibility of truncation is obtained as:

$$1/\gamma_T = 1 - \frac{8}{\pi^2} \sum_{n=0}^{N-1} \frac{1}{(2n+1)^2}$$

   Including the equivalent flexibility [4.74] in the model, and adopting $\gamma_L = \pi/4$ and $N = 4$, the following natural frequencies are obtained: $\varpi_1 = 1.9532$; $\varpi_2 = 4.8725$; $\varpi_3 = 7.9531$; $\varpi_4 = 11.068$, which are found to be very close to the exact values. Even if the basis is reduced to the two first modes only, the results are still sufficiently accurate for most applications: $\gamma_e = 0.7285$ and $\varpi_1 = 1.9538$; $\varpi_2 = 4.8755$. Finally, if the end spring stiffness is so large as to be practically equivalent to a fixed condition, satisfactory results can be obtained by selecting $N = 6$, which gives $\gamma_e = 29.67$.

4.4.1.2 Truncation stiffness for a free-free modal basis

Here, the beam is supported by a spring K_L at one end and left free at the other. Again, we want to compute the longitudinal vibration modes of the supported beam by using a truncated basis of the vibration modes in the free-free configuration, i.e. without end springs. As unnatural as it may appear, the procedure is useful in the context of the substructuring method, as made clear in subsection 4.4.3.

If a direct calculation is performed by modelling the elastic support as an elastic impedance, the following characteristic equation for the natural frequencies is $\gamma_L = \varpi_n \tan(\varpi_n)$. Selecting for instance $\gamma_L = 1$, the first reduced frequencies are $\varpi_1 = 0,86035$; $\varpi_2 = 3,4256$; $\varpi_3 = 6,4373$. Now if the modal projection method is used, a difficulty arises in determining the truncation flexibility since it is a priori infinite, due to the presence of the free rigid mode. However, let us discuss the problem a little further. The constrained beam is governed by:

$$-\bar{X}'' - \varpi^2 \bar{X} = \gamma_L \bar{X} \delta(\xi); \quad \bar{X}'(1) = 0$$

By setting $\gamma_L = 0$, the free-free modal basis is obtained:

$$\varpi_n^{(ff)} = n\pi; \quad a_n = 0 \Rightarrow \varphi_n(\xi) = \cos(n\pi\xi); \quad n = 0, 1, 2 \ldots$$

If four modes are retained in the projection, the following homogeneous system to be solved is:

$$\begin{bmatrix} 2(\gamma_L + \varpi^2) & 2\gamma_L & 2\gamma_L & 2\gamma_L \\ 2\gamma_L & 2\gamma_L + \pi^2 - \varpi^2 & 2\gamma_L & 2\gamma_L \\ 2\gamma_L & 2\gamma_L & 2\gamma_L + 4\pi^2 - \varpi^2 & 2\gamma_L \\ 2\gamma_L & 2\gamma_L & 2\gamma_L & 2\gamma_L + 9\pi^2 - \varpi^2 \end{bmatrix} \begin{bmatrix} q_1 \\ q_2 \\ q_3 \\ q_4 \end{bmatrix} = \begin{bmatrix} 0 \\ 0 \\ 0 \\ 0 \end{bmatrix}$$

The first row of the modal equation is the projection of the constrained local equation onto the rigid body mode $n = 1$. By solving numerically the system for $\gamma_L = 1$, the following natural frequencies are obtained:

$$\varpi_1 = 0.87901; \quad \varpi_2 = 3.4412; \quad \varpi_3 = 6.4770; \quad \varpi_4 = 9.5387$$

Although accounting for the truncation stiffness is not formally feasible, nothing prevents us in practice from providing the system with a fictitious spring of stiffness coefficient K_f, see Figure 4.31. It is appropriate to select for K_f a value much smaller than the modal stiffness related to the first non-rigid mode of the free-free beam, in such a way that the perturbation remains negligibly small, except in so far as the frequency of the rigid mode is concerned. The static response to an axial force is thus performed with this support, artificially introduced in the system to make the calculation possible. The erroneous contribution of the rigid mode can be eliminated afterwards. The procedure can be described starting from the equivalent flexibility as determined in the initial and then in the perturbed system. For the initial system:

$$\frac{1}{\gamma_e} = \text{``}\left(\frac{1}{0}\right)\text{''} + \frac{2}{\pi^2} \sum_{n=1}^{\infty} \frac{1}{n^2} = \text{``}\left(\frac{1}{0}\right)\text{''} + \frac{1}{3}$$

where "1/0" stands for the free rigid mode singular contribution.

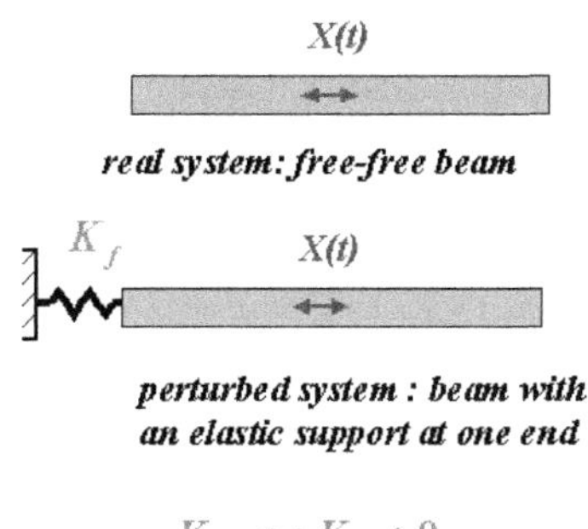

Figure 4.31. *Beam provided with an elastic support at one end*

For the perturbed system, in which perturbation of the non-rigid modes is assumed to be negligible:

$$\frac{1}{\gamma'} = \frac{1}{\gamma_f} + \frac{1}{3}$$

where $1/\gamma_f$ is the finite flexibility of the fictitious support.

Then the truncation flexibility is given by:

$$\frac{1}{\gamma_T} = \frac{1}{\gamma'} - \frac{1}{\gamma_f} - \frac{2}{\pi^2}\sum_{n=2}^{N}\frac{1}{n^2} = \frac{1}{3} - \frac{2}{n^2}\sum_{n=2}^{N}\frac{1}{n^2}$$

where $n = 2$ is the order of the first non-rigid mode of the free-free beam.

Using this corrective term, a truncation to $N = 5$ gives:

$$\gamma_e = 0.9465; \quad \varpi_1 = 0.8607; \quad \varpi_2 = 3.4263; \quad \varpi_3 = 6.4386; \quad \varpi_4 = 9.5322$$

which are very close to the exact values.

4.4.1.3 Bending modes of an axially prestressed beam

A beam clamped at one end and provided with a sliding support at the other, is prestressed by a compressive load P_0. The buckling load $P_c = (\pi/L)^2 EI$ was already determined in subsection 4.2.5.2. The related buckling mode shape was found to be:

$$\psi_n(x) = \frac{1}{2}\left(\cos\left(\frac{n\pi x}{L}\right) - 1\right)$$

However, as already indicated, the vibration modes cannot be expressed in a closed analytical form. So the modal projection method is applied to solve the problem in terms of a truncated series by using the modal basis of the unloaded beam. The equation of the prestressed beam is thus written as:

$$\rho S\ddot{Z} + P_0 \frac{\partial^2 Z}{\partial x^2} + EI \frac{\partial^4 Z}{\partial x^4} = 0$$

provided with the boundary condition: $\left(P_0(\partial Z/\partial x) + EI(\partial^3 Z/\partial x^3)\right)\big|_L = 0$

It is emphasized again that this condition depends on the axial force, which is not accounted for in the modal basis. Accordingly, the projection is carried out directly as a weighting integral, producing a non-symmetrical prestress operator denoted $K^{(0)}$, which couples the modal degrees of freedom:

$$\int_0^L \varphi_n(x) \left(\rho S\ddot{Z} + P_0 \frac{\partial^2 Z}{\partial x^2} + EI \frac{\partial^4 Z}{\partial x^4}\right) dx = 0$$

leading to:

$$(K_n - \omega^2 M_n) q_n + P_0 \sum K_{n,j}^{(0)} q_j = 0$$

The results obtained by using the first four modes are fairly accurate, as shown in Figure 4.32. Here, the mode shapes are normalized with respect to the largest deflection magnitude. It is noted that the analytical buckling mode shape is nicely fitted by the modal approximation as soon as the preload is sufficiently large.

4.4.2 *Additional inertia*

The methods presented in the last subsection can be transposed without any difficulty in the case of 'inertial connections' i.e. when concentrated masses are added to the original system. To illustrate this point we take the example often encountered in rotating machinery of a shaft provided with a fly wheel of large inertia.

Figure 4.33 is a simple model of a rotor which includes a shaft simply supported at both ends (pinned-pinned conditions) and a fly wheel, modelled as a rigid circular disk, located at the distance x_0 from the left end. The bending moment of inertia I_b of the shaft is small in comparison with that of the fly wheel, denoted I_w. The equilibrium equation could be obtained by adding the kinetic energy of the disk induced by bending and axial displacements to the Bernoulli–Euler Lagrangian. Another means to derive such an equation is to add directly the inertia forces and

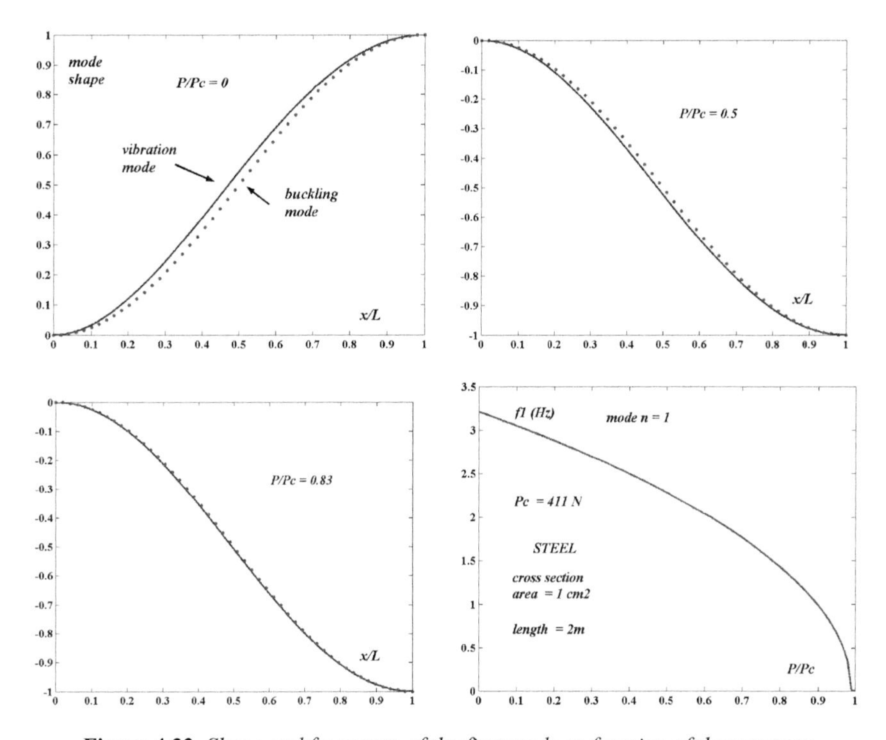

Figure 4.32. *Shape and frequency of the first mode as function of the prestress*

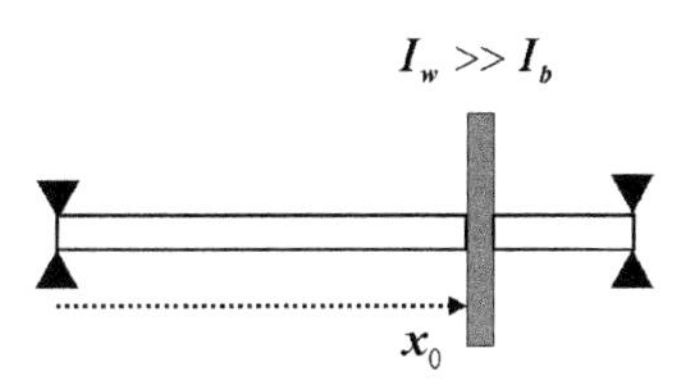

Figure 4.33. *Shaft and fly wheel*

moments due to the disk in the force and moment balance equations [2.18], [2.21]. Following the last procedure, the equations are:

$$\frac{\partial \mathcal{Q}_z}{\partial x} + M_w \ddot{Z} \delta(x - x_0) + \rho S \ddot{Z} = 0$$
$$\frac{\partial \mathcal{M}_z}{\partial x} + \mathcal{Q}_z - I_w \left. \frac{\partial \ddot{Z}}{\partial x} \right|_{x_0} \delta(x - x_0) \qquad [4.75]$$

which leads to the force balance:

$$EI_b \frac{\partial^4 Z}{\partial x^4} + (\rho S + M_w \delta(x - x_0))\, \ddot{Z} + \rho I_w \left. \frac{\partial \ddot{Z}}{\partial x} \right|_{x_0} \delta'(x - x_0) = 0 \qquad [4.76]$$

The modal basis is described by the following quantities:

$$\varphi_n = \sin(n\pi\xi); \quad K_n = n^4 K_0; \quad K_0 = \frac{EI_b \pi^4}{2L^3};$$
$$M_n = \frac{\rho SL}{2} = \frac{M_b}{2}; \quad \omega_n^2 = n^4 \omega_0^2; \quad \omega_0^2 = \frac{K_0}{M_n}$$

Projecting [4.76] on this basis, a coupled system of equations is obtained, which is written in terms of reduced quantities as:

$$(n^4 - \varpi^2) q_n - \varpi^2 \mu_R \sum_{j=1}^{\infty} \left(j \cos\left(\frac{n\pi x_0}{L}\right) \cos\left(\frac{j\pi x_0}{L}\right) q_j \right)$$
$$- \varpi^2 \mu_Z \sum_{j=1}^{\infty} \left(\sin\left(\frac{n\pi x_0}{L}\right) \sin\left(\frac{j\pi x_0}{L}\right) q_j \right) = 0 \qquad [4.77]$$

$$\text{where } \varpi = \frac{\omega}{\omega_0}; \quad \mu_R = \frac{2\pi^2 I_w}{M_b L^2}; \quad \mu_Z = \frac{M_w}{M_b}$$

Inertia of the disk induces coupling both in translation and rotation. If the disk is at mid-span the odd order modes are sensitive to translation only, whereas the even order modes are sensitive to rotation only. So all the modes are perturbed by the presence of the flywheel. The shapes of the two first modes are shown in Figure 4.34, in the case $\mu_z = 2$ and $\mu_R = \mu_R/4$. The solid lines refer to the flywheel mounted on the shaft and the dashed lines to the shaft alone. As expected, the effect of the wheel is important both on the mode shapes and on the natural frequencies. Here, the modal basis is truncated to $N = 23$ but $N = 6$ would produce very similar results.

4.4.3 *Substructures by using modal projection*

4.4.3.1 Basic ideas of the method

Many mechanical structures encountered in practice may be conveniently described as an assembly of more simple substructures that are denoted here (S_1), (S_2) etc., attached to each other at various places by connecting elements,

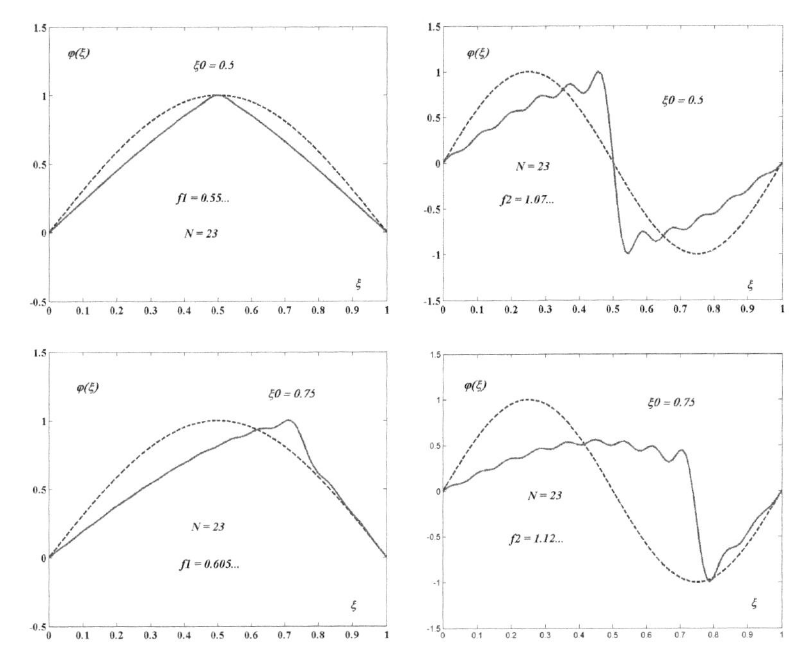

Figure 4.34. *The first two modes of the shaft with a flywheel computed by using the modal projection method*

for instance linear or nonlinear springs, as depicted schematically in Figure 4.35. Numerical studies of the dynamic behaviour of such structures, taken as a whole, may need a large number of DOF, especially if the finite element method is used. In particular, when building the finite element model of the whole structure, it is often difficult to take full advantage of the peculiarities of each substructure to simplify the model. Furthermore, it is often difficult to optimize the size of the mesh with respect to the space resolution and frequency range of physical interest. In many applications, the alternative method described here has the decisive advantage of saving a large number of DOF and time steps and providing the analyst with pertinent information on the physical features of the problem to be solved. The method consists of projecting the equations of the whole system on a modal basis made up of a set of modes which refer to each substructure taken individually, that is in the absence of the elements connecting it to the others. The modes to be retained in the model can be judiciously selected in close relation to the physical particularities of the problem to be solved, by using the criteria already discussed in subsection 4.3.3. The procedure can be conveniently described by considering an assembly of two substructures only. In the absence of connecting elements, the substructure (1) satisfies the equation:

$$\left.\begin{aligned} K_1\vec{X} + C_1\dot{\vec{X}} + M_1\ddot{\vec{X}} = \vec{F}_1^{(e)}(\vec{r},t) \\ + \text{elastic } B.C. \end{aligned}\right\} \qquad [4.78]$$

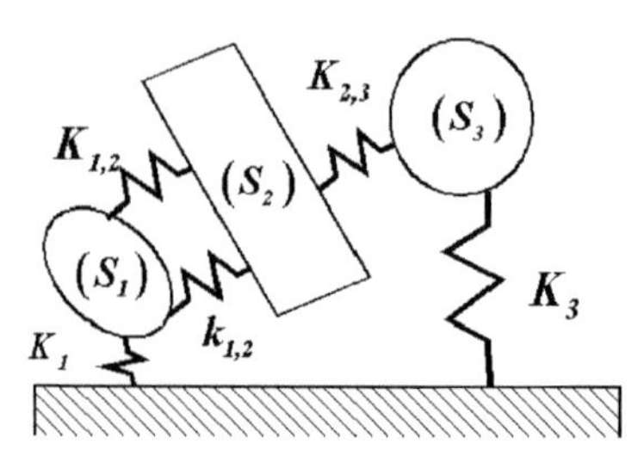

Figure 4.35. *Assembly of distinct substructures*

The related modal basis is:

$$\{\varphi_n^{(1)}\}; \quad \omega_n; \quad M_n; \quad K_n \tag{4.79}$$

The substructure (2) satisfies the equation:

$$\left.\begin{aligned} &K_2\vec{Y} + C_2\dot{\vec{Y}} + M_2\ddot{\vec{Y}} = \vec{F}_2^{(e)}(\vec{r},t) \\ &+ \text{elastic } B.C. \end{aligned}\right\} \tag{4.80}$$

The related modal basis is:

$$\{\varphi_n^{(2)}\}; \quad \omega_n'; \quad M_n'; \quad K_n' \tag{4.81}$$

The connection between two points of the two substructures is modelled as interaction forces, which can depend, linearly or not, on displacements, velocities and accelerations of the connected points and be explicitly time dependent. The coupled system of equations is thus written as:

$$\begin{aligned} &K_1\vec{X} + C_1\dot{\vec{X}} + M_1\ddot{\vec{X}} = \vec{F}_1^{(e)}(\vec{r};t) + \sum_{j=1}^{J} \vec{F}_j^L(\vec{X},\vec{Y},\dot{\vec{X}},\dot{\vec{Y}},\ddot{\vec{X}},\ddot{\vec{Y}};t) \\ &+ \text{elastic} B.C. \\ &K_2\vec{Y} + C_2\dot{\vec{Y}} + M_2\ddot{\vec{Y}} = \vec{F}_2^{(e)}(\vec{r};t) - \sum_{j=1}^{J} \vec{F}_j^L(\vec{X},\vec{Y},\dot{\vec{X}},\dot{\vec{Y}},\ddot{\vec{X}},\ddot{\vec{Y}};t) \\ &+ \text{elastic } B.C. \end{aligned} \tag{4.82}$$

where F_j^L stands for the force induced by the j-th connection.

The projection of [4.82] on the basis $\{\Phi_n\} = \{\varphi_n^{(1)}\} \cup \{\varphi_n^{(2)}\}$ produces a differential system of modal oscillators coupled by the connecting forces. Depending on their nature, different techniques can be used to solve the problem. The method is further described in the next subsections by taking a few examples.

4.4.3.2 Vibration modes of an assembly of two beams linked by a spring

Two identical beams, pinned at both ends are considered; the object is to determine the bending vibration modes of the beams coupled by a spring acting at mid-span, see Figure 4.36. The formulation of the problem is as follows:

1. *Local equilibrium equations for harmonic vibration*:

$$EI\frac{d^4Z_1}{dx^4} - \omega^2\rho SZ_1 = -K_L(Z_1 - Z_2)\delta(x - L/2)$$
$$EI\frac{d^4Z_2}{dx^4} - \omega^2\rho SZ_2 = -K_L(Z_2 - Z_1)\delta(x - L/2)$$

2. *Projection on the modal bases:*
   Here $\{\varphi_n^{(1)}\} = \{\varphi_n^{(2)}\}$ are relative to the pinned-pinned configuration. The projected equations are written as:

$$[n^4 - \varpi^2]q_n = \gamma \sin(n\pi/2)\sum_j \sin(j\pi/2)(p_j - q_j)$$
$$[n^4 - \varpi^2]p_n = \gamma \sin(n\pi/2)\sum_j \sin(j\pi/2)(q_j - p_j)$$

$\varpi = \omega/\omega_0$ where $\omega_0 = \sqrt{2K_0/\rho SL}$ and $K_0 = EI\pi^4/2L^3$. It is recalled that ω_0 is the first natural pulsation of the pinned-pinned beams without connection and K_0 the corresponding modal stiffness coefficient. $\gamma = K_L/K_0$ is the dimensionless coupling coefficient and q_j, p_j are the modal displacements of the two beams.

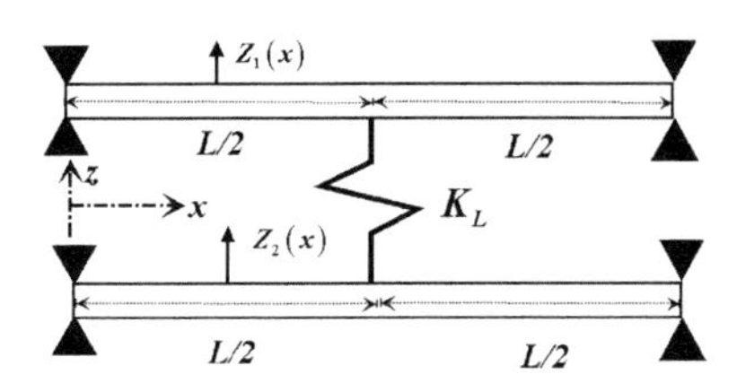

Figure 4.36. *Two identical beams connected at mid-span by a linear spring*

As is easily understood, only the odd order modes give rise to coupling terms, since the even modes have a node at mid-span. Then the system is reduced to :

$$[(2k+1)^4 - \varpi^2] q_{2k+1} = \gamma_{kj} \sum_j (p_j - q_j)$$

$$[(2k+1)^4 - \varpi^2] p_{2k+1} = \gamma_{kj} \sum_j (q_j - p_j)$$

$$\gamma_{kj} = \gamma \sin((2k+1)\pi/2) \sin((2j+1)\pi/2) = \gamma(-1)^{k+j}$$

Furthermore, it is noted that only the out-of-phase modes of the beam assembly are modified by the coupling. Thus, the modal system can be further simplified by using the variable transformation:

$$d_{(2k+1)} = q_{(2k+1)} - p_{(2k+1)}$$

Then the linear system to be solved is written as a modal system, of current row $n = 2j + 1$:

$$\left([(2k+1)^4 - \varpi^2]\right)(d_{(2k+1)}) - 2\gamma_{kj} \sum_j d_{2j+1} = 0$$

The results obtained by solving numerically such a set of equations, projected on the first ten odd modes of the pinned-pinned beams, are shown in Figures 4.37 and 4.38. In Figure 4.37, the natural frequency of the first out-of-phase mode is plotted versus the reduced value γ of the stiffness coefficient of the connecting spring. As expected, its effect is rather negligible in the range $\gamma \ll 1$. To the contrary, if $\gamma \gg 1$ the spring acts as a rigid connection and the natural frequency converges to $\varpi_\infty \simeq 2\pi$. As could be anticipated, this asymptotic value corresponds to the first natural frequency of an equivalent beam of half length which would be pinned at $x = 0$ and clamped at $x = L/2$ $\varpi_\infty = 4\,(3.9266/\pi)^2 \simeq 1.99\pi$.

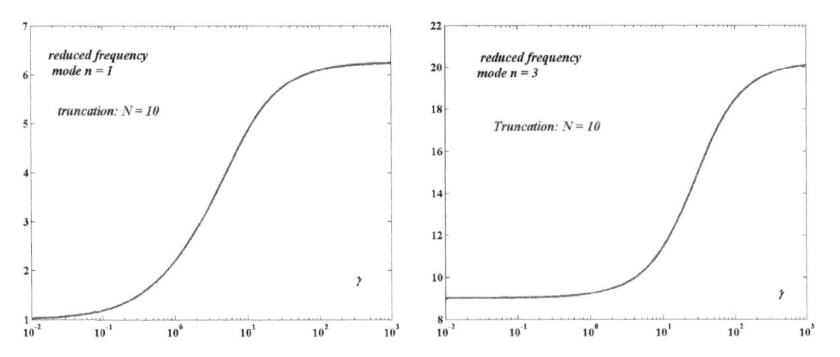

Figure 4.37. *Natural frequency of the first out-of-phase mode versus spring stiffness coefficient*

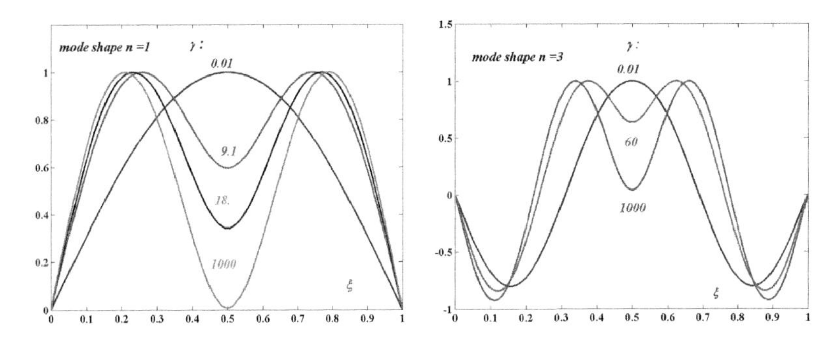

Figure 4.38. *Mode shape of the first out-of-phase mode versus spring stiffness coefficient (deflection of only one beam is displayed for sake of clarity)*

Finally, in Figure 4.38 the shape of the first and third modes are plotted for a few values of γ. Similar computation performed with only four odd modes lead to nearly the same results, which indicates that convergence is pretty fast, avoiding the necessity of including a quasi-static corrective term in the computation.

4.4.3.3 Multispan beams

The preceding exercise is also suitable for discussing briefly the vibration modes of multispan beams which are of practical importance in various industrial components, such as the tube and shell heat exchangers of fossil fuel or nuclear power plants. Each span can be considered as a subsystem coupled to the next one through the rotation and/or the transverse displacement at each support. As an example, we consider the periodic case of a uniform straight beam provided with equally spaced pinned supports. According to the results of the last subsection, it is realized that each individual support may act either as a nearly clamped or a nearly pinned condition, depending whether rotation is essentially zero or not at the support for the specific mode considered. Then, referring to the data of Table 4.2, the generalized stiffness of the first mode of a single span is found to be roughly proportional either to 10, 15, or 20 for pinned-pinned, pinned-clamped, or clamped-clamped equivalent support conditions, respectively. The generalized stiffness of the second mode is significantly higher, being roughly proportional either to 40, 50, or 60. Accordingly, the first modes of the multi-span beam are marked by mode shapes with a single antinode per span, which can be either in-phase, or out-of-phase from one span to the next. Generalized stiffness – and so natural frequencies – are thus found to increase much more progressively than in the single span case, starting from the lowest frequency for which the mode shape corresponds to a pinned condition at each support, up to the highest frequency for which the mode shape corresponds to a clamped condition at each support, except of course the end pinned supports. As a consequence, the number of modes per frequency interval, or *modal density*, of multi-span beams is found to be proportional to the number of spans.

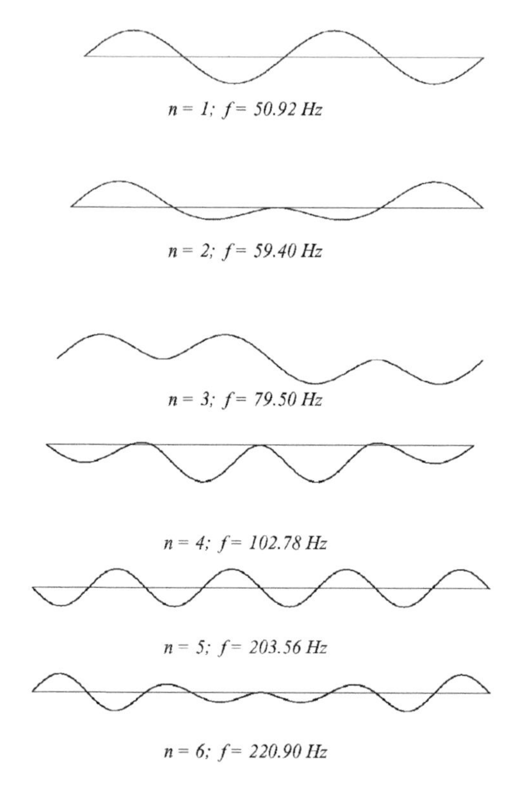

Figure 4.39. *Multispan beam comprising four equal spans*

Figure 4.39 refers to a typical heat exchanger tube made of stainless steel: span-length $L = 1$ m, external radius $R = 1$ cm, wall thickness $h = 2$ mm, number of spans $N = 4$. It displays the first 6 mode shapes and natural frequencies as computed by using a FEM program. The first mode vibrates at $f_1 \simeq 51$ Hz and the highest mode of the first family with a single antinode at each span corresponds to $n = 4$. It vibrates at 103 Hz, which is less than the analytical approximation $f_4 \simeq f_1(4.73/\pi)^2 \simeq 115$ Hz. The remaining discrepancy is easily understood by inspecting the mode shape which is marked by a smaller magnitude of the side lobes in comparison with that of the central lobes. Significant variations of the lobe magnitude from one span to the next are also conspicuous on the mode shape $n = 2$. They are a consequence of differences in the equivalent support conditions at the ends of the span. Finally, the $n = 5$ mode is the lowest mode of the next family, marked by two antinodes at each span. It vibrates at a natural frequency close to the analytical value $f_5 \simeq 4f_1 \simeq 203$ Hz.

To conclude on the subject, let us mention that, in reality, some departure from periodicity cannot be avoided because of manufacturing tolerances and variable working conditions in service. As emphasized for instance in [KIS 91], [LAN 94],

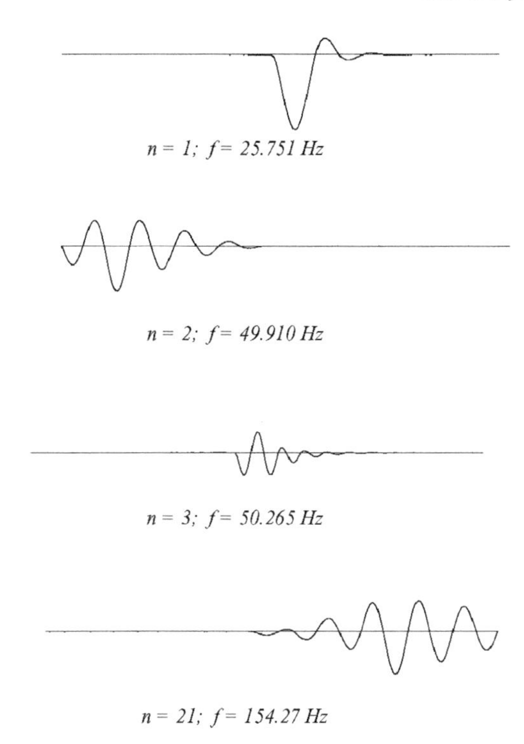

Figure 4.40. *Localized modes of a beam comprising 20 spans with 4% random error in the location of the intermediate supports*

[BOU 96], [BAN 97], [CET 99], the presence of small irregularities in nominally periodic structures may significantly affect their dynamics, the natural modes of vibration in particular. Major effects include the so called *localization* of the mode shapes to zones of relatively small extent, typically a single or a few spans in the case of a multispan beam, as shown in Figure 4.40 which refers to $N = 20$ and a random error of about 4 per cent on the location of the intermediate supports.

4.4.4 *Nonlinear connecting elements*

The modal projection method, which has been illustrated above by solving a few linear problems, is also very useful for dealing with many nonlinear problems where the nonlinearities are concentrated at specific locations. In such cases, the connecting linear springs are replaced by nonlinear connecting elements. As an illustration of the method, the procedure is applied here to a few impact problems, which have been selected for their simplicity. They can be solved analytically providing thus a valuable insight into some basic physical aspects of impact problems in elastic structures. It is worth mentioning that the method proved to be efficient in much more complicated structures than those considered here. For instance, it has been applied by one of the authors to compute piping systems and tube arrays provided with loose supports against which the structures experience intermittent contacts; the shocks are essentially elastic but accompanied by dry friction (see [ANT 90a and 90b], [AXI 84, 88, 92]). Another example of industrial application is concerned with the numerical simulation of car crash tests of various types, leading to large plastic deformations and large displacements [MOU 04].

By using a modal model instead of a finite element model to perform dynamic analyses, it is often possible to reduce drastically both the size of the model, i.e. the number of DOF, and the number of time steps, or internal iterations, required to compute the time-histories of the responses. This is mainly because the physical information included in the natural modes of vibration can be easily compared to the space distribution and the spectral content of the excitation on one side, and to the space and time resolution of practical interest, on the other side. Then, the modal basis can be truncated in a consistent way, providing an optimal model. Furthermore, if the time histories of the nonlinear responses are computed by using an explicit algorithm, such as the finite difference method, the time step can be conveniently adjusted to the smallest natural period present in the model, which in turn is controlled by the number of modes retained in the modal basis.

4.4.4.1 Axial impact of a beam on a rigid wall

For discussing impact problems involving elastically deformable bodies, it is advisable to start from the case of discrete systems. In the present problem a single DOF system is appropriate. As already discussed in [AXI 04] Chapter 5, the elastic impact of a particle against a rigid wall can be modelled in two distinct ways, namely either by using a kinematically constrained model, or by using a truly elastic model, which accounts for the elastic deformation of the actual body during the shock.

According to the constrained model depicted in Figure 4.41, the impact is infinitely short and the system is constrained by the conservation conditions of linear momentum and kinetic energy:

$$M_1 \dot{X}_1 = -M_1 \dot{X}'_1; \quad \tfrac{1}{2} M_1 \dot{X}_1^2 = \tfrac{1}{2} M_1 \dot{X}_1'^2 \qquad [4.83]$$

which of course reduces to the condition of perfect reflection with sign change $\dot{X}'_1 = -\dot{X}_1$, where $\dot{X}_1$ and $\dot{X}'_1$ denote the velocities of the particle just before and just after the impact, respectively.

However, for integrating numerically the equation of motion step by step in time, it is found more convenient to adopt the truly elastic model of Figure 4.42, which accounts for the elastic deformation of the actual system (body plus wall)

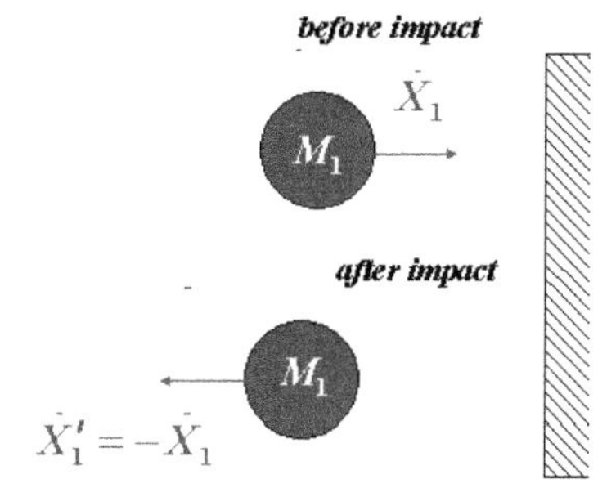

Figure 4.41. *Elastic impact of a particle against a rigid wall: constrained model*

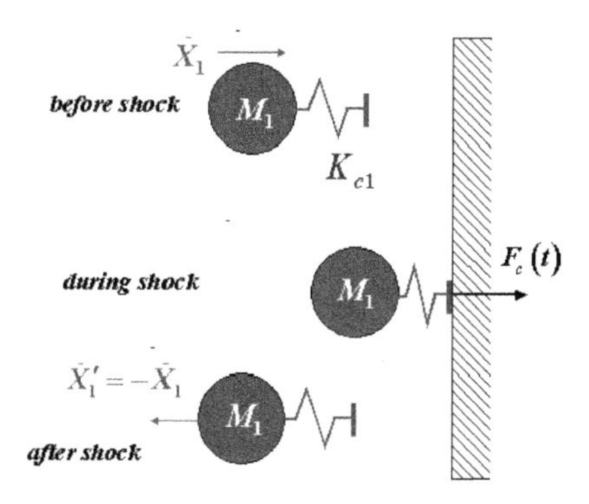

Figure 4.42. *The truly elastic shock model of a particle against a wall*

during the shock. According to the truly elastic model, shock duration is finite and equal to:

$$\tau_c = \pi/\omega_c; \quad \text{where } \omega_c = \sqrt{\frac{K_c}{M_1}} \qquad [4.84]$$

The contact force $F_c(t)$ behaves as a half-sine signal of duration τ_c, the action of which being equal to $2M_1\dot{X}_1$, is appropriate. Furthermore the value of the impact stiffness coefficient K_{c1} can be related in practice either to the equivalent stiffness of the real impacting system, or be adjusted by using [4.84] to fit appropriately the time resolution required by the analyst to perform the numerical simulation.

Turning now to the case of a deformable solid body, let us consider a straight uniform beam moving at a uniform axial speed $-\dot{X}_0$ and impinging on a rigid wall. The shock is supposed perfectly elastic. As the motion is purely axial, the use of a traction-compression model is justified. Furthermore, in the simple approach adopted here, the Poisson ratio effect is neglected. As a consequence, the waves triggered by the impact are not dispersive. The equation and the initial and boundary conditions are:

$$
\begin{aligned}
&-ES\frac{\partial^2 X}{\partial x^2}+\rho S\ddot{X}=0 \\
&\left.\frac{\partial X}{\partial x}\right|_L=0;\quad \dot{X}(x;0)=-\dot{X}_0;\quad \dot{X}(0;t_c)=0
\end{aligned}
\qquad [4.85]
$$

The problem is solved by extending to the continuous case the constrained model of Figure 4.41. The shock starts at the contact time t_c and position $x=0$ complying with the conditions $X(0;t)=0$ and $\dot{X}(0;t)=0$. The system [4.85] is solved by using the Laplace transformation $\tilde{X}(x;s)=TL[X(x;t')]$ where $t'=t-t_c$. The Laplace transform of [4.85] is written as:

$$
\begin{aligned}
&-c^2\frac{d^2\tilde{X}}{dx^2}+s^2\tilde{X}=-\dot{X}_0;\quad \text{where } c=\sqrt{\frac{E}{\rho}} \\
&\tilde{X}(0;s)=0;\quad \left.\frac{d\tilde{X}}{dx}\right|_L=0
\end{aligned}
\qquad [4.86]
$$

The solution is of the form:

$$
\tilde{X}(x;s)=-\frac{\dot{X}_0}{s^2}+ae^{-(xs/c)}+be^{+(xs/c)} \qquad [4.87]
$$

where the constants a and b are adjusted to the boundary conditions:

$$
\begin{aligned}
&b=ae^{-\tau};\quad \text{where } \tau=\frac{2L}{c} \\
&a=\frac{\dot{X}_0}{s^2}\frac{1}{1+e^{-\tau}}=\frac{\dot{X}_0}{s^2}\left\{1-e^{-\tau}+e^{-2\tau}-e^{-3\tau}+\ldots\right\}
\end{aligned}
\qquad [4.88]
$$

where τ designates the time delay taken by the longitudinal wave to travel forth and back along the beam.

The Laplace transform of the displacement field is thus found to be:

$$
\begin{aligned}
&\tilde{X}(x;s)=\frac{\dot{X}_0}{s^2}\left\{-1+(e^{-s\theta}+e^{s(\theta-\tau)})(1-e^{-\tau}+e^{-2\tau}-e^{-3\tau}+\ldots)\right\}; \\
&\text{where } \theta=\frac{x}{c}
\end{aligned}
\qquad [4.89]
$$

The Laplace series is inverted as:

$$
X(x,t)=t\dot{X}_0\left\{-\mathcal{U}(t)+\mathcal{U}\left(t-t_c-\frac{x}{c}\right)+\mathcal{U}\left(t-t_c-\frac{2L-x}{c}\right)-\cdots\right\} \qquad [4.90]
$$

where $\mathcal{U}(t)$ is the Heaviside step function.

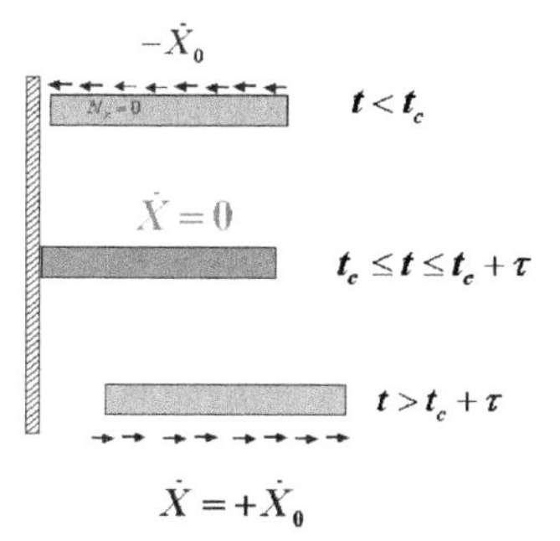

Figure 4.43. *The three consecutive steps of beam motion*

It is easy to verify that the contact time between the beam and the wall is equal to τ. At time $t \geq t_c + \tau$, the beam rebounds towards $x > 0$ with the uniform and constant speed $+\dot{X}_0$. So, no other retarded terms than those present in [4.90] are necessary. In Figure 4.43 the beam motion during the shock is sketched; as the displacement is known, the force during the shock follows as:

$$F_c(t) = ES\frac{\partial X}{\partial x}\bigg|_{x=0} \qquad [4.91]$$

or, by using [4.90] :

$$F_c(t) = \lim_{x\to 0}\left\{\frac{ESt\dot{X}_0}{x}\left(\mathcal{U}\left(t - t_c - \frac{x}{c}\right) - \mathcal{U}(t - t_c)\right)\right\} = \frac{ES\dot{X}_0}{c} \qquad [4.92]$$

As expected, the force is constant during the impact time and its action is equal to the change of the linear momentum before and after the shock:

$$ES\frac{\dot{X}_0}{c}\tau = 2\rho SL\dot{X}_0 = 2M\dot{X}_0 \qquad [4.93]$$

Figure 4.44 shows how the stress and displacement fields vary during the shock. The stress wave is shaped as a rectangular pulse which is reflected with sign change at the rigid wall and without sign change at the free end. At a given time during the shock, the stressed part remains motionless and the unstressed part is moving at a uniform speed $-\dot{X}_0$, the material stops when reached by the wave reflected from the wall; the speed becomes $+\dot{X}_0$ when the stressed material is relaxed by the reflected wave originating from the free end. As soon as the stress is relaxed in the whole beam, the latter leaves the wall at uniform speed $+\dot{X}_0$. This analytical solution is used as a reference to investigate the features of the truncated series solution obtained by using the modal projection method. As a preliminary and gross approximation, the beam can be described by using a single oscillator of mass $M = \rho SL$ and stiffness coefficient $K_\ell = ES/L$, that is the equivalent stiffness of a fixed-free beam. This 1-DOF model, which corresponds to the truly elastic shock model of

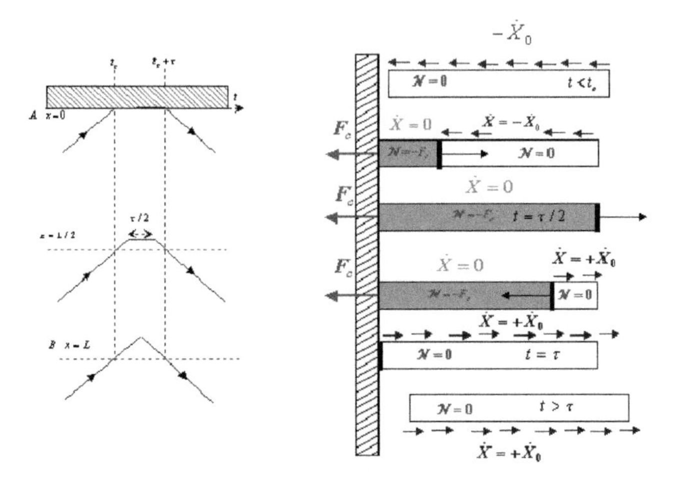

Figure 4.44. *Displacement and stress fields in the beam*

Figure 4.42, is sufficient to describe the motion before and after the shock, but is clearly unsuitable to describe the shock itself, as the contact force is found to vary as a half-sine and the shock time value $\tau' = \pi L/c$ is overestimated by 60 per cent.

We turn now to the more refined model which retains several modes of vibration of the free-free beam and which accounts for the neglected modes by a truncation stiffness coefficient K_T, connected to the impacted end of the beam. In the present problem, K_T is found to be:

$$\frac{1}{K_T} = \frac{1}{K_1}\left\{\sum_{n=1}^{\infty}\frac{1}{n^2} - \sum_{n=1}^{N}\frac{1}{n^2}\right\} = \frac{1}{K_1}\left\{\frac{\pi^2}{6} - \sum_{n=1}^{N}\frac{1}{n^2}\right\} \qquad [4.94]$$

$$\text{where } K_1 = \pi^2\frac{ES}{2L}$$

Then the problem is formulated as:

$$-ES\frac{\partial^2 X}{\partial x^2} + \rho S\ddot{X} = F_c(x;t)\delta(x)$$
$$\left.\frac{\partial X}{\partial x}\right|_L = 0; \quad X(0;0) = X_0; \quad \dot{X}(x;0) = -\dot{X}_0 \qquad [4.95]$$

As shown in Figure 4.45, the gap X_0 is the initial distance between the wall and the beam end that will impinge on the wall. The shock force acting on the beam during the impact is modelled as the nonlinear force:

$$F_c(x;t) = \begin{cases} -K_T(X(0;t) - X_0); & \text{if } (X(0;t) - X_0) < 0 \\ 0; & \text{otherwise} \end{cases} \qquad [4.96]$$

where the equivalent impact stiffness coefficient K_c is identified with K_T.

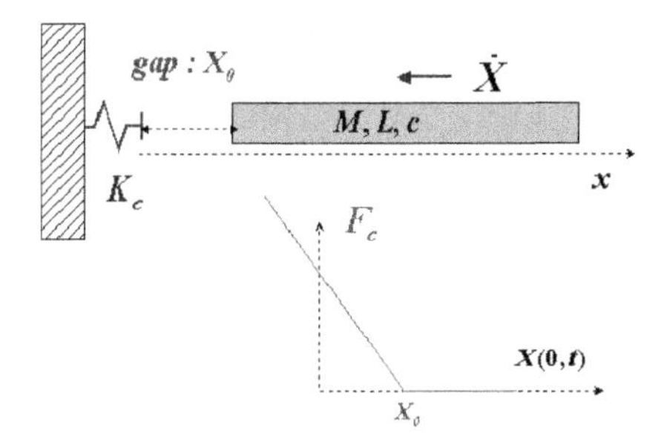

Figure 4.45. *Elastic impact model*

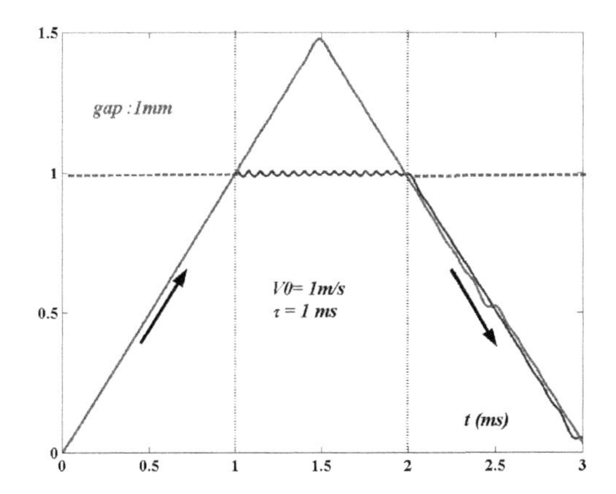

Figure 4.46. *Displacement of the end beam sections. The impacted end is oscillating about the wall position (i.e. the gap) at a frequency corresponding to that of the first mode neglected in the modal basis. According to the motion of the other end, maximum compressive deformation of the beam is nearly 0.002*

Modal projection of the system [4.95] is straightforward. As the impact force [4.96] is assumed to depend upon displacement variables only, it is found very convenient to use an explicit algorithm - such as the central differences scheme – to integrate step by step the equation of motion. According to such an algorithm, described for instance in [AXI 04] Chapter 5, the modal displacements $q_n(k+1)$ at the $(k+1)$-th time step of duration h, are extrapolated from the values taken at the previous time steps k and $k-1$. To compute the contact force at the $(k+1)$-th time step it is necessary to derive the physical displacement of the beam impacted end from the known values $q_nt(k+1)$, and then to project the contact force on the modal basis to compute the modal impact forces $Q_n(k+1)$ which serve as an input to the recursive sequence giving $q_n(k+2)$. Results of such a numerical simulation are illustrated in Figures 4.46 and 4.47. The modal basis is restricted to the first 20 modes, including the rigid mode which course is necessary to account for the free flight motion before and after impact. The beam parameters are: length $L_1 = 2.5$ m, full circular section, *radius* 5 cm. mass per unit volume 8000 kg/m^3. Young's modulus $E = 2.10^{11}$ Pa. wave speed $c = 5000$ m/sec; wave travel time

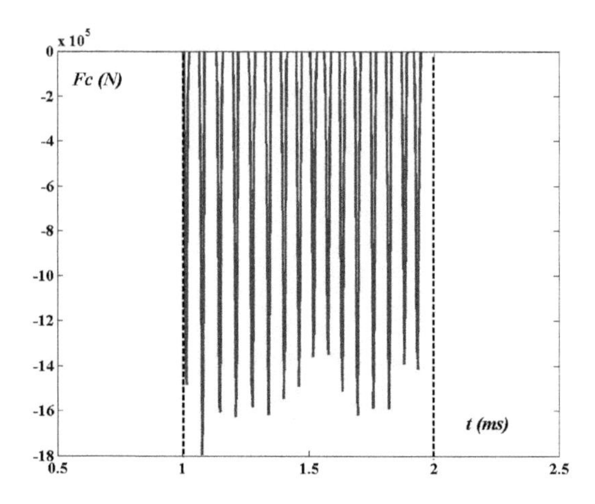

Figure 4.47. *Contact force*

$\tau = 1$ ms. The beam is hurled against the wall with an initial velocity $V_0 = 1$ m/sec from a distance of 1 mm.

As shown in Figure 4.46, the computed time history of the end displacements of the beam are satisfactorily reproduced. The effect of the modal basis truncation is best evidenced by looking at the time-history of the shock force signal. As shown in Figure 4.47, in place of a rectangular pulse, a series of short lived peaks are obtained at the cut-off frequency of the model. However, such a discrepancy may be considered to be of minor importance since the action of the computed force signal still agrees with the analytical result [4.93], which is the important point in most engineering applications. Furthermore, the shock duration can be made as close to the analytical value τ as desired, by increasing the size of the modal basis.

4.4.4.2 Beam motion initiated by a local impulse followed by an impact on a rigid wall

Another interesting problem is that of a beam initially at rest and loaded by an impulsive load at its free end, resulting typically from a stiff hammer impact, see Figure 4.48.

An analytic calculation similar to the previous one starts from the free-free beam motion which is described before the shock by the equation:

$$\begin{aligned} & -ES\frac{\partial^2 X}{\partial x^2} + \rho S\ddot{X} = M\dot{X}_0\delta(t)\delta(x-L) \\ & \left.\frac{\partial X}{\partial x}\right|_{x=0,L} = 0; \quad X(x;0) = \dot{X}(x;0) = 0 \end{aligned} \qquad [4.97]$$

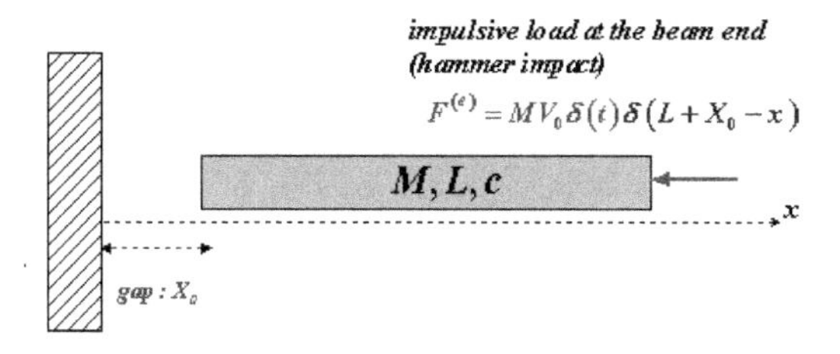

Figure 4.48. *Beam motion initiated by a hammer impact at the free end opposite to the wall*

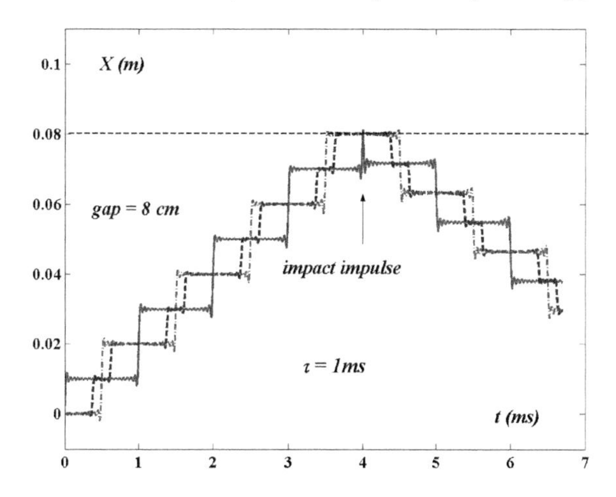

Figure 4.49. *Displacement calculated with the 20 first beam modes*

The Laplace transform gives the solution before the shock:

$$\tilde{X}(x;s) = \frac{c}{s}\frac{M\dot{X}_0}{ES}\left\{\frac{e^{-sx/c} + e^{s(x/c-\tau)}}{1-e^{-s\tau}}\right\} \qquad [4.98]$$

The motion given by [4.98] is superposing a rigid body translation at constant velocity and a train of waves shaped as a series of rectangular pulses of duration $\tau = 2L/c$, whence the staircase signal:

$$X(x;t) = \frac{cM\dot{X}_0}{ES}\sum_{n=0}^{\infty}\{\mathcal{U}(t-x/c-n\tau) + \mathcal{U}(t+x/c-(n+1)\tau)\} \qquad [4.99]$$

The results of the modal projection method are illustrated in Figure 4.49, which superposes the time histories of displacement at three positions $x = 0$, $x = L/4$, $x = L$ along the beam. The travel delay of the wave from one position to the next can be clearly seen. The contact takes place during the transition between two steps. As the incident wave is reflected with sign change, contact is lost as soon as the impact occurs. The reflected wave is propagated along the beam exactly in the same way as the incident wave. Therefore, the contact force signal is impulsive in nature and its action is twice that of the impulsive elastic stress occurring at a rectangular wave front, incident, or reflected.

4.4.4.3 Elastic collision between two beams

To deal with the problem shown in Figure 4.50 of two beams impacting elastically against each other, the substructuring method described in subsection 4.4.3 is used together with the nonlinear elastic force model described above in the case of a single beam. Each substructure is described by its own truncated modal basis. The stiffness coefficients K_{c1}, K_{c2} mounted in series are related to the quasi-static mode of each truncated basis.

The simplest model of this kind is reduced to the rigid mode of each free beam, the problem being therefore reduced to the elastic impact of two particles. The constrained model for this particular case is shown in Figure 4.51 and the truly elastic model is shown in Figure 4.52. According to the constrained model, the impact is again infinitely short and described by the conservation conditions of linear momentum and kinetic energy:

$$\begin{aligned} M_1\dot{X}_1 + M_2\dot{X}_2 &= M_1\dot{X}_1' + M_2\dot{X}_2' \\ \tfrac{1}{2}M_1\dot{X}_1^2 + \tfrac{1}{2}M_2\dot{X}_2^2 &= \tfrac{1}{2}M_1\dot{X}_1'^2 + \tfrac{1}{2}M_2\dot{X}_2'^2 \end{aligned} \qquad [4.100]$$

where $\dot{X}_1$, $\dot{X}_2$ and $\dot{X}_1'$, $\dot{X}_2'$ denote the velocities of the two particles just before and just after the impact, respectively.

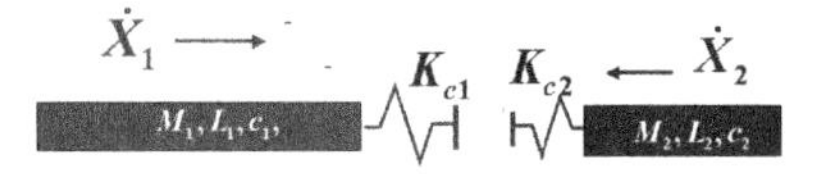

Figure 4.50. *Two beams impacting against each other*

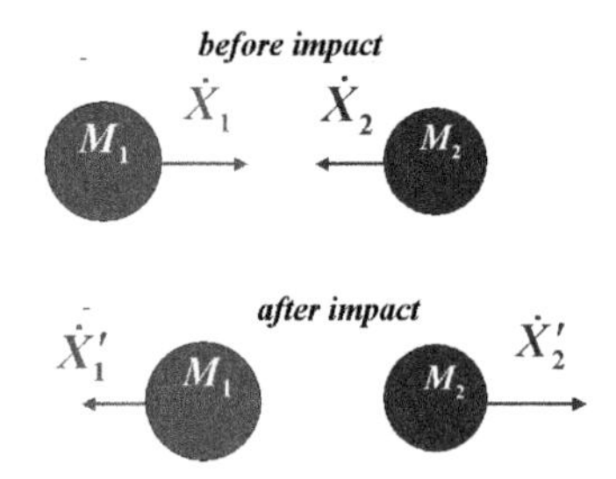

Figure 4.51. *Two particles impacting against each other: constrained model*

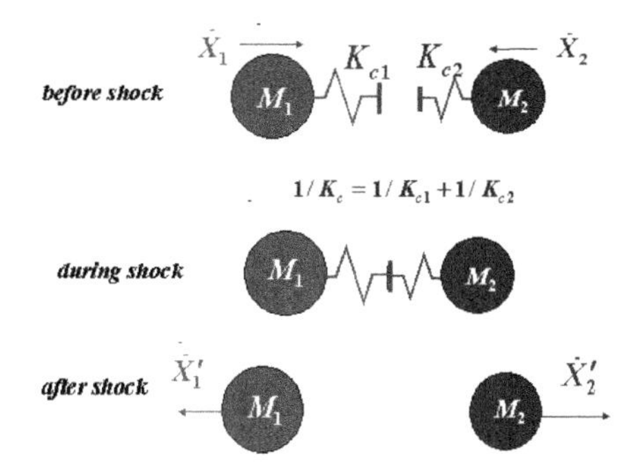

Figure 4.52. *Two particles impacting against each other: truly elastic model*

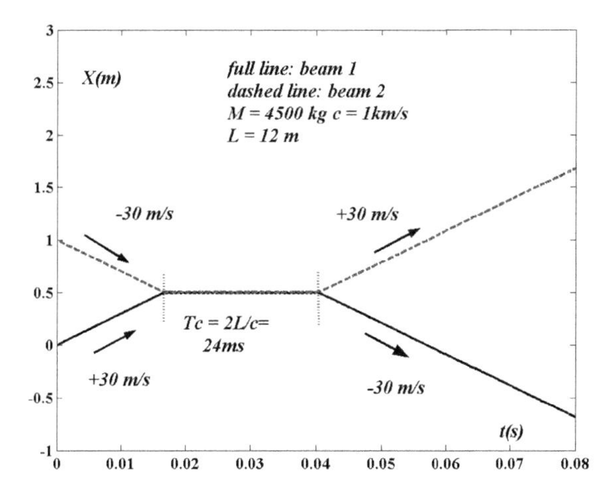

Figure 4.53. *Frontal collision of two identical beams, symmetrical case*

Again, according to the truly elastic model, the impacting particles remain in contact during a finite time interval and the contact force varies as a half-sine signal of pulsation:

$$\omega_c = \sqrt{\frac{K_g}{M_g}} \qquad [4.101]$$

where K_g and M_g are the modal stiffness and mass coefficients of the out-of-phase mode of the 2-DOF system during contact. The velocities before and after impact still comply with the conditions [4.100].

Finally, Figures 4.53 and 4.54 refer to two distinct cases of impact, as computed retaining the first 15 modes of vibration of each colliding beam. The first corresponds to a symmetrical system of two identical beams with an initial gap,

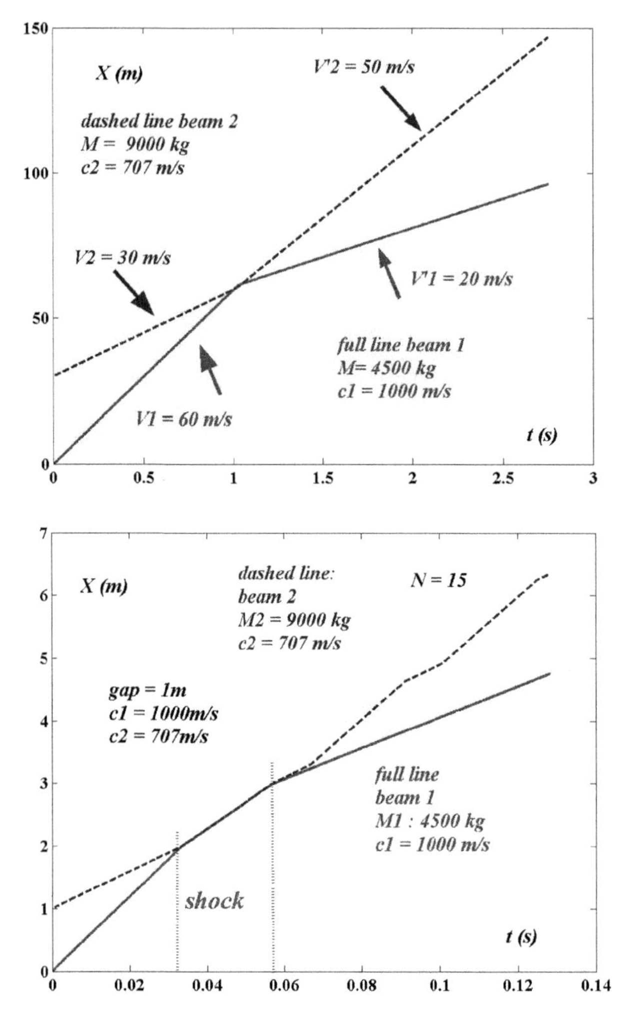

Figure 4.54. *Collision of two beams, asymmetrical case*

which start with equal and opposite velocities at $t = 0$. The contact time is equal to the back and forth travelling time along an equivalent beam of length $2L$. No waves are excited and the velocities of the beams are simply interchanged by the collision. The last case corresponds to an asymmetrical system. The upper plot of Figure 4.54 shows a global view of the impact, which agrees satisfactorily with the conditions [4.100] and the lower plot focuses on the collision itself. Contact time depends on the beam properties and it may be noticed that waves are excited by the impact, which are clearly visible in the trajectory of the second beam (dashed line in Figure 4.54).

Chapter 5

Plates: in-plane motion

Plates are structural components used as walls, roofs, panels, windows, etc. They are modelled as two-dimensional structures characterized by a plane geometry bounded by a contour comprising straight and/or curved lines. Plates are intended to resist various load conditions, broadly classified as in-plane and out-of-plane loads, that is forces either parallel or perpendicular to the plane of the plate, respectively. It is found convenient to study their response properties by separating first the in-plane and the out-of-plane motions. In-plane motions of plates are generally marked by the coupling between the two in-plane components of the displacement field. Such a coupling gives rise to new interesting features of in-plane motions with respect to the longitudinal motions of straight beams. Coupling between in-plane and out-of-plane motions occurs in the presence of in-plane preloads and will be considered in the next chapter. On the other hand, for mathematical convenience in using Cartesian coordinates, presentation will focus first on rectangular plates. Then use of curvilinear coordinates will be introduced to consider plates bounded by curved contours.

5.1. Introduction

5.1.1 *Plate geometry*

As shown in Figure 5.1, structural elements called *plates* are characterized by the two following geometrical properties:

1. One dimension, termed the thickness, denoted h, is much smaller than the other two (length L_1 and width L_2 for a rectangular shape, diameter D for a circular one etc.). This allows one to model the mechanical properties of plates by using a two-dimensional solid medium.
2. The geometrical support of the 2D model is the midsurface, taken at $h/2$ if the material is homogeneous. In contrast to the case of shells to be studied later, this surface is flat and called the *midplane* of the plate. The border lines of the contour of the midplane are called *edges*, which can be straight or/and curved lines.

A few other definitions concerning plate geometry are useful. As a three-dimensional body, the plate is bounded by closed surfaces comprising the so called *faces* and *edge surfaces*. If the plate thickness is uniform, the faces are flat and parallel to the midplane, whereas the edge surfaces can be cylindrical or plane, depending on the plate geometry. Intersection of two edge surfaces defines a corner edge. Finally, it is also found convenient to orientate the contour by defining a positive unit vector normal to the midplane, which points from the lower face to the upper face of the plate. The usual positive orientation is in the anticlockwise direction.

5.1.2 *Incidence of plate geometry on the mechanical response*

As a preliminary, it is useful to outline a few generic features of the response properties of plates which serve as a guideline to organize the presentation of this

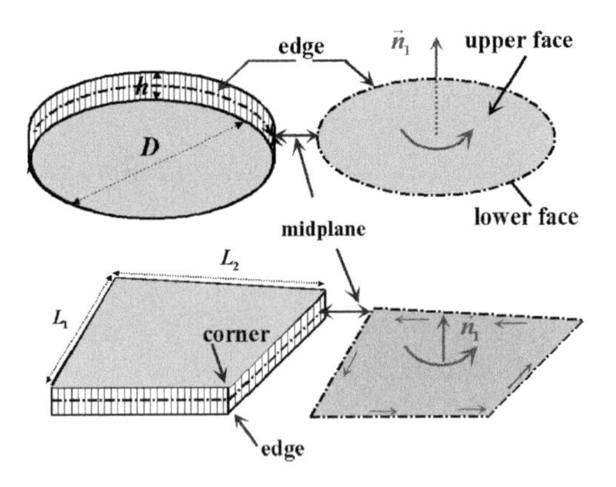

Figure 5.1. *Plates: geometrical definitions*

and the next chapters. As will be verified 'a posteriori', they arise as a consequence of the concept of midplane.

Considering first linear motions about an equilibrium position with zero or negligible stress level, it can be stated that:

1. In-plane loads, i.e. parallel to the midplane, are balanced by in-plane stresses, which can be normal (compressive or tensile forces) and/or tangential (in-plane shear forces).
2. Out-of-plane loads, i.e. perpendicular to the midplane, are balanced by bending and torsional moments and by transverse shear forces.
3. If the load is oblique, all the previous stresses participate with the equilibrium. However, in the absence of initial stresses, coupling between the in-plane and the out-of-plane motions of the plate can be discarded. So the response to an oblique load is obtained by superposing the responses on the in-plane and to the out-of-plane components of the load, which can be studied separately.

Considering then linear motions about a prestressed equilibrium position, it can be stated that:

1. In-plane initial stresses contribute to the out-of-plane equilibrium of a plate through restoring, or alternatively, destabilizing forces, similar to those arising in straight beams when prestressed axially.
2. As a limit case, the forces that contribute to the out-of-plane equilibrium of suitably stretched skin structures originate almost completely from the in-plane stresses which are prescribed initially. Such skin structures are referred to as membranes. A membrane can be seen as an idealized two-dimensional medium which has no flexural rigidity, in contrast with plates and shells.
3. Like cables, or strings, membranes belong to the general class of the so called tension structures which can support mechanical loads by tensile stresses only.

In agreement with these preliminary remarks, it is found appropriate to investigate first the motions of plates which involve in-plane components of force and displacement fields only. Such in-plane fields are termed *membrane components*, or *fields*, as their nature is basically the same as those induced when a skin is stretched in its own plane. On the other hand, it is also found convenient to study first the case of rectangular plates as they can be described using Cartesian coordinates. Non rectangular plates have to be analysed by using oblique or curved coordinates, depending on the specificities of their shapes, which are more difficult to manipulate mathematically.

5.2. Kirchhoff–Love model

5.2.1 *Love simplifications*

Let us consider a rectangular plate of uniform thickness h. The coordinates of a material point are expressed in a direct Cartesian frame *Oxyz*, where *Oxy* is the midplane of the plate, the origin *O* is at a corner, the axes *Ox* and *Oy* are parallel to the edges of lengths L_x and L_y respectively. The unit vectors of the *Ox*, *Oy*, *Oz* axes are denoted $\vec{i}, \vec{j}, \vec{k}$ respectively. A *main cross-section* is obtained by cutting mentally the plate with a plane which is perpendicular to the midplane and parallel either to *Ox* or to *Oy*. An ordinary cross-section is neither parallel to *Ox* nor to *Oy*, *P(x, y, 0)* or $P(x, y)$ designating a point of the midplane; the intersection of two cross-sections passing through P defines a *normal fibre* (see Figure 5.2). As h is very small with respect to L_x and to L_y, the following simplifying assumptions can be made, which hold in the case of small deformations:

1. The normal fibres behave as a rigid body.
2. As the midplane is deformed, the normal fibres remain perpendicular to it.
3. The normal stresses acting on the planes parallel to the midplane are consistent with the first hypothesis of rigidity of the normal fibres.

These assumptions can be understood as a natural extension to the two-dimensional case of the basic beam model. In particular, the second hypothesis means that there is no shear between two neighbouring normal fibres as the plate bends, in agreement with the Bernoulli–Euler model.

5.2.2 *Degrees of freedom and global displacements*

According to the first assumption made above, the motion of a normal fibre can be described by using five independent parameters, termed *global displacements*,

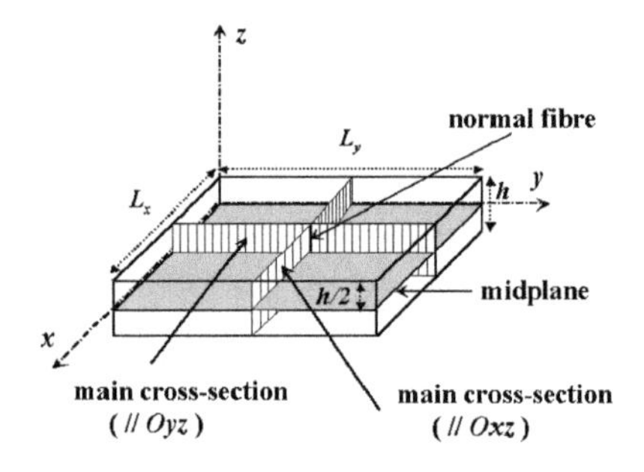

Figure 5.2. *Cross-sections and normal fibre in a plate*

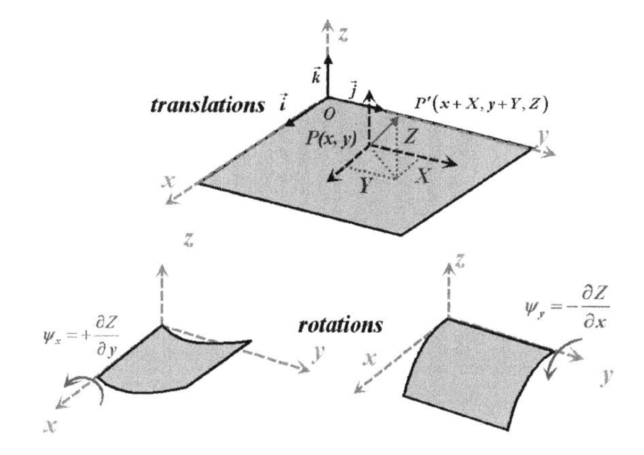

Figure 5.3. *Global displacements: translation and rotation variables*

which are referred to a current point $P(x, y)$ of the midplane. They are shown in Figure 5.3 and defined as follows:

1. the longitudinal displacement in the direction Ox: $X(x, y; t)\vec{i}$
2. the lateral displacement in the direction Oy: $Y(x, y; t)\vec{j}$
3. the transverse displacement in the direction Oz: $Z(x, y; t)\vec{k}$
4. the rotation around the Ox axis: $\psi_x(x, y; t)\vec{i}$
5. the rotation around the Oy axis: $\psi_y(x, y; t)\vec{j}$

After deformation, the point $P(x, y, 0)$ is transformed into $P(x + X, y + Y, Z)$.

5.2.3 *Membrane displacements, strains and stresses*

5.2.3.1 Global and local displacements

The components X, Y are termed *membrane displacements*, which are sufficient to describe the in-plane motions of the plate. In such motions, all the material points lying on a same normal fibre have the same displacements. So, global and local displacements are also the same, see Figure 5.4.

$$\vec{\xi}(x, y; t) = \vec{X} = X(x, y; t)\vec{i} + Y(x, y; t)\vec{j} \qquad [5.1]$$

5.2.3.2 Global and local strains

In agreement with [5.1], the local and the global strains are also the same: $\bar{\bar{\varepsilon}} = \bar{\bar{\eta}}$. Restricting the study to the case of small deformations, by substituting [5.1] into

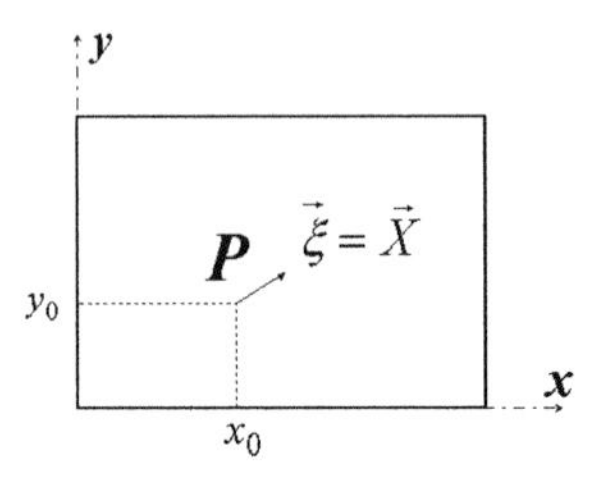

Figure 5.4. *Local and global displacements for in-plane motion*

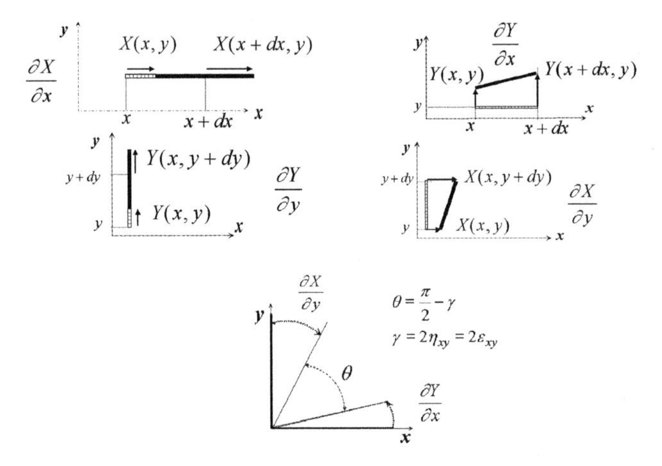

Figure 5.5. *Membrane strains*

the small strain tensor [1.25], we get:

$$\eta_{xx} = \frac{\partial X}{\partial x}; \quad \eta_{yy} = \frac{\partial Y}{\partial y}; \quad \eta_{xy} = \eta_{yx} = \frac{1}{2}\left(\frac{\partial X}{\partial y} + \frac{\partial Y}{\partial x}\right) \qquad [5.2]$$

The geometrical meaning of these quantities is illustrated in Figure 5.5. They comprise two normal and one shear component. η_{xx} is the relative longitudinal elongation (*Ox* direction), η_{yy} is the relative lateral elongation (*Oy* direction) and $\eta_{xy} = \eta_{yx} = \gamma/2$ is the in-plane shear strain which is also expressed in terms of the shear angle γ. Finally, the strain tensor [5.2] can be conveniently written in the following matrix form:

$$[\varepsilon] = [\eta] = \begin{bmatrix} \dfrac{\partial X}{\partial x} & \dfrac{1}{2}\left(\dfrac{\partial Y}{\partial x} + \dfrac{\partial X}{\partial y}\right) \\ \dfrac{1}{2}\left(\dfrac{\partial X}{\partial y} + \dfrac{\partial Y}{\partial x}\right) & \dfrac{\partial Y}{\partial y} \end{bmatrix} \qquad [5.3]$$

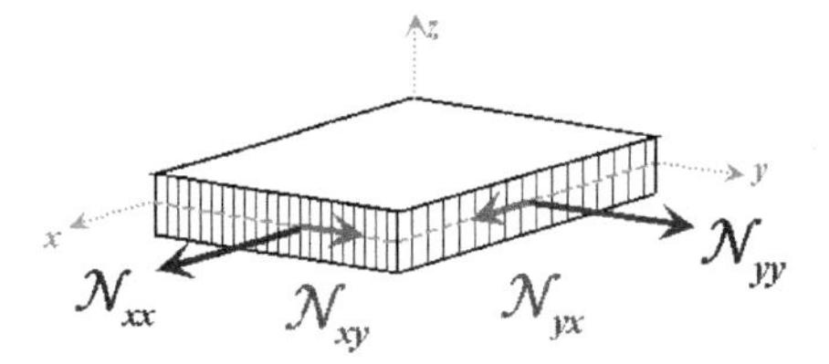

Figure 5.6. *Global stress components*

5.2.3.3 Membrane stresses

The membrane strains induce the *membrane local stresses* $\sigma_{xx}, \sigma_{yy}, \sigma_{xy}$ which are independent of z. These quantities are integrated over the plate thickness to produce the *global membrane stresses* which are suitably defined as forces per unit length:

$$\begin{bmatrix} \mathcal{N}_{xx} \\ \mathcal{N}_{yy} \\ \mathcal{N}_{xy} \end{bmatrix} = \int_{-h/2}^{h/2} \begin{bmatrix} \sigma_{xx} \\ \sigma_{yy} \\ \sigma_{xy} \end{bmatrix} dz = h \begin{bmatrix} \sigma_{xx} \\ \sigma_{yy} \\ \sigma_{xy} \end{bmatrix} \qquad [5.4]$$

In [5.4] the stress components are used to form a stress vector instead of a stress tensor. To avoid confusion when using the matrix notation, the local and global stress vectors are denoted $[\vec{\sigma}]$ and $[\vec{\mathcal{N}}]$ respectively, whereas the local and global stress tensors are denoted $[\sigma]$ and $[\mathcal{N}]$.

As shown in Figure 5.6, they are consistent with the sign convention adopted for the three-dimensional stresses (see Figure 1.6). $\mathcal{N}_{xx}$ and $\mathcal{N}_{yy}$ are the normal stresses, directed along the normal vectors of the edge surfaces parallel to Oy and Ox respectively, and $\mathcal{N}_{xy} = \mathcal{N}_{yx}$ are the in-plane shear stresses directed in the tangential directions Oy and Ox respectively. The global stress tensor is written in matrix notation as:

$$[\mathcal{N}] = \begin{bmatrix} \mathcal{N}_{xx} & \mathcal{N}_{yx} \\ \mathcal{N}_{xy} & \mathcal{N}_{yy} \end{bmatrix} \qquad [5.5]$$

5.3. Membrane equilibrium of rectangular plates

5.3.1 *Equilibrium in terms of generalized stresses*

Because the geometry of rectangular plates is very simple, there is no difficulty in deriving the equilibrium equations using the Newtonian approach i.e. direct balancing of the forces acting on an infinitesimal rectangular element, as shown in the next subsection. However, it is also of interest to solve the problem by using Hamilton's principle, which deals with scalar instead of vector quantities. It is applied first to the case of rectangular plates in subsection 5.3.1.2, and then to the case of orthogonal curvilinear coordinates in section 5.4.

5.3.1.1 *Local balance of forces*

Let us consider a rectangular plate loaded by the external force field:

$$\vec{f}^{(e)}(x, y; t) = f_x^{(e)}\vec{i} + f_y^{(e)}\vec{j} \qquad [5.6]$$

where $\vec{f}^{(e)}$ is a force density per unit area of the midplane surface.

In Figure 5.7, the balances of the longitudinal and lateral forces acting on an elementary rectangle *dx, dy* are sketched, which extend to the two-dimensional case the longitudinal force balance of straight beams shown in Figure 2.9. Of course, in the 2D case, it is appropriate to distinguish between a longitudinal and a lateral force balance, giving rise thus to two distinct equilibrium equations. Furthermore, the contribution of the tangential stresses due to the in-plane shear must be added to the in-plane normal stresses. As shear stresses are symmetric $\mathcal{N}_{xy} = \mathcal{N}_{yx}$, the moments are automatically balanced, not requiring any additional condition to describe the equilibrium of the rectangle.

The two equilibrium equations, in the *Ox, Oy* directions respectively, are written as:

$$\begin{aligned} \rho h\ddot{X} - \left(\frac{\partial \mathcal{N}_{xx}}{\partial x} + \frac{\partial \mathcal{N}_{yx}}{\partial y}\right) &= f_x^{(e)} \\ \rho h\ddot{Y} - \left(\frac{\partial \mathcal{N}_{yy}}{\partial y} + \frac{\partial \mathcal{N}_{xy}}{\partial x}\right) &= f_y^{(e)} \end{aligned} \qquad [5.7]$$

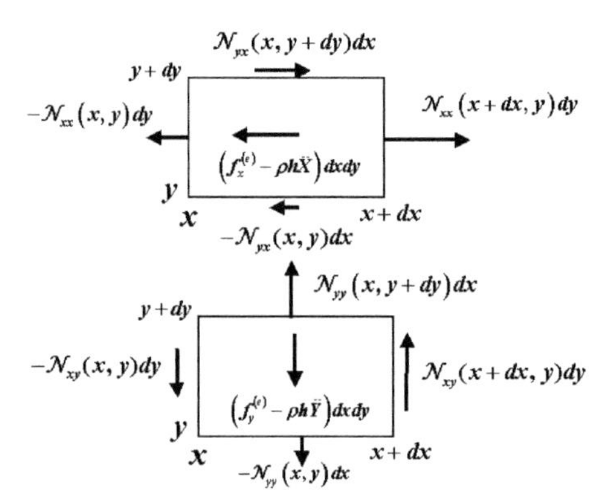

Figure 5.7. *In-plane forces acting on an infinitesimal element of the rectangular plate*

In tensor notation, the system [5.7] is expressed as:

$$\rho h \ddot{\vec{X}} \cdot \vec{\ell} - \left(\mathrm{div}\overline{\overline{\mathcal{N}}}\right) \cdot \vec{\ell} = \vec{f}^{(e)} \cdot \vec{\ell} \qquad [5.8]$$

where $\vec{\ell}$ denotes a unit vector in the plane of the plate.

It may be noticed that equation [5.8] is similar to the general 3D equation [1.32] and is independent of the coordinate system (intrinsic form), in contrast with [5.7] which holds in the case of Cartesian coordinates only.

5.3.1.2 Hamilton's principle

Here, in addition to the surface force density defined in [5.7], an edge force density is also introduced:

$$\vec{t}^{(e)}(x, y; t) = \left(t_x^{(e)}\vec{i} + t_y^{(e)}\vec{j}\right) \qquad [5.9]$$

The Lagrangian is written as:

$$\mathcal{L} = \int_0^{L_x}\int_0^{L_y} \{e_\kappa - e_s + w_F\}\, dx\, dy + \oint_{(\mathcal{C})} w_T\, ds \qquad [5.10]$$

where e_κ is the kinetic and e_s the strain energy densities per unit area; w_F is the work density of an external force field acting within the plate midplane and w_T is the work density of an external force field acting on the contour $(\mathcal{C})$ of the plate and s is the curvilinear abscissa along $(\mathcal{C})$.

Hamilton's principle is first written as:

$$\delta[\mathcal{A}] = \int_{t_1}^{t_2}\left[\int_0^{L_x}\int_0^{L_y} \{\delta[e_\kappa] - \delta[e_s] + \delta[w_F]\} dx\, dy + \oint_{C} \delta[w_T] ds\right] dt = 0 \qquad [5.11]$$

The remaining task is to evaluate the terms of [5.11] in a suitable way, as detailed just below.

The kinetic energy density is found to be:

$$e_\kappa = \frac{1}{2}\rho h(\dot{X}^2 + \dot{Y}^2) \qquad [5.12]$$

Its variation is integrated with respect to time, giving:

$$\int_{t_1}^{t_2}\left[\int_0^{L_x}\int_0^{L_y}\{\delta[e_\kappa]\}\,dx\,dy\right]dt = -\int_{t_1}^{t_2}\left[\int_0^{L_x}\int_0^{L_y}\rho h\{\ddot{X}\delta X + \ddot{Y}\delta Y\}\,dx\,dy\right]dt \qquad [5.13]$$

As long as the material law is not specified, it is not possible to express the strain energy density analytically. Nevertheless, its virtual variation is still given by (cf. the 3D equation [1.47]):

$$\delta[e_s] = \overline{\overline{\mathcal{N}}} : \delta\overline{\overline{\eta}} \iff \delta[e_s] = \mathcal{N}_{ij}\delta\eta_{ij} \qquad [5.14]$$

So, in Cartesian coordinates,

$$\delta[e_s] = \mathcal{N}_{xx}\frac{\partial\,\delta X}{\partial x} + \mathcal{N}_{xy}\frac{\partial\delta Y}{\partial x} + \mathcal{N}_{yx}\frac{\partial\delta X}{\partial y} + \mathcal{N}_{yy}\frac{\partial\delta Y}{\partial y} \qquad [5.15]$$

Therefore, we arrive at:

$$\begin{aligned}
&-\int_{t_1}^{t_2}\left[\int_0^{L_x}\int_0^{L_y}\{\delta[e_s]\}\,dx\,dy\right]dt \\
&= +\int_{t_1}^{t_2}\left[\int_0^{L_x}\int_0^{L_y}\left\{\left(\frac{\partial\mathcal{N}_{xx}}{\partial x}+\frac{\partial\mathcal{N}_{yx}}{\partial y}\right)\delta X + \left(\frac{\partial\mathcal{N}_{xy}}{\partial x}+\frac{\partial\mathcal{N}_{yy}}{\partial y}\right)\delta Y\right\}dx\,dy\right]dt \\
&\quad - \int_{t_1}^{t_2}dt\int_0^{L_y}[\mathcal{N}_{xx}\delta X+\mathcal{N}_{xy}\delta Y]_0^{L_x}dy - \int_{t_1}^{t_2}dt\int_0^{L_x}[\mathcal{N}_{yx}\delta X+\mathcal{N}_{yy}\delta Y]_0^{L_y}dx
\end{aligned} \qquad [5.16]$$

Turning finally to the virtual work of external loading, for the surface forces we get immediately:

$$\int_{t_1}^{t_2}\left[\int_0^{L_x}\int_0^{L_y}\{\delta[w_F]\}\,dx\,dy\right]dt = \int_{t_1}^{t_2}\left[\int_0^{L_x}\int_0^{L_y}\{f_x^{(e)}\delta X + f_y^{(e)}\delta Y\}\,dx\,dy\right]dt \qquad [5.17]$$

and for the edge forces,

$$\int_{t_1}^{t_2}\left[\oint_{(C)}\{\delta[w_T]\}\,ds\right]dt = \int_{t_1}^{t_2}\left[\oint_{(C)}\{t_x^{(e)}\delta X + t_y^{(e)}\delta Y\}\,ds\right]dt \qquad [5.18]$$

By cancelling the δX and δY cofactors in the surface and contour integrals arising in [5.13] and [5.16] to [5.18], the equations [5.7] and the following boundary conditions are readily derived. The contour integral [5.18] is developed as:

$$
\begin{aligned}
&\oint_{(C)} \{t_x^{(e)}\delta X + t_y^{(e)}\delta Y\}\,ds \\
&= \int_0^{L_x} \left[t_x^{(e)}\delta X + t_y^{(e)}\delta Y\right]_{y=0} dx + \int_0^{L_y} \left[t_x^{(e)}\delta X + t_y^{(e)}\delta Y\right]_{x=L_x} dy \\
&\quad + \int_0^{L_x} \left[t_x^{(e)}\delta X + t_y^{(e)}\delta Y\right]_{y=L_y} dx + \int_0^{L_y} \left[t_x^{(e)}\delta X + t_y^{(e)}\delta Y\right]_{x=0} dy
\end{aligned}
\qquad [5.19]
$$

Equation [5.19] used with the contour integrals of [5.16] gives:

$$
\begin{aligned}
&\int_0^{L_x} \left[\left(t_x^{(e)} + \mathcal{N}_{yx}\right)\delta X + \left(t_y^{(e)} + \mathcal{N}_{yy}\right)\delta Y\right]_{y=0} dx = 0 \\
&\quad \Rightarrow\; t_x^{(e)} + \mathcal{N}_{yx}\Big|_{y=0} = 0; \quad t_y^{(e)} + \mathcal{N}_{yy}\Big|_{y=0} = 0 \\
&\int_0^{L_x} \left[\left(t_x^{(e)} - \mathcal{N}_{yx}\right)\delta X + \left(t_y^{(e)} - \mathcal{N}_{yy}\right)\delta Y\right]_{y=L_y} dx = 0 \\
&\quad \Rightarrow\; t_x^{(e)} - \mathcal{N}_{yx}\Big|_{y=L_y} = 0; \quad t_y^{(e)} - \mathcal{N}_{yy}\Big|_{y=L_y} = 0 \\
&\int_0^{L_y} \left[\left(t_x^{(e)} + \mathcal{N}_{xx}\right)\delta X + \left(t_y^{(e)} + \mathcal{N}_{xy}\right)\delta Y\right]_{x=0} dy = 0 \\
&\quad \Rightarrow\; t_x^{(e)} + \mathcal{N}_{xx}\Big|_{x=0} = 0; \quad t_y^{(e)} + \mathcal{N}_{xy}\Big|_{x=0} = 0 \\
&\int_0^{L_y} \left[(t_x^{(e)} - \mathcal{N}_{xx})\delta X + (t_y^{(e)} - \mathcal{N}_{xy})\delta Y\right]_{x=L_x} dy = 0 \\
&\quad \Rightarrow\; t_x^{(e)} - \mathcal{N}_{xx}\Big|_{x=L_x} = 0; \quad t_y^{(e)} - \mathcal{N}_{xy}\Big|_{x=L_x} = 0
\end{aligned}
\qquad [5.20]
$$

It may be checked that the signs of the different stress terms which appear in the boundary conditions [5.20] are consistent with the general definition of stresses given in Chapter 1, subsection 1.2.3. This is illustrated in Figure 5.8 in which positive edge forces are applied to the four edges of the plate. For instance, according to [5.20] $\mathcal{N}_{xx}$ is found to be equal to the external edge force $t_x^{(e)}$ at $x = L_x$ and to $-t_x^{(e)}$ at $x = 0$. A slightly different way to check the signs consists of claiming that a positive external force $t_x^{(e)}$ applied along $x = L_x$ must induce tensile, hence positive, stresses and the same force when applied along $x = 0$ must induce compressive, hence negative stresses. Of course, the signs of the stresses will also agree with those found in a 3D rectangular parallelepiped and in a straight beam.

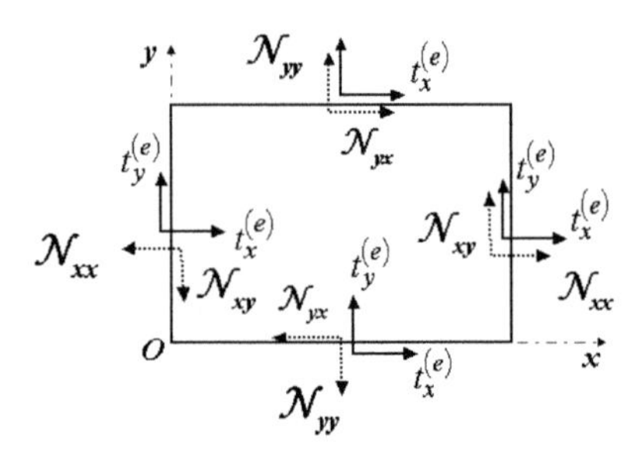

Figure 5.8. *Stresses induced by a uniform edge loading*

5.3.1.3 Homogeneous boundary conditions

If $\vec{t}^{(e)} \equiv 0$, the following standard homogeneous boundary conditions arise:

1. Fixed edge:

$$X \text{ and } Y = 0 \tag{5.21}$$

2. Sliding edge:

$$\begin{cases} \text{parallel to } O\vec{x}: & Y = 0; \quad \mathcal{N}_{yx} = 0 \\ \text{parallel to } O\vec{y}: & X = 0; \quad \mathcal{N}_{xy} = 0 \end{cases} \tag{5.22}$$

3. Free edge:

$$\begin{cases} \text{parallel to } O\vec{x}: & \mathcal{N}_{yy} = 0; \quad \mathcal{N}_{yx} = 0 \\ \text{parallel to } O\vec{y}: & \mathcal{N}_{xx} = 0; \quad \mathcal{N}_{xy} = 0 \end{cases} \tag{5.23}$$

5.3.1.4 Concentrated loads

As in the case of beams, it is found convenient to describe concentrated loads by using singular distributions. For example a force per unit length distributed along the line $x = x_0$ is written as:

$$\vec{f}^{(e)} = \left(f_x^{(e)}(y;t)\vec{i} + f_y^{(e)}(y;t)\vec{j}\right)\delta(x - x_0) \tag{5.24}$$

Then the equilibrium equations are written in terms of distributions as:

$$\begin{aligned} \rho h\ddot{X} - \left(\frac{\partial \mathcal{N}_{xx}}{\partial x} + \frac{\partial \mathcal{N}_{yx}}{\partial y}\right) &= f_x^{(e)}(y;t)\delta(x - x_0) \\ \rho h\ddot{Y} - \left(\frac{\partial \mathcal{N}_{xy}}{\partial x} + \frac{\partial \mathcal{N}_{yy}}{\partial y}\right) &= f_y^{(e)}(y;t)\delta(x - x_0) \end{aligned} \tag{5.25}$$

The integration in the domain $[x_0 - \varepsilon, x_0 + \varepsilon]$ where ε is arbitrarily small gives the equations:

$$\rho h\ddot{X} - \left(\frac{\partial \mathcal{N}_{xx}}{\partial x} + \frac{\partial \mathcal{N}_{yx}}{\partial y}\right) = 0; \quad \rho h\ddot{Y} - \left(\frac{\partial \mathcal{N}_{yy}}{\partial y} + \frac{\partial \mathcal{N}_{xy}}{\partial x}\right) = 0$$
$$\mathcal{N}_{xx}|_{x0-} - \mathcal{N}_{xx}|_{x0+} = f_x^{(e)}(y,t); \quad \mathcal{N}_{xy}|_{x0-} - \mathcal{N}_{xy}|_{x0+} = f_y^{(e)}(y,t) \qquad [5.26]$$

In the same way, a force concentrated at the point x_0, y_0 is defined by the distribution:

$$\vec{f}^{(e)} = \left(f_x^{(e)}(t)\vec{i} + f_y^{(e)}(t)\vec{j}\right)\delta(x - x_0) \cap \delta(y - y_0) \qquad [5.27]$$

In [5.27] the symbol $\cap$ means that the application point of the loading is located at the intersection of the lines $x = x_0$ and $y = y_0$. Accordingly, the equilibrium equations are written in terms of distributions as:

$$\rho h\ddot{X} - \left(\frac{\partial \mathcal{N}_{xx}}{\partial x} + \frac{\partial \mathcal{N}_{yx}}{\partial y}\right) = f_x^{(e)}(t)\delta(x - x_0) \cap \delta(y - y_0)$$
$$\rho h\ddot{Y} - \left(\frac{\partial \mathcal{N}_{xy}}{\partial x} + \frac{\partial \mathcal{N}_{yy}}{\partial y}\right) = f_y^{(e)}(t)\delta(x - x_0) \cap \delta(y - y_0) \qquad [5.28]$$

By integrating [5.28] over the intervals $[x_0 - \varepsilon, x_0 + \varepsilon; y_0 - \eta, y_0 + \eta]$ where ε and η are arbitrarily small, the equations of motion can be written as:

$$\rho h\ddot{X} - \left(\frac{\partial \mathcal{N}_{xx}}{\partial x} + \frac{\partial \mathcal{N}_{yx}}{\partial y}\right) = 0; \quad \rho h\ddot{Y} - \left(\frac{\partial \mathcal{N}_{xy}}{\partial x} + \frac{\partial \mathcal{N}_{yy}}{\partial y}\right) = 0$$
$$\mathcal{N}_{xx}(x_{0-}, y_0) - \mathcal{N}_{xx}(x_{0+}, y_0) + \mathcal{N}_{yx}(x_0, y_{0-}) - \mathcal{N}_{yx}(x_0, y_{0+}) = f_x^{(e)}(t)$$
$$\mathcal{N}_{yy}(x_0, y_{0-}) - \mathcal{N}_{yy}(x_0, y_{0+}) + \mathcal{N}_{xy}(x_{0-}, y_0) - \mathcal{N}_{xy}(x_{0+}, y_0) = f_y^{(e)}(t) \qquad [5.29]$$

The stress discontinuities are easily understood by turning back to Figure 5.8 and letting the plate dimensions tend to zero. For example, the balance of forces along the edge Ox is:

$$\mathcal{N}_{xx}(x_0 + \varepsilon, y) - \mathcal{N}_{xx}(x_0 - \varepsilon, y) + \mathcal{N}_{yx}(x, y_0 + \eta)$$
$$- \mathcal{N}_{yx}(x, y_0 - \eta) + f_x^{(e)}(t) = 0 \qquad [5.30]$$

where the inertia forces are neglected, since they vanish when ε and η tend to zero. So, the limit of [5.30] identifies with the second equation [5.29]. In the same manner, calculation performed on a plate of infinitesimal width 2ε would result in equation [5.26].

5.3.2 *Elastic stresses*

The three-dimensional law of elasticity as given by equations [1.36] to [1.38] is particularized here to the case of plane stresses. In agreement with the Kirchhoff–Love assumptions, the out-of-plane components of the strain tensor and of the elastic stress tensor are set to zero. Substituting such simplifications into [1.38], the following strain-stress relationships are derived:

$$\begin{aligned} \varepsilon_{xx} &= \frac{1+\nu}{E}\sigma_{xx} - \frac{\nu}{E}(\sigma_{xx}+\sigma_{yy}) = \frac{1}{E}\sigma_{xx} - \frac{\nu}{E}\sigma_{yy} \\ \varepsilon_{yy} &= \frac{1+\nu}{E}\sigma_{yy} - \frac{\nu}{E}(\sigma_{xx}+\sigma_{yy}) = \frac{1}{E}\sigma_{yy} - \frac{\nu}{E}\sigma_{xx} \\ \varepsilon_{xy} &= \frac{1+\nu}{E}\sigma_{xy} \end{aligned} \qquad [5.31]$$

The stress–strain relationships can be conveniently written in matrix form as:

$$\begin{bmatrix} \varepsilon_{xx} \\ \varepsilon_{yy} \\ 2\varepsilon_{xy} \end{bmatrix} = \frac{1}{E}\begin{bmatrix} 1 & -\nu & 0 \\ -\nu & 1 & 0 \\ 0 & 0 & 2(1+\nu) \end{bmatrix}\begin{bmatrix} \sigma_{xx} \\ \sigma_{yy} \\ \sigma_{xy} \end{bmatrix}$$

$$\Longleftrightarrow \begin{bmatrix} \sigma_{xx} \\ \sigma_{yy} \\ \sigma_{xy} \end{bmatrix} = \frac{E}{1-\nu^2}\begin{bmatrix} 1 & \nu & 0 \\ \nu & 1 & 0 \\ 0 & 0 & \dfrac{1-\nu}{2} \end{bmatrix}\begin{bmatrix} \varepsilon_{xx} \\ \varepsilon_{yy} \\ 2\varepsilon_{xy} \end{bmatrix} \qquad [5.32]$$

where the local strain and stress components are used to form a strain and a stress vector denoted $[\vec{\varepsilon}]$ and $[\vec{\sigma}]$ respectively.

In terms of global quantities the relation [5.32] reads as:

$$\begin{bmatrix} \eta_{xx} \\ \eta_{yy} \\ 2\eta_{xy} \end{bmatrix} = \frac{1}{Eh}\begin{bmatrix} 1 & -\nu & 0 \\ -\nu & 1 & 0 \\ 0 & 0 & 2(1+\nu) \end{bmatrix}\begin{bmatrix} \mathcal{N}_{xx} \\ \mathcal{N}_{yy} \\ \mathcal{N}_{xy} \end{bmatrix}$$

$$\Longleftrightarrow \begin{bmatrix} \mathcal{N}_{xx} \\ \mathcal{N}_{yy} \\ \mathcal{N}_{xy} \end{bmatrix} = \frac{Eh}{1-\nu^2}\begin{bmatrix} 1 & \nu & 0 \\ \nu & 1 & 0 \\ 0 & 0 & \dfrac{1-\nu}{2} \end{bmatrix}\begin{bmatrix} \eta_{xx} \\ \eta_{yy} \\ 2\eta_{xy} \end{bmatrix} \qquad [5.33]$$

where the global strain and stress components are used to form a strain vector and a stress vector denoted $[\vec{\eta}]$ and $[\vec{\mathcal{N}}]$ respectively.

NOTE. – *way of writing the stress–strain relationships*

Because of the symmetry of the strain and stress tensors it follows that $2\sigma_{xy}\varepsilon_{xy} = \sigma_{xy}\varepsilon_{xy} + \sigma_{yx}\varepsilon_{yx}$. Hence, in [5.32] the notation $2\varepsilon_{xy} = \gamma$ in place of ε_{xy} allows one to write the elastic energy density as the scalar product of a stress and a strain vectors:

$$2e_e = \overline{\overline{\sigma}} : \overline{\overline{\varepsilon}} = [\vec{\sigma}]^T [\vec{\varepsilon}];$$
$$\text{where } [\vec{\varepsilon}]^T = \begin{bmatrix} \varepsilon_{xx} & \varepsilon_{yy} & 2\varepsilon_{xy} \end{bmatrix} \quad [\vec{\sigma}]^T = \begin{bmatrix} \sigma_{xx} & \sigma_{yy} & \sigma_{xy} \end{bmatrix}$$

NOTE. – *3D elasticity versus plane stress and strain approximations*

According to the 3D elasticity law, there is an incompatibility between a plane stress model and a plane strain model, because if σ_{zz} is assumed to vanish, then $\varepsilon_{zz} = -\nu(\sigma_{xx} + \sigma_{yy})/E$ differs from zero as a result of the Poisson effect. Nevertheless, in the Kirchhoff–Love model the strains in the direction of plate thickness are discarded and at the same time a plane stress model is assumed to hold. Again, such a simplification is of the same nature as that made in beams by assuming that the cross-sections are rigid.

5.3.3 *Equations and boundary conditions in terms of displacements*

Substituting the strain–displacement relationships [5.3] into the stress–strain relationships [5.33], we obtain the global membrane stresses:

$$\mathcal{N}_{xx} = \frac{Eh}{1-\nu^2}\left(\frac{\partial X}{\partial x} + \nu\frac{\partial Y}{\partial y}\right); \quad \mathcal{N}_{yy} = \frac{Eh}{1-\nu^2}\left(\frac{\partial Y}{\partial y} + \nu\frac{\partial X}{\partial x}\right)$$
$$\mathcal{N}_{xy} = \frac{Eh}{2(1+\nu)}\left(\frac{\partial X}{\partial y} + \frac{\partial Y}{\partial x}\right) = Gh\left(\frac{\partial X}{\partial y} + \frac{\partial Y}{\partial x}\right) \qquad [5.34]$$

Substituting the stresses [5.34] into the equilibrium equations [5.7], we obtain the vibration equations:

$$\rho h\ddot{X} - \frac{Eh}{1-\nu^2}\frac{\partial}{\partial x}\left(\frac{\partial X}{\partial x} + \nu\frac{\partial Y}{\partial y}\right) - Gh\frac{\partial}{\partial y}\left(\frac{\partial X}{\partial y} + \frac{\partial Y}{\partial x}\right) = f_x^{(e)}(x, y; t)$$
$$\rho h\ddot{Y} - \frac{Eh}{1-\nu^2}\frac{\partial}{\partial y}\left(\frac{\partial Y}{\partial y} + \nu\frac{\partial X}{\partial x}\right) - Gh\frac{\partial}{\partial x}\left(\frac{\partial Y}{\partial x} + \frac{\partial X}{\partial y}\right) = f_y^{(e)}(x, y; t)$$
$$[5.35]$$

On the left-hand side of [5.35] it is appropriate to make the distinction between at least two stiffness operators. The first one, proportional to E, describes the stretching forces, and the second one, proportional to G, describes the in-plane shear forces. Furthermore, both the stretching and the shear operators are found to couple the longitudinal and the lateral motions together, since a displacement in the Ox direction is found to induce a displacement in the Oy direction and vice versa. The coupling terms appearing in the stretching operator is clearly due to the Poisson effect. The coupling terms appearing in the shear operator are inherent in 2D shear effect. They can be put in evidence even more clearly by writing the equations [5.35] in matrix form:

$$\begin{bmatrix} \rho h & 0 \\ 0 & \rho h \end{bmatrix}\begin{bmatrix} \ddot{X} \\ \ddot{Y} \end{bmatrix} + \begin{bmatrix} \left(-\frac{Eh}{1-\nu^2}\left(\frac{\partial^2}{\partial x^2}\right) - Gh\left(\frac{\partial^2}{\partial y^2}\right)\right) & -\left(\left(\frac{Eh\nu}{1-\nu^2}+Gh\right)\frac{\partial^2}{\partial x \partial y}\right) \\ -\left(\left(\frac{Eh\nu}{1-\nu^2}+Gh\right)\frac{\partial^2}{\partial x \partial y}\right) & \left(-\frac{Eh}{1-\nu^2}\left(\frac{\partial^2}{\partial y^2}\right) - Gh\left(\frac{\partial^2}{\partial x^2}\right)\right) \end{bmatrix} \times \begin{bmatrix} X \\ Y \end{bmatrix} = \begin{bmatrix} f_x^{(e)} \\ f_y^{(e)} \end{bmatrix} \qquad [5.36]$$

In [5.36] we recognize the canonical form $[M][\ddot{q}] + [K][q] = [Q^{(e)}]$ of the linear equations of any forced conservative mechanical system. In contrast with discrete systems, the stiffness coefficients are partial derivative linear operators. The mass matrix is diagonal and the non diagonal terms of the stiffness matrix are symmetric and made up of even operators, so as to comply with the condition of self-adjointness of $[M]$ and $[K]$.

The homogeneous boundary conditions describing edge elastic supports can be inferred from the inhomogeneous conditions [5.20], by assimilating first the support forces to an "external" load applied to the edges:

$$\begin{aligned} t_x^{(e)}(0,y) &= -K_{xx}X(0,y); \quad t_x^{(e)}(L_x,y) = -K_{xx}X(L_x,y) \\ t_y^{(e)}(0,y) &= -K_{xy}Y(0,y); \quad t_y^{(e)}(L_x,y) = -K_{xy}Y(L_x,y) \\ t_y^{(e)}(x,0) &= -K_{yy}Y(x,0); \quad t_y^{(e)}(x,L_y) = -K_{yy}Y(x,L_y) \\ t_x^{(e)}(x,0) &= -K_{yx}Y(x,0); \quad t_x^{(e)}(x,L_y) = -K_{yx}Y(x,L_y) \end{aligned}$$

The supports are described by the stiffness coefficients $K_{xx}, K_{yy}, K_{xy}, K_{yx}$ defined as forces per unit area (Nm^{-2} or Pa using the S.I. units). If the edge equilibrium conditions are considered as force balances between the 'external' loads and the internal stresses, then they are written as:

$$\begin{aligned}
&\frac{Eh}{1-\nu^2}\left(\frac{\partial X}{\partial x}+\nu\frac{\partial Y}{\partial y}\right)-K_{xx}X\bigg|_{x=0}=0; \\
&\frac{Eh}{1-\nu^2}\left(\frac{\partial X}{\partial x}+\nu\frac{\partial Y}{\partial y}\right)+K_{xx}X\bigg|_{x=L_x}=0 \\
&Gh\left(\frac{\partial X}{\partial y}+\frac{\partial Y}{\partial x}\right)-K_{xy}Y\bigg|_{x=0}=0; \\
&Gh\left(\frac{\partial X}{\partial y}+\frac{\partial Y}{\partial x}\right)+K_{xy}Y\bigg|_{x=L_x}=0 \\
&\frac{Eh}{1-\nu^2}\left(\frac{\partial Y}{\partial y}+\nu\frac{\partial X}{\partial x}\right)-K_{yy}Y\bigg|_{y=0}=0; \\
&\frac{Eh}{1-\nu^2}\left(\frac{\partial Y}{\partial y}+\nu\frac{\partial X}{\partial x}\right)+K_{yy}Y\bigg|_{y=L_y}=0 \\
&Gh\left(\frac{\partial X}{\partial y}+\frac{\partial Y}{\partial x}\right)-K_{yx}X\bigg|_{y=0}=0; \\
&Gh\left(\frac{\partial X}{\partial y}+\frac{\partial Y}{\partial x}\right)+K_{yx}X\bigg|_{y=L_y}=0
\end{aligned} \qquad [5.37]$$

Of course, the standard boundary conditions [5.21] to [5.23] may be recovered as limit cases of [5.37] where the appropriate stiffness coefficients tend either to zero or to infinity.

5.3.4 *Examples of application in elastostatics*

5.3.4.1 Sliding plate subject to a uniform longitudinal load at the free edge

The rectangular plate (h, L_x, L_y) shown in Figure 5.9, is loaded along the edge $x = L_x$ by a uniform longitudinal force density $t_x^{(e)}\vec{i}$. The edge $x = 0$ slides freely

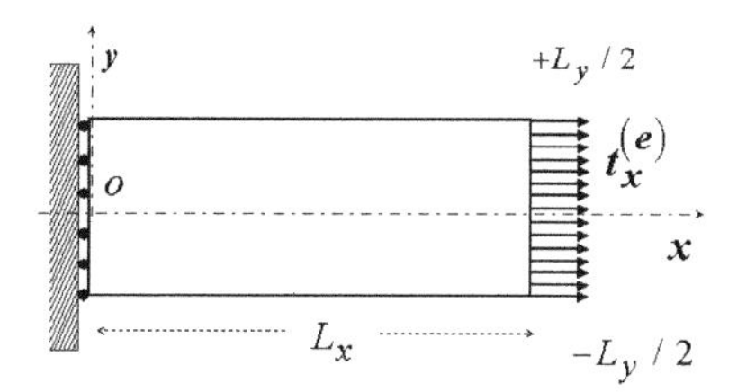

Figure 5.9. *Plate subject to a uniform longitudinal force density at one edge*

in the lateral direction Oy. The other edges are free. In terms of distributions the equilibrium equations are:

$$-\frac{Eh}{1-\nu^2}\left(\frac{\partial^2 X}{\partial x^2}+\nu\frac{\partial^2 Y}{\partial x\partial y}\right)-Gh\left(\frac{\partial^2 X}{\partial y^2}+\frac{\partial^2 Y}{\partial x\partial y}\right)=t_x^{(e)}\delta(L_x-x)$$

$$-\frac{Eh}{1-\nu^2}\left(\frac{\partial^2 Y}{\partial y^2}+\nu\frac{\partial^2 X}{\partial x\partial y}\right)-Gh\left(\frac{\partial^2 Y}{\partial x^2}+\frac{\partial^2 X}{\partial x\partial y}\right)=0$$

The boundary conditions are:

1. *Sliding edge* $x=0$:

$$X(0,y)=0;\quad \mathcal{N}_{xy}=0 \iff \frac{\partial X(0,y)}{\partial y}+\frac{\partial Y(0,y)}{\partial x}=0$$

2. *Free edge* $x=L_x$:

$$\mathcal{N}_{xx}=0 \iff \frac{\partial X(L_x,y)}{\partial x}+\nu\frac{\partial Y(L_x,y)}{\partial y}=0$$

$$\mathcal{N}_{xy}=0 \iff \frac{\partial X(L_x,y)}{\partial y}+\frac{\partial Y(L_x,y)}{\partial x}=0$$

3. *Free edges* $y=\pm L_y/2$:

$$\mathcal{N}_{yy}=0 \iff \frac{\partial Y(x,\pm L_y/2)}{\partial y}+\nu\frac{\partial X(x,\pm L_y/2)}{\partial x}=0$$

$$\mathcal{N}_{xy}=0 \iff \frac{\partial X(x,\pm L_y/2)}{\partial y}+\frac{\partial Y(x,\pm L_y/2)}{\partial x}=0$$

As a first step to solving the problem, the longitudinal equilibrium equation, written in terms of distributions, is replaced by the homogeneous equation:

$$-\frac{Eh}{1-\nu^2}\left(\frac{\partial^2 X}{\partial x^2}+\nu\frac{\partial^2 Y}{\partial x\partial y}\right)-Gh\left(\frac{\partial^2 X}{\partial y^2}+\frac{\partial^2 Y}{\partial x\partial y}\right)=0$$

and the related homogeneous boundary condition is replaced by the non homogeneous condition:

$$\mathcal{N}_{xx}|_{L_x-}=t_x^{(e)} \iff \frac{Eh}{1-\nu^2}\left(\frac{\partial X(L_x,y)}{\partial x}+\nu\frac{\partial Y(L_x,y)}{\partial y}\right)=t_x^{(e)}$$

As is rather obvious, any polynomial which is linear in x and y satisfies the homogeneous partial derivative equations. So the general solution may be written as:

$$X=a_1x+b_1y+c_1;\quad Y=a_2x+b_2y+c_2$$

where the coefficients a_1, b_1, etc. are determined by the boundary conditions, as detailed below.

1. *Sliding edge $x = 0$*

$$X(0, y) = 0 \;\Rightarrow\; c_1 = 0,\; b_1 = 0;$$

$$\frac{\partial X(0, y)}{\partial y} + \frac{\partial Y(0, y)}{\partial x} = 0 \;\Rightarrow\; b_1 + a_2 = 0$$

It is worth pointing out that the second condition implies that $\mathcal{N}_{xy} \equiv 0$ is zero everywhere in the plate, including at the edges.

2. *Loaded edge $x = L_x$*

$$a_1 + \nu b_2 = \frac{t_x^{(e)}(1 - \nu^2)}{Eh}$$

3. *Free edges $y = \pm L_y/2$*

$$\mathcal{N}_{yy} = 0 \;\Rightarrow\; b_2 + \nu a_1 = 0$$

Hence, $\mathcal{N}_{yy}$ as well as $\mathcal{N}_{xy}, \mathcal{N}_{yx}$ are found to vanish everywhere in the plate, including at the edges; so the stress field is uniaxial and uniform $\mathcal{N}_{xx} = t_x^{(e)}$. The displacement field is found to be:

$$X = \frac{t_x^{(e)}}{Eh} x; \quad Y = \frac{-\nu t_x^{(e)}}{Eh} y + c_2$$

where c_2 is arbitrary since the plate can move freely along the Oy axis. The problem can be further particularized by prescribing $c_2 = 0$, which means that the solution is symmetric about the Ox axis:

$$X = \frac{t_x^{(e)}}{Eh} x; \quad Y = \frac{-\nu t_x^{(e)}}{Eh} y \qquad [5.38]$$

The only difference between the plate solution [5.38] and the corresponding beam solution lies in the lateral contraction proportional to Poisson's ratio, which is related to the longitudinal stretching. Finite element solution of the problem is shown in colour plate 4, the load being applied to the $x = L_x$ and to the $x = L_x/2$ line successively. In the last case, only the part of the plate $0 \le x \le L_x/2$ is deformed. The finite jump of $\mathcal{N}_{xx}$ across the loaded line is smoothed out by the finite element discretization process.

5.3.4.2 Fixed instead of sliding condition at the supported edge

The condition along the edge $x = 0$ becomes $X(0, y) = 0; Y(0, y) = 0$. A priori, it could be surmised that the plate response does not differ much from the former one, though solution must be quite different from the mathematical standpoint since the relation $\mathcal{N}_{xy} \equiv 0$ does not hold anymore. Actually, the problem has no analytical solution in closed form. Numerical solution obtained by using the finite element method is shown in colour plates 5. It can be verified that by fixing the supported edge, the lateral contraction is prevented and shear stresses arise. Nevertheless, such effects are concentrated in the vicinity of the supported edge, in agreement with the Saint-Venant principle (cf. Chapter 1, subsection 1.5.1).

5.3.4.3 Three sliding edges: plate in uniaxial strain configuration

Here, the lateral contraction related to the longitudinal stretching of the plate is prevented by prescribing the sliding conditions $X = 0$ at $x = 0$ and $Y = 0$ at $y = \pm L_y/2$. Accordingly, it seems natural to anticipate a solution of the type $X = a_1 x; Y \equiv 0$ which, after some elementary algebra, is found to be:

$$X = \frac{t_x^{(e)}}{Eh}(1 - \nu^2)x; \quad Y \equiv 0$$

$$\mathcal{N}_{xx} = t_x^{(e)}; \quad \mathcal{N}_{yy} = \nu \mathcal{N}_{xx}; \quad \mathcal{N}_{xy} = 0; \quad \eta_{xx} = \frac{t_x^{(e)}}{Eh}; \quad \eta_{yy} = \eta_{xy} = 0 \qquad [5.39]$$

5.3.4.4 Uniform plate stretching

As sketched in Figure 5.10, the plate is stretched in the longitudinal and the lateral directions by a tensile load distributed uniformly on the four edges. The plate centre is supposed fixed, so central symmetry holds. Again, the displacements X and Y are linear functions and because of the boundary conditions, there is no shear. Thus, we get:

$$\mathcal{N}_{xy} = 0 \;\Rightarrow\; b_1 + a_2 = 0$$

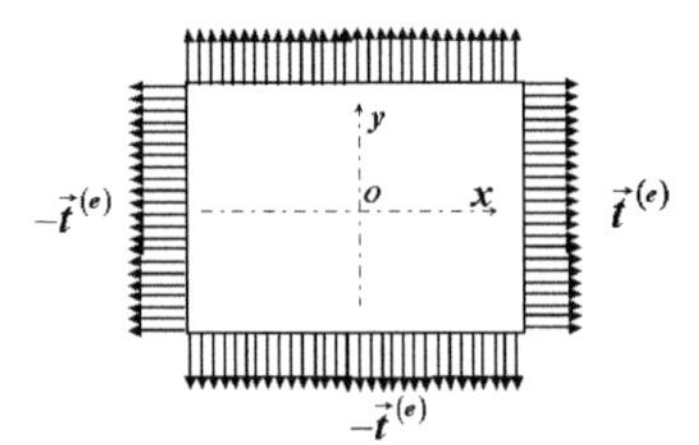

Figure 5.10. *Rectangular plate stretched in the Ox and Oy directions*

The normal stresses verify:

$$\begin{cases} \mathcal{N}_{xx} = \dfrac{Eh}{1-\nu^2}(a_1 + \nu b_2) = t_x = t^{(e)} \\ \mathcal{N}_{yy} = \dfrac{Eh}{1-\nu^2}(b_2 + \nu a_1) = t_y = t^{(e)} \end{cases} \Rightarrow a_1 = b_2$$

$$b_2 = \frac{t^{(e)}(1-\nu)}{Eh}$$

and because of the central symmetry:

$$X(x,y) = \frac{t^{(e)}(1-\nu)}{Eh}x$$
$$Y(x,y) = \frac{t^{(e)}(1-\nu)}{Eh}y \qquad [5.40]$$

5.3.4.5 In-plane uniform shear loading

Again, the centre of the rectangular plate is fixed. It is loaded by shear forces uniformly distributed on the four edges, as shown in Figure 5.11. Of course, the resultant of the loading vanishes. The moment of the longitudinal forces is $\mathcal{M}_{z1} = (t^{(e)}L_x)L_y$, where $L_y/2$ is the lever arm. Similarly, the moment of lateral forces is $\mathcal{M}_{z2} = -(t^{(e)}L_y)L_x$ where the lever arm is $L_x/2$. Thus, the resultant moment is also null and the global condition for plate equilibrium is fulfilled. Because of the central symmetry and the absence of normal stresses, the solution in terms of displacements is necessarily of the type:

$$X = by \qquad Y = ax \qquad [5.41]$$

The shear stress field follows as:

$$\mathcal{N}_{xy} = \mathcal{N}_{yx} = Gh(a+b) = \gamma Gh = t^{(e)}h \qquad [5.42]$$

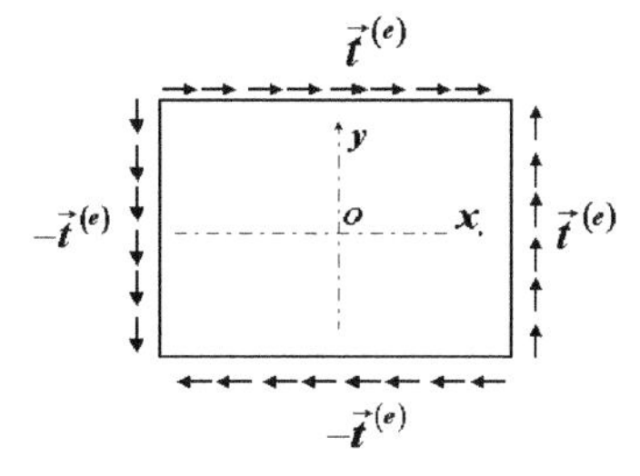

Figure 5.11. *Plate uniformly loaded by shear forces*

However, a and b are not single valued because the plate can rotate freely around the Oz axis, normal to the plate.

5.3.4.6 In-plane shear and bending

A rectangular plate is supported along the edge $x = 0$ and loaded by a uniform shear force along the edge $x = L$, the resultant of which is denoted F, see Figure 5.12. A priori, it would be desirable to prescribe a fixed support condition at $x = 0$, in order to compare the plate response to the in-plane shear load and the bending response to F of the same plate, modelled as a cantilevered beam. Such a comparative study is expected to provide us with enlightening information concerning both the relative importance of shear and normal stresses in the plate and the hypothesis of cross-sectional rigidity in beam models, as a function of L/ℓ. Unfortunately, solving analytically the problem stated just above is far from being a simple task, as difficulties arise in complying with the fixed boundary conditions. It is thus found more convenient to approximate the boundary conditions, based on physical reasoning.

At first, the plate solution is expected to be not very far from the beam solution (Bernoulli–Euler model) which, for the present problem, is written as:

$$Y_b(x) = \frac{FL^3}{6EI}\left(3\left(\frac{x}{L}\right)^2 - \left(\frac{x}{L}\right)^3\right)$$

As Y_b is independent of y, the stress field related to the bending solution is of the type:

$$\sigma_{xx} = \alpha(L - x)y, \quad \sigma_{yy} = 0, \quad \sigma_{xy} = 0$$

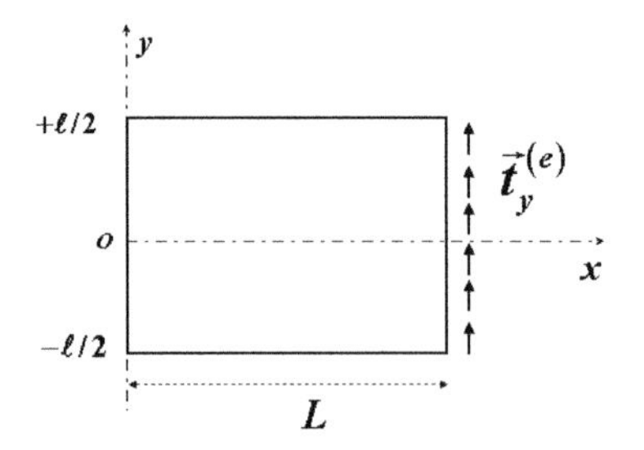

Figure 5.12. *Lateral shear loading*

In the plate model shear differs from zero; so a plate stress field not far from the beam model could be anticipated as:

$$\mathcal{N}_{xx} = \alpha(x-L)y; \quad \mathcal{N}_{yy} = 0; \quad \mathcal{N}_{xy} = \beta\left(\left(\frac{\ell}{2}\right)^2 - y^2\right)$$

The normal stresses are identical to those obtained from the beam theory and the form postulated for the shear stress $\mathcal{N}_{xy}$ agrees with the boundary condition $\mathcal{N}_{xy} = 0$ at the lateral edges $y = \pm\ell/2$. Furthermore, to be compatible with the edge loading, the shear stress field along the edge $x = L$ needs to be an even function of y. As a first approximation it is legitimate to adopt a parabolic distribution, leading to the resultant:

$$F = \int_{-\ell/2}^{\ell/2} t_y^{(e)}(L,y)\,dy = \int_{-\ell/2}^{\ell/2} \beta h\left(\left(\frac{\ell}{2}\right)^2 - y^2\right)dy$$

It follows that $\beta = 6F/(h\ell^3) = F/(2I)$

The moment associated with F is:

$$\mathcal{M}(x) = (x-L)F = \alpha(x-L)h\int_{-\ell/2}^{\ell/2} y^2 dy, \quad \text{so } \alpha = \frac{12F}{h\ell^3} = \frac{F}{I}$$

Using the stress–strain and the strain–displacement relationships, the following system of partial derivative equations is obtained:

$$\frac{\partial X}{\partial x} = \frac{1}{E}\left(\mathcal{N}_{xx} - \nu\mathcal{N}_{yy}\right) \Rightarrow \frac{\partial X}{\partial x} = \frac{F}{EI}(x-L)y;$$

$$\frac{\partial Y}{\partial y} = \frac{1}{E}\left(\mathcal{N}_{yy} - \nu\mathcal{N}_{xx}\right) \Rightarrow \frac{\partial Y}{\partial y} = -\nu\frac{F}{EI}(x-L)y$$

$$\frac{\partial X}{\partial y} + \frac{\partial Y}{\partial x} = \frac{1}{G}\mathcal{N}_{xy} \Rightarrow \frac{\partial X}{\partial y} + \frac{\partial Y}{\partial x} = \frac{F}{2GI}\left(\left(\frac{\ell}{2}\right)^2 - y^2\right)$$

By integrating the two first equations, it follows that:

$$X(x,y) = \frac{Fxy(x-2L)}{2EI} + X_0(y); \qquad Y(x,y) = -\frac{\nu F(x-L)y^2}{2EI} + Y_0(x)$$

Substitution of these results into the shear equation leads to:

$$\frac{F\ell^2}{8GI} = \left(\frac{dX_0}{dy} + \frac{Fy^2}{2GI} - \frac{\nu F y^2}{2EI}\right) + \left(\frac{dY_0}{dx} + \frac{Fx(x-2L)}{2EI}\right)$$

so,

$$\frac{dX_0}{dy} = a - \frac{(2+\nu)Fy^2}{2EI}; \quad \frac{dY_0}{dx} = b - \left(\frac{Fx(x-2L)}{2EI}\right)$$

$$\text{where } a + b = \frac{F\ell^2}{8GI}$$

The resulting displacement field is:

$$X_0(y) = ay + c - \frac{F(2+\nu)}{6EI}y^3; \quad Y_0(x) = bx + d + \frac{F}{6EI}(3Lx^2 - x^3)$$

$$X(x,y) = \frac{Fxy(x-2L)}{2EI} - \frac{F(2+\nu)}{6EI}y^3 + ay + c$$

$$Y(x,y) = -\frac{\nu F(x-L)y^2}{2EI} + \frac{F}{6EI}(3Lx^2 - x^3) + bx + d$$

However, if such a solution is adopted, it is impossible to fix the edge $x = 0$ in its integrity, so the boundary condition is modified by fixing the point O only. This new condition is clearly less rigid that the former one. It leads to:

$$X(0,0) = Y(0,0) = 0 \Rightarrow c = d = 0$$

Supplementary conditions are still needed to determine the coefficients a and b. They can be selected in such a way as to rigidify the support. A suitable condition which prevents shear strain at the fixed point is:

$$\left.\frac{\partial X}{\partial y}\right|_{0,0} - \left.\frac{\partial Y}{\partial x}\right|_{0,0} = 0 \Rightarrow a - b = 0 \Rightarrow a = b = \frac{F\ell^2}{16GI}$$

Therefore, the final solution is written as:

$$X(x,y) = \frac{Fx(x-2L)y}{2EI} - \frac{F(2+\nu)}{6EI}y^3 + \frac{F\ell^2}{16GI}y$$
$$Y(x,y) = +\frac{F}{6EI}(3Lx^2 - x^3) - \frac{\nu F(x-L)y^2}{2EI} + \frac{F\ell^2}{16GI}x \quad [5.43]$$

The displacement field [5.43] must be understood as an approximate solution to the "real" problem which comply with the fixed edge condition. However, according to Saint-Venant's principle, it may be expected that it provides satisfactory results, except in the vicinity of the fixed edge. This important point can be checked by comparison with the beam results, as outlined below.

1. In [5.43], the first term of the lateral displacement is precisely the beam solution $Y_b(x)$.

2. The first term of the longitudinal displacement fully agrees with the axial component of the local displacement of beam theory: $\xi_x = -\psi_z y = -(\partial Y_b/\partial x)y = Fx(x-2L)y/(2EI)$.
3. The other terms of [5.43] are corrective terms induced by plate shearing. As expected, their relative importance decreases quickly as the aspect ratio $\eta = L/\ell$ increases. In particular, considering the displacements at a corner of the plate the following characteristic ratios are found:

$$\frac{X_{\text{plate}}(L,\ell/2)}{\xi_{x_{\text{beam}}}(L,\ell/2)} = 1 + \frac{4+5\nu}{3}\left(\frac{\ell}{2L}\right)^2$$
$$\frac{Y_{\text{plate}}(L,\ell/2)}{Y_{\text{beam}}(L,\ell/2)} = 1 + \frac{3(1+\nu)}{2}\left(\frac{\ell}{2L}\right)^2 \quad [5.44]$$

Finally, finite element solutions of the plate fixed along the edge $x = 0$ and loaded by a uniform in-plane shear force along the edge $x = L$ are shown in colour plates 6 and 7. They are found to be in good agreement with the approximate solution [5.43]. In plate 6, which refers to a small aspect ratio, warping of the plate edge section is clearly evidenced at $x = L$. The longitudinal stress distribution along the fixed edge departs also significantly from the values given by the beam model. As expected, results shown in plate 7 which refers to a much slender plate are in much better agreement with those given by the beam model.

5.3.5 *Examples of application in thermoelasticity*

5.3.5.1 Thermoelastic law

Thermal strains due to a temperature change in a plate have the following expressions:

$$\tilde{\varepsilon}_{xx} = \tilde{\eta}_{xx} = \alpha\Delta\theta; \quad \tilde{\varepsilon}_{yy} = \tilde{\eta}_{yy} = \alpha\Delta\theta; \quad \tilde{\varepsilon}_{xy} = \tilde{\eta}_{xy} = 0 \quad [5.45]$$

α is the thermal dilatation coefficient and $\Delta\theta$ the temperature change. In thermoelasticity, elastic and thermal strains are added, so:

$$\varepsilon_{xx} = \eta_{xx} = \hat{\varepsilon}_{xx} + \tilde{\varepsilon}_{xx} = \frac{\hat{\sigma}_{xx}}{E} - \frac{\nu\hat{\sigma}_{yy}}{E} + \alpha\Delta\theta$$
$$\varepsilon_{yy} = \eta_{yy} = \hat{\varepsilon}_{yy} + \tilde{\varepsilon}_{yy} = \frac{\hat{\sigma}_{yy}}{E} - \frac{\nu\hat{\sigma}_{xx}}{E} + \alpha\Delta\theta \quad [5.46]$$
$$\varepsilon_{xy} = \eta_{xy} = \hat{\varepsilon}_{xy}$$

$\bar{\bar{\tilde{\varepsilon}}}$ and $\bar{\bar{\tilde{\sigma}}}$ are the strain and the stress isotherm tensors. Since the elastic stresses are related to the elastic strains alone, the thermoelastic law is written as:

$$\begin{bmatrix} \sigma_{xx} \\ \sigma_{yy} \\ \sigma_{xy} \end{bmatrix} = \frac{E}{(1-\nu^2)} \begin{bmatrix} 1 & \nu & 0 \\ \nu & 1 & 0 \\ 0 & 0 & \dfrac{1-\nu}{2} \end{bmatrix} \begin{bmatrix} \varepsilon_{xx} - \tilde{\varepsilon}_{xx} \\ \varepsilon_{yy} - \tilde{\varepsilon}_{yy} \\ 2\varepsilon_{xy} \end{bmatrix}$$

$$\Longleftrightarrow \begin{bmatrix} \mathcal{N}_{xx} \\ \mathcal{N}_{yy} \\ \mathcal{N}_{xy} \end{bmatrix} = \frac{Eh}{(1-\nu^2)} \begin{bmatrix} 1 & \nu & 0 \\ \nu & 1 & 0 \\ 0 & 0 & \dfrac{1-\nu}{2} \end{bmatrix} \begin{bmatrix} \eta_{xx} - \tilde{\eta}_{xx} \\ \eta_{yy} - \tilde{\eta}_{yy} \\ 2\tilde{\eta}_{xy} \end{bmatrix} \qquad [5.47]$$

In terms of displacements [5.47] becomes:

$$\mathcal{N}_{xx} = \frac{Eh}{1-\nu^2}\left(\frac{\partial X}{\partial x} + \nu\frac{\partial Y}{\partial y}\right) - \frac{\alpha Eh\Delta\theta}{1-\nu}$$

$$\mathcal{N}_{yy} = \frac{Eh}{1-\nu^2}\left(\frac{\partial Y}{\partial y} + \nu\frac{\partial X}{\partial x}\right) - \frac{\alpha Eh\Delta\theta}{1-\nu} \qquad [5.48]$$

$$\mathcal{N}_{xy} = \hat{N}_{xy} = Gh\left(\frac{\partial X}{\partial y} + \frac{\partial Y}{\partial x}\right)$$

5.3.5.2 *Thermal stresses*

A square plate $(2L, h)$ is considered. The edges are on sliding supports. The initial temperature θ of the plate is changed into $\theta + \Delta\theta$. We want to determine the stresses induced by this temperature change. As the edges are fixed along the normal direction, the corresponding strains are null: $\varepsilon_{xx} = \varepsilon_{yy} = 0$. Furthermore, no shear is induced by the temperature change, as shown in [5.48], so the plate is not deformed at all, $\varepsilon_{xx} = \varepsilon_{yy} = \varepsilon_{xy} = 0$ everywhere in the plate. Then, by using [5.47]:

$$\begin{bmatrix} \tilde{\mathcal{N}}_{xx} \\ \tilde{\mathcal{N}}_{yy} \\ \tilde{\mathcal{N}}_{xy} \end{bmatrix} = \frac{Eh}{(1-\nu^2)} \begin{bmatrix} 1 & \nu & 0 \\ \nu & 1 & 0 \\ 0 & 0 & \dfrac{1-\nu}{2} \end{bmatrix} \begin{bmatrix} -\tilde{\eta}_{xx} \\ -\tilde{\eta}_{yy} \\ 0 \end{bmatrix}$$

$$\Longleftrightarrow \begin{bmatrix} \tilde{\mathcal{N}}_{xx} \\ \tilde{\mathcal{N}}_{yy} \\ \tilde{\mathcal{N}}_{xy} \end{bmatrix} = \frac{-\alpha\Delta\theta Eh}{(1-\nu^2)} \begin{bmatrix} 1 & \nu & 0 \\ \nu & 1 & 0 \\ 0 & 0 & \dfrac{1-\nu}{2} \end{bmatrix} \begin{bmatrix} 1 \\ 1 \\ 0 \end{bmatrix} \qquad [5.49]$$

Depending on the sign of $\Delta\theta$, the stresses are positive (tension) or negative (compression). The thermal forces exerted on the edges are:

$$F_x|_{x=L} = F_y|_{y=L} - \frac{2\alpha EhL\Delta\theta}{1-\nu}; \quad F_x|_{x=-L} = F_y|_{y=-L}\frac{2\alpha EhL\Delta\theta}{1-\nu} \qquad [5.50]$$

A numerical example is useful to emphasize that such thermal stresses may be very large:

$$L = 0.5\,\text{m}; \quad h = 2\,\text{mm}; \quad E = 2.10^{11}\,\text{Pa}; \quad \nu = 0.3$$

$$\Rightarrow \; F = \left|\frac{2\alpha EhL\Delta\theta}{1-\nu}\right| \cong 280\,\text{kN} \quad \alpha \cong 10^{-5}/°\text{C}, \quad \Delta\theta = 100°\text{C}$$

5.3.5.3 Expansion joints

The same plate is supported now by using expansion joints modelled as a normal and uniform stiffness coefficient per unit edge length noted K. The tangential stiffness of the joints is neglected. Because of the symmetry of the structure it is advantageous to use the centre of the plate as the origin of the axes. According to the thermoelastic stresses [5.48], the equilibrium equations are:

$$\begin{aligned} &\frac{Eh}{1-\nu^2}\left(\frac{\partial^2 X}{\partial x^2} + \nu\frac{\partial^2 Y}{\partial x \partial y}\right) + Gh\left(\frac{\partial^2 X}{\partial y^2} + \frac{\partial^2 Y}{\partial x \partial y}\right) = -\frac{\alpha Eh}{1-\nu}\frac{\partial \Delta\theta}{\partial x} \\ &\frac{Eh}{1-\nu^2}\left(\frac{\partial^2 Y}{\partial y^2} + \nu\frac{\partial^2 X}{\partial x \partial y}\right) + Gh\left(\frac{\partial^2 Y}{\partial x^2} + \frac{\partial^2 X}{\partial x \partial y}\right) = -\frac{\alpha Eh}{1-\nu}\frac{\partial \Delta\theta}{\partial y} \end{aligned} \qquad [5.51]$$

However, as $\Delta\theta$ is uniform, the right-hand side of [5.48] vanishes. The general solution of [5.51] is thus

$$X(x,y) = a_1 x + b_1 y + c_1; \quad Y(x,y) = a_2 x + b_2 y + c_2$$

Because of the symmetry it follows that $c_1 = c_2 = 0$. On the other hand the following conditions hold at the edges,

$$\mathcal{N}_{xy}|_C = 0 \iff \frac{\partial Y}{\partial x} + \frac{\partial X}{\partial y} = 0$$

$$\mathcal{N}_{xx}|_{x=\pm L} \mp KX(\pm L, y) = 0;$$

$$\Rightarrow \quad \frac{Eh}{1-\nu^2}\left(\frac{\partial X}{\partial x} + \nu\frac{\partial Y}{\partial y}\right)\bigg|_{x=\pm L} \mp KhX(L,y) = \frac{\alpha Eh\Delta\theta}{1-\nu}$$

$$\mathcal{N}_{yy}|_{y=\pm L} \mp KX(x, \pm L) = 0$$

$$\Rightarrow \quad \frac{Eh}{1-\nu^2}\left(\frac{\partial Y}{\partial y} + \nu\frac{\partial X}{\partial x}\right)\bigg|_{y=\pm L} \mp KhY(x,\pm L) = \frac{\alpha Eh\Delta\theta}{1-\nu}$$

The shear stresses are null along the edges and as they must be constant, they vanish everywhere:

$$\frac{\partial Y}{\partial x}+\frac{\partial X}{\partial y}=a_2+b_1;\qquad \left.\frac{\partial Y}{\partial x}+\frac{\partial X}{\partial y}\right|_C=0\ \Rightarrow\ \frac{\partial Y}{\partial x}+\frac{\partial X}{\partial y}\equiv 0$$

The normal stresses verify the following equations:

$$\frac{Eh}{1-\nu^2}\left.\left(\frac{\partial X}{\partial x}+\nu\frac{\partial Y}{\partial y}\right)\right|_{x=\pm L}\mp KhX(\pm L,y)=\frac{\alpha Eh\Delta\theta}{1-\nu}$$

$$\frac{Eh}{1-\nu^2}\left.\left(\frac{\partial Y}{\partial y}+\nu\frac{\partial X}{\partial x}\right)\right|_{y=\pm L}\mp KhY(x,\pm L)=\frac{\alpha Eh\Delta\theta}{1-\nu}$$

Finally, because of the central symmetry of the problem, the following conditions hold:

$$X(x,y)=Y(y,x)\iff a_1=a_2=a\quad b_1=b_2=0$$

The coefficient a can be expressed in terms of the dimensionless stiffness coefficient:

$$\kappa=\frac{KL(1-\nu^2)}{E}\tag{5.52}$$

then,

$$\begin{aligned}&X(x)=\frac{\alpha E(1+\nu)\Delta\theta}{(1+\nu+\kappa)}x;\quad Y(y)=\frac{\alpha E(1+\nu)\Delta\theta}{(1+\nu+\kappa)}y\\&\mathcal{N}_{xx}=\mathcal{N}_{yy}=-\frac{\alpha Eh\kappa\Delta\theta}{(1-\nu)(1+\nu+\kappa)};\quad \mathcal{N}_{xy}=\mathcal{N}_{yx}=0\end{aligned}\tag{5.53}$$

The limit cases of sliding and free plate are recovered as follows:

$$\begin{aligned}&\kappa\to 0\ \Rightarrow\ \begin{cases}X(x)\to(\alpha E\Delta\theta)x;\ Y(y)\to(\alpha E\Delta\theta)y\\ \mathcal{N}_{xx}=\mathcal{N}_{yy}\to 0\end{cases}\\&\kappa\to\infty\ \Rightarrow\ \begin{cases}X(x)\to 0;\ Y(y)\to 0\\ \mathcal{N}_{xx}=\mathcal{N}_{yy}\to-\dfrac{\alpha Eh\Delta\theta}{1-\nu}\end{cases}\end{aligned}\tag{5.54}$$

5.3.5.4 Uniaxial plate expansion

A rectangular plate $(L,2\ell,h)$ is considered. The edge $x=0$ is a sliding support. The problem consists in determining the response to the linear temperature distribution $\Delta\theta(x)=\Delta\theta_0 x/L$, see Figure 5.13. It is governed by the

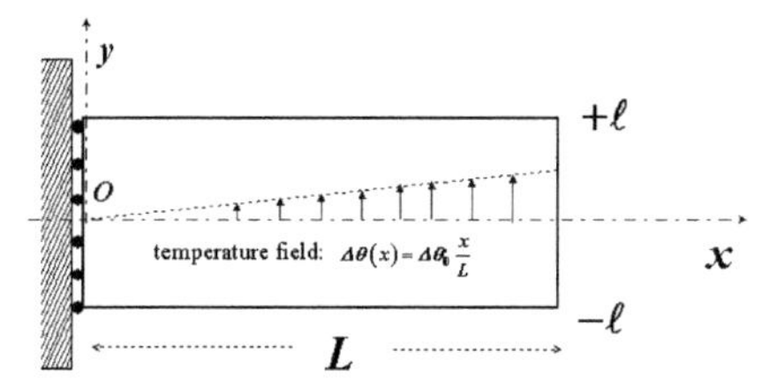

Figure 5.13. *Plate loaded by a uniform thermal gradient*

following system of equations:

$$\frac{Eh}{1-\nu^2}\left(\frac{\partial^2 X}{\partial x^2}+\nu\frac{\partial^2 Y}{\partial x \partial y}\right)+Gh\left(\frac{\partial^2 X}{\partial y^2}+\frac{\partial^2 Y}{\partial x \partial y}\right)=\frac{\alpha Eh}{1-\nu}\frac{\Delta\theta_0}{L}$$

$$\frac{Eh}{1-\nu^2}\left(\frac{\partial^2 Y}{\partial y^2}+\nu\frac{\partial^2 X}{\partial x \partial y}\right)+Gh\left(\frac{\partial^2 Y}{\partial x^2}+\frac{\partial^2 X}{\partial x \partial y}\right)=0$$

The solution has the general form:

$$X = a_1x^2 + b_1xy + c_1y^2 + d_1x + e_1y + f_1$$

$$Y = a_2x + b_2y + c_2$$

The constants are determined by using the boundary conditions and the symmetry of the problem with respect to Ox axis, implying thus the following set of relations:

$$X(0,y)=0; \quad Y(x,0)=0$$

$$\mathcal{N}_{xx}|_{x=L}=0 \Rightarrow \frac{Eh}{1-\nu^2}\left(\frac{\partial X}{\partial x}+\nu\frac{\partial Y}{\partial y}\right)\bigg|_{x=L}=\frac{\alpha Eh}{1-\nu}\Delta\theta_0$$

$$\mathcal{N}_{yy}\big|_{y=\pm\ell}=0 \Rightarrow \frac{Eh}{1-\nu^2}\left(\frac{\partial Y}{\partial y}+\nu\frac{\partial X}{\partial x}\right)\bigg|_{y=\pm\ell}=\frac{\alpha Eh}{1-\nu}\Delta\theta_0\frac{x}{L};$$

$$\mathcal{N}_{xy}\big|_{C}=0 \iff \frac{\partial Y}{\partial x}+\frac{\partial X}{\partial y}=0$$

The conditions on displacement give:

$$X(0,y)=0 \Rightarrow c_1y^2+e_1y+f_1=0 \Rightarrow c_1=e_1=f_1=0$$

$$Y(x,0)=0 \Rightarrow a_2x+c_2=0 \Rightarrow a_2=c_2=0$$

Changing the name of the constants, the solution is rewritten as

$$X = A_1x^2 + B_1xy + C_1x; \qquad Y = A_2y$$

The condition of no shear at the edges gives:

$$\frac{\partial X}{\partial y} + \frac{\partial Y}{\partial x} = B_1 = 0$$

As a consequence, shear vanishes everywhere in the plate. The condition of zero normal stresses at the edges gives:

$$\begin{aligned}
&\frac{Eh}{1-\nu^2}\left(\frac{\partial X}{\partial x} + \nu\frac{\partial Y}{\partial y}\right)\bigg|_{x=L} = \frac{\alpha Eh\Delta\theta_0}{1-\nu} \\
&\Rightarrow \quad \frac{E}{1-\nu^2}(2A_1L + C_1 + \nu A_2) = \frac{\alpha E}{1-\nu}\Delta\theta_0 \\
&\frac{Eh}{1-\nu^2}\left(\frac{\partial Y}{\partial y} + \nu\frac{\partial X}{\partial x}\right)\bigg|_{y=\pm\ell} = \frac{\alpha Eh}{1-\nu}\frac{\Delta\theta_0}{L}x \\
&\Rightarrow \quad \frac{E}{1-\nu^2}(A_2 + \nu(2A_1x + C_1)) = \frac{\alpha E}{1-\nu}\frac{\Delta\theta_0}{L}x
\end{aligned}$$

then,

$$\begin{aligned}
&\frac{E}{1-\nu^2}(A_2 + \nu(2A_1x + C_1)) = \frac{\alpha E}{1-\nu}\frac{\Delta\theta_0}{L}x \\
&\Rightarrow \; A_2 + \nu C_1 = 0; \quad A_1 = \frac{\alpha\Delta\theta_0(1+\nu)}{2L} \\
&\frac{E}{1-\nu^2}(2A_1L + C_1 + \nu A_2) = \frac{\alpha E}{1-\nu}\Delta\theta_0 \\
&\Rightarrow \; C_1 + \nu A_2 = 0 \; \Rightarrow \; C_1 = A_2 = 0
\end{aligned}$$

leading finally to the longitudinal displacement field:

$$X(x) = \frac{\alpha\Delta\theta_0(1+\nu)}{2L}x^2; \quad Y = 0 \qquad [5.55]$$

Such a result indicates that in the present problem the contraction due to the Poisson effect is exactly balanced by the thermal expansion.

5.3.6 *In-plane, or membrane, natural modes of vibration*

5.3.6.1 Solutions of the modal equations by variable separation

The modal equations derived from the vibration equations [5.35] can be written as:

$$\begin{aligned}
&\left(\frac{\omega}{c}\right)^2 X + \frac{\partial^2 X}{\partial x^2} + \frac{1+\nu}{2}\frac{\partial^2 Y}{\partial x \partial y} + \frac{1-\nu}{2}\frac{\partial^2 X}{\partial y^2} = 0 \\
&\left(\frac{\omega}{c}\right)^2 Y + \frac{\partial^2 Y}{\partial y^2} + \frac{1+\nu}{2}\frac{\partial^2 X}{\partial x \partial y} + \frac{1-\nu}{2}\frac{\partial^2 Y}{\partial x^2} = 0 \\
&\text{where} \quad c = \sqrt{\frac{E}{\rho(1-\nu^2)}}
\end{aligned} \qquad [5.56]$$

Incidentally, it may be noted that the speed of membrane waves is intermediate between those of the longitudinal 1D and 3D waves, in agreement with the discussion of Chapter 4, subsection 4.2.1.1:

$$c_{1D} = \sqrt{\frac{E}{\rho}} < c_{2D} = \sqrt{\frac{E}{\rho(1-\nu^2)}} < c_{3D} = \sqrt{\frac{E(1-\nu)}{\rho(1+\nu)(1-2\nu)}}$$

For instance in steel ($E = 2.110^{11}$ Pa; $\rho = 7800\,\text{kgm}^{-3}$; $\nu = 0.3$), $c_{1D} \simeq 5200$; $c_{2D} = 5450$; $c_{3D} = 6034$ (m/s).

The solution of [5.56] is assumed to be of the type,

$$X(x, y) = A(x)B(y); \quad Y(x, y) = C(x)D(y) \qquad [5.57]$$

Substitution of [5.57] into [5.56] gives:

$$\begin{aligned}
&\left(\frac{\omega}{c}\right)^2 + \frac{A''}{A} + \left(\frac{1-\nu}{2}\right)\frac{B''}{B} + \left(\frac{1+\nu}{2}\right)\frac{C'D'}{AB} = 0 \\
&\left(\frac{\omega}{c}\right)^2 + \frac{D''}{D} + \left(\frac{1-\nu}{2}\right)\frac{C''}{C} + \left(\frac{1+\nu}{2}\right)\frac{A'B'}{CD} = 0
\end{aligned} \qquad [5.58]$$

Because x and y may vary independently, the first equation leads to:

$$\begin{aligned}
C' &= \alpha A; & D' &= B \\
A'' &= -k_x^2 A; & B'' &= -k_y^2 B
\end{aligned} \qquad [5.59]$$

where the signs of the constant are suitably chosen to obtain positive frequency values.

The general solutions are thus of the type:

$$\begin{aligned} A(x) &= a_1 \sin(k_x x) + a_2 \cos(k_x x) \\ B(y) &= b_1 \sin(k_y y) + b_2 \cos(k_y y) \\ C(x) &= \frac{\alpha}{k_x}(a_2 \sin(k_x x) - a_1 \cos(k_x x)) \\ D(y) &= \frac{1}{k_y}(b_2 \sin(k_y y) - b_1 \cos(k_y y)) \end{aligned} \quad [5.60]$$

Substituting the solutions [5.60] into [5.56], two equations of dispersion are obtained, which relate the wave pulsation to the longitudinal k_x and lateral k_y wave numbers.

5.3.6.2 *Natural modes of vibration for a plate on sliding supports*

As for three-dimensional dilatation waves, separated variable solutions of the problem hold in the particular case of sliding boundary conditions. Though such conditions are of little interest in practice, they are very convenient from the mathematical standpoint and suitable for discussing the general features of the membrane modes. The sliding boundary conditions impose:

$$X(0, y) = X(L_x, y) = Y(x, 0) = Y(0, L_y) = 0$$

which lead to the mode shapes:

$$\begin{cases} A_n(x) = \sin\left(\dfrac{n\pi x}{L_x}\right) \\ D_m(x) = \dfrac{L_y}{m\pi} \sin\left(\dfrac{m\pi y}{L_y}\right) \end{cases}$$

$$\Rightarrow \begin{cases} C_n(x) = -\dfrac{\alpha L_x}{n\pi} \cos\left(\dfrac{n\pi x}{L_x}\right); \quad n = 0, 1, 2, \ldots \\ B_m(y) = \cos\left(\dfrac{m\pi y}{L_y}\right); \quad m = 0, 1, 2, \ldots \end{cases} \quad [5.61]$$

It is easily checked that they verify also the sliding conditions of vanishing shear stresses at the edges,

$$\mathcal{N}_{xy}\big|_{(\mathcal{C})} = 0 \iff \left.\frac{\partial X}{\partial y} + \frac{\partial Y}{\partial x}\right|_{(\mathcal{C})} = 0 \quad [5.62]$$

The only parameter which remains unknown is α; it controls the relative amplitude of the displacements X and Y. To find the α values, the shapes [5.61] are substituted into equations [5.59] to give:

$$\left(\frac{\omega}{c}\right)^2 - \left(\frac{n\pi}{L_x}\right)^2 - \left(\frac{1-\nu}{2}\right)\left(\frac{m\pi}{L_y}\right)^2 + \left(\frac{1+\nu}{2}\right)\alpha = 0$$

$$\left(\frac{\omega}{c}\right)^2 - \left(\frac{m\pi}{L_y}\right)^2 - \left(\frac{1-\nu}{2}\right)\left(\frac{n\pi}{L_x}\right)^2 + \left(\frac{1+\nu}{2}\right)\left(\frac{n\pi}{L_x}\right)^2\left(\frac{m\pi}{L_y}\right)^2\frac{1}{\alpha} = 0 \quad [5.63]$$

Elimination of the pulsation between the two equations is immediate. It produces a quadratic equation which determines the modal values of α. The result is:

$$\alpha_{n,m}^{(1)} = \left(\frac{n\pi}{L_x}\right)^2; \quad \alpha_{n,m}^{(2)} = -\left(\frac{m\pi}{L_y}\right)^2 \quad [5.64]$$

There are infinitely many modes which can be enumerated by using two integer indices, namely the longitudinal index n and the lateral index m. The particular values $n = 0$, or $m = 0$, stand for rectilinear modes, with the following natural frequencies and mode shapes:

$$\begin{aligned} f_{0,m} &= \frac{mc}{2L_y}; \quad X_{0,m} = 0; \quad Y_{0,m} = \sin\left(\frac{m\pi y}{L_y}\right) \\ f_{n,0} &= \frac{nc}{2L_x}; \quad X_{n,0} = \sin\left(\frac{n\pi x}{L_x}\right); \quad Y_{n,0} = 0 \end{aligned} \quad [5.65]$$

Of course, such modes are similar to the longitudinal modes in straight beams. The sole difference lies in the value of wave speed c.

The two-dimensional modes $n \neq 0, m \neq 0$ can be classified into two distinct families according to the sign of the coupling coefficient $\alpha_{n,m}$. Modes related to $\alpha_{n,m}^{(1)} > 0$ are named in-phase modes. They have the following natural frequencies and mode shapes:

$$f_{n,m} = \frac{c}{2\pi}\sqrt{\left(\frac{n\pi}{L_x}\right)^2 + \left(\frac{m\pi}{L_y}\right)^2}; \quad n, m = 1, 2, \ldots$$

$$X_{n,m} = \sin\left(\frac{n\pi x}{L_x}\right)\cos\left(\frac{m\pi y}{L_y}\right); \quad Y_{n,m} = \frac{mL_x}{nL_y}\cos\left(\frac{n\pi x}{L_x}\right)\sin\left(\frac{m\pi y}{L_y}\right) \quad [5.66]$$

Modes related to $\alpha_{n,m}^{(2)} < 0$ are the out-of-phase modes. The natural frequencies and mode shapes are:

$$f_{n,m} = \frac{c}{2\pi}\sqrt{\left(\frac{1-\nu}{2}\right)\left(\left(\frac{n\pi}{L_x}\right)^2 + \left(\frac{m\pi}{L_y}\right)^2\right)}; \quad n,m = 1,2,\ldots$$

$$X_{n,m} = \sin\left(\frac{n\pi x}{L_x}\right)\cos\left(\frac{m\pi y}{L_y}\right); \quad Y_{n,m} = -\frac{nL_y}{mL_x}\cos\left(\frac{n\pi x}{L_x}\right)\sin\left(\frac{m\pi y}{L_y}\right) \tag{5.67}$$

It may be noted that the frequencies of the out-of-phase modes are less than those of the in-phase modes indicating that shear deformation is also smaller, as easily checked by direct calculation. In Figure 5.14 the shapes of the pair of modes $n = 1, m = 1$ are shown, as computed by using the finite element software CASTEM 2000 [CAS 92]), for a rectangular steel plate ($L_x = 4,4$ m, $L_y = 4$ m). The out-of-phase mode is depicted in the left-hand side plot (natural frequency $\simeq$ 533 Hz) and the in-phase mode is depicted in the right-hand side plot (natural frequency $\simeq$ 898 Hz). Within this frequency interval four other membrane modes exist. In Figure 5.15 the natural frequencies of the in-phase modes are plotted versus the longitudinal index n, the lateral index m being taken as a parameter varied from one to ten. The natural frequencies of the out-of-phase modes are plotted in the same way in Figure 5.16. It is immediately apparent from such plots that the *modal density*, defined as the number of modes lying within a given frequency interval, is much higher in the case of plates than in the case of beams.

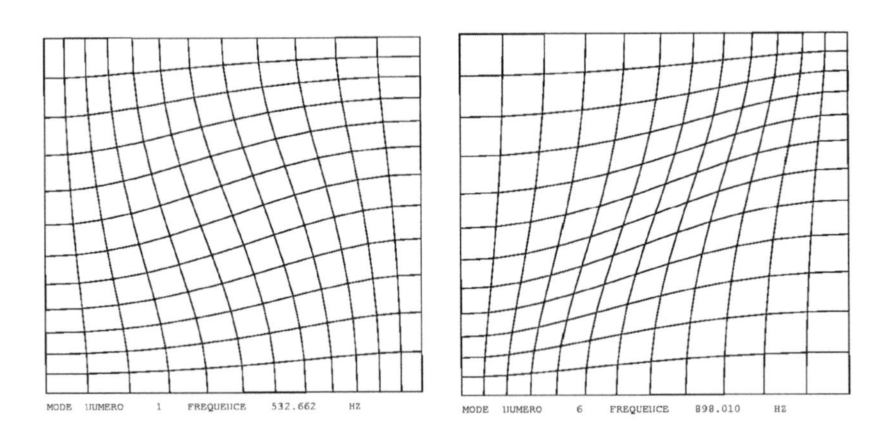

Figure 5.14. *Mode shapes for $m = n = 1$*

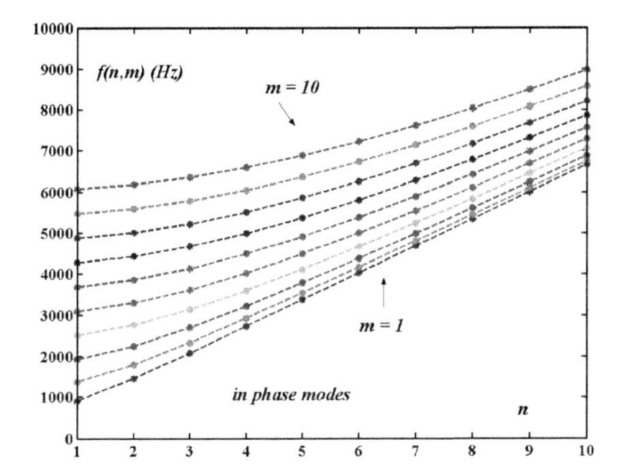

Figure 5.15. *Natural frequencies of the in-phase modes*

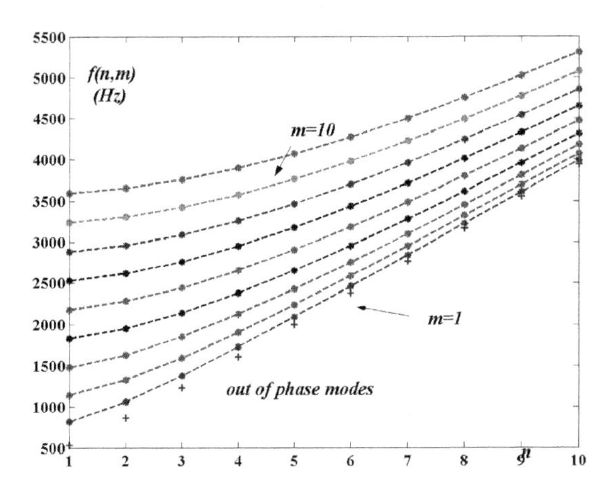

Figure 5.16. *Natural frequencies of the out-of-phase modes*

5.3.6.3 Semi-analytical approximations: Rayleigh–Ritz and Galerkin discretization methods

Beside the finite element and the modal expansion methods, there exist also two semi-analytical techniques of interest to deal with boundary problems known as the Rayleigh–Ritz and the Galerkin methods, which are closely related to each other and leading to the same semi-analytical procedure, as will be outlined here in the context of modal problems. In both methods, the basic idea is to search for an approximate solution by expanding first the displacement field in terms of suitably selected trial functions, in finite number N, as:

$$\vec{\Psi}(\vec{r}) = \sum_{1}^{N} a_n \vec{\psi}_n(\vec{r}) \qquad [5.68]$$

Starting from the linear manifold [5.68], where the unknown coefficients a_n constitute a generalized displacement vector $[a]$, a variational principle is then used to solve for them, producing thus an approximate solution to the original problem, restricted to the subspace of the manifold. In contrast to the modal expansion method described in Chapter 4, orthogonality of the trial functions is not necessarily required.

The variational principle invoked in the Rayleigh–Ritz method is the Rayleigh minimum principle which states that the Rayleigh quotient is a minimum with respect to any variations of the mode shapes:

$$(\omega_n')^2 = (\omega_n)^2 + \delta[\omega^2] = \frac{\langle \vec{\varphi}_n + \delta\vec{\varphi}, K[\vec{\varphi}_n + \delta\vec{\varphi}]\rangle}{\langle \vec{\varphi}_n + \delta\vec{\varphi}, M[\vec{\varphi}_n + \delta\vec{\varphi}]\rangle} > (\omega_n)^2$$

$$= \frac{\langle \vec{\varphi}_n, K[\vec{\varphi}_n]\rangle}{\langle \vec{\varphi}_n, M[\vec{\varphi}_n]\rangle}; \quad \forall \delta\vec{\varphi} \quad \text{(admissible)} \qquad [5.69]$$

First, $\delta\vec{\varphi}$ has to be admissible, that is, it must comply with the same boundary conditions as the exact mode shape $\vec{\varphi}_n$. As K and M are self-adjoint operators and $\delta\vec{\varphi}$ is admissible, we have:

$$(\omega_n')^2 = \frac{\langle \vec{\varphi}_n + \delta\vec{\varphi}, K[\vec{\varphi}_n + \delta\vec{\varphi}]\rangle}{\langle \vec{\varphi}_n + \delta\vec{\varphi}, M[\vec{\varphi}_n + \delta\vec{\varphi}]\rangle} = \frac{K_n + 2\langle \delta\vec{\varphi}, K[\vec{\varphi}_n]\rangle + \langle \delta\vec{\varphi}, K[\delta\vec{\varphi}]\rangle}{M_n + 2\langle \delta\vec{\varphi}, M[\vec{\varphi}_n]\rangle + \langle \delta\vec{\varphi}, M[\delta\vec{\varphi}]\rangle} \qquad [5.70]$$

On the other hand, projecting the exact modal equation $(K - \omega_n^2 M)\vec{\varphi}_n = 0$ on $\delta\vec{\varphi}$, we get:

$$\langle \delta\vec{\varphi}, (K - (\omega_n)^2 M)\vec{\varphi}_n\rangle = 0 \;\Rightarrow\; \langle \delta\vec{\varphi}, K[\vec{\varphi}_n]\rangle = (\omega_n)^2 \langle \delta\vec{\varphi}, M[\delta\vec{\varphi}]\rangle \qquad [5.71]$$

Substituting [5.71] into [5.70], we arrive at:

$$(\omega_n')^2 = \frac{(\omega_n)^2 (M_n + 2\langle \delta\vec{\varphi}, M[\vec{\varphi}_n]\rangle) + \langle \delta\vec{\varphi}, K[\delta\vec{\varphi}]\rangle}{M_n + 2\langle \delta\vec{\varphi}, M[\vec{\varphi}_n]\rangle + \langle \delta\vec{\varphi}, M[\delta\vec{\varphi}]\rangle} \qquad [5.72]$$

Further, $\delta\vec{\varphi}$ is assumed to be 'reasonably small', in such a way that $\langle \delta\vec{\varphi}, M[\delta\vec{\varphi}_n]\rangle$ and $\langle \delta\vec{\varphi}, K[\delta\vec{\varphi}_n]\rangle$ are small positive quantities of the order of $\|\delta\vec{\varphi}\|^2$. So, from

[5.72] it can be stated that the Rayleigh quotient is stationary, since up to the first order $(\omega'_n)^2$ is found to be equal to $(\omega_n)^2$; which explains why it can provide a satisfactory estimate of the exact eigenvalue $(\omega_n)^2$ even if some error is made concerning the mode shape. Using the Rayleigh principle as a stationary principle, the $(\omega'_n)^2$ can be calculated through a discretization procedure, which turns out to be analogous to that performed in the finite element method, except that it is carried out on the structure taken as a whole instead of an assembly of small elements and by using suitable trial functions sufficiently close to the exact solution instead of low order polynomials. The energy functionals calculated by using the manifold [5.68] produce two quadratic forms in terms of the unknown vector $[a]$:

$$\left\langle \vec{\Psi}, K[\vec{\Psi}] \right\rangle = [a]^T[K_d][a]; \quad \langle \vec{\Psi}, M[\vec{\Psi}] \rangle = [a]^T[M_d][a] \qquad [5.73]$$

$[K_d]$ is the stiffness matrix of the discretized system, which is symmetric and positive. $[M_d]$ is the mass matrix which is symmetric and positive definite. Such forms are stationary with respect to virtual variations $[\delta a]$, so the following discretized modal system is obtained:

$$\left[[K_d] - (\omega'_n)^2 [M_d] \right] [a] = 0 \qquad [5.74]$$

which is expected to give satisfactory estimates of the true $(\omega_n)^2$ and mode shapes, corresponding to the manifold spanned by the trial functions $\psi_n(\vec{r})$.

The final step would be to establish that $(\omega'_n)^2 > (\omega_n)^2$. However, the mathematical proof will be omitted in this book, as it requires rather lengthy developments which can be found for instance in [STA 70], Vol 2, Chapter 8. On the other hand, the Galerkin discretization method proceeds in a quite similar way, except that the trial functions are required to comply only with the so called *essential boundary conditions* i.e. those boundary conditions which concern the displacement variables solely. If such relaxed conditions are made, it can be shown that the Rayleigh quotient is still stationary but the minimum property is lost.

The Rayleigh–Ritz method is illustrated here by taking the example of a rectangular plate fixed at the four edges. An admissible displacement field which fulfils the fixed conditions is given by the following manifold:

$$\psi^{(X)}_{n,m} = \alpha_{n,m} \sin\left(\frac{n\pi x}{L_x}\right) \sin\left(\frac{m\pi y}{L_y}\right)$$

$$\psi^{(Y)}_{n,m} = \beta_{n,m} \sin\left(\frac{m\pi x}{L_x}\right) \sin\left(\frac{n\pi y}{L_y}\right); \quad n, m = 1, 2, \ldots$$

The algebra is further alleviated by assuming $\nu = 0$. The elastic energy functional may be written as:

$$\left\langle \vec{\Psi}_{n,m}, K[\vec{\Psi}_{n,m}] \right\rangle$$

$$= -Eh \int_0^{L_y} \int_0^{L_x} \left[\psi_{n,m}^{(X)}, \psi_{n,m}^{(Y)} \right] \begin{bmatrix} \frac{\partial^2}{\partial x^2} + \frac{1}{2}\frac{\partial^2}{\partial y^2} & \frac{1}{2}\frac{\partial^2}{\partial x \partial y} \\ \frac{1}{2}\frac{\partial^2}{\partial x \partial y} & \frac{\partial^2}{\partial y^2} + \frac{1}{2}\frac{\partial^2}{\partial x^2} \end{bmatrix} \begin{bmatrix} \psi_{n,m}^{(X)} \\ \psi_{n,m}^{(Y)} \end{bmatrix} dx\, dy$$

which results in:

$$\langle \vec{\Psi}_{n,m}, K[\vec{\Psi}_{n,m}] \rangle$$

$$= \begin{cases} \text{if } n \text{ and } m \text{ are both even or odd} \\ \frac{Eh\pi^2}{4} \left(\left(\frac{n^2}{\eta} + \frac{m^2 \eta}{2} \right) \alpha_{n,m}^2 + \left(n^2 \eta + \frac{m^2}{2\eta} \right) \beta_{n,m}^2 \right) \\ \text{otherwise} \\ \frac{Eh\pi^2}{4} \left(\left(\frac{n^2}{\eta} + \frac{m^2 \eta}{2} \right) \alpha_{n,m}^2 + \left(n^2 \eta + \frac{m^2}{2\eta} \right) \beta_{n,m}^2 + \frac{8n^2 m^2}{\pi^2 (n^2 - m^2)^2} \alpha_{n,m} \beta_{n,m} \right) \end{cases}$$

where $\eta = L_x / L_y$ and the following identity is used:

$$\int_0^{\pi} \sin(nu)\cos(mu)\, du = \begin{cases} -\left[\frac{\cos(n-m)u}{2(n-m)} + \frac{\cos(n+m)u}{2(n+m)} \right]_0^{\pi}; & \text{if } n \neq m \\ 0; & \text{if } n = m \end{cases}$$

which gives the final result:

$$\int_0^{\pi} \sin(nu)\cos(mu)\, du = \begin{cases} \frac{2n}{n^2 - m^2}; & \text{if } n \text{ and } m \text{ are both even or odd} \\ 0; & \text{otherwise} \end{cases}$$

The functional of kinetic energy is:

$$\left\langle \psi_{n,m}^{(X)}, \rho h \psi_{n,m}^{(X)} \right\rangle + \left\langle \psi_{n,m}^{(Y)}, \rho h \psi_{n,m}^{(Y)} \right\rangle = \frac{M}{4} \left(\alpha_{n,n}^2 + \beta_{n,n}^2 \right)$$

where $M = \rho h L_x L_y$

If n and m are both even or odd integers, the modal equation is:

$$\frac{4Eh}{M}\begin{bmatrix} \frac{\pi^2}{4}\left(n^2+\frac{m^2\eta^2}{2}\right)-\left(\omega'_{n,m}\right)^2 & 0 \\ 0 & \frac{\pi^2}{4}\left(n^2\eta^2+\frac{m^2}{2}\right)-\left(\omega'_{n,m}\right)^2 \end{bmatrix} \times \begin{bmatrix} \alpha_{n,m} \\ \beta_{n,m} \end{bmatrix} = \begin{bmatrix} 0 \\ 0 \end{bmatrix}$$

Solution is immediate, giving two orthogonal families of one-directional modes, vibrating in the x, and the y direction respectively, with exactly the same properties, as expected due to the symmetry of the problem. The natural frequencies may be expressed as:

$$f^{(X)}_{n,m} = \frac{c}{2L_x}\sqrt{n^2+0.5(m\eta)^2}; \quad f^{(Y)}_{n,m} = \frac{c}{2L_x}\sqrt{(n\eta)^2+0.5m^2}$$

The mode shapes

$$\vec{\Psi}_{n,m} = \begin{cases} \sin\left(\frac{n\pi x}{L_x}\right)\sin\left(\frac{m\pi y}{L_y}\right)\vec{i} \\ \sin\left(\frac{m\pi x}{L_x}\right)\sin\left(\frac{n\pi y}{L_y}\right)\vec{j} \end{cases}$$

are not exact, as they do not verify the exact modal equations, because of the mixed derivative in the coupling term.

The other modes are approximated as the solutions of the following coupled modal system:

$$\frac{4Eh}{M}\begin{bmatrix} \frac{\pi^2}{4}\left(n^2+\frac{m^2\eta^2}{2}\right)-\left(\omega'_{n,m}\right)^2 & \frac{2n^2m^2\eta^2}{(n^2-m^2)^2} \\ \frac{2n^2m^2\eta^2}{(n^2-m^2)^2} & \frac{\pi^2}{4}\left(n^2\eta^2+\frac{m^2\eta^2}{2}\right)-\left(\omega'_{n,m}\right)^2 \end{bmatrix} \times \begin{bmatrix} \alpha_{n,m} \\ \beta_{n,m} \end{bmatrix} = \begin{bmatrix} 0 \\ 0 \end{bmatrix}$$

In the particular case of a square plate, they are immediately obtained as:
in-phase modes:

$$f_{n,m} = \frac{c}{2L}\sqrt{n^2 + \frac{m^2}{2} + \frac{8n^2m^2}{\pi^2(n^2 - m^2)^2}}; \quad \alpha_{n,m} = \beta_{n,m}$$

out-of-phase modes:

$$f_{n,m} = \frac{c}{2L}\sqrt{n^2 + \frac{m^2}{2} - \frac{8n^2m^2}{\pi^2(n^2 - m^2)^2}}; \quad \alpha_{n,m} = -\beta_{n,m}$$

Figure 5.17 shows the mode shapes (1, 1), (1, 2) and (2, 2) of a steel plate (though assuming $\nu = 0$) $L_x = 4.04\,\text{m}, L_y = 4\,\text{m}$, as computed by the finite

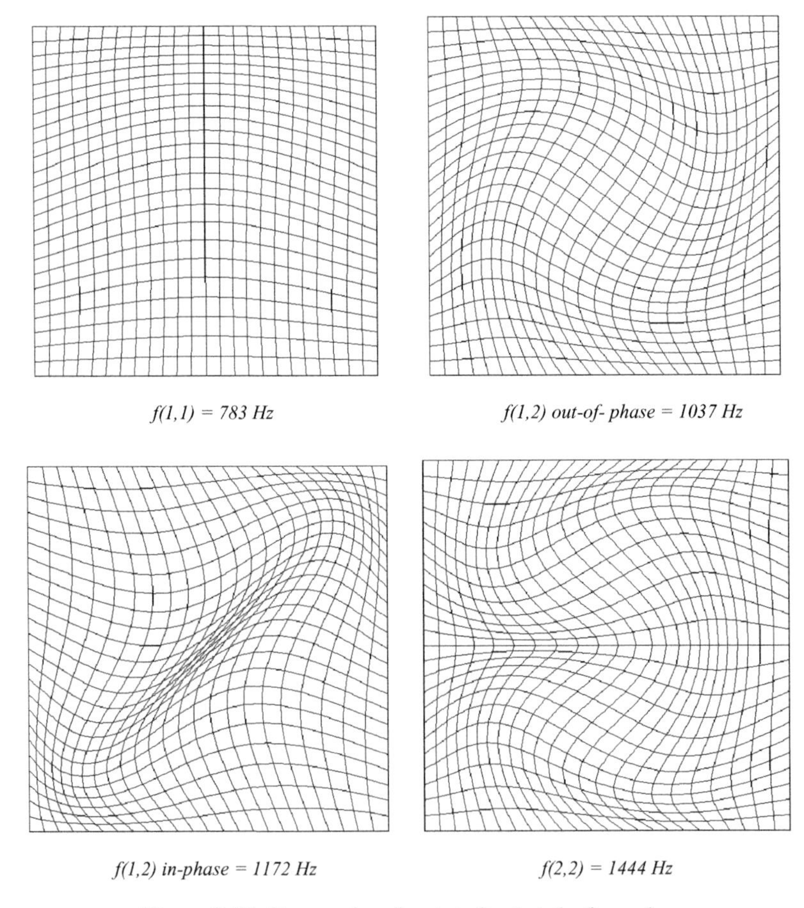

Figure 5.17. *First modes of a plate fixed at the four edges*

element method. The computed natural frequencies are found to be less than the Rayleigh–Ritz values: $f(1,1) = 789\,\text{Hz}$, $f(1,2)$ *out-of phase* $= 1048\,\text{Hz}$, $f(1,2)$ *in-phase* $= 1180\,\text{Hz}$, $f(2,2) = 1578\,\text{Hz}$, as expected from the Rayleigh minimum principle. The unsatisfactory result for the mode (2,2) concerning both the value of the natural frequency and the mode shape, which is clearly marked by a coupling between the x and the y directions, could be corrected by selecting more complicated trial functions such as:

$$\psi_{2,2}^{(X)} = \alpha_{2,2} \sin\left(\frac{2\pi x}{L_x}\right) \sin\left(\frac{2\pi y}{L_y}\right) + \alpha_{1,3} \sin\left(\frac{\pi x}{L_x}\right) \sin\left(\frac{3\pi y}{L_y}\right)$$

$$\psi_{2,2}^{(Y)} = \beta_{2,2} \sin\left(\frac{2\pi x}{L_x}\right) \sin\left(\frac{2\pi y}{L_y}\right) + \beta_{1,3} \sin\left(\frac{3\pi x}{L_x}\right) \sin\left(\frac{\pi y}{L_y}\right)$$

which would produce non-vanishing coupling terms between $\alpha_{2,2}$ and $\beta_{1,3}$ as well as between $\beta_{2,2}$ and $\alpha_{1,3}$.

5.3.6.4 Plate loaded by a concentrated in-plane force: spatial attenuation of the local response

A rectangular plate (L_x, L_y, h) is loaded by an in-plane force concentrated at $P(x_0, y_0)$, see Figure 5.18. The four edges are on sliding supports. We are interested in studying the field of the normal longitudinal stress $\mathcal{N}_{xx}(x, y)$. Colour plate 8 illustrates the results of a finite element computation, referring to a longitudinal load applied to the plate centre. It is worth noticing that magnitude of displacement and stress fields are sharply peaked in the close vicinity of the loaded point, in agreement with Saint-Venant's principle. It is also noted that the support reactions are essentially normal to the edges and not distributed uniformly. As expected, longitudinal reactions are negative along the edge $x = L$, and positive along the edge $x = 0$. Their magnitude is maximum at $y = L/2$ in agreement with the longitudinal stress field $\mathcal{N}_{xx}(x, y)$. Lateral reactions are found to be antisymmetric about the middle lateral axis, in agreement with the lateral stress field $\mathcal{N}_{yy}(x, y)$. Such behaviour is clearly related to the Poisson effect.

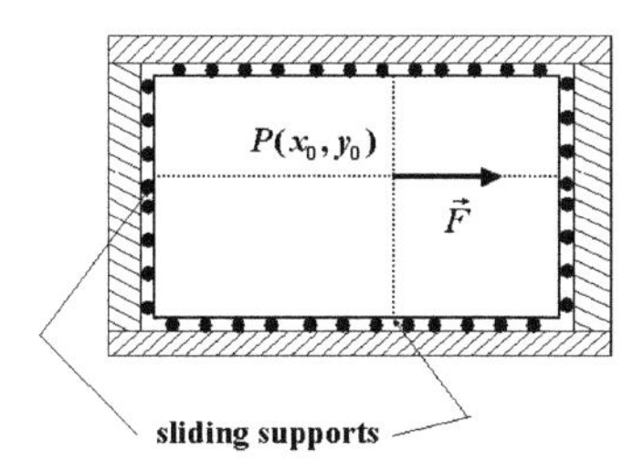

Figure 5.18. *Rectangular plate on sliding support loaded by a longitudinal force at point P*

Here the selected boundary conditions allow us to expand the solution as a modal series by using [5.67]. Furthermore, by making the distinction between the contribution of the rectilinear modes and that of the membrane modes $(n, m \neq 0)$ it becomes possible to separate the global and the local responses. The displacement fields are expanded as:

$$X(x,y) = \sum_{m=0}^{\infty}\sum_{n=1}^{\infty} q_{n,m} \sin\left(\frac{n\pi x}{L_x}\right)\cos\left(\frac{m\pi y}{L_y}\right)$$
$$Y(x,y) = \sum_{m=1}^{\infty}\sum_{n=0}^{\infty} p_{n,m} \cos\left(\frac{n\pi x}{L_x}\right)\sin\left(\frac{m\pi y}{L_y}\right) \quad [5.75]$$

The problem is further analysed by assuming a longitudinal axial load F_x applied at a point $P(x_0, L_y/2)$ located on the middle longitudinal axis.

1. *Rectilinear mode contribution and global response*
   A simple calculation gives:

$$q_{n,0} = \frac{2(1-\nu^2)F_x}{EhL_xL_y}\left(\frac{L_x}{n\pi}\right)^2 \sin\left(\frac{n\pi x_0}{L_x}\right); \quad p_{n,0} = 0 \quad [5.76]$$

It follows the displacements in terms of rectilinear modes,

$$X_{(m=0)}(x) = F_x \frac{2(1-\nu^2)}{EhL_xL_y}\sum_{n=1}^{\infty}\left(\frac{L_x}{n\pi}\right)^2 \sin\left(\frac{n\pi x_0}{L_x}\right)\sin\left(\frac{n\pi x}{L_x}\right);$$
$$Y_{(m=0)} = 0 \quad [5.77]$$

This field is independent of the lateral distribution of the load and characterises the global response of the plate. The longitudinal normal stress field is given by:

$$\mathcal{N}_{xx}^{(m=0)}(x) = \frac{Eh}{1-\nu^2}\frac{\partial X_{(m=0)}}{\partial x}$$
$$= \left(\frac{F_x}{L_y}\right)\left(\frac{2}{\pi}\right)\sum_{n=1}^{\infty}\frac{1}{n}\sin\left(\frac{n\pi x_0}{L_x}\right)\cos\left(\frac{n\pi x}{L_x}\right) \quad [5.78]$$

It could be shown that, provided x_0 differs from zero, or L_x, the modal expansion [5.78] is the Fourier series on the interval $0 \le x \le L_x$ of the following step function:

$$+\frac{F_x}{L_y}\left(\frac{L_x - x_0}{L_x}\right); \quad \text{if } 0 \le x \le x_0$$
$$-\frac{F_x}{L_y}\left(\frac{x_0}{L_x}\right); \quad \text{if } x_0 \le x \le L_x \quad [5.79]$$

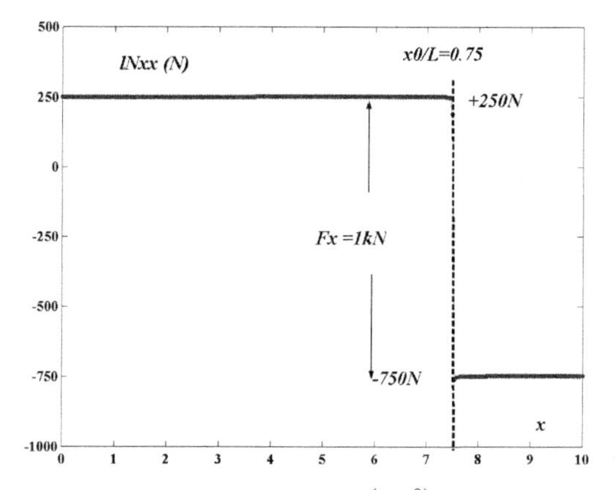

Figure 5.19. *Distribution of the normal stress $\mathcal{N}_{xx}^{(m=0)}$ along a longitudinal line of the plate*

This is illustrated in Figure 5.19 which refers to a plate $L = 10\,\text{m}$; $\ell = 2\,\text{m}$ loaded by a longitudinal force of 1 kN applied at midwidth $y_0 = \ell/2$ and three quarter length $x_0 = 0.75\,\text{L}$. The series [5.78] is truncated to $N = 400$ in order to minimize the Gibbs oscillations near the discontinuity. As expected, the step magnitude is F_x/L_y, the plate is compressed in the domain $[x_0, L_x]$ and stretched in the domain $[0, x_0]$, the external load is exactly balanced by the longitudinal component of the support reactions: $L_y(\mathcal{N}_{xx}^{(m=0)}(L_x) - \mathcal{N}_{xx}^{(m=0)}(0)) = F_x$. On the other hand, the series [5.78] vanishes if the loading is applied at the sliding edges. The shear stresses $\mathcal{N}_{xy}^{(m=0)}$ are null and the lateral normal stress field is given by the relation $\mathcal{N}_{yy}^{(m=0)} = \nu\mathcal{N}_{xx}^{(m=0)}$.

2. *Contribution of the membrane modes $(n, m \neq 0)$ and local response*
   The modal displacements are given by:

$$q_{n,m} = \frac{8(1+\nu)F_x}{EhL_xL_y}\frac{\left\{\left(\frac{n\pi}{L_x}\right)^2 + \left(\frac{1-\nu}{2}\right)\left(\frac{m\pi}{L_y}\right)^2\right\}\sin\left(\frac{n\pi x_0}{L_x}\right)\cos\left(\frac{m\pi y_0}{L_y}\right)}{\left\{\left(\frac{n\pi}{L_x}\right)^2 + \left(\frac{m\pi}{L_y}\right)^2\right\}^2}$$

[5.80]

$$p_{n,m} = -\frac{8(1+\nu)F_x}{EhL_xL_y}\frac{\left(\frac{1+\nu}{2}\right)\left(\frac{n\pi}{L_x}\right)\left(\frac{m\pi}{L_y}\right)\sin\left(\frac{n\pi x_0}{L_x}\right)\cos\left(\frac{m\pi y_0}{L_y}\right)}{\left\{\left(\frac{n\pi}{L_x}\right)^2 + \left(\frac{m\pi}{L_y}\right)^2\right\}^2}$$

[5.81]

The membrane stresses related to the membrane modes are:

$$\mathcal{N}_{xx}^{(n,m)} = \frac{Eh}{1-\nu^2}\sum_{n=1}^{\infty}\sum_{m=1}^{\infty}\left\{q_{n,m}\frac{n\pi}{L_x} + \nu p_{n,m}\frac{m\pi}{L_y}\right\}\cos\left(\frac{n\pi x}{L_x}\right)\cos\left(\frac{m\pi y}{L_y}\right)$$

$$\mathcal{N}_{yy}^{(n,m)} = \frac{Eh}{1-\nu^2}\sum_{n=1}^{\infty}\sum_{m=1}^{\infty}\left\{\nu q_{n,m}\frac{n\pi}{L_x} + p_{n,m}\frac{m\pi}{L_y}\right\}\cos\left(\frac{n\pi x}{L_x}\right)\cos\left(\frac{m\pi y}{L_y}\right)$$

$$\mathcal{N}_{xy}^{(n,m)} = \frac{Eh}{1-\nu^2}\sum_{n=1}^{\infty}\sum_{m=1}^{\infty}\left\{q_{n,m}\frac{m\pi}{L_y} + p_{n,m}\frac{n\pi}{L_x}\right\}\sin\left(\frac{n\pi x}{L_x}\right)\sin\left(\frac{m\pi y}{L_y}\right)$$

[5.82]

These describe the local response of the plate in the vicinity of the loaded point P. In numerical evaluation of the series some difficulties arise related to the singularity of the stress distribution at P. Along the lateral direction the $\mathcal{N}_{xx}^{(n,m)}$ profile comprises a Dirac component $\delta(y - y_0)$ and in the longitudinal direction it comprises a Dirac dipole $\delta'(x - x_0)$. It is thus appropriate to calculate [5.82] over a rectangular grid of elementary size $\Delta x, \Delta y$. Then the series are truncated in such a way that the smallest wavelengths of the modes are roughly a few tenths of $\Delta x, \Delta y$. This is illustrated in Figure 5.20 which refers to a square plate $L = \ell = 2$ m loaded by a longitudinal force of 1 kN applied at $y_0 = 1$ m and $x_0 = 1.5$ m. The elementary lengths are $\Delta x = \Delta y = 7$ cm and the series are truncated up to the smallest wavelengths $\lambda_x = \lambda_y = 5$ cm. As expected, the larger the distance from P, the smaller is the response and the less is the number of modes which are necessary to compute the series [5.82].

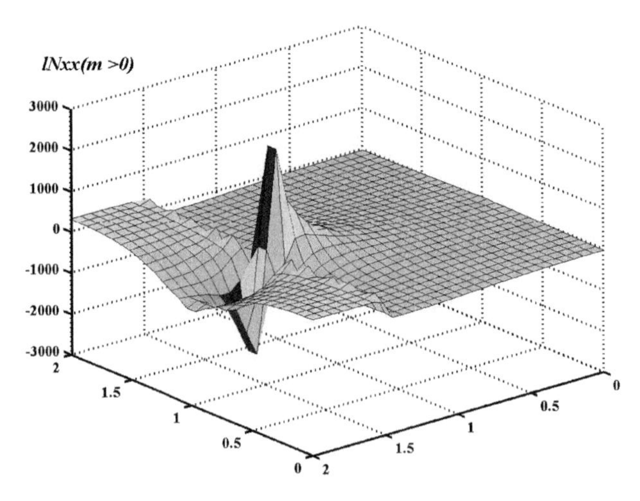

Figure 5.20. *Distribution of the local component of $\mathcal{N}_{xx}$ over the plate*

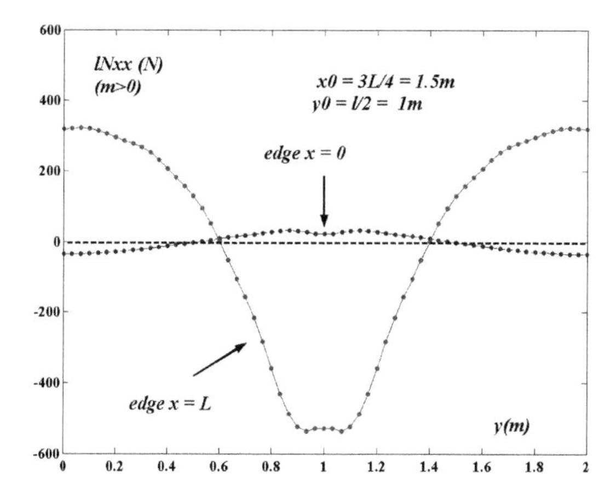

Figure 5.21. *Distribution of the local component of $\mathcal{N}_{xx}$ over the plate*

In the same way, the larger the distance of the loaded point from one of the lateral edges, the more uniform is the stress distribution along this edge. This is illustrated in Figures 5.21 where the lateral distribution of $L\mathcal{N}_{xx}^{(n,m)}$ is plotted along the lateral edges $x = 0$, $x = L$. On such plots, the singular values at y_0 are replaced by the nearest value, which is convenient to focus on the regular part of the local stress field. The local normal stress is very important at the edge nearest to the loading and remains significant at the other edge, though of much less magnitude. One observes a stress peak centred about y_0, which is rather broad and of the same sign as the 'global' field $\mathcal{N}_{xx}^{(m=0)}$, at about $y = L/4$, the sign is reversed. Such a sign reversal is a very necessary feature of the local field $\mathcal{N}_{xx}^{(n,m)}$, since it must vanish when integrated over the lateral edge. Similar calculations carried out on rectangular plates show that the local response vanishes with a characteristic length of the order of a fraction of the width of the plate. Such a result indicates that Saint-Venant's principle can be applied to plate in-plane problems.

As a final remark it is worth mentioning that the use of the finite element method to study the local effects induced by a concentrated load gives rise to the same kind of difficulties as the semi-analytical method described here. In both cases, the singular component of the stress field is smoothed out by the discretization procedure and no very reliable values of the stress can be obtained at the loaded point and even along the y_0 line, see Colour Plate 8.

5.4. Curvilinear coordinates

If the plates are limited by curved edges, a mathematical difficulty arises as the boundary conditions of the problem cannot be expressed in a tractable way by using Cartesian coordinates. Fortunately, the use of curvilinear coordinates is found appropriate, to deal with rather simple geometries at least, for instance circular and elliptical plates, and more generally when an orthogonal curvilinear coordinate system can be fitted to the edge geometry.

5.4.1 *Linear strain tensor*

Let us define the position of a point lying on the midplane of a plate by curvilinear coordinates denoted α and β as shown in Figure 5.22. The curves $(\mathcal{C}_\alpha)$ defined by α = constant are orthogonal to the curves $(\mathcal{C}_\beta)$ defined by β = constant. The unit vectors tangent to these curves are denoted $\vec{t}_\alpha, \vec{t}_\beta$. Transformation to Cartesian coordinates is defined as:

$$x = x(\alpha, \beta); \quad y = y(\alpha, \beta) \qquad [5.83]$$

The length of any segment drawn in the midplane is independent of the coordinate system, then for any infinitesimal segment of length ds, the following relationship holds:

$$ds^2 = dx^2 + dy^2 = g_\alpha^2 d\alpha^2 + g_\beta^2 d\beta^2 \qquad [5.84]$$

where:

$$g_\alpha^2 = \left(\frac{\partial x}{\partial \alpha}\right)^2 + \left(\frac{\partial y}{\partial \alpha}\right)^2; \quad g_\beta^2 = \left(\frac{\partial x}{\partial \beta}\right)^2 + \left(\frac{\partial y}{\partial \beta}\right)^2$$

g_α, g_β are termed the Lamé parameters of the plane surface.

On the other hand, according to [5.84], in the α, β orthonormal system, the area of the elementary surface must be defined as:

$$g_\alpha g_\beta d\alpha d\beta \qquad [5.85]$$

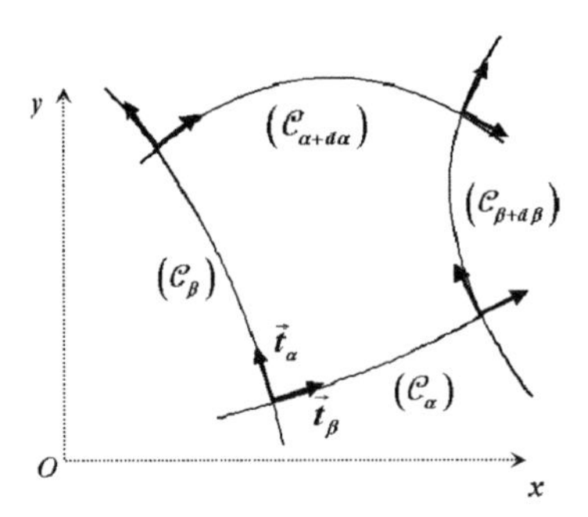

Figure 5.22. *Curvilinear orthogonal coordinates*

The linear membrane strain tensor is

$$[\eta] = \frac{1}{2}\left(\left[\nabla(X_\alpha \vec{t}_\alpha + X_\beta \vec{t}_\beta)\right] + \left[\nabla\left(X_\alpha \vec{t}_\alpha + X_\beta \vec{t}_\beta\right)\right]^T\right) \qquad [5.86]$$

As detailed in Appendix 3 (formula [A.3.13]), it can be written as:

$$[\eta] = \begin{bmatrix} \frac{1}{g_\alpha}\left(\frac{\partial X_\alpha}{\partial \alpha} + \frac{X_\beta}{g_\beta}\frac{\partial g_\alpha}{\partial \beta}\right) & \frac{1}{2}\left(\frac{g_\alpha}{g_\beta}\frac{\partial}{\partial \beta}\left(\frac{X_\alpha}{g_\alpha}\right) + \frac{g_\beta}{g_\alpha}\frac{\partial}{\partial \alpha}\left(\frac{X_\beta}{g_\beta}\right)\right) \\ \frac{1}{2}\left(\frac{g_\alpha}{g_\beta}\frac{\partial}{\partial \beta}\left(\frac{X_\alpha}{g_\alpha}\right) + \frac{g_\beta}{g_\alpha}\frac{\partial}{\partial \alpha}\left(\frac{X_\beta}{g_\beta}\right)\right) & \frac{1}{g_\beta}\left(\frac{\partial X_\beta}{\partial \beta} + \frac{X_\alpha}{g_\alpha}\frac{\partial g_\beta}{\partial \alpha}\right) \end{bmatrix} \qquad [5.87]$$

The components of $[\eta]$ can also be used to define a strain vector $[\vec{\eta}]^T = [\eta_{\alpha\alpha} \quad 2\eta_{\alpha\beta} \quad \eta_{\beta\beta}]$.

5.4.2 *Equilibrium equations and boundary conditions*

The kinetic energy has the form:

$$\mathcal{E}_k = \frac{1}{2}\int_{\alpha 1}^{\alpha 2}\int_{\beta 1}^{\beta 2} \rho h \left\{\dot{X}_\alpha^2 + \dot{X}_\beta^2\right\} g_\alpha g_\beta \, d\alpha \, d\beta \qquad [5.88]$$

The strain energy is also invariant in any coordinate transformation, so its variation is written as:

$$\begin{aligned} \delta[\mathcal{E}_s] &= \int_{\alpha 1}^{\alpha 2}\int_{\beta 1}^{\beta 2} [\vec{\mathcal{N}}]^T \, \delta[[\vec{\eta}]] g_\alpha g_\beta \, d\alpha \, d\beta \\ &= \int_{\alpha 1}^{\alpha 2}\int_{\beta 1}^{\beta 2} \{\mathcal{N}_{\alpha\alpha}\delta[\eta_{\alpha\alpha}] + \mathcal{N}_{\alpha\beta}\delta[\eta_{\alpha\beta}] + \mathcal{N}_{\beta\beta}\delta[\eta_{\beta\beta}]\} g_\alpha g_\beta \, d\alpha \, d\beta \end{aligned} \qquad [5.89]$$

where the factor $g_\alpha g_\beta$ comes from the definition [5.85] of the elementary area in curvilinear coordinates.

Using [5.87] and integrating once by parts to express all the variations in terms of δX_α and δX_β, we arrive at the following expressions, written in a suitable form to apply Hamilton's principle, as detailed below:

$$\begin{aligned} &\int_{\alpha 1}^{\alpha 2}\int_{\beta 1}^{\beta 2} \{\mathcal{N}_{\alpha\alpha}\delta[\eta_{\alpha\alpha}]\} g_\alpha g_\beta \, d\alpha \, d\beta \\ &= \int_{\alpha 1}^{\alpha 2}\int_{\beta 1}^{\beta 2} \left\{\frac{\mathcal{N}_{\alpha\alpha}}{g_\alpha}\left(\frac{\partial(\delta X_\alpha)}{\partial \alpha} + \frac{\delta X_\beta}{g_\beta}\frac{\partial g_\alpha}{\partial \beta}\right)\right\} g_\alpha g_\beta \, d\alpha \, d\beta \end{aligned}$$

$$= \int_{\beta 1}^{\beta 2} [g_\beta \mathcal{N}_{\alpha\alpha} \delta X_\alpha]_{\alpha 1}^{\alpha 2} d\beta$$

$$+ \int_{\alpha 1}^{\alpha 2} \int_{\beta 1}^{\beta 2} \left\{ \left(-\frac{\partial (g_\beta \mathcal{N}_{\alpha\alpha})}{\partial \alpha} \delta X_\alpha + \frac{\partial g_\alpha}{\partial \beta} \delta X_\beta \right) \right\} d\alpha \, d\beta$$

and,

$$\int_{\alpha 1}^{\alpha 2} \int_{\beta 1}^{\beta 2} \{\mathcal{N}_{\beta\beta} \delta[\eta_{\beta\beta}]\} g_\alpha g_\beta \, d\alpha \, d\beta$$

$$= \int_{\alpha 1}^{\alpha 2} \int_{\beta 1}^{\beta 2} \left\{ \frac{\mathcal{N}_{\beta\beta}}{g_\beta} \left(\frac{\partial (\delta X_\beta)}{\partial \beta} + \frac{\delta X_\alpha}{g_\alpha} \frac{\partial g_\beta}{\partial \alpha} \right) \right\} g_\alpha g_\beta \, d\alpha \, d\beta$$

$$= \int_{\alpha 1}^{\alpha 2} [g_\alpha \mathcal{N}_{\beta\beta} \delta X_\beta]_{\beta 1}^{\beta 2} \, d\alpha$$

$$+ \int_{\alpha 1}^{\alpha 2} \int_{\beta 1}^{\beta 2} \left\{ \left(-\frac{\partial (g_\alpha \mathcal{N}_{\beta\beta})}{\partial \beta} \delta X_\beta + \frac{\partial g_\beta}{\partial \alpha} \delta X_\alpha \right) \right\} d\alpha \, d\beta$$

$$2 \int_{\alpha 1}^{\alpha 2} \int_{\beta 1}^{\beta 2} \{\mathcal{N}_{\alpha\beta} \delta[\eta_{\alpha\beta}]\} g_\alpha g_\beta \, d\alpha \, d\beta$$

$$= \int_{\alpha 1}^{\alpha 2} \int_{\beta 1}^{\beta 2} \left\{ \mathcal{N}_{\alpha\beta} \left(\frac{g_\alpha}{g_\beta} \frac{\partial}{\partial \beta} \left(\frac{\partial X_\alpha}{g_\alpha} \right) + \frac{g_\beta}{g_\alpha} \frac{\partial}{\partial \alpha} \left(\frac{\delta X_\beta}{g_\beta} \right) \right) \right\} g_\alpha g_\beta \, d\alpha \, d\beta$$

$$= \int_{\alpha 1}^{\alpha 2} [g_\beta \mathcal{N}_{\alpha\beta} \delta X_\beta]_{\beta 1}^{\beta 2} \, d\alpha + \int_{\beta 1}^{\beta 2} [g_\alpha \mathcal{N}_{\alpha\beta} \delta X_\alpha]_{\alpha 1}^{\alpha 2} \, d\beta$$

$$- \int_{\alpha 1}^{\alpha 2} \int_{\beta 1}^{\beta 2} \left\{ \left(\frac{1}{g_\alpha} \frac{\partial (g_\alpha^2) \mathcal{N}_{\alpha\beta})}{\partial \beta} \delta X_\alpha + \frac{1}{g_\beta} \frac{\partial (g_\beta^2) \mathcal{N}_{\alpha\beta})}{\partial \alpha} \delta X_\beta \right) \right\} d\alpha \, d\beta$$

So, the two following equations of motion are found:

$$\rho h \ddot{X}_\alpha - \frac{1}{g_\alpha g_\beta} \left\{ \frac{\partial (g_\beta \mathcal{N}_{\alpha\alpha})}{\partial \alpha} - \mathcal{N}_{\beta\beta} \frac{\partial g_\beta}{\partial \alpha} + \frac{1}{g_\alpha} \frac{\partial (g_\alpha^2 \mathcal{N}_{\beta\alpha})}{\partial \beta} \right\} = f_\alpha^{(e)}$$

$$\rho h \ddot{X}_\beta - \frac{1}{g_\alpha g_\beta} \left\{ \frac{\partial (g_\alpha \mathcal{N}_{\beta\beta})}{\partial \beta} - \mathcal{N}_{\alpha\alpha} \frac{\partial g_\alpha}{\partial \beta} + \frac{1}{g_\beta} \frac{\partial (g_\beta^2 \mathcal{N}_{\alpha\beta})}{\partial \alpha} \right\} = f_\beta^{(e)} \quad [5.90]$$

where $f_\alpha^{(e)}$ and $f_\beta^{(e)}$ are the external forces per unit length applied along the $\vec{t}_\alpha$ and $\vec{t}_\beta$ directions, respectively.

The associated elastic boundary conditions are,

$$\begin{aligned}
&\mathcal{N}_{\alpha\alpha} - K_{\alpha\alpha} X_\alpha = 0; \quad \mathcal{N}_{\beta\beta} - K_{\beta\beta} X_\beta = 0 \\
&\mathcal{N}_{\alpha\beta} - K_{\alpha\beta} X_\beta = 0; \quad \mathcal{N}_{\beta\alpha} - K_{\beta\alpha} X_\alpha = 0
\end{aligned} \qquad [5.91]$$

where $K_{\alpha\alpha}, K_{\beta\beta}, K_{\alpha\beta}$ are the stiffness coefficients of the supports, acting in the normal and tangential directions to the boundary lines $(\mathcal{C}_\alpha)$ and $(\mathcal{C}_\beta)$.

5.4.3 *Elastic law in curvilinear coordinates*

The invariance of the strain energy with respect to any coordinate transformation implies that the strain–stress relationship [5.33] is not changed and is written as:

$$\begin{bmatrix} \mathcal{N}_{\alpha\alpha} \\ \mathcal{N}_{\beta\beta} \\ \mathcal{N}_{\alpha\beta} \end{bmatrix} = \frac{Eh}{1-\nu^2} \begin{bmatrix} 1 & \nu & 0 \\ \nu & 1 & 0 \\ 0 & 0 & \dfrac{1-\nu}{2} \end{bmatrix} \begin{bmatrix} \eta_{\alpha\alpha} \\ \eta_{\beta\beta} \\ 2\eta_{\alpha\beta} \end{bmatrix} \qquad [5.92]$$

5.4.4 *Circular cylinder loaded by a radial pressure*

As an interesting application of plate in-plane equations in curvilinear coordinates, let us consider the problem sketched in Figure 5.23, which deals with a circular cylinder loaded by an external pressure P_e and an internal pressure P_i. Both P_e and P_i are assumed to be uniform and $P_e \neq P_i$. A priori, the reader could be surprised to find here a shell instead of a plate problem. The shell is conveniently described by using the cylindrical coordinate system r, θ, z. Nevertheless, as here the pressure is assumed to be independent of z, the dimension of the problem can be reduced to two dimensions described by the polar coordinates r and θ. By doing so, the cylinder is reduced to an annular plate of unit thickness. The Lamé parameters are found to be:

$$ds^2 = dr^2 + (rd\theta)^2 \;\Rightarrow\; g_r = 1; \quad g_\theta = r \qquad [5.93]$$

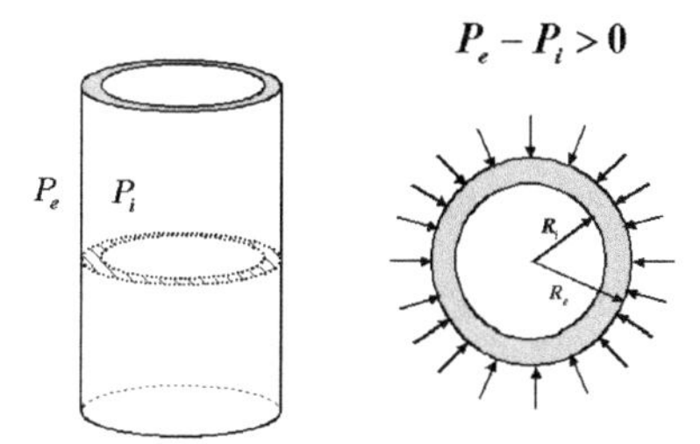

Figure 5.23. *Circular cylinder loaded by a uniform radial pressure*

The radial and tangential equilibrium equations are:

$$\frac{1}{r}\frac{\partial(r\mathcal{N}_{rr})}{\partial r}+\frac{\partial(\mathcal{N}_{r\theta})}{r\partial\theta}-\frac{\mathcal{N}_{\theta\theta}}{r}=0;\qquad \frac{\partial(\mathcal{N}_{\theta\theta})}{r\partial\theta}+\frac{1}{r^2}\frac{\partial(r^2\mathcal{N}_{\theta r})}{\partial r}=0 \qquad [5.94]$$

U denoting the radial and V the tangential displacements, [5.87] and [5.92] give:

$$\eta_{rr}=\frac{\partial U}{\partial r};\quad \eta_{r\theta}=\frac{1}{2}\left(\frac{1}{r}\frac{\partial U}{\partial \theta}+r\frac{\partial}{\partial r}\left(\frac{V}{r}\right)\right);\quad \eta_{\theta\theta}=\frac{1}{r}\frac{\partial V}{\partial\theta}+\frac{U}{r}$$

$$\mathcal{N}_{rr}=\frac{Eh}{1-\nu^2}\left(\frac{\partial U}{\partial r}+\nu\left(\frac{1}{r}\frac{\partial V}{\partial\theta}+\frac{U}{r}\right)\right);\quad \mathcal{N}_{r\theta}=Gh\left(\frac{1}{r}\frac{\partial U}{\partial\theta}+r\frac{\partial}{\partial r}\left(\frac{V}{r}\right)\right)$$

$$\mathcal{N}_{\theta\theta}=\frac{Eh}{1-\nu^2}\left(\frac{1}{r}\frac{\partial V}{\partial\theta}+\frac{U}{r}+\nu\frac{\partial U}{\partial r}\right) \qquad [5.95]$$

Obviously the solution is independent of θ and the equations [5.94] reduce to:

$$\frac{1}{r}\frac{\partial(r\mathcal{N}_{rr})}{\partial r}-\frac{\mathcal{N}_{\theta\theta}}{r}=0;\qquad \frac{1}{r^2}\frac{\partial(r^2\mathcal{N}_{r\theta})}{\partial r}=0 \qquad [5.96]$$

as r is strictly positive, the second equation implies $r^2\mathcal{N}_{r\theta}=constant$. The boundary conditions imply $\mathcal{N}_{r\theta}(R_1)=\mathcal{N}_{r\theta}(R_2)=0$, then $\mathcal{N}_{r\theta}\equiv 0$ for any r in the interval $R_1\le r\le R_2$. However, from [5.95] it is also found that:

$$\mathcal{N}_{r\theta}=\frac{Eh}{1-\nu^2}(1-\nu)\eta_{r\theta}=Ghr\frac{\partial}{\partial r}\left(\frac{V}{r}\right) \qquad [5.97]$$

which means

$$\frac{\partial}{\partial r}\left(\frac{V}{r}\right)=0\ \Rightarrow\ V=\alpha r.$$

The constant α is arbitrary; as the physical meaning of V is a rotation of the plate (considered as a rigid solid) around the Oz axis, α may be chosen equal to zero. This result confirms the intuition of a pure radial displacement of the cylinder which is a contraction if the pressure differential P_e-P_i is positive. Figure 5.24 visualizes the equilibrium of a plate sector, delimited by the radii r and $r+dr$ and by an angular sector $d\theta$. The condition of radial equilibrium is readily found to be:

$$\begin{aligned} dF &= \mathcal{N}_{rr}(r+dr)\{(r+dr)\,d\theta\}-\mathcal{N}_{rr}(r)\{r\,d\theta\}\\ &\simeq\left(\mathcal{N}_{rr}(r)+\tfrac{\partial\mathcal{N}_{rr}}{\partial r}dr\right)\{(r+dr)\,d\theta\}-\mathcal{N}_{rr}(r)\{r\,d\theta\}\\ &\simeq\left(\mathcal{N}_{rr}(r)+\tfrac{\partial\mathcal{N}_{rr}}{\partial r}\right)\{dr\,d\theta\}=\tfrac{\partial r\mathcal{N}_{rr}}{\partial r}\{dr\,d\theta\}=F\{dr\,d\theta\}\end{aligned}$$

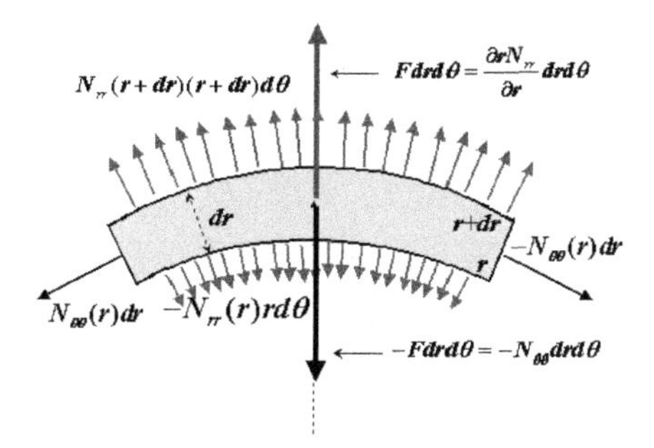

Figure 5.24. *Plate sector equilibrium*

The resultant of the tangential forces is also radial $-dF = -\mathcal{N}_{\theta\theta}\{drd\theta\}$. As suitable, this force balance is consistent with the first equation [5.96]. It must be emphasized that the curvature of the sector induces a coupling between the radial and tangential stresses; this is a very important feature which holds for any curved structures (beams, plates, shells) as will be stressed several times in Chapters 7 and 8. To alleviate to some extent the algebra, without changing the physics of the problem, P_i is assumed hereafter to be negligible in comparison with P_e. The radial equation is thus associated with the boundary conditions:

$$\mathcal{N}_{rr}(R_1) = 0; \quad \mathcal{N}_{rr}(R_2) = -P_e h \qquad [5.98]$$

Strains and stresses reduce to

$$\eta_{rr} = \frac{dU}{dr}; \quad \eta_{\theta\theta} = \frac{U}{r}$$
$$\mathcal{N}_{rr} = \frac{Eh}{1-\nu^2}\left(\frac{dU}{dr} + \nu\left(\frac{U}{r}\right)\right); \quad \mathcal{N}_{\theta\theta} = \frac{Eh}{1-\nu^2}\left(\frac{U}{r} + \nu\frac{dU}{dr}\right) \qquad [5.99]$$

and using [5.96], we arrive at the differential equation:

$$r\frac{d^2U}{dr^2} + \frac{dU}{dr} - \frac{U}{r} = 0 \qquad [5.100]$$

The solutions have the general form r^m with the values $m = 1$ and $m = -1$, so

$$U = ar + b\frac{1}{r} \qquad [5.101]$$

a and b are determined by the boundary conditions and the solution is

$$U = -\frac{P_e}{E}\frac{R_2^2}{R_2^2 - R_1^2}\left(r(1-\nu) + \frac{(1+\nu)R_1^2}{r}\right); \quad \mathcal{N}_{\theta r} = 0$$

$$\mathcal{N}_{rr} = \frac{-P_e h}{1-(R_1/R_2)^2}\left(1 - \left(\frac{R_1}{r}\right)^2\right); \quad \mathcal{N}_{\theta\theta} = \frac{-P_e h}{1-(R_1/R_2)^2}\left(1 + \left(\frac{R_1}{r}\right)^2\right) \quad [5.102]$$

The tangential stress is always greater than the pressure load; it is maximum for $r = R_1$. The sum of the stresses (radial and tangential) is constant; which is written in terms of local stresses as:

$$\sigma_{rr} + \sigma_{\theta\theta} = -2P_e\left(1 - \left(\frac{R_1}{R_2}\right)^2\right) \quad [5.103]$$

If the cylinder thickness is small $R_1 \cong R_2 \cong R$; $R_2 = R + e$, with $e/R \ll 1$, then a first order approximation gives

$$U \cong -\frac{P_e R^2}{Ee}; \quad \sigma_{rr} \cong -P_e\frac{(r-R)}{e}; \quad \sigma_{\theta\theta} \cong -P_e\frac{R}{e} \quad [5.104]$$

The tangential stress is much greater than the radial stress:

$$\sigma_{\theta\theta} \cong -P_e\frac{R}{e} \gg \sigma_{rr}|_{\max} \cong -P_e \quad [5.105]$$

Because of the curvature the radial loading can be equilibrated essentially by the tangential stresses.

Chapter 6

Plates: out-of-plane motion

Out-of-plane, or transverse, motions of plates is of paramount importance because, as in the case of straight beams and for the same reasons, plates are much less rigid when solicited in the out-of-plane than in the in-plane direction. Furthermore, the transverse response of a plate is sensitive to the presence of in-plane stresses, just as the bending of a straight beam is sensitive to the presence of a longitudinal stress. Modelling of the out-of-plane motions of plates is based on the so called Kirchhoff–Love hypotheses which extends the simplifying assumptions used to establish the Bernoulli–Euler model of straight beam bending to the two-dimensional case. Accordingly, the transverse motion can be described in terms of a single displacement field, the so called transverse displacement denoted Z. As Z depends on two coordinates (x, y in the case of a rectangular plate), new interesting features arise in plate bending and torsion with respect to straight beam bending, both from the mathematical and physical viewpoints. The Kirchhoff–Love model is found appropriate to deal with thin plates where the thickness is small compared with either the other dimensions of the plate or to the modal wavelengths of the highest modal frequency of interest. Though this book is strictly restricted to the case of thin plates and thin shells, it may be worth mentioning that the Kirchhoff–Love model can be improved by accounting for rotatory inertia and out-of-plane shear deformation, in a similar manner as the Bernoulli–Euler model can be improved, to give rise to the Rayleigh–Timoshenko model.

6.1. Kirchhoff–Love hypotheses

6.1.1 *Local displacements*

As already stated in Chapter 5 section 5.2, the local displacement field of a material point M, located at the distance z from the plate midplane is (see Figure 6.1):

$$\vec{\xi}(x, y, z; t) = \vec{X} + \vec{\psi} \times \vec{r} \qquad [6.1]$$

where

$$\vec{X} = X\vec{i} + Y\vec{j} + Z\vec{k}; \quad \vec{r} = z\vec{k}; \quad \vec{\psi} = \psi_x\vec{i} + \psi_y\vec{j} \qquad [6.2]$$

Then the Cartesian components of the local displacements are:

$$\xi_x = X + z\psi_y; \quad \xi_y = Y - z\psi_x; \quad \xi_z = Z \qquad [6.3]$$

According to the second hypothesis of the Kirchhoff–Love model, the rotations are related to the transverse displacements through the derivatives:

$$\psi_x = +\frac{\partial Z}{\partial y}; \quad \psi_y = -\frac{\partial Z}{\partial x} \qquad [6.4]$$

The signs of these derivatives are consistent with a direct frame of reference. Using [6.4], [6.3] becomes:

$$\xi_x = X - z\frac{\partial Z}{\partial x}; \quad \xi_y = Y - z\frac{\partial Z}{\partial y}; \quad \xi_z = Z \qquad [6.5]$$

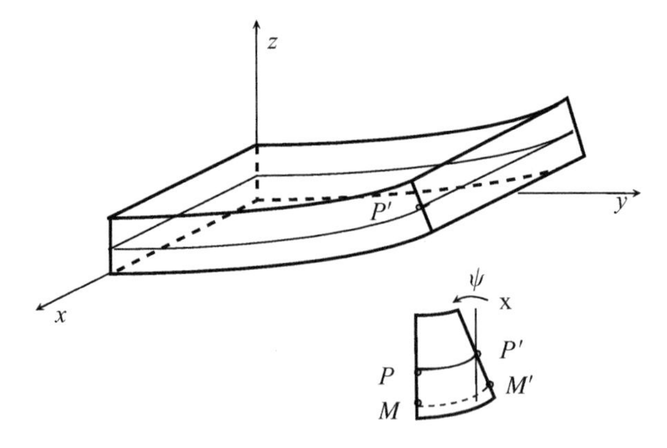

Figure 6.1. *Local (M point) and global (P point) displacements*

6.1.2 *Local and global strains*

6.1.2.1 Local strains

Starting from [6.5], the local strains are found to be:

$$\varepsilon_{xx} = \frac{\partial \xi_x}{\partial x} = \frac{\partial X}{\partial x} - z\frac{\partial^2 Z}{\partial x^2}; \quad \varepsilon_{yy} = \frac{\partial \xi_y}{\partial y} = \frac{\partial Y}{\partial y} - z\frac{\partial^2 Z}{\partial y^2}$$
$$\varepsilon_{xy} = \varepsilon_{yx} = \frac{1}{2}\left(\frac{\partial \xi_x}{\partial y} + \frac{\partial \xi_y}{\partial x}\right) = \frac{1}{2}\left(\frac{\partial X}{\partial y} + \frac{\partial Y}{\partial x} - 2z\frac{\partial^2 Z}{\partial x \partial y}\right) \qquad [6.6]$$

As a statement of the Kirchhoff–Love model, the other components (in particular the shear strains ε_{xz} and ε_{yz}) are nil, leading thus to a flexure model without shear.

6.1.2.2 Global flexure and torsional strains

As in the case of beams, the local strains [6.6] can be viewed as the sum of two distinct components, namely the membrane strains independent of z, and the flexure and torsional strains, proportional to z. Thus, the tensor form of [6.6] is conveniently written as:

$$\overline{\overline{\varepsilon}} = \overline{\overline{\eta}} + z\overline{\overline{\chi}} \qquad [6.7]$$

where $\overline{\overline{\eta}}$ stands for the membrane strain tensor, already introduced in the preceding chapter, and $\overline{\overline{\chi}}$ stands for the flexure stress tensor, written in the matrix form as:

$$[\chi] = \begin{bmatrix} \chi_{xx} & \chi_{xy} \\ \chi_{yx} & \chi_{yy} \end{bmatrix} = \begin{bmatrix} -\dfrac{\partial^2 Z}{\partial x^2} & -\dfrac{\partial^2 Z}{\partial y \partial x} \\ -\dfrac{\partial^2 Z}{\partial x \partial y} & -\dfrac{\partial^2 Z}{\partial y^2} \end{bmatrix} \qquad [6.8]$$

The coefficients of $[\chi]$ are interpreted as small curvatures of the deformed mid-plane; the diagonal terms are the flexure curvatures (similar to the beam bending curvatures) whereas the non diagonal terms represent a torsion of the plate which induces shear of the cross-sections. It is also worth mentioning that the tensor form [6.7] can be replaced by the vector form:

$$[\vec{\varepsilon}] = [\vec{\eta}] + z[\vec{\chi}] \qquad [6.9]$$

where,

$$[\vec{\eta}]^T = \begin{bmatrix}\eta_{xx} & \eta_{yy} & 2\eta_{xy}\end{bmatrix} = \begin{bmatrix}\frac{\partial X}{\partial x} & \frac{\partial Y}{\partial y} & \frac{\partial X}{\partial y} + \frac{\partial Y}{\partial x}\end{bmatrix}$$

$$[\vec{\chi}]^T = \begin{bmatrix}\chi_{xx} & \chi_{yy} & 2\chi_{xy}\end{bmatrix} = \begin{bmatrix}-\frac{\partial^2 Z}{\partial x^2} & -\frac{\partial^2 Z}{\partial y^2} & -2\frac{\partial^2 Z}{\partial y \partial x}\end{bmatrix}$$

6.1.3 *Local and global stresses: bending and torsion*

The global stresses are obtained by integrating the local stresses through the plate thickness. If the material is isotropic (which is the limitation adopted in this book), the Kirchhoff–Love model restricts the elastic stresses to the following three components, σ_{xx}, σ_{xy}, σ_{yy} which depend linearly on the coordinate z. Using the matrix form [5.32] and [6.9], in Cartesian coordinates we obtain:

$$[\vec{\sigma}] = \begin{bmatrix}\sigma_{xx} \\ \sigma_{yy} \\ \sigma_{xy}\end{bmatrix} = [C][\vec{\varepsilon}] = \frac{E}{1-\nu^2}\begin{bmatrix}1 & \nu & 0 \\ \nu & 1 & 0 \\ 0 & 0 & (1-\nu)/2\end{bmatrix}\begin{bmatrix}\varepsilon_{xx} \\ \varepsilon_{yy} \\ 2\varepsilon_{xy}\end{bmatrix} = [C]([\vec{\eta}] + z[\vec{\chi}]) \quad [6.10]$$

Starting from [6.10], integration through the thickness of the bending terms $z[\vec{\chi}]$ gives zero, whereas integration of the bending stress moments $z^2[\vec{\chi}]$ gives:

$$[\vec{\mathcal{M}}]^T = \begin{bmatrix}\mathcal{M}_{xx} & \mathcal{M}_{yy} & \mathcal{M}_{xy}\end{bmatrix} = \int_{-h/2}^{h/2} z\vec{\sigma}^T \, dz = \int_{-h/2}^{h/2} z^2[\vec{\chi}^T] \, dz = \frac{h^3}{12}[C][\vec{\chi}^T] \quad [6.11]$$

The moments $\mathcal{M}_{xx}$ and $\mathcal{M}_{yy}$ are termed bending moments, and $\mathcal{M}_{xy} = \mathcal{M}_{yx}$ is a torsion moment. Actually, these quantities are dimensioned as a moment, or torque, per unit length, i.e. as a force. Of course [6.11] can also be written as a tensor $\overline{\overline{\mathcal{M}}}$ written in matrix form as:

$$[\mathcal{M}] = \begin{bmatrix}\mathcal{M}_{xx} & \mathcal{M}_{xy} \\ \mathcal{M}_{yx} & \mathcal{M}_{yy}\end{bmatrix} \quad [6.12]$$

In Figures 6.2a and 6.2b the components of $\overline{\overline{\mathcal{M}}}$ acting on a rectangular plate element are sketched in relation to the local stresses, which helps to make clear the appropriate sign. As will be shown in subsection 6.2.1, it is also useful to visualize these moments by equivalent torques whose force components are parallel to the Oz axis.

In a similar manner to that for beams, the equilibrium of plate elements loaded by external transverse forces requires the presence of internal transverse forces,

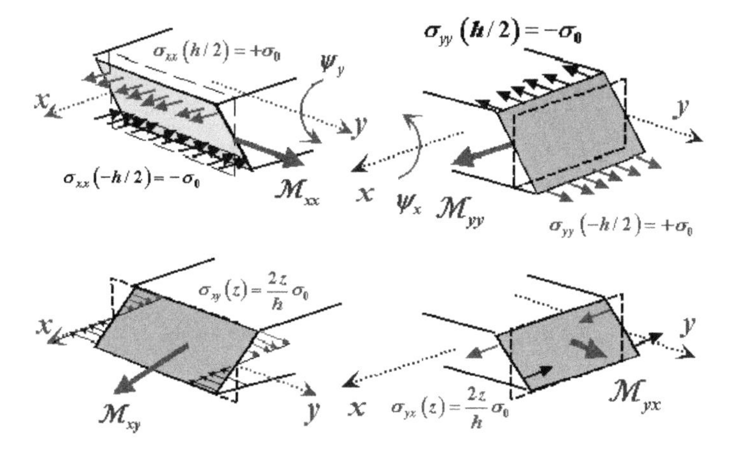

Figure 6.2a. *Bending moments*

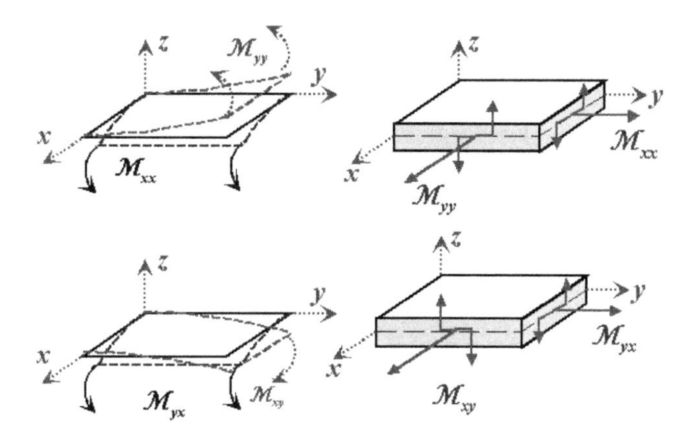

Figure 6.2b. *Torsion moments*

as shown in Figure 6.3. They result from the shear local stresses, according to the formulas:

$$\mathcal{Q}_{xz} = \int_{-h/2}^{+h/2} \sigma_{xz}\, dz; \quad \mathcal{Q}_{yz} = \int_{-h/2}^{+h/2} \sigma_{yz}\, dz \tag{6.13}$$

However, in the same way as in the Bernoulli–Euler model for beams, the transverse shear stresses cannot be introduced through a material law, since according to the Kirchhoff–Love model $\varepsilon_{xz} = \varepsilon_{yz} = 0$. So they must be introduced conceptually via the Lagrange multipliers related to the conditions $\varepsilon_{xz} = \varepsilon_{yz} = 0$ and derived in practice by balancing the transverse forces applied to a plate element. Another way to give a physical interpretation of the Kirchoff–Love model is to

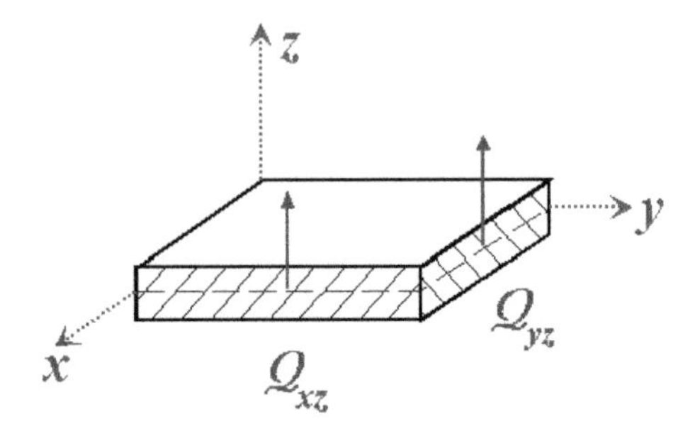

Figure 6.3. *Transverse shear forces Q_{xz}, Q_{yz}*

consider that the plate is made of an orthotropic material, which is characterized by a finite Young modulus E and Poisson's ratio ν in the in-plane directions, whereas E is assumed to be infinitely large and ν is zero in the out-of-plane direction.

6.2. Bending equations

In the absence of in-plane loading, the small transverse displacements of a plate depend on the bending and torsion terms only. The present section is restricted to this case. Study of the effect of in-plane stressed plates is postponed to section 6.4.

6.2.1 *Formulation in terms of stresses*

As in the case of beams, the local equilibrium equations could be derived by using again the Newtonian approach. However difficulties would arise in writing down the proper boundary conditions, as further discussed in subsection 6.2.2. Hence, in the present problem it is found more appropriate to use a variational principle such as Hamilton's principle. Because the detailed calculation is rather cumbersome and tedious and because some boundary conditions give rise to a few interesting subtleties, we will proceed step by step in the analysis. At first, a plate without external loading of any kind will be considered. Then the homogeneous and inhomogeneous boundary conditions will be discussed. Finally surface and concentrated loads applied on lines or points on the midplane will be included in the analysis.

6.2.1.1 Variation of the inertia terms

If the rotatory inertia of the transverse plate fibres is neglected, the kinetic energy density is:

$$de_\kappa = \tfrac{1}{2}\rho h \dot{Z}^2 \qquad [6.14]$$

The variation of the kinetic energy is:

$$\rho h \int_{t_1}^{t^2} \left[\int_0^{L_x} \int_0^{L_y} \dot{Z}\delta[\dot{Z}]\, dx\, dy \right] dt = -\int_{t_1}^{t_2} \left[\int_0^{L_x} \int_0^{L_y} \rho h \ddot{Z}\delta Z\, dx\, dy \right] dt \qquad [6.15]$$

6.2.1.2 Variation of the strain energy

Using a tensor formulation, the variation of the strain energy density is written as:

$$\delta[e_s] = \overline{\overline{\mathcal{M}}} : \delta[\overline{\overline{\chi}}] = [\vec{\mathcal{M}}]^T \delta[\vec{\chi}] \qquad [6.16]$$

Using [6.8] and [6.12] we get:

$$-\delta[e_s] = \mathcal{M}_{xx} \frac{\partial^2 \delta Z}{\partial x^2} + (\mathcal{M}_{xy} + \mathcal{M}_{yx}) \frac{\partial^2 \delta Z}{\partial x \partial y} + \mathcal{M}_{yy} \frac{\partial^2 \delta Z}{\partial y^2} \qquad [6.17]$$

Despite being equal to each other, the terms $\mathcal{M}_{xy}$ and $\mathcal{M}_{yx}$ are still considered separately because they do not act in the same direction. As the time dependent variations are not considered, the calculus of variations can be restricted to the space domain:

$$\int_0^{L_x} \int_0^{L_y} \mathcal{M}_{xx} \frac{\partial^2 \delta Z}{\partial x^2}\, dx\, dy = \int_0^{L_x} \int_0^{L_y} \frac{\partial^2 \mathcal{M}_{xx}}{\partial x^2} \delta Z\, dx\, dy + \int_0^{L_y} \left[\mathcal{M}_{xx} \frac{\partial \delta Z}{\partial x} - \frac{\partial \mathcal{M}_{xx}}{\partial x} \delta Z \right]_0^{L_x} dy \qquad [6.18]$$

$$\int_0^{L_y} \int_0^{L_x} \mathcal{M}_{yy} \frac{\partial^2 \delta Z}{\partial y^2}\, dx\, dy = \int_0^{L_x} \int_0^{L_y} \frac{\partial^2 \mathcal{M}_{yy}}{\partial y^2} \delta Z\, dx\, dy + \int_0^{L_x} \left[\mathcal{M}_{yy} \frac{\partial \delta Z}{\partial y} - \frac{\partial \mathcal{M}_{yy}}{\partial y} \delta Z \right]_0^{L_y} dx \qquad [6.19]$$

and for the torsion terms,

$$\int_0^{L_y} \int_0^{L_x} \mathcal{M}_{xy} \frac{\partial^2 \delta Z}{\partial x \partial y}\, dx\, dy = \int_0^{L_x} \int_0^{L_y} \frac{\partial^2 \mathcal{M}_{xy}}{\partial x \partial y} \delta Z\, dx\, dy + \int_0^{L_x} \left[\mathcal{M}_{xy} \frac{\partial \delta Z}{\partial x} \right]_0^{L_y} dx - \int_0^{L_y} \left[\frac{\partial \mathcal{M}_{xy}}{\partial y} \delta Z \right]_0^{L_x} dy \qquad [6.20]$$

$$\int_0^{L_x}\int_0^{L_y} \mathcal{M}_{yx}\frac{\partial^2 \delta Z}{\partial x \partial y}\,dx\,dy = \int_0^{L_x}\int_0^{L_y} \frac{\partial^2 \mathcal{M}_{yx}}{\partial x \partial y}\delta Z\,dx\,dy + \int_0^{L_y}\left[\mathcal{M}_{yx}\frac{\partial \delta Z}{\partial y}\right]_0^{L_x} dy - \int_0^{L_x}\left[\frac{\partial \mathcal{M}_{yx}}{\partial x}\delta Z\right]_0^{L_y} dx \tag{6.21}$$

6.2.1.3 *Local equilibrium without external loads*

Putting together all the surface integrals, we obtain:

$$-\delta[\mathcal{E}_s] = \int_0^{L_x}\int_0^{L_y}\left\{\left(\frac{\partial^2 \mathcal{M}_{xx}}{\partial x^2} + 2\frac{\partial^2 \mathcal{M}_{xy}}{\partial x \partial y} + \frac{\partial^2 \mathcal{M}_{yy}}{\partial y^2}\right)\delta Z\right\} dx\,dy \tag{6.22}$$

Then, in the absence of any external loading, Hamilton's principle is found to reduce to:

$$\int_{t_1}^{t_2}\int_0^{L_x}\int_0^{L_y}\left[-\rho h\ddot{Z} + \left(\frac{\partial^2 \mathcal{M}_{xx}}{\partial x^2} + 2\frac{\partial^2 \mathcal{M}_{xy}}{\partial x \partial y} + \frac{\partial^2 \mathcal{M}_{yy}}{\partial y^2}\right)\right]\delta Z\,dx\,dy\,dt = 0 \tag{6.23}$$

which gives the local equilibrium equation at a current point of the midplane of the plate:

$$\rho h\ddot{Z} - \left(\frac{\partial^2 \mathcal{M}_{xx}}{\partial x^2} + 2\frac{\partial^2 \mathcal{M}_{xy}}{\partial x \partial y} + \frac{\partial^2 \mathcal{M}_{yy}}{\partial y^2}\right) = 0 \tag{6.24}$$

$\vec{k}$ denoting the unit vector normal to the plate midplane, [6.24] is written in intrinsic form as:

$$\rho h\ddot{\vec{X}} \cdot \vec{k} - \operatorname{div}(\operatorname{div} \overline{\overline{\mathcal{M}}}) = 0 \tag{6.25}$$

NOTE. – *Shear loads and gradient of the moments*

Equation [6.25] represents the equilibrium of the internal forces. Accordingly, it may be written as:

$$\rho h\ddot{\vec{X}} \cdot \vec{n} - \operatorname{div}\vec{\mathcal{Q}} = 0 \quad \text{where } \vec{\mathcal{Q}} = \operatorname{div} \overline{\overline{\mathcal{M}}} \tag{6.26}$$

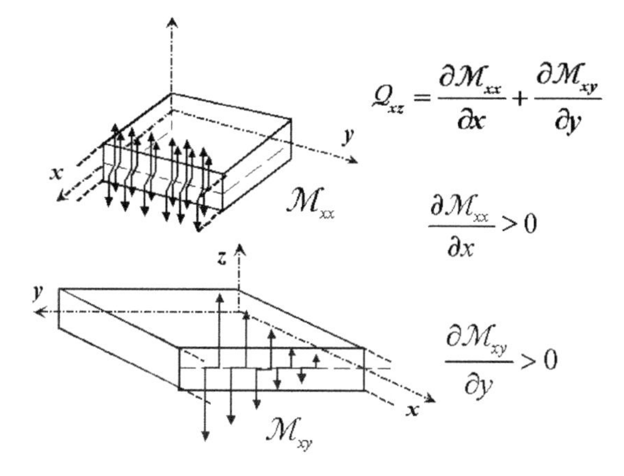

Figure 6.4. *Shear loads related to the moment derivatives*

$\vec{\mathcal{Q}}$ is the shear force vector which equilibrates the inertia forces. Its components are given by:

$$\mathcal{Q}_{xz} = \frac{\partial \mathcal{M}_{xx}}{\partial x} + \frac{\partial \mathcal{M}_{yx}}{\partial y}; \quad \mathcal{Q}_{yz} = \frac{\partial \mathcal{M}_{yy}}{\partial y} + \frac{\partial \mathcal{M}_{xy}}{\partial x} \qquad [6.27]$$

These components are represented in Figure 6.4 in the same way as in Figure 2.21. Their orientation is consistent with the orientation of the local stresses acting on the facets of a cuboid, or of the global stresses acting on a straight beam. As $\mathcal{M}_{xy} = \mathcal{M}_{yx}$, [6.24] can also be expressed as:

$$\rho h \ddot{Z} - \left(\frac{\partial \mathcal{Q}_{xz}}{\partial x} + \frac{\partial \mathcal{Q}_{yz}}{\partial y} \right) = 0 \qquad [6.28]$$

6.2.2 *Boundary conditions*

6.2.2.1 Kirchhoff effective shear forces and corner forces

In [6.17] to [6.21], cancellation of the integral variations on the edges of the plate gives the boundary conditions. A correct formulation of these conditions is not simple and concerning it one might be reminded of an historic story. At the beginning of the nineteen century, the physicist and acoustician E. Chladni (1756–1829) displayed many complex and beautiful geometrical figures when, using a bow, he excited a plate covered with a thin sand layer. This experiment was presented to the emperor Napoléon Bonaparte who was very puzzled and decided in 1809 to transfer 3000 francs to the French Academy of Sciences to be given as a prize for anybody who could be able to explain the vibrations of the plate. After several attempts, Miss S. Germain won the prize in 1816; she produced a correct differential equation but derived erroneous boundary conditions. Though Lagrange, Cauchy and Poisson participated in her research work, it was not until about 1850 that the consistent boundary conditions were established by G.R. Kirchhoff (1824–1887).

1. Components of the moment densities. Let us consider a plate edge parallel to *Oy*. The terms involving the moments in formulas [6.18] to [6.21] are:

$$\int_0^{L_y} \left[\mathcal{M}_{xx}\delta\left(\frac{\partial Z}{\partial x}\right) + \mathcal{M}_{yx}\delta\left(\frac{\partial Z}{\partial y}\right) \right]_0^{L_x} dy \qquad [6.29]$$

Before Kirchhoff, the rotations were considered as two independent variables. Accordingly, at a free edge the following boundary conditions were assumed to occur: $\mathcal{M}_{xx} = \mathcal{M}_{yx} = 0$. Such a result cannot be true because an additional boundary condition has still to be complied with, which concerns the shear forces, as further discussed in the next subsection. Hence, it would turn out that three conditions ought to be fulfilled at a plate edge, producing thus an oversized system of equations since shear forces and moments are already related through equation [6.28]. The difficulty was overcome by Kirchhoff who correctly accounted for the fact that rotations and displacements are dependent variables by integrating the torsion term to express the variation in terms of δZ solely. So, in place of [6.29] he obtained:

$$\int_0^{L_y} \left[\mathcal{M}_{xx}\delta\left(\frac{\partial Z}{\partial x}\right) + \mathcal{M}_{yx}\delta\left(\frac{\partial Z}{\partial y}\right) \right]_0^{L_x} dy$$
$$= \int_0^{L_y} \left[\mathcal{M}_{xx}\delta\left(\frac{\partial Z}{\partial x}\right) - \frac{\partial \mathcal{M}_{yx}}{\partial y}\delta Z \right]_0^{L_x} dy + \left[[\mathcal{M}_{yx}\delta Z]_0^{L_x} \right]_0^{L_y} \qquad [6.30]$$

Then, in [6.30] the sole condition to be fulfilled concerning the moment is:

$$\int_0^{L_y} \left[\mathcal{M}_{xx}\delta\left(\frac{\partial Z}{\partial x}\right) \right]_0^{L_x} dy = 0 \qquad [6.31]$$

Similarly along an edge parallel to the Ox axis we get:

$$\int_0^{L_x} \left[\mathcal{M}_{yy}\delta\left(\frac{\partial Z}{\partial y}\right) + \mathcal{M}_{xy}\delta\left(\frac{\partial Z}{\partial x}\right) \right]_0^{L_y} dx$$
$$= \int_0^{L_x} \left[\mathcal{M}_{yy}\delta\left(\frac{\partial Z}{\partial x}\right) - \frac{\partial \mathcal{M}_{xy}}{\partial x}\delta Z \right]_0^{L_y} dx + \left[[\mathcal{M}_{xy}\delta Z]_0^{L_x} \right]_0^{L_y} \qquad [6.32]$$

$$\int_0^{L_x} \left[\mathcal{M}_{yy} \delta \left(\frac{\partial Z}{\partial y} \right) \right]_0^{L_y} dx = 0 \qquad [6.33]$$

The other terms are to be included in the boundary condition concerning the shear forces, as detailed below.

2. Components of the shear force densities. By gathering the pertinent terms of the relations [6.18] to [6.21] together with the Kirchhoff term in [6.30] along an edge parallel to Oy, we obtain:

$$\int_0^{L_y} \left[-\left\{ \frac{\partial \mathcal{M}_{xx}}{\partial x} + \frac{\partial \mathcal{M}_{yx}}{\partial y} + \frac{\partial \mathcal{M}_{xy}}{\partial y} \right\} \right]_0^{L_x} \delta Z \, dy = 0 \qquad [6.34]$$

and for an edge parallel to Ox:

$$\int_0^{L_x} \left[-\left\{ \frac{\partial \mathcal{M}_{yy}}{\partial y} + \frac{\partial \mathcal{M}_{xy}}{\partial x} + \frac{\partial \mathcal{M}_{yx}}{\partial x} \right\} \right]_0^{L_y} \delta Z \, dx = 0 \qquad [6.35]$$

The terms between brackets are interpreted as shear force densities which are known as *effective Kirchhoff shear forces* per unit length, expressed as:

$$\mathcal{V}_{xz} = \frac{\partial \mathcal{M}_{xx}}{\partial x} + \frac{\partial \mathcal{M}_{yx}}{\partial y} + \frac{\partial \mathcal{M}_{xy}}{\partial y} = \mathcal{Q}_{xz} + \frac{\partial \mathcal{M}_{yx}}{\partial y} = \frac{\partial \mathcal{M}_{xx}}{\partial x} + 2\frac{\partial \mathcal{M}_{xy}}{\partial y}$$

$$\mathcal{V}_{yz} = \frac{\partial \mathcal{M}_{yy}}{\partial y} + \frac{\partial \mathcal{M}_{xy}}{\partial x} + \frac{\partial \mathcal{M}_{yx}}{\partial x} = \mathcal{Q}_{yz} + \frac{\partial \mathcal{M}_{xy}}{\partial x} = \frac{\partial \mathcal{M}_{yy}}{\partial y} + 2\frac{\partial \mathcal{M}_{yx}}{\partial x} \qquad [6.36]$$

where use is made of the equality $\mathcal{M}_{xy} = \mathcal{M}_{yx}$.

3. Corner forces. Corner forces originate from the Kirchhoff relations [6.30] and [6.32] which imply the boundary condition at a corner:

$$\left[[(\mathcal{M}_{xy} + \mathcal{M}_{yx})\delta Z]_0^{L_x} \right]_0^{L_y} = 0 \qquad [6.37]$$

If the corner is supported in such a way that $\delta Z = 0$, a corner force $\mathcal{M}_{xy} + \mathcal{M}_{yx} \neq 0$ arises in the transverse direction, as further discussed in subsection 6.2.2.3.

6.2.2.2 *Elastic boundary conditions*

When applied to elastic stresses, the preceding relations lead to the following homogeneous boundary conditions:

1. *Clamped edge*

$$\begin{cases} \text{parallel to } Ox: & Z(x, y = 0 \text{ or } L_y) = 0; \quad \partial Z/\partial y|_{y=0 \text{ or } L_y} = 0 \\ \text{parallel to } Ox: & Z(x = 0 \text{ or } L_x, y) = 0; \quad \partial Z/\partial x|_{x=0 \text{ or } L_x} = 0 \end{cases} \quad [6.38]$$

2. *Sliding edge*

$$\begin{cases} \text{parallel to } Ox: & Z(x, y = 0, \text{ or } x, L_y) = 0; \quad \mathcal{V}_{yz}|_{y=0 \text{ or } L_y} = 0 \\ \text{parallel to } Oy: & Z(x = 0, \text{ or } L_x, y) = 0; \quad \mathcal{V}_{xz}|_{x=0 \text{ or } L_x} = 0 \end{cases} \quad [6.39]$$

3. *Hinged edge*

$$\begin{cases} \text{parallel to } Ox: & Z(x, y = 0 \text{ or } x, L_y) = 0; \quad \mathcal{M}_{yy}|_{y=0 \text{ or } L_y} = 0 \\ \text{parallel to } Oy: & Z(x = 0 \text{ or } L_x, y) = 0; \quad \mathcal{M}_{xx}|_{x=0 \text{ or } L_x} = 0 \end{cases} \quad [6.40]$$

4. *Free edge*

$$\begin{cases} \text{parallel to } Ox: & \mathcal{V}_{yz}|_{x,o \text{ or } L_y} = 0; \quad \mathcal{M}_{yy}|_{y=0 \text{ or } L_y} = 0 \\ \text{parallel to } Oy: & \mathcal{V}_{xz}|_{x=0 \text{ or } L_x,y} = 0; \quad \mathcal{M}_{xx}|_{x=0 \text{ or } L_x} = 0 \end{cases} \quad [6.41]$$

5. *Free corner*

$$(\mathcal{M}_{yx} = \mathcal{M}_{xy})|_{0,0 \text{ or } 0,L_y \text{ or } L_x,0 \text{ or } L_x,L_y} = 0 \quad [6.42]$$

6.2.2.3 *External loading of the edges and inhomogeneous boundary conditions*

The external loading is comprised of forces or/and moments per unit length applied to the edges of the plate, the latter defining the plate contour denoted $(\mathcal{C})$:

$$\vec{\mathcal{Q}}^{(e)}(x, y; t) = \vec{\mathcal{Q}}_z^{(e)}\vec{k}; \quad \vec{\mathcal{M}}^{(e)}(x, y; t) = \left(\mathcal{M}_x^{(e)}\vec{i} + \mathcal{M}_y^{(e)}\vec{j}\right) \quad x, y \in (\mathcal{C}) \quad [6.43]$$

The analysis is restrained to the simplest configuration which supposes the moment $\vec{\mathcal{M}}^{(e)}$ parallel to the edge; then the virtual work is:

$$\oint_{(\mathcal{C})} \left\{ \mathcal{Q}_z^{(e)} \delta Z + \mathcal{M}_x^{(e)} \frac{\partial \delta Z}{\partial y} - \mathcal{M}_y^{(e)} \frac{\partial \delta Z}{\partial x} \right\} ds \quad [6.44]$$

The expression [6.44] is expanded as follows:

$$\oint_{(C)} \left\{ \mathcal{Q}_z^{(e)} \delta Z + \mathcal{M}_x^{(e)} \frac{\partial \delta Z}{\partial y} - \mathcal{M}_y^{(e)} \frac{\partial \delta Z}{\partial x} \right\} ds$$
$$= \int_0^{L_x} \left(\mathcal{Q}_z^{(e)} \delta Z + \left[\mathcal{M}_x^{(e)} \frac{\partial \delta Z}{\partial y} \right]_{y=L_y} + \left[\mathcal{Q}_z^{(e)} \delta Z + \mathcal{M}_x^{(e)} \frac{\partial \delta Z}{\partial y} \right]_{y=0} \right) dx$$
$$+ \int_0^{L_y} \left(\left[\mathcal{Q}_z^{(e)} \delta Z + \mathcal{M}_y^{(e)} \frac{\partial \delta Z}{\partial y} \right]_{x=L_x} + \left[\mathcal{Q}_z^{(e)} \delta Z - \mathcal{M}_y^{(e)} \frac{\partial \delta Z}{\partial x} \right]_{x=0} \right) dy \quad [6.45]$$

The inhomogeneous boundary conditions are determined by collecting appropriately the variations of the external edge loading and those of the edge and corner stresses (formulas [6.30] to [6.37]). Letting the coefficients of the variation $\delta(\partial Z/\partial x)$ and $\delta(\partial Z/\partial y)$ vanish, the following moment balances are produced:

$$\begin{aligned}
\text{edge } x = 0: &\quad \mathcal{M}_{xx} = -\mathcal{M}_y^{(e)} \\
\text{edge } x = L_x: &\quad \mathcal{M}_{xx} = +\mathcal{M}_y^{(e)} \\
\text{edge } y = 0: &\quad \mathcal{M}_{yy} = +\mathcal{M}_x^{(e)} \\
\text{edge } y = L_y: &\quad \mathcal{M}_{yy} = -\mathcal{M}_x^{(e)}
\end{aligned} \quad [6.46]$$

Letting the coefficients of the variation δZ vanish, the following force balances are produced:

$$\begin{aligned}
\text{edge } x = 0: &\quad \mathcal{V}_{xz} = -\mathcal{Q}_z^{(e)}(0, y; t) \\
\text{edge } x = L_x: &\quad \mathcal{V}_{xz} = +\mathcal{Q}_z^{(e)}(L_x, y; t) \\
\text{edge } y = 0: &\quad \mathcal{V}_{yz} = -\mathcal{Q}_z^{(e)}(x, 0; t) \\
\text{edge } y = L_y: &\quad \mathcal{V}_{yz} = +\mathcal{Q}_z^{(e)}(x, L_y; t)
\end{aligned} \quad [6.47]$$

Similarly, the following corner force balance is obtained:

$$\begin{aligned}
2\mathcal{M}_{yx}(L_x, L_y) &= -F_{zC}^{(e)}(\delta(x - L_x) \cap \delta(y - L_y)); \\
2\mathcal{M}_{yx}(0, 0) &= -F_{zC}^{(e)}(\delta(x) \cap \delta(y)) \\
2\mathcal{M}_{yx}(L_x, 0) &= +F_{zC}^{(e)}(\delta(x - L_x) \cap \delta(y)); \\
2\mathcal{M}_{yx}(0, L_y) &= +F_{zC}^{(e)}(\delta(x) \cap \delta(y - L_y))
\end{aligned} \quad [6.48]$$

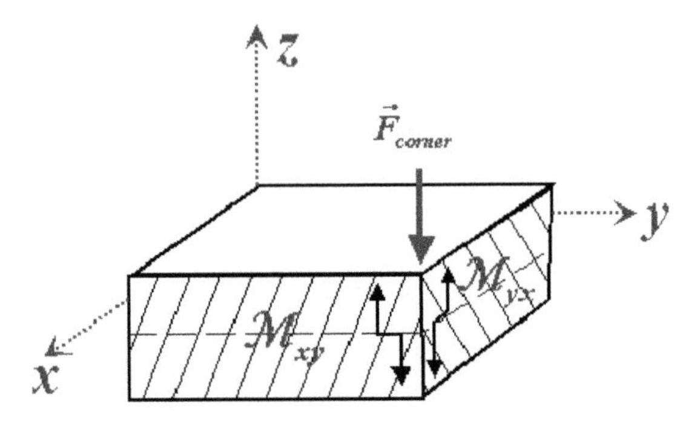

Figure 6.5. *Representation of the corner forces*

The force balance at the corner (L_x, L_y) is sketched in Figure 6.5. It makes clear that the internal torsion moments induce an internal corner force as the result of the discontinuity in the edge directions.

6.2.3 *Surface and concentrated loadings*

6.2.3.1 Loading distributed over the midplane surface

The surface loads are defined by the two vectors:

$$\vec{f}^{(e)}(x,y;t) = f_z^{(e)}\vec{k}; \quad \vec{\mathfrak{M}}^{(e)}(x,y;t) = \mathfrak{M}_x^{(e)}\vec{i} + \mathfrak{M}_y^{(e)}\vec{j} \tag{6.49}$$

To avoid any redundancy with edge loading, surface loads are assumed to vanish at the plate edges. The related virtual work is written as:

$$\delta[\mathcal{W}^{(e)}] = \int_0^{L_x}\int_0^{L_y}\left(f_z^{(e)}\delta Z + \mathfrak{M}_x^{(e)}\frac{\partial \delta Z}{\partial y} - \mathfrak{M}_y^{(e)}\frac{\partial \delta Z}{\partial x}\right)dx\,dy \tag{6.50}$$

Integrating by parts [6.50], the terms related to the variation δZ are:

$$\delta[\mathcal{W}^{(e)}] = \int_0^{L_x}\int_0^{L_y}\left(f_z^{(e)} - \frac{\partial \mathfrak{M}_x^{(e)}}{\partial y} + \frac{\partial \mathfrak{M}_y^{(e)}}{\partial x}\right)\delta Z\,dx\,dy \tag{6.51}$$

Then Hamilton's principle leads to:

$$\int_{t_1}^{t_2}\int_0^{L_x}\int_0^{L_y}\left[-\rho h\ddot{Z} + \left(\frac{\partial^2 \mathcal{M}_{xx}}{\partial x^2} + 2\frac{\partial^2 \mathcal{M}_{xy}}{\partial x \partial y} + \frac{\partial^2 \mathcal{M}_{yy}}{\partial y^2}\right)\right.$$
$$\left. + \left(f_z^{(e)} - \frac{\partial \mathfrak{M}_x^{(e)}}{\partial y} + \frac{\partial \mathfrak{M}_y^{(e)}}{\partial x}\right)\right]\delta Z\,dx\,dydt = 0$$

from which the equilibrium equation is obtained as:

$$\rho h\ddot{Z} - \left(\frac{\partial^2 \mathcal{M}_{xx}}{\partial x^2} + 2\frac{\partial^2 \mathcal{M}_{xy}}{\partial x \partial y} + \frac{\partial^2 \mathcal{M}_{yy}}{\partial y^2}\right) = f_z^{(e)} - \frac{\partial \mathfrak{M}_x^{(e)}}{\partial y} + \frac{\partial \mathfrak{M}_y^{(e)}}{\partial x} \quad [6.52]$$

or in intrinsic form:

$$\rho h\ddot{\vec{X}} \cdot \vec{k} - \mathrm{div}(\mathrm{div}\, \bar{\bar{\mathcal{M}}}) = (\vec{f}^{(e)} + \mathrm{curl}\, \vec{\mathfrak{M}}^{(e)}) \cdot \vec{k}$$

$$\mathrm{div}(\mathrm{div}\, \bar{\bar{\mathcal{M}}}) = \vec{\nabla} \cdot \vec{\nabla} \cdot \bar{\bar{\mathcal{M}}} = \frac{\partial}{\partial x_i}\left(\frac{\partial \mathcal{M}_{ij}}{\partial x_j}\right) \quad \text{and} \quad \mathrm{curl}\, \vec{\mathfrak{M}}^{(e)} = \vec{\nabla} \times \vec{\mathfrak{M}}^{(e)}$$

$$\text{where} \quad \vec{\nabla} = \left[\frac{\partial}{\partial x} \quad \frac{\partial}{\partial y} \quad \frac{\partial}{\partial z}\right] \quad [6.53]$$

6.2.3.2 Load distributed along a straight line parallel to an edge

A transverse external force, concentrated along $x = x_0$, is described by the singular distribution:

$$\vec{f}^{(e)} = f_z^{(e)}(y;t)\delta(x - x_0)\vec{k} \quad [6.54]$$

Using [6.52], the equilibrium is given in terms of distributions as:

$$\rho h\ddot{Z} - \left(\frac{\partial^2 \mathcal{M}_{xx}}{\partial x^2} + 2\frac{\partial^2 \mathcal{M}_{xy}}{\partial x \partial y} + \frac{\partial^2 \mathcal{M}_{yy}}{\partial y^2}\right) = f_z^{(e)}(y;t)\delta(x - x_0) \quad [6.55]$$

If the integration is performed in the interval $[x_0 - \varepsilon, x_0 + \varepsilon]$ where $\varepsilon \to 0$, it is found that:

$$\mathcal{Q}_{xz}|_{x_0-} - \mathcal{Q}_{xz}|_{x_0+} = f_z^{(e)}(y;t) \quad [6.56]$$

This result generalizes the inhomogeneous boundary conditions [6.47]. On the other hand, equation [6.55] is equivalent to the system:

$$\rho h\ddot{Z} - \left(\frac{\partial^2 \mathcal{M}_{xx}}{\partial x^2} + 2\frac{\partial^2 \mathcal{M}_{xy}}{\partial x \partial y} + \frac{\partial^2 \mathcal{M}_{yy}}{\partial y^2}\right) = 0$$

$$\left[\frac{\partial \mathcal{M}_{xx}}{\partial x} + 2\frac{\partial \mathcal{M}_{xy}}{\partial y}\right]\Bigg|_{x_0-} - \left[\frac{\partial \mathcal{M}_{xx}}{\partial x} + 2\frac{\partial \mathcal{M}_{xy}}{\partial y}\right]\Bigg|_{x_0+} = \mathcal{Q}_{xz}|_{x_0-} - \mathcal{Q}_{xz}|_{x_0+} = f_z^{(e)}(y;t) \quad [6.57]$$

Similarly, for external moments distributed according to

$$\mathfrak{M}_y^{(e)}(x, y; t) = \mathfrak{M}_y^{(e)}(y; t)\delta(x - x_0)\vec{j} \quad [6.58]$$

we obtain:

$$\rho h\ddot{Z} - \left(\frac{\partial^2 \mathcal{M}_{xx}}{\partial x^2} + 2\frac{\partial^2 \mathcal{M}_{xy}}{\partial x \partial y} + \frac{\partial^2 \mathcal{M}_{yy}}{\partial y^2}\right) = -\mathfrak{M}_y^{(e)}(y; t)\delta'(x - x_0) \quad [6.59]$$

which is equivalent to the system:

$$\begin{aligned} &\rho h\ddot{Z} - \left(\frac{\partial^2 \mathcal{M}_{xx}}{\partial x^2} + 2\frac{\partial^2 \mathcal{M}_{xy}}{\partial x \partial y} + \frac{\partial^2 \mathcal{M}_{yy}}{\partial y^2}\right) = 0 \\ &[\mathcal{M}_{xx}]|_{x_0+} - [\mathcal{M}_{xx}]|_{x_0-} = -\mathcal{M}_y^{(e)}(x_0, y; t) \end{aligned} \quad [6.60]$$

The result [6.60] generalizes the inhomogeneous boundary conditions [6.46].

6.2.3.3 *Point loads*

An external force, concentrated at x_0, y_0, is described by the singular distribution:

$$\vec{f}^{(e)} = f_z(t)(\delta(x - x_0) \cap \delta(y - y_0))\vec{k} \quad [6.61]$$

So the equilibrium equation is written in terms of distributions as:

$$\rho h\ddot{Z} - \left(\frac{\partial^2 \mathcal{M}_{xx}}{\partial x^2} + 2\frac{\partial^2 \mathcal{M}_{xy}}{\partial x \partial y} + \frac{\partial^2 \mathcal{M}_{yy}}{\partial y^2}\right) = f_z^{(e)}(t)\delta(x - x_0) \cap \delta(y - y_0) \quad [6.62]$$

Again, an integration over the domain $[x_0 - \varepsilon, x_0 + \varepsilon; y_0 - \eta, y_0 + \eta]$ leads one to describe this loading by a finite stress discontinuity, as follows:

$$\begin{aligned} &\int_{x_0-\varepsilon}^{x_0+\varepsilon} \int_{y_0-\eta}^{y_0+\eta} -\left\{\frac{\partial^2 \mathcal{M}_{xx}}{\partial x^2} + \frac{\partial^2 \mathcal{M}_{xy}}{\partial x \partial y} + \frac{\partial^2 \mathcal{M}_{yy}}{\partial y^2}\right\} dx\, dy \\ &= \int_{y_0-\eta}^{y_0+\eta} -\left[\frac{\partial \mathcal{M}_{xx}}{\partial x}\right]_{x_0-\varepsilon}^{x_0+\varepsilon} dy + \int_{y_0-\eta}^{y_0+\eta} -\left[\frac{\partial \mathcal{M}_{xy}}{\partial y}\right]_{x_0-\varepsilon}^{x_0+\varepsilon} dy \\ &+ \int_{x_0-\varepsilon}^{x_0+\varepsilon} -\left[\frac{\partial \mathcal{M}_{yy}}{\partial y}\right]_{y_0-\eta}^{y_0+\eta} dx + \int_{x_0-\varepsilon}^{x_0+\varepsilon} -\left[\frac{\partial \mathcal{M}_{xy}}{\partial x}\right]_{y_0-\eta}^{y_0+\eta} dx \end{aligned} \quad [6.63]$$

When ε and η tend to zero, the sole non-vanishing components are those related to the torsion moments. Then it is found that [6.63] is equivalent to:

$$\rho h\ddot{Z} - \left(\frac{\partial^2 \mathcal{M}_{xx}}{\partial x^2} + 2\frac{\partial^2 \mathcal{M}_{xy}}{\partial x \partial y} + \frac{\partial^2 \mathcal{M}_{yy}}{\partial y^2}\right) = 0$$
$$-2\left[[\mathcal{M}_{xy}]_{x_{0-}}^{x_{0+}}\right]_{y_{0-}}^{y_{0+}} = f_z^{(e)}(t) \qquad [6.64]$$

6.2.4 *Elastic vibrations*

6.2.4.1 *Global stresses*

The bending and torsion moments are given by:

$$\begin{bmatrix} \mathcal{M}_{xx} \\ \mathcal{M}_{yy} \\ \mathcal{M}_{xy} \end{bmatrix} = \int_{-h/2}^{+h/2} \frac{Ez^2}{1-\nu^2} \begin{bmatrix} 1 & \nu & 0 \\ \nu & 1 & 0 \\ 0 & 0 & (1-\nu)/2 \end{bmatrix} \begin{bmatrix} \chi_{xx} \\ \chi_{yy} \\ 2\chi_{xy} \end{bmatrix} dz \qquad [6.65]$$

By substituting [6.8] into [6.65], it becomes:

$$\mathcal{M}_{xx} = -D\left(\frac{\partial^2 Z}{\partial x^2} + \nu\frac{\partial^2 Z}{\partial y^2}\right); \quad \mathcal{M}_{yy} = -D\left(\frac{\partial^2 Z}{\partial y^2} + \nu\frac{\partial^2 Z}{\partial x^2}\right);$$
$$\mathcal{M}_{xy} = -D(1-\nu)\frac{\partial^2 Z}{\partial x \partial y} \quad \text{where } D = \frac{E}{1-\nu^2}\int_{-h/2}^{+h/2} z^2 dz = \frac{Eh^3}{12(1-\nu^2)} \qquad [6.66]$$

D is known as the *bending stiffness coefficient* of the plate.

6.2.4.2 *Vibration equations*

The shear forces needed to ensure the local force equilibrium are obtained by substituting [6.66] into [6.25]; which gives the equation of transverse motion:

$$D\left\{\frac{\partial^4 Z}{\partial x^4} + 2\frac{\partial^4 Z}{\partial x^2 \partial y^2} + \frac{\partial^4 Z}{\partial y^4}\right\} + \rho h\frac{\partial^2 Z}{\partial t^2} = f_z^{(e)} - \frac{\partial \mathfrak{M}_x^{(e)}}{\partial y} + \frac{\partial \mathfrak{M}_y^{(e)}}{\partial x} \qquad [6.67]$$

The intrinsic form of this equation is written as:

$$D\Delta[\Delta[\vec{X}\cdot\vec{k}]] + \rho h\left(\ddot{\vec{X}}\cdot\vec{k}\right) = \vec{f}^{(e)}\cdot\vec{k} + \text{curl}\,\vec{\mathfrak{M}}^{(e)}\cdot\vec{k}$$
$$\Delta = \text{div grad} = \vec{\nabla}\cdot\vec{\nabla}[\,] \qquad [6.68]$$

6.2.4.3 Elastic boundary conditions

The boundary conditions of subsection 6.2.2 take the following forms:

1. *Inhomogeneous edge conditions*

$$x = L_x \begin{cases} D\left[\dfrac{\partial^2 Z}{\partial x^2} + \nu\dfrac{\partial^2 Z}{\partial y^2}\right]_{x=L_x} = -\mathcal{M}_y^{(e)}(L_x, y; t) \\ D\left(\dfrac{\partial^3 Z}{\partial x^3} + (2-\nu)\dfrac{\partial^3 Z}{\partial x \partial y^2}\right)\Bigg|_{x=L_x} = -\mathcal{Q}_z^{(e)}(L_x, y; t) \end{cases} \quad [6.69]$$

$$y = 0 \begin{cases} D\left[\dfrac{\partial^2 Z}{\partial y^2} + \nu\dfrac{\partial^2 Z}{\partial x^2}\right]_{y=0} = -\mathcal{M}_x^{(e)}(x, 0; t) \\ D\left(\dfrac{\partial^3 Z}{\partial y^3} + (2-\nu)\dfrac{\partial^3 Z}{\partial x^2 \partial y}\right)\Bigg|_{y=0} = +\mathcal{Q}_z^{(e)}(x, 0; t) \end{cases} \quad [6.70]$$

$$y = L_y \begin{cases} D\left[\dfrac{\partial^2 Z}{\partial y^2} + \nu\dfrac{\partial^2 Z}{\partial x^2}\right]_{y=L_y} = -\mathcal{M}_x^{(e)}(L_x, y; t) \\ D\left(\dfrac{\partial^3 Z}{\partial x^3} + (2-\nu)\dfrac{\partial^3 Z}{\partial x \partial y^2}\right)\Bigg|_{x=L_x} = -\mathcal{Q}_z^{(e)}(L_x, y; t) \end{cases} \quad [6.71]$$

2. *Inhomogeneous corner conditions*

$$\begin{aligned} &x = L_x, y = L_y: \quad +2D(1-\nu)\frac{\partial^2 Z}{\partial x \partial y}\bigg|_{x=L_x;\, y=L_y} = F_{corner}^{(e)} \\ &x = 0, y = 0: \quad +2D(1-\nu)\frac{\partial^2 Z}{\partial x \partial y}\bigg|_{x=0;\, y=0} = F_{corner}^{(e)} \\ &x = 0, y = L_y: \quad -2D(1-\nu)\frac{\partial^2 Z}{\partial x \partial y}\bigg|_{x=0;\, y=L_y} = F_{corner}^{(e)} \\ &x = L_x, y = 0: \quad -2D(1-\nu)\frac{\partial^2 Z}{\partial x \partial y}\bigg|_{x=L_x;\, y=0} = F_{corner}^{(e)} \end{aligned} \quad [6.72]$$

3. *Usual homogeneous conditions*
Clamped edge

$$\begin{cases} \text{parallel to } Ox\text{: } Z(x, y = 0 \text{ or } L_y) = 0; \ \ \partial Z/\partial y|_{y=0 \text{ or } L_y} = 0 \\ \text{parallel to } Oy\text{: } Z(x = 0 \text{ or } L_x, y) = 0; \ \ \partial Z/\partial y|_{x=0 \text{ or } L_x} = 0 \end{cases} \quad [6.73]$$

Sliding edge

$$\begin{cases} \text{parallel to } Ox\text{: } Z(x, y = 0 \text{ or } L_y) = 0; & \left.\dfrac{\partial^3 Z}{\partial y^3}\right|_{y=0 \text{ or } L_y} = 0 \\ \text{parallel to } Oy\text{: } Z(x = 0 \text{ or } L_x, y) = 0; & \left.\dfrac{\partial^3 Z}{\partial y^3}\right|_{x=0 \text{ or } L_x} = 0 \end{cases} \qquad [6.74]$$

It must be pointed out that the disappearance of the rotation implies automatically that of the mixed derivative in the shear forces.
Hinged edge

$$\begin{cases} \text{parallel to } Ox\text{: } Z(x, y = 0 \text{ or } L_y) = 0; & \left.\dfrac{\partial^2 Z}{\partial y^2}\right|_{y=0 \text{ or } L_y} = 0 \\ \text{parallel to } Oy\text{: } Z(x = 0 \text{ or } L_x, y) = 0; & \left.\dfrac{\partial^2 Z}{\partial x^2}\right|_{x=0 \text{ or } L_x} = 0 \end{cases} \qquad [6.75]$$

It must be pointed out that the disappearance of the displacement implies automatically that of the mixed derivative in the moment.
Free edge

$$\begin{aligned} &\text{parallel to } Ox: & \left.\left(\frac{\partial^3 Z}{\partial y^3} + (2-\nu)\frac{\partial^3 Z}{\partial y \partial x^2}\right)\right|_{y=0 \text{ or } L_y} &= 0 \\ & & \left.\left(\frac{\partial^2 Z}{\partial y^2} + \nu\frac{\partial^2 Z}{\partial x^2}\right)\right|_{y=0 \text{ or } L_y} &= 0 \\ &\text{parallel to } Oy: & \left.\left(\frac{\partial^3 Z}{\partial x^3} + (2-\nu)\frac{\partial^3 Z}{\partial x \partial y^2}\right)\right|_{x=0 \text{ or } L_x} &= 0 \\ & & \left.\left(\frac{\partial^2 Z}{\partial x^2} + \nu\frac{\partial^2 Z}{\partial y^2}\right)\right|_{x=0 \text{ or } L_x} &= 0 \end{aligned} \qquad [6.76]$$

Free corner

$$\left.\left(\frac{\partial^2 Z}{\partial x \partial y}\right)\right|_{0,0 \text{ or } 0,L_y \text{ or } L_x,0 \text{ or } L_x,L_y} = 0 \qquad [6.77]$$

6.2.5 *Application to a few problems in statics*

6.2.5.1 Bending of a plate loaded by edge moments

Let us consider a rectangular plate loaded by a uniform moment density $-\mathcal{M}_y^{(e)}$ along the edge $x = -L/2$ and by $+\mathcal{M}_y^{(e)}$ along the edge $x = L/2$, the two other

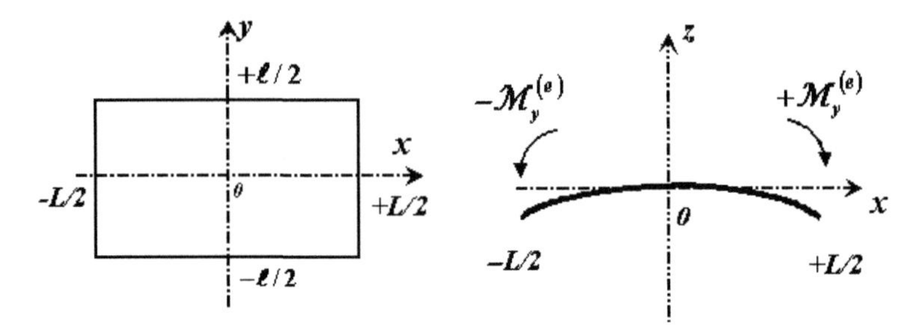

Figure 6.6. *Bending of a rectangular plate*

edges being left free. The choice of the global frame of Figure 6.6 is clearly guided by the symmetries of the problem. The equilibrium equations are:

$$D\left(\frac{\partial^4 Z}{\partial x^4} + 2\frac{\partial^4 Z}{\partial x^2 \partial y^2} + \frac{\partial^4 Z}{\partial y^4}\right) = 0$$

$$D\left(\frac{\partial^2 Z}{\partial x^2} + \nu\frac{\partial^2 Z}{\partial y^2}\right)\bigg|_{x=\pm L/2} = -\mathcal{M}_y^{(e)}$$

$$\left(\frac{\partial^2 Z}{\partial y^2} + \nu\frac{\partial^2 Z}{\partial x^2}\right)\bigg|_{y=\pm\ell/2} = 0; \quad \frac{\partial^2 Z}{\partial x \partial y}\bigg|_{x=\pm L/2;\, y=\pm\ell/2} = 0$$

$$\left(\frac{\partial^3 Z}{\partial x^3} + (2-\nu)\frac{\partial^3 Z}{\partial x \partial y^2}\right)\bigg|_{x=\pm L/2} = \left(\frac{\partial^3 Z}{\partial y^3} + (2-\nu)\frac{\partial^3 Z}{\partial x^2 \partial y}\right)\bigg|_{y=\pm\ell/2} = 0 \quad [6.78]$$

with the following symmetry conditions:

$$\frac{\partial Z}{\partial x}\bigg|_{x=y=0} = \frac{\partial Z}{\partial y}\bigg|_{x=y=0} = 0 \quad [6.79]$$

The solution is a second degree polynomial of the general type,

$$Z(x, y) = ax^2 + by^2 + cx + dy + e \quad [6.80]$$

Because of the symmetries $c = d = 0$; the cross-term xy is also zero since there is no corner force. Finally, the rigid displacement mode accounted for by e may be discarded, so the solution is rewritten as:

$$Z(x, y) = ax^2 + by^2 \quad [6.81]$$

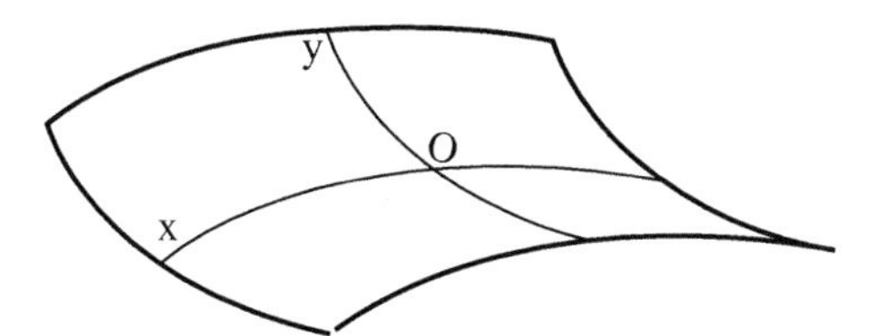

Figure 6.7. *Anticlastic deformation of the plate*

The coefficients a and b are deduced from the following boundary conditions:

$$D\left(\frac{\partial^2 Z}{\partial x^2}+\nu\frac{\partial^2 Z}{\partial y^2}\right)\bigg|_{x=\pm L/2}=-\mathcal{M}_y^{(e)}\ \Rightarrow\ -2D(a+\nu b)=\mathcal{M}_y^{(e)}$$

$$\left(\frac{\partial^2 Z}{\partial y^2}+\nu\frac{\partial^2 Z}{\partial x^2}\right)\bigg|_{y=\pm\ell/2}=0\ \Rightarrow\ (b+\nu a)=0$$

Then, the final solution is found to be:

$$Z(x,y)=\frac{-6\mathcal{M}_y^{(e)}}{Eh^3}(x^2-\nu y^2) \qquad [6.82]$$

The shape of the deformed plate is an hyperboloid; the saddle point is the centre of the plate. The two principal curvatures have opposite signs and this kind of deformation is termed *anticlastic* (see Figure 6.7). The contour is made of hyperbolic lines with asymptotes crossing each other at an angle θ defined by $\tan(\theta)=1/\sqrt{\nu}$. Colour plates 9 and 10 show the bending deflection, the stresses and the reactions along the supports for two different loadings and boundary conditions. Plate 9 refers to a rectangular plate hinged at the lateral edges and loaded by a uniform pressure. Plate 10 refers to a rectangular plate clamped at the lateral edges and loaded by a transverse force concentrated at the plate centre.

6.2.5.2 Torsion by corner forces

The problem is sketched in Figure 6.8 together with the solution. The equilibrium equations are:

$$D\left(\frac{\partial^4 Z}{\partial x^4}+2\frac{\partial^4 Z}{\partial x^2\partial y^2}+\frac{\partial^4 Z}{\partial y^4}\right)=0$$

$$\left(\frac{\partial^2 Z}{\partial x^2}+\nu\frac{\partial^2 Z}{\partial y^2}\right)\bigg|_{x=\pm L/2}=\left(\frac{\partial^2 Z}{\partial y^2}+\nu\frac{\partial^2 Z}{\partial x^2}\right)\bigg|_{y=\pm\ell/2}=0$$

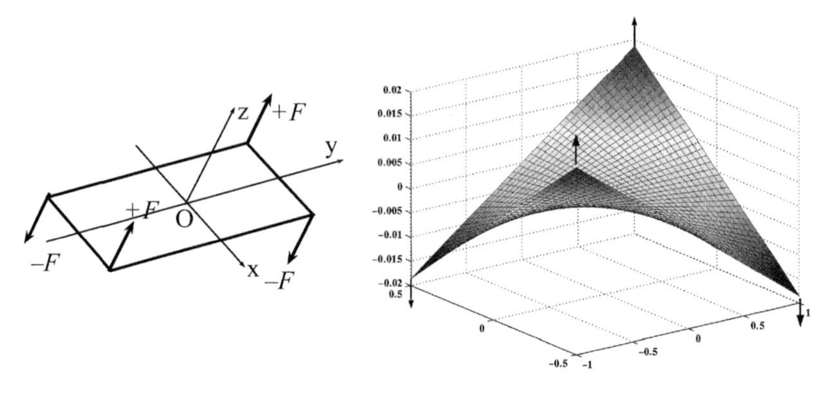

Figure 6.8. *Torsion of a rectangular steel plate (L = 2 m, l = 1 m, h = 1 cm) loaded at its corners F = 1kN*

$$2D(1-\nu)\left.\frac{\partial^2 Z}{\partial x \partial y}\right|_{x=\pm L/2;\, y=\pm \ell/2} = -F$$

$$\left.\left(\frac{\partial^3 Z}{\partial x^3} + (2-\nu)\frac{\partial^3 Z}{\partial x \partial y^2}\right)\right|_{x=\pm L/2} = \left.\left(\frac{\partial^3 Z}{\partial y^3} + (2-\nu)\frac{\partial^3 Z}{\partial x^2 \partial y}\right)\right|_{y=\pm \ell/2} = 0 \qquad [6.83]$$

The polynomial solution is of the type $Z = axy$, and the coefficient a is determined by using the corner boundary condition , which leads to:

$$Z(x,y) = -\frac{3Fxy}{Gh^3} \qquad [6.84]$$

6.3. Modal analysis

6.3.1 *Natural modes of vibration*

6.3.1.1 Flexure equation of a plate prestressed in its own plane

A rectangular plate is assumed to be initially loaded by in-plane forces distributed uniformly along the edges: $F_{xx}^{(0)}, F_{xy}^{(0)}$ on $x = L_x$ and $F_{yy}^{(0)}, F_{yx}^{(0)}$ on $y = L_y$ (see Figure 6.9). The consequence of such a preloading on the flexure modes of vibration of the plate is analysed here. This problem extends that of an axially preloaded beam already discussed in Chapter 4, subsection 4.2.5 to the 2D case. The analysis is somewhat heavier and some interesting differences in the results occur when passing from the 1D beam case to the 2D plate case.

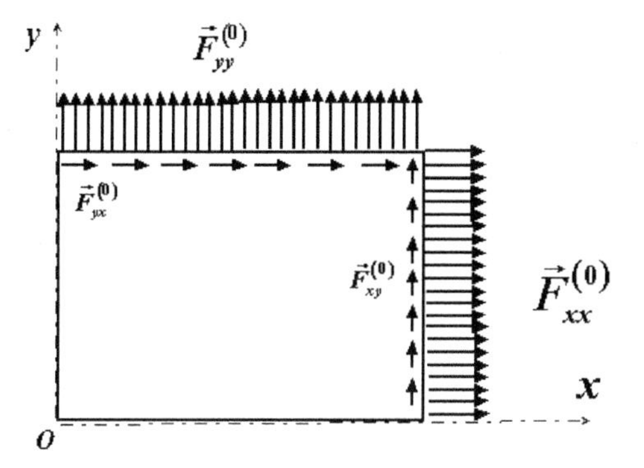

Figure 6.9. *Plate prestressed by in-plane loads applied to the edges*

The local stress field is written as the sum of a prestress field $\overline{\overline{\sigma^{(0)}}}$ and of an elastic stress field $\overline{\overline{\sigma^{(1)}}}$:

$$\overline{\overline{\sigma}} = \overline{\overline{\sigma^{(0)}}} + \overline{\overline{\sigma^{(1)}}} \quad \Rightarrow$$

$$\begin{bmatrix} \sigma_{xx} & \sigma_{yx} \\ \sigma_{xy} & \sigma_{yy} \end{bmatrix} = \frac{1}{h}\begin{bmatrix} F_{xx}^{(0)} & F_{yx}^{(0)} \\ F_{xy}^{(0)} & F_{yy}^{(0)} \end{bmatrix} + \frac{E}{1-\nu^2}\begin{bmatrix} \varepsilon_{xx} + \nu\varepsilon_{yy} & (1-\nu)\varepsilon_{yx} \\ (1-\nu)\varepsilon_{xy} & \varepsilon_{yy} + \nu\varepsilon_{xx} \end{bmatrix} \quad [6.85]$$

The strain energy density per unit plate volume is thus:

$$e_s = \overline{\overline{\sigma^{(0)}}} : \overline{\overline{\varepsilon}} + \tfrac{1}{2}\overline{\overline{\sigma^{(1)}}} : \overline{\overline{\varepsilon}} \quad [6.86]$$

The prestress energy is calculated starting from the following expression:

$$e_s = \overline{\overline{\sigma^{(0)}}} : \overline{\overline{\varepsilon}} = \sigma_{xx}^{(0)}\varepsilon_{xx} + 2\sigma_{xy}^{(0)}\varepsilon_{xy} + \sigma_{yy}^{(0)}\varepsilon_{yy}$$

$$\text{where} \quad \sigma_{xx}^{(0)} = \frac{F_{xx}^{(0)}}{h}, \; \sigma_{xy}^{(0)} = \frac{F_{xy}^{(0)}}{h}, \sigma_{yy}^{(0)} = \frac{F_{yy}^{(0)}}{h} \quad [6.87]$$

The strains are determined by using the Green–Lagrange strain tensor:

$$\varepsilon_{ij} = \frac{1}{2}\left\{\frac{\partial\xi_i}{\partial x_j} + \frac{\partial\xi_j}{\partial x_i} + \frac{\partial\xi_k}{\partial x_i}\frac{\partial\xi_k}{\partial x_j}\right\} \quad \Rightarrow$$

$$\varepsilon_{xx} = \frac{\partial\xi_x}{\partial x} + \frac{1}{2}\left\{\left(\frac{\partial\xi_x}{\partial x}\right)^2 + \left(\frac{\partial\xi_y}{\partial x}\right)^2 + \left(\frac{\partial\xi_z}{\partial x}\right)^2\right\}$$

$$\varepsilon_{xy} = \frac{1}{2}\left\{\frac{\partial \xi_x}{\partial y} + \frac{\partial \xi_y}{\partial x} + \frac{\partial \xi_x}{\partial x}\frac{\partial \xi_x}{\partial y} + \frac{\partial \xi_y}{\partial x}\frac{\partial \xi_y}{\partial y} + \frac{\partial \xi_z}{\partial x}\frac{\partial \xi_z}{\partial y}\right\}$$

$$\varepsilon_{yy} = \frac{\partial \xi_y}{\partial y} + \frac{1}{2}\left\{\left(\frac{\partial \xi_x}{\partial y}\right)^2 + \left(\frac{\partial \xi_y}{\partial y}\right)^2 + \left(\frac{\partial \xi_z}{\partial y}\right)^2\right\}$$

[6.88]

Substituting into [6.88] the local displacements field [6.5] related to the transverse global displacement Z, the following results are found:

$$\xi_x = -z\frac{\partial Z}{\partial x}; \quad \xi_y = -z\frac{\partial Z}{\partial y}; \quad \xi_z = Z$$

$$\varepsilon_{xx} = -z\frac{\partial^2 Z}{\partial x^2} + \frac{1}{2}\left(\frac{\partial Z}{\partial x}\right)^2 + \frac{z^2}{2}\left\{\left(\frac{\partial^2 Z}{\partial x^2}\right)^2 + \left(\frac{\partial^2 Z}{\partial x \partial y}\right)^2\right\}$$

$$\varepsilon_{xy} = -z\frac{\partial^2 Z}{\partial x \partial y} + \frac{z^2}{2}\frac{\partial^2 Z}{\partial x \partial y}\left(\frac{\partial^2 Z}{\partial x^2} + \frac{\partial^2 Z}{\partial y^2}\right) + \frac{1}{2}\left(\frac{\partial Z}{\partial x}\frac{\partial Z}{\partial y}\right)$$

$$\varepsilon_{yy} = -z\frac{\partial^2 Z}{\partial y^2} + \frac{1}{2}\left(\frac{\partial Z}{\partial y}\right)^2 + \frac{z^2}{2}\left\{\left(\frac{\partial^2 Z}{\partial y^2}\right)^2 + \left(\frac{\partial^2 Z}{\partial x \partial y}\right)^2\right\}$$

[6.89]

In the same way as in the case of beams, it suffices to retain the terms of the first and second order only, where the distance z of a material point from the midplane is considered as a first order infinitesimal term. Hence, when using [6.89] hereafter, terms proportional to z^2 are dropped out. Accordingly, the prestress energy is found to be:

$$\int_{-h/2}^{h/2}\int_0^{L_x}\int_0^{L_y}\left(\overline{\overline{\sigma^{(0)}}} : \overline{\overline{\varepsilon}}\right) dx\, dy\, dz$$
$$= F_{xx}^{(0)}\int_0^{L_x}\int_0^{L_y}\frac{1}{2}\left(\frac{\partial Z}{\partial x}\right)^2 dx\, dy + F_{yy}^{(0)}\int_0^{L_x}\int_0^{L_x}\frac{1}{2}\left(\frac{\partial Z}{\partial x}\right)^2 dx\, dy$$
$$+ \left(F_{xy}^{(0)} + F_{yx}^{(0)}\right)\int_0^{L_x}\int_0^{L_y}\left(\frac{\partial Z}{\partial x}\frac{\partial Z}{\partial y}\right) dx\, dy$$

[6.90]

The variation of [6.90] is expressed as:

$$- F_{xx}^{(0)}\int_0^{L_x}\int_0^{L_y}\frac{\partial^2 Z}{\partial x^2}\delta Z\, dx\, dy - F_{yy}^{(0)}\int_0^{L_x}\int_0^{L_y}\frac{\partial^2 Z}{\partial y^2}\delta Z\, dx\, dy$$
$$- \left(F_{xy}^{(0)} + F_{yx}^{(0)}\right)\int_0^{L_x}\int_0^{L_y}\left(\frac{\partial^2 Z}{\partial x \partial y}\right)\delta Z\, dx\, dy$$

$$+\left(F_{xx}^{(0)}+F_{yx}^{(0)}\right)\int_0^{L_y}\left[\frac{\partial Z}{\partial x}\delta Z\right]_0^{L_x}dx\,dy$$

$$+\left(F_{yy}^{(0)}+F_{xy}^{(0)}\right)\int_0^{L_x}\left[\frac{\partial Z}{\partial y}\delta Z\right]_0^{L_y}dx\,dy \qquad [6.91]$$

The transverse vibration equation is obtained by adding the appropriate terms of [6.91] to the elastic equation [6.67], where the external loading is assumed to be zero. The result is:

$$\rho h\ddot{Z}+D\left(\frac{\partial^4 Z}{\partial x^4}+2\frac{\partial^4 Z}{\partial x^2\partial y^2}+\frac{\partial^4 Z}{\partial y^4}\right)$$
$$-\left(F_{xx}^{(0)}\frac{\partial^2 Z}{\partial x^2}+F_{yy}^{(0)}\frac{\partial^2 Z}{\partial y^2}+\left(F_{xy}^{(0)}+F_{yx}^{(0)}\right)\frac{\partial^2 Z}{\partial x\partial y}\right)=0 \qquad [6.92]$$

The boundary conditions which are eventually modified with respect to the non-stressed case are found to be:

$$\left[\left(\left(F_{xx}^{(0)}+F_{yx}^{(0)}\right)\frac{\partial Z}{\partial x}-\mathcal{Q}_{xz}\right)\right]_{x=0;\,x=L_x}$$
$$=0\left[\left(\left(F_{yy}^{(0)}+F_{xy}^{(0)}\right)\frac{\partial Z}{\partial y}-\mathcal{Q}_{yz}\right)\right]_{y=0;\,x=L_x}=0 \qquad [6.93]$$

They extend to the 2D case the beam boundary conditions [3.52] and they hold when $Z(0)$ and $Z(L)$ are different from zero.

6.3.1.2 *Natural modes of vibration and buckling load*

A closed-form solution does not exist for the natural modes of vibration of a rectangular plate with various standard boundary conditions such as those defined by the relations [6.73] to [6.77]. It is thus found convenient to discuss the modal properties of rectangular plates by restricting first the analysis to the particular case of four hinged edges, for which the analytical solution is easily derived. In subsection 6.3.1.4, an approximate solution based on the Rayleigh–Ritz method will be presented to deal with other boundary conditions.

Let us consider a rectangular plate with hinged edges and subjected to a compressive force $-F_{xx}^{(0)}$. The modal equation is thus written as:

$$D\left(\frac{\partial^4 Z}{\partial x^4}+2\frac{\partial^4 Z}{\partial x^2\partial y^2}+\frac{\partial^4 Z}{\partial y^4}\right)+F_{xx}^{(0)}\frac{\partial^2 Z}{\partial x^2}-\omega^2\rho hZ=0 \qquad [6.94]$$

all edges: $Z=0$; lateral edges: $\partial^2 Z/\partial x^2=0$; longitudinal edges: $\partial^2 Z/\partial y^2=0$

The system [6.94] can be solved by separating the variables, so $Z(x,y) = A(x)B(y)$, which gives:

$$\left(\frac{1}{A}\frac{d^4A}{dx^4} + 2\left(\frac{1}{A}\frac{d^2A}{dx^2}\right)\left(\frac{1}{B}\frac{d^2B}{dy^2}\right) + \frac{1}{B}\frac{d^4B}{dy^4}\right) - F_{xx}^{(0)}\left(\frac{1}{A}\frac{d^2A}{dx^2}\right) = \frac{\rho h}{D}\omega^2$$

$$A(x)|_{x=0,L} = 0; \quad B(y)|_{y=0,L} = 0; \quad \left.\frac{d^2A}{dx^2}\right|_{x=0,L} = 0; \quad \left.\frac{d^2B}{dy^2}\right|_{y=0,L} = 0$$

A priori, a difficulty arises here, since the second term is still an x and y function. However, by looking to the prestressed term, it is recognized that the method can be applied, because the condition:

$$\frac{1}{A}\frac{d^2A}{dx^2} = k_x^2$$

where k_x^2 is a constant implies necessarily that:

$$\frac{1}{B}\frac{d^2B}{dy^2} = k_y^2$$

where k_y^2 is another constant.

The general solution of equation [6.94] is thus found to be of the type:

$$A(x) = ae^{k_x x} + be^{-k_x x}; \quad B(y) = ae^{k_y y} + be^{-k_y y}$$

By using the hinged edge conditions, they take the form:

$$A_n(x) = \alpha_n \sin\left(\frac{n\pi x}{L_x}\right); \quad B_m(y) = \beta_m \sin\left(\frac{m\pi y}{L_y}\right), \qquad n, m = 1, 2, 3, \ldots, \infty$$

which leads immediately to the normalized mode shapes:

$$\varphi_{nm}(x,y) = \sin\left(\frac{n\pi x}{L_x}\right)\sin\left(\frac{m\pi y}{L_y}\right), \quad n, m = 1, 2, 3, \ldots, \infty \qquad [6.95]$$

and to the related natural pulsations:

$$\omega_{nm} = \left(\frac{1}{\rho h}\left\{D\left(\left(\frac{n\pi}{L_x}\right)^2 + \left(\frac{m\pi}{L_y}\right)^2\right)^2 - F_{xx}^{(0)}\left(\frac{n\pi}{L_x}\right)^2\right\}\right)^{1/2} \qquad [6.96]$$

From [6.96] it can be immediately concluded that the mode indexed by n and m buckles as soon as the compressive load exceeds the critical value:

$$F_{n,m}^{(c)}\left(\frac{L_y}{m\pi}\right)^2 = D\left(\frac{n}{m\eta}+\frac{m\eta}{n}\right)^2 \quad \text{where } \eta=\frac{L_x}{L_y} \qquad [6.97]$$

Obviously, if the compression load is increased from zero, the buckling load corresponds to the lowest modal critical load. In contrast with the beam case, the buckling load is found to depend upon the aspect ratio η of the plate and does not necessarily correspond to the first n (or m) mode, as illustrated in Figure 6.10, which refers to $m = 1$ and where L_x is varied and L_y left constant. The buckling mode shape is clearly dependent on η. In this simple example, it is possible to find the less stable mode by cancelling the derivative of [6.97] with respect to n (or m); for instance if $\eta = 5$, the plate will buckle according to the $m = 1, n = 4$ mode, see Figure 6.11. Such features are clearly a consequence of the 2D nature of the problem.

Thermal buckling is of practical interest in many applications. Let us consider the case of a square plate subjected to a uniform temperature increase $\Delta\theta$. The formula [6.96] can be easily adapted to the present problem by using the results established in Chapter 5, subsection 5.3.5.2. We obtain:

$$\omega_{nm} = \left(\frac{E\pi^2(n^2+m^2)}{\rho(1-\nu)L^2}\left\{\frac{\pi^2(n^2+m^2)h^2}{12(1+\nu)L^2}-\alpha\Delta\theta\right\}\right)^{1/2} \qquad [6.98]$$

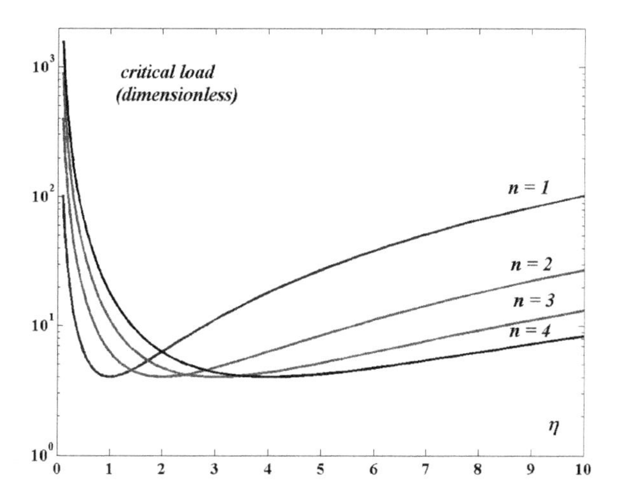

Figure 6.10. *Buckling load as function of the plate aspect ratio and of the mode order*

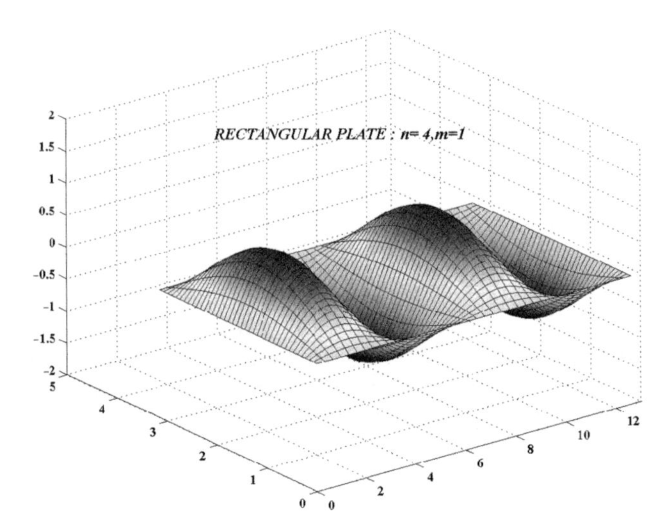

Figure 6.11. *Mode shape: $n = 4$, $m = 1$*

Here the plate is found to buckle according to the first mode $n = m = 1$. It corresponds to the critical temperature increase:

$$\Delta\theta_c = \frac{1}{6\alpha(1+\nu)}\left(\frac{\pi h}{L}\right)^2 \qquad [6.99]$$

Figure 6.12 refers to a steel plate $L = 1$ m. The natural frequency of the (1,1) mode is plotted versus $\Delta\theta$ together with the variation of the critical temperature increase as a function of the plate thickness.

6.3.1.3 Modal density and forced vibrations near resonance

Of course, the formula [6.96] also holds in the particular case of unstressed plates. The point here is to investigate a few consequences of the n, m dependency of the natural frequencies which is governed by the coefficient:

$$\left(\frac{n\pi}{L_x}\right)^2 + \left(\frac{m\pi}{L_y}\right)^2$$

It is easy to check that it can take similar values for several pairs of mode indices. This is illustrated in Figure 6.13, where all the natural frequencies are plotted for m and n varying from 1 to 10. The results refer to a steel square plate $L = 1$ m, $h = 2$ mm. It clearly shows that the plate can vibrate according to several flexure modes, whose natural frequencies are very close to each other. So, if a plate is excited by a 'nearly resonant' force, even if the excitation spectrum is

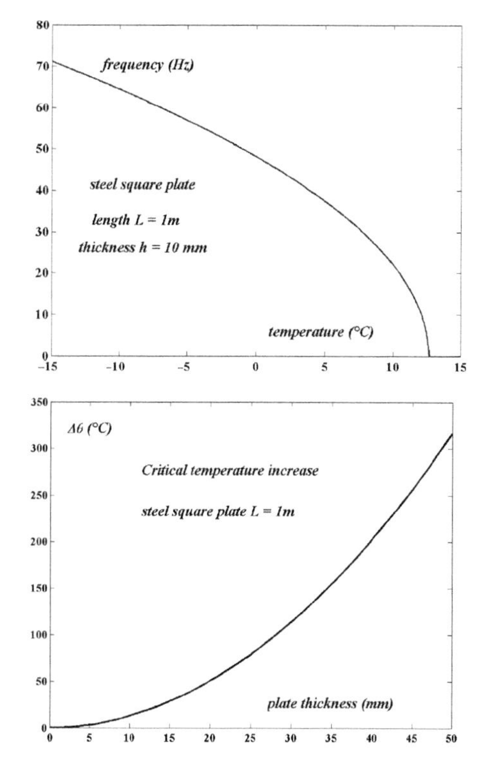

Figure 6.12. *Thermal buckling of a square steel plate*

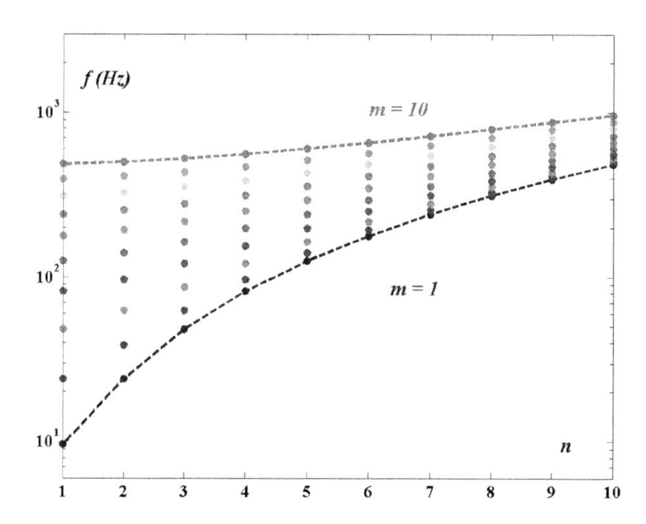

Figure 6.13. *Natural frequencies of the flexure modes of a square plate*

limited to a fairly narrow frequency range, the plate response is not restricted to a single resonant mode but is made up of a linear superposition of all the nearly resonant modes whose frequencies lie within the frequency range of the excitation signal. Furthermore, the coefficients entering in the superposition are very sensitive to 'small' defects in the plate geometry, support conditions and material. As a consequence, the experimental determination of the modal properties of plates is often a difficult task. On the other hand, the multi modal response can explain the complexity of the figures obtained by Chladni, already evoked in subsection 6.2.2.

As an example, the response of a square plate, hinged at the four edges and excited by a point harmonic force $F_0\delta(x - x_0) \cap \delta(y - y_0)e^{i\omega_0 t}$ is considered here. Using the modal expansion [4.51] of the transfer function, the Fourier transform of the response is found to be:

$$Z(x, y, x_0, y_0; \omega_0) = \frac{F_0}{M_G} \sum_{n=1}^{\infty} \sum_{m=1}^{\infty} \frac{\sin(n\pi x/L)\sin(m\pi y/L)\sin(n\pi x_0/L)\sin(m\pi y_0/L)}{(\omega_{n,m}^2 - \omega_0^2 + 2i\omega_0\omega_{n,m}\varsigma_{n,m})}$$

where $M_G = \rho h L^2/4$

Consider, for instance, the case of a resonant excitation at frequency $f_{1,2} = f_{2,1}$. If the excitation point lies on the nodal line of the mode (1,2) $y = L/2$, the response is marked by a nodal line $x = L/2$. Conversely, if it lies on the nodal line of the mode (2,1) $x = L/2$, the response is marked by a nodal line $y = L/2$. Finally, if the excitation is applied at $x = y = L/4$ the two modes are excited with the same efficiency and the resulting nodal line is $x = -y$. Much more complicated nodal patterns can be obtained by increasing the frequency of excitation, as illustrated in colour plate 11, where the red colour corresponds to vibration levels equal to or less than 10% of the maximum vibration magnitude, representing thus the zones where the sand would accumulate in a Chladni experiment.

6.3.1.4 *Natural modes of vibration of a stretched plate*

As already mentioned in subsection 6.3.1.2, the modal problem cannot be solved analytically in closed form for various boundary conditions. This is the case for instance of a plate with hinged supports along the lateral edges and left free along the longitudinal edges. Such a configuration is of practical importance in many industrial applications, such as the rolling process of thin strips of paper or metal, see Figure 6.14. In so far as such strips are stretched uniformly in the longitudinal

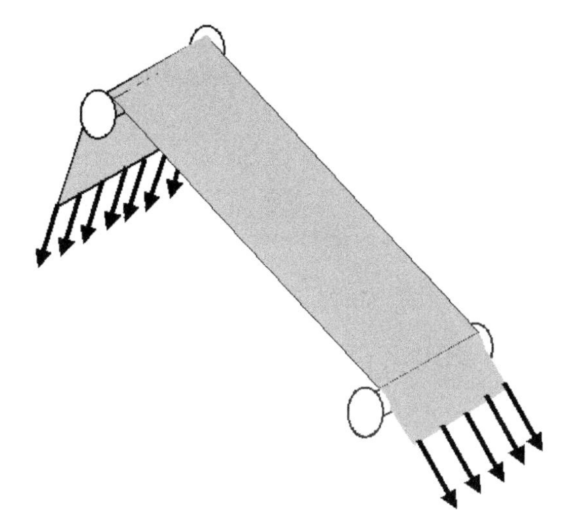

Figure 6.14. *Stretching of a rolled strip of metal*

direction, the modal problem is formulated as follows:

$$D\left(\frac{\partial^4 Z}{\partial x^4} + 2\frac{\partial^4 Z}{\partial x^2 \partial y^2} + \frac{\partial^4 Z}{\partial y^4}\right) - F_{xx}^{(0)}\frac{\partial^2 Z}{\partial x^2} - \omega^2 \rho h Z = 0$$

$$\text{lateral edges: } Z(0,y) = Z(L,y) = 0; \quad \left.\frac{\partial^2 Z}{\partial x^2}\right|_{x=0,L} = 0$$

longitudinal edges:

$$\left.\left(\frac{\partial^2 Z}{\partial y^2} + \nu\frac{\partial^2 Z}{\partial x^2}\right)\right|_{y=\pm\ell/2} = 0; \quad \left.\left(\frac{\partial^3 Z}{\partial y^3} + (2-\nu)\frac{\partial^3 Z}{\partial x^2 \partial y}\right)\right|_{y=\pm\ell/2} = 0$$

[6.100]

If an exact separate variables solution is attempted as in subsection 6.3.1.2, the longitudinal mode shapes $\varphi_n(x) = \sin(n\pi x/L)$ arise necessarily as admissible functions. However, no suitable lateral mode shapes can be found. Then, an approximate solution can be attempted based on the Rayleigh–Ritz method described in Chapter 5 subsection 5.3.6.3. As trial functions for the mode shapes, it seems reasonable and convenient to adopt a linear manifold of products of the natural modes of bending vibration of beams. Furthermore, because of the orthogonality properties of such functions, to approximate suitably the shape of the mode (n, m) a single product will suffice:

$$\psi_{n,m}(x,y) - \varphi_n(x)\psi_m(y) \qquad [6.101]$$

where $\varphi_n(x)$ stands for the beam modes complying with the lateral support conditions and $\psi_m(y)$ for the beam modes complying with the longitudinal support conditions:

$$\begin{aligned}
&\varphi_n(x) = \sin\frac{n\pi x}{L} \\
&\psi_1(y) = 1, \quad \psi_2(y) = \frac{2y}{\ell} - 1 \\
&m > 2, \quad \psi_m(y) = a_m \sin\left(\frac{\varpi_m y}{\ell}\right) + b_m \sinh\left(\frac{\varpi_m y}{\ell}\right) \\
&\qquad + c_m \cos\left(\frac{\varpi_m y}{\ell}\right) + b_m \cosh\left(\frac{\varpi_n y}{\ell}\right)
\end{aligned} \quad [6.102]$$

where $m = 1, 2$ refer to the rigid modes of pure translation and of pure rotation.

So, in this example the Rayleigh–Ritz procedure is reduced to that of Rayleigh's quotient. Accordingly, the natural frequencies are evaluated by:

$$\omega_{n,m}^2 = \frac{\left\langle \mathcal{E}_{n,m}^{(p)} \right\rangle}{\left\langle \mathcal{E}_{n,m}^{(\kappa)} \right\rangle} = \frac{K_{n,m}}{M_{n,m}} \quad [6.103]$$

where the functional of potential and kinetic energies are calculated by using the postulated mode shapes. Since the trial functions [6.102] do not comply with the boundary conditions on the longitudinal edges of the plate, it is appropriate to calculate the functional of potential energy by starting from the symmetric form:

$$\left\langle \mathcal{E}_{n,m}^{(p)} \right\rangle = \int_0^\ell dy \int_0^L \left(\overline{\overline{\mathcal{M}_{nm}}} : \overline{\overline{\chi_{nm}}} + F_{xx}^{(0)} \left(\frac{\partial \phi_{nm}}{\partial x}\right)^2 \right) dx \quad [6.104]$$

By using the relations [6.66] and [6.102] it is found that:

$$\begin{aligned}
\left\langle \mathcal{E}_{n,m}^{(p)} \right\rangle = &\frac{DL}{2} \left\{ \left(\frac{n\pi}{L}\right)^4 \mu_1 - 2\nu \left(\frac{n\pi}{L}\right)^2 \mu_2 + \mu_3 + 2(1-\nu)\left(\frac{n\pi}{L}\right)^2 \mu_4 \right\} \\
&+ \frac{F_{xx}^{(0)} L}{2} \left(\frac{n\pi}{L}\right)^2 \mu_1
\end{aligned}$$

where

$$\begin{aligned}
\mu_1 &= \int_0^\ell (\psi_m(y))^2 dy, \quad \mu_2 = \int_0^\ell \psi_m(y)\psi_m^n(y) dy \\
\mu_3 &= \int_0^\ell (\psi_m^n(y))^2 dy, \quad \mu_4 = \int_0^\ell (\psi_m'(y))^2 dy
\end{aligned} \quad [6.105]$$

the functional of kinetic energy is readily found to be:

$$\left\langle \mathcal{E}_{n,m}^{(\kappa)} \right\rangle = \frac{\rho h L \mu_1}{2} \quad [6.106]$$

As an example we consider a strip of steel $L = 10\,\text{m}, l = 1\,\text{m}, h = 0.7\,\text{mm}$ subjected to a uniform tensile stress σ_{xx} which is varied from 0 to 100 Mpa. In Figure 6.15 the natural frequencies of the modes (1,1), (1,2) and (1,3) are plotted versus σ_{xx}. The values in full lines refer to the Rayleigh quotient method and those marked by upward triangles refer to the finite element method. Both kinds of results are found to agree with each other within a few percent. As expected, when σ_{xx} is sufficiently large, most of the stiffness is provided by the prestress term and the natural frequencies of the three modes become essentially the same and equal to

$$f_1 = \frac{1}{2L}\frac{h}{\ell}\sqrt{\frac{\sigma_{xx}}{\rho}}$$

On the other hand, Figure 6.16 shows the mode shapes, obtained by using the finite element method, as viewed from two distinct points of view (see the reference frames below the views). As expected, the mode shape (1,1) is very close to the first bending mode of the equivalent beam of length L. The mode shape (1,2) is very close to the first torsional mode.

Figures 6.17 and 6.18 refer to a strip of steel of low aspect ratio, $L = 1$ m, $l = 1$ m, $h = 0.7$ mm. The mode shape (1,1) is clearly marked by an anticlastic

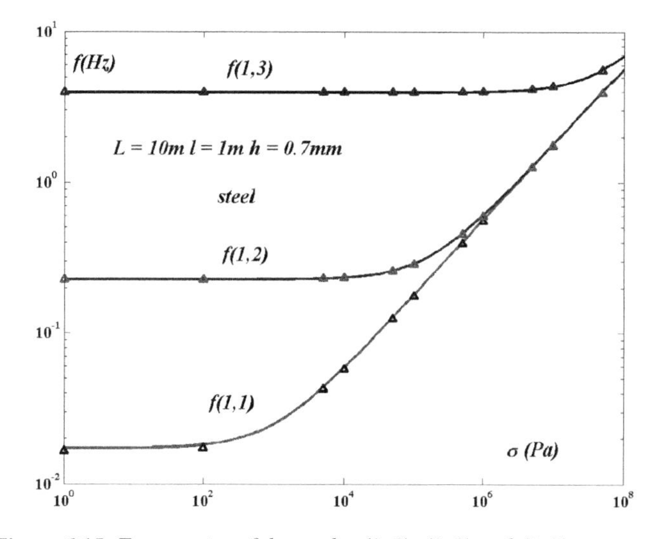

Figure 6.15. *Frequencies of the modes (1, 1), (1, 2) and (1, 3) versus* σ_{xx}

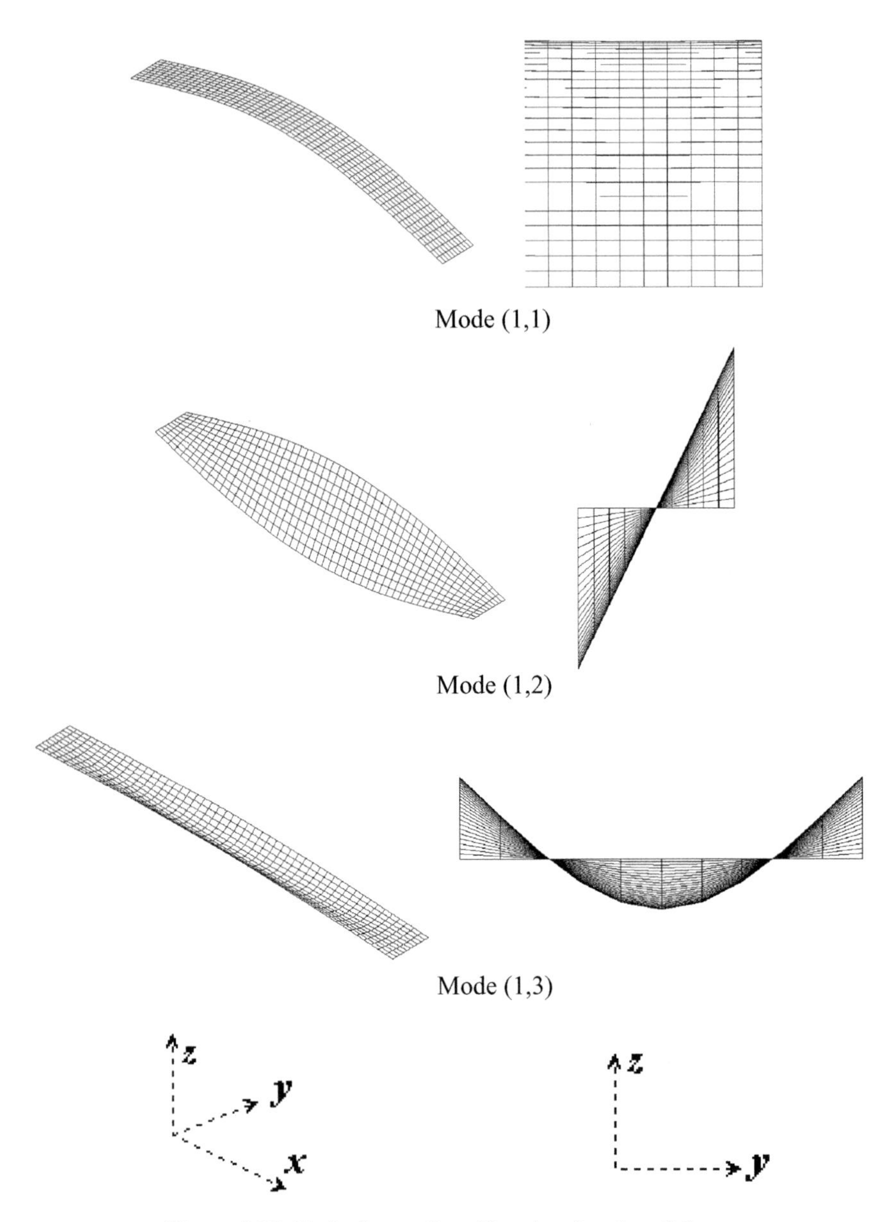

Figure 6.16. *Mode shapes: $L = 10$ m, $l = 1$ m, $h = 0.7$ mm*

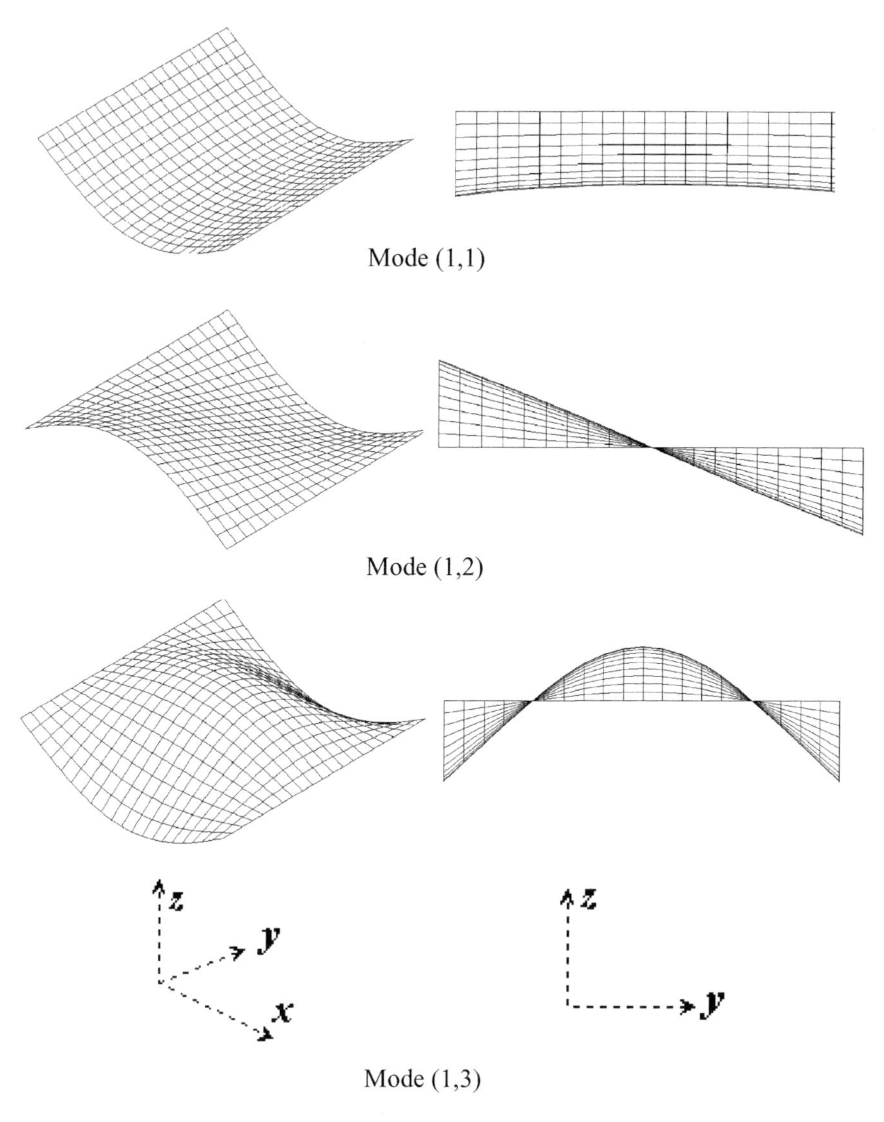

Figure 6.17. *Mode shapes: L = 1 m, l = 1 m, h = 0.7 mm*

bending along the lateral direction (cf. Figure 6.7). Again, the Rayleigh quotient and the finite element results compare satisfactorily.

As a final remark, it may be noted that the natural frequencies of the (1,1) and (1,2) modes can be also obtained by modelling the plates as an equivalent pinned-pinned beam. According to the results of Chapters 2 and 3, the beam equation for

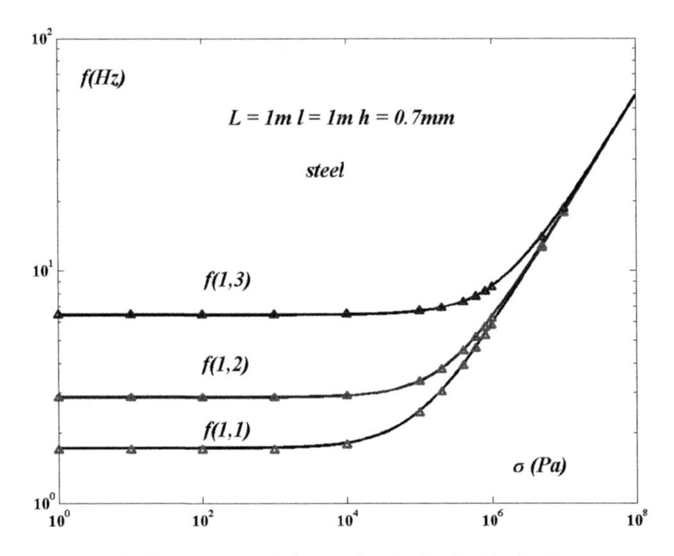

Figure 6.18. *Frequencies of the modes (1, 1), (1, 2), (1, 3) versus* σ_{xx}

the mode (1,1) is found to be:

$$\frac{Eh^3\ell}{12}\frac{\partial^4 Z}{\partial x^4} - \sigma h\ell \frac{\partial^2 Z}{\partial x^2} - \omega^2 \rho h \ell Z = 0; \quad Z(0) = Z(L) = 0;$$

$$\left.\frac{\partial^2 Z}{\partial x^2}\right|_{x=0} = \left.\frac{\partial^2 Z}{\partial x^2}\right|_{x=L} = 0$$

and the beam model for the mode (1,2) is:

$$-\left(GJ_T + \sigma \frac{h\ell^3}{12}\right)\frac{\partial^2 \psi_x}{\partial x^2} - \omega^2 \rho J \psi_x = 0 \quad \text{where } J_T = \frac{1}{3}\left(\frac{h^3\ell^3}{\ell^2 + h^2}\right)\frac{h^3\ell}{3}$$

$$\psi_x(0) = \psi_x(L) = 0$$

the prestress term is obtained by determining the prestress potential related to ψ_x. The transverse displacement induced by the rotation about the beam axis $y = 0$ is:

$$Z(x, y) = y\psi_x(x) \qquad -\ell/2 \le y \le \ell/2$$

Consequently, the prestress potential is:

$$\mathcal{E}_p = \sigma h \left(\frac{\partial \psi_x}{\partial x}\right)^2 \int_{-\ell/2}^{-\ell/2} y^2 dy \int_0^L \left(\sin\left(\frac{n\pi x}{L}\right)\right)^2 dx = \sigma \frac{h\ell^3 L}{24}\left(\frac{\partial \psi_x}{\partial x}\right)^2$$

6.3.1.5 Warping of a beam cross-section: membrane analogy

In Chapter 2 subsection 2.2.3.7, it was established that the warping function Φ of the torsion theory of Barré de Saint-Venant is the solution of the following boundary problem (cf. equations [2.61]):

$$\begin{aligned} &\Delta\Phi(y,z) = 0 \\ &\overrightarrow{\mathrm{grad}\Phi}\cdot\vec{n} = zn_y - yn_z = \vec{n}\times\vec{r}\cdot\vec{i} \quad \forall\,\vec{r}\in(\mathcal{C}) \end{aligned} \qquad [6.107]$$

The solution for a rectangular cross-section was established as the following series:

$$\Phi(y,z) = yz - \frac{8}{a}\sum_{n=0}^{\infty}\frac{(-1)^n}{k_n^3\cosh(k_n b/2)}\sin(k_n y)\sinh(k_n z) \qquad [6.108]$$

Here it is of interest to reconsider the problem by noticing the analogy between the problem [2.61] and that of the transverse displacement of a stretched membrane. Let us consider a rectangular plate stretched uniformly in the longitudinal and lateral direction by an in-plane force $F^{(0)}$ normal to the plate edges. Furthermore, the plate is assumed to be so thin that the flexure terms can be neglected in comparison with the prestressed terms. Thus the equation [6.92] reduces to:

$$\rho h\ddot{Z} - F^{(0)}\left(\frac{\partial^2 Z}{\partial x^2} + \frac{\partial^2 Z}{\partial y^2}\right) = 0 \qquad [6.109]$$

The problem analogue to [6.107] is expressed as:

$$\begin{cases} \left(\dfrac{\partial^2 Z}{\partial x^2} + \dfrac{\partial^2 Z}{\partial y^2}\right) = 0 \\ Z(y,0) = Z(0,z) = 0 \\ \left.\dfrac{\partial Z}{\partial x}\right|_{x=a/2} = y; \quad \left.\dfrac{\partial Z}{\partial y}\right|_{y=b/2} = -x \end{cases} \Leftrightarrow \begin{cases} \left(\dfrac{\partial^2 Z}{\partial x^2} + \dfrac{\partial^2 Z}{\partial y^2}\right) = y\delta\left(x-\dfrac{a}{2}\right) - x\delta\left(y-\dfrac{b}{2}\right) \\ \\ Z(y,0) = Z(0,z) = 0 \end{cases} \qquad [6.110]$$

where use is made of the central symmetry of the problem to deal with a quarter of plate of length $a/2$ and width $b/2$. The solution of [6.110] can be expressed as a modal expansion of the type:

$$Z(x,y) = \sum_n \sum_m q_{n,m} \varphi_{n,m}(x,y)$$

where $\varphi_{n,m}(x,y)$ are the mode shapes of the following modal problem:

$$\left(\frac{\partial^2 Z}{\partial x^2} + \frac{\partial^2 Z}{\partial y^2}\right) + \omega^2 Z = 0$$
$$Z(x,0) = Z(0,y) = 0$$
$$\left.\frac{\partial Z}{\partial x}\right|_{x=a/2} = 0; \qquad \left.\frac{\partial Z}{\partial y}\right|_{y=b/2} = 0$$

After some straightforward algebra, we arrive at:

$$\varphi_{n,m}(x,y) = \sin(k_n x)\sin(\eta k_m y)$$
$$q_{n,m} = \frac{16}{ab}(-1)^{n+m} \frac{(\eta k_m)^2 - (k_n)^2}{(\eta k_m k_n)^2 \left((\eta k_m)^2 + (k_n)^2\right)}$$
$$k_n = \frac{(2n+1)\pi}{a}, \quad n = 0,1,2\ldots; \quad k_m = \frac{(2m+1)\pi}{a}, \quad m = 0,1,2\ldots$$

where $\eta = a/b$ is the aspect ratio of the plate.

Numerical evaluation of this modal expansion shows that convergence rate is significantly slower than that of the series [6.108].

6.4. Curvilinear coordinates

6.4.1 *Bending and torsion displacements and strains*

As shown in the preceding chapter, the use of curvilinear coordinates is appropriate to deal with plates limited by curved edges of simple geometries. It is recalled that an orthonormal curvilinear coordinate system is defined by the coordinates α, β and the Lamé parameters g_α, g_β. Dropping the in-plane components in [6.5], the displacement of a point at a distance z from the midplane is given by:

$$\xi_\alpha = -\frac{z}{g_\alpha}\frac{\partial Z}{\partial \alpha} = -z\psi_\beta; \quad \xi_\beta = -\frac{z}{g_\beta}\frac{\partial Z}{\partial \beta} = -z\psi_\alpha$$
$$\text{where} \quad \psi_\alpha = \frac{1}{g_\beta}\frac{\partial Z}{\partial \beta}; \quad \psi_\beta = \frac{1}{g_\alpha}\frac{\partial Z}{\partial \alpha} \qquad [6.111]$$

Using the gradient vector expressed in curvilinear coordinates given in appendix A.3 (formula [A.3.13]), the strain tensor components are:

$$\chi_{\alpha\alpha} = -\left\{ \frac{1}{g_\alpha}\left(\frac{\partial}{\partial\alpha}\left(\frac{1}{g_\alpha}\frac{\partial Z}{\partial\alpha}\right) + \frac{1}{g_\beta^2}\frac{\partial Z}{\partial\beta}\frac{\partial g_\alpha}{\partial\beta}\right)\right\}$$

$$\chi_{\beta\beta} = -\left\{ \frac{1}{g_\beta}\left(\frac{\partial}{\partial\beta}\left(\frac{1}{g_\beta}\frac{\partial Z}{\partial\beta}\right) + \frac{1}{g_\alpha^2}\frac{\partial Z}{\partial\alpha}\frac{\partial g_\beta}{\partial\alpha}\right)\right\} \qquad [6.112]$$

$$\chi_{\alpha\beta} = -\frac{1}{2}\left\{ \frac{g_\alpha}{g_\beta}\left(\frac{\partial}{\partial\beta}\left(\frac{1}{g_\alpha^2}\frac{\partial Z}{\partial\alpha}\right)\right) + \frac{g_\beta}{g_\alpha}\left(\frac{\partial}{\partial\alpha}\left(\frac{1}{g_\beta^2}\frac{\partial Z}{\partial\beta}\right)\right)\right\}$$

6.4.2 *Equations of motion*

The variation of strain energy density per unit volume is:

$$\delta e_s = -(\mathcal{M}_{\alpha\alpha}\delta\chi_{\alpha\alpha} + \mathcal{M}_{\alpha\beta}\delta\chi_{\alpha\beta} + \mathcal{M}_{\beta\alpha}\delta\chi_{\beta\alpha} + \mathcal{M}_{\beta\beta}\delta\chi_{\beta\beta}) \qquad [6.113]$$

After some rather tedious but straightforward manipulations detailed in appendix A4, we arrive at:

$$\rho h\ddot{Z} - \left\{\left(\frac{\partial}{\partial\alpha}\left(\frac{1}{g_\alpha}\frac{\partial(g_\beta\mathcal{M}_{\alpha\alpha})}{\partial\alpha} + \frac{1}{g_\alpha^2}\frac{\partial\left(g_\alpha^2\mathcal{M}_{\alpha\beta}\right)}{\partial\beta} - \frac{\mathcal{M}_{\beta\beta}}{g_\alpha}\frac{\partial g_\beta}{\partial\alpha}\right)\right)\right\}$$

$$-\left\{\frac{\partial}{\partial\beta}\left(\frac{1}{g_\beta}\frac{\partial\left(g_\alpha\mathcal{M}_{\beta\beta}\right)}{\partial\beta} + \frac{1}{g_\beta^2}\frac{\partial\left(g_\beta^2\mathcal{M}_{\beta\alpha}\right)}{\partial\alpha} - \frac{\mathcal{M}_{\alpha\alpha}}{g_\beta}\frac{\partial g_\alpha}{\partial\beta}\right)\right\} \qquad [6.114]$$

$$= f_z^{(e)}(\alpha,\beta;t) + \frac{\partial\mathfrak{M}_\alpha^{(e)}}{g_\alpha\partial\alpha} + \frac{\partial\mathfrak{M}_\beta^{(e)}}{g_\beta\partial\beta}$$

NOTE. – *Vector form of the equation of motion*

Equation [6.114] may also be obtained starting from [6.25] or [6.26] and using the formulas given in appendix A3. Nevertheless, attention must be paid to the sign changes introduced by the rotation angles, as α and β definitions are not consistent with the convention of direct frame used in Cartesian coordinates. Equation [6.26] is written as:

$$\rho h\ddot{Z} - \frac{1}{g_\alpha g_\beta}\left(\frac{\partial(g_\beta Q_{\alpha z})}{\partial\alpha} + \frac{\partial(g_\alpha Q_{\beta z})}{\partial\beta}\right) = f_z^{(e)}(\alpha,\beta;t) + \frac{\partial\mathfrak{M}_\alpha^{(e)}}{g_\alpha\partial\alpha} + \frac{\partial\mathfrak{M}_\beta^{(e)}}{g_\beta\partial\beta} \qquad [6.115]$$

According to the calculation detailed in appendix A3, the shear forces are:

$$Q_{\alpha z} = \frac{1}{g_\alpha g_\beta}\left(\frac{\partial(g_\beta \mathcal{M}_{\alpha\alpha})}{\partial\alpha} + \frac{1}{g_\alpha}\frac{\partial\left(g_\alpha^2 \mathcal{M}_{\alpha\beta}\right)}{\partial\beta} - \mathcal{M}_{\beta\beta}\frac{\partial g_\beta}{\partial\alpha}\right)$$

$$Q_{\beta z} = \frac{1}{g_\alpha g_\beta}\left(\frac{\partial(g_\alpha \mathcal{M}_{\beta\beta})}{\partial\beta} + \frac{1}{g_\beta}\frac{\partial\left(g_\beta^2 \mathcal{M}_{\beta\alpha}\right)}{\partial\alpha} - \mathcal{M}_{\alpha\alpha}\frac{\partial g_\alpha}{\partial\beta}\right)$$

[6.116]

6.4.3 *Boundary conditions*

The boundary conditions are given again by gathering together the appropriate edge terms arising in the variational calculus. The following non homogeneous conditions are:

1. *Corner condition*

$$2\mathcal{M}_{\alpha\beta} = -F_{corner}$$ [6.117]

2. *Edge moments*

$$\mathcal{M}_{\alpha\alpha} = M_\beta^{(e)}; \quad \mathcal{M}_{\beta\beta} = \mathcal{M}_\alpha^{(e)}$$ [6.118]

3. *Effective Kirchhoff shear forces*

$$\text{boundary } \alpha = \text{cst:} \quad \mathcal{V}_{\alpha z} = Q_{\alpha z} + \frac{\partial \mathcal{M}_{\beta\alpha}}{g_\beta \partial\beta} = t_z^{(e)};$$

$$\text{boundary } \beta = \text{cst:} \quad \mathcal{V}_{\beta z} = Q_{\beta z} + \frac{\partial \mathcal{M}_{\alpha\beta}}{g_\alpha \partial\alpha} = t_z^{(e)}$$

[6.119]

Particularization of equations [6.115] to [6.119] to elasticity is straightforward as the strain-stress relationship [6.65] still holds in curvilinear coordinates, the subscripts x, y being replaced by the subscripts α, β as illustrated by an example in the next subsection.

6.4.4 *Circular plate loaded by a uniform pressure*

A uniform pressure p_0 is applied on the lower face of a circular plate of radius R and thickness h. The Lamé parameters are readily found to be $g_\alpha = 1, g_\beta = r$. In statics, equation [6.114] is written as:

$$-\left(\frac{1}{r}\frac{\partial(r\mathcal{M}_{rr})}{\partial r} + 2\frac{\partial}{r\partial\theta}\left(\frac{\partial \mathcal{M}_{\theta r}}{\partial r} + \frac{\mathcal{M}_{\theta r}}{r}\right) + \frac{\partial^2 \mathcal{M}_{\theta\theta}}{r^2\partial\theta^2} - \frac{\partial \mathcal{M}_{\theta\theta}}{r\partial r}\right) = p_0$$

[6.120]

Then [6.112] takes the form:

$$\chi_{rr} = -\frac{\partial^2 Z}{\partial r^2}; \quad \chi_{\theta\theta} = -\left\{\frac{1}{r}\frac{\partial}{\partial \theta}\left(\frac{1}{r}\frac{\partial Z}{\partial \theta}\right) + \frac{1}{r}\frac{\partial Z}{\partial r}\right\};$$

$$\chi_{\theta r} = -\frac{1}{2}\left\{\frac{1}{r}\frac{\partial}{\partial \theta}\left(\frac{\partial Z}{\partial r}\right) + r\frac{\partial}{\partial r}\left(\frac{1}{r^2}\frac{\partial Z}{\partial \theta}\right)\right\} \qquad [6.121]$$

The elastic moments are:

$$\mathcal{M}_{rr} = D(\chi_{rr} + \nu\chi_{\theta\theta}) = -D\left\{\frac{\partial^2 Z}{\partial r^2} + \nu\left(\frac{1}{r}\frac{\partial Z}{\partial r} + \frac{1}{r}\frac{\partial}{\partial \theta}\left(\frac{1}{r}\frac{\partial Z}{\partial \theta}\right)\right)\right\}$$

$$\mathcal{M}_{\theta\theta} = D(\chi_{\theta\theta} + \nu\chi_{rr}) = -D\left\{\frac{1}{r}\frac{\partial Z}{\partial r} + \frac{1}{r}\frac{\partial}{\partial \theta}\left(\frac{1}{r}\frac{\partial Z}{\partial \theta}\right) + \nu\frac{\partial^2 Z}{\partial r^2}\right\}$$

$$\mathcal{M}_{r\theta} = D(1-\nu)\chi_{r\theta} = -\frac{1}{2}D(1-\nu)\left\{\frac{1}{r}\frac{\partial}{\partial \theta}\left(\frac{\partial Z}{\partial r}\right) + r\frac{\partial}{\partial r}\left(\frac{1}{r^2}\frac{\partial Z}{\partial \theta}\right)\right\}$$

[6.122]

and the shear forces are:

$$Q_{rz} = \frac{\partial \mathcal{M}_{rr}}{\partial r} + \frac{1}{r}\frac{\partial \mathcal{M}_{\theta r}}{\partial \theta} + \frac{\mathcal{M}_{rr} - \mathcal{M}_{\theta\theta}}{r}; \quad Q_{rz} = \frac{\partial \mathcal{M}_{r\theta}}{\partial r} + \frac{1}{r}\frac{\partial \mathcal{M}_{\theta\theta}}{\partial \theta} + \frac{2\mathcal{M}_{r\theta}}{r}$$

[6.123]

Equation [6.115] takes a form similar to [6.68],

$$D\Delta[\Delta[Z]] = p_0 \Rightarrow \left(\frac{\partial^2}{\partial r^2} + \frac{1}{r}\frac{\partial}{\partial r} + \frac{1}{r^2}\frac{\partial^2}{\partial \theta^2}\right)\left(\frac{\partial^2 Z}{\partial r^2} + \frac{1}{r}\frac{\partial Z}{\partial r} + \frac{1}{r^2}\frac{\partial^2 Z}{\partial \theta^2}\right) = \frac{p_0}{D}$$

[6.124]

The problem is independent of θ, so the corresponding derivatives are null and [6.124] is reduced to the ordinary differential equation:

$$\left(\frac{d^2}{dr^2} + \frac{1}{r}\frac{d}{dr}\right)\left(\frac{d^2 Z}{dr^2} + \frac{1}{r}\frac{dZ}{dr}\right) = \frac{p_0}{D} \qquad [6.125]$$

It may be noticed that

$$\frac{d^2 Z}{dr^2} + \frac{1}{r}\frac{dZ}{dr} = \frac{1}{r}\left(r\frac{dZ}{dr}\right)$$

then [6.125] takes the form:

$$\frac{d}{dr}\left\{r\frac{d}{dr}\left(\frac{1}{r}\frac{d}{dr}\left(\frac{dZ}{dr}\right)\right)\right\} = \frac{p_0 r}{D} \qquad [6.126]$$

The successive integrations of [6.126] lead to:

$$\left\{r\frac{d}{dr}\left(\frac{1}{r}\frac{d}{dr}\left(r\frac{dZ}{dr}\right)\right)\right\} = \frac{p_0 r^2}{2D} + a$$

$$\frac{d}{dr}\left(\frac{1}{r}\frac{d}{dr}\left(r\frac{dZ}{dr}\right)\right) = \frac{p_0 r}{2D} + \frac{a}{r} \Rightarrow \frac{1}{r}\frac{d}{dr}\left(r\frac{dZ}{dr}\right) = \frac{p_0 r^2}{4D} + a\ln r + b \qquad [6.127]$$

$$\frac{d}{dr}\left(r\frac{dZ}{dr}\right) = \frac{p_0 r^3}{4D} + ar\ln r + br \Rightarrow \frac{dZ}{dr} = \frac{p_0 r^3}{16D} + a\left(\frac{r}{2}\ln r - \frac{r}{4}\right) + \frac{br}{2} + c$$

$$Z = \frac{p_0 r^4}{64D} + a\left(\frac{r^2}{4}\operatorname{lng} r - \frac{r^2}{4}\right) + \frac{br^2}{4} + c\ln r + d \qquad [6.128]$$

As the plate is assumed to be complete (no central hole) the coefficients of the logarithmic terms must vanish since otherwise they would induce stress or displacement singularities at $r = 0$. On the other hand, if the plate is assumed to be clamped along its outer edge, we arrive at the following solution:

$$Z(R) = 0 \Rightarrow Z(r) = \frac{p_0(R^2 - r^2)^2}{64D} \qquad [6.129]$$

The maximum displacement is obviously obtained at $r = 0$. The moment variation is parabolic:

$$\mathcal{M}_{rr} = \frac{p_0 R^2}{16}\left\{(1+\nu) - \left(\frac{r}{R}\right)^2(3+\nu)\right\};$$

$$\mathcal{M}_{\theta\theta} = \frac{p_0 R^2}{16}\left\{(1+\nu) - \left(\frac{r}{R}\right)^2(1+3\nu)\right\}; \quad \mathcal{M}_{r\theta} = 0$$

and finally the shear force is,

$$Q_{rz}(r) = -\frac{p_0 r}{2} \quad \Rightarrow \quad \int_0^{2\pi} Q_{rz}(R)R\,d\theta = -\pi R^2 p_0$$

as appropriate to fulfil the condition of global equilibrium.

It is also of interest to evaluate the influence of the boundary conditions on the plate deflection. For a hinged outer edge the result is:

$$Z(R) = 0; \quad \mathcal{M}_{rr}(R) = 0 \Rightarrow$$

$$Z\left(\frac{r}{R}\right) = \frac{p_0 R^4}{64D}\left(\left(\frac{r}{R}\right)^4 - \frac{2(3+\nu)}{1+\nu}\left(\frac{r}{R}\right)^2 + \frac{(5+\nu)}{1+\nu}\right)$$

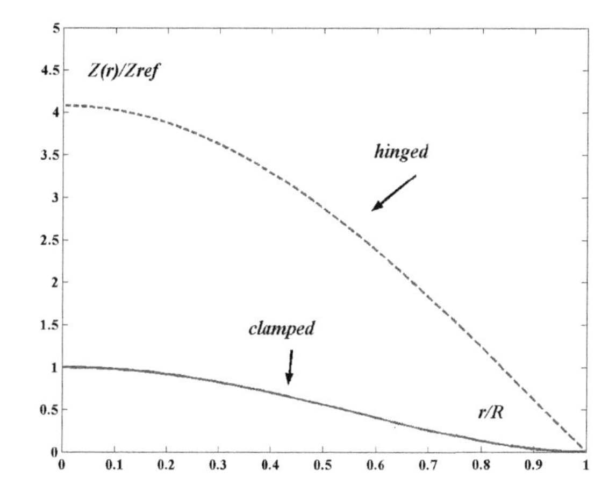

Figure 6.19. *Deflection of a circular hinged or clamped plate*

Figure 6.19 shows the deflection of a steel plate in the case of a clamped and hinged support conditions at the edge. The plot is made dimensionless by using the plate radius to scale r and the central deflection of the clamped plate to scale Z. This puts clearly in evidence that the hinged plate maximum deflection is about four times larger than in the clamped case. Finally, in colour plate 12 the deformation of a clamped circular plate is represented together with the isovalues of the radial and tangential moments $\mathcal{M}_{rr}, \mathcal{M}_{\theta\theta}$ and the edge reaction forces for two cases of loading (a) uniform pressure and (b) antisymmetrical pressure field.

Chapter 7

Arches and shells: string and membrane forces

Curved beams and shells are often preferable to straight beams and plates, as exemplified by most vegetable, animal and even mineral creations, as well as by a large variety of human architectural and engineering works. Amongst several other reasons – convenience, streamlining and aesthetics among them, the basic advantage of curved geometry is to provide the structural elements with a highly improved resistance to external loads for the same quantity of material required. This is because, in a curved structure, most of the transverse load is often balanced by tensile or compressive stresses. This can be conveniently emphasized by considering first simplified arch and shell models where bending and torsion terms are entirely discarded. They extend to the curved geometry, the models used to describe the longitudinal motion of straight beams and the in-plane motion of plates. Though the range of validity of such models is clearly limited to certain load conditions, it is appropriate to present and discuss them in a rather detailed manner, for sake of clarity at least, before embarking on the more elaborate models presented in the next chapter, which account for string or membrane stresses as well as for bending and torsion stresses.

7.1. Introduction: why curved structures?

7.1.1 *Resistance of beams to transverse loads*

As pointed out in Chapter 3 in connection with beam assemblies, there is a large difference between the flexure and the longitudinal stiffness scale factors of a straight beam element, the ratio being given by:

$$\frac{K_b}{K_\ell} = \left(\frac{EI}{L^3}\right)\left(\frac{L}{ES}\right) \approx \eta^{-2} \qquad [7.1]$$

where η stands for the slenderness ratio of the beam.

As a consequence, a straight beam resists much better a longitudinal than a transverse load. This can be further illustrated by considering the elastic response of a cantilevered straight beam loaded by a point force at its free end, as sketched in Figure 7.1. In traction, the displacement field is $X(x) = F_x x/ES$ and the local stress field $\sigma_{xx} = F_x/S$ is uniformly distributed on S and along the beam. In flexure, the displacement field is $Z(x) = F_z(3Lx^2 - x^3)/6EI$ and the local stress field is $\sigma_{xx} = F_z(L-x)z/I_y$. Then, its maximum magnitude is $|\sigma_{xx}|_{\max} = (F_z hL)/2I$, where h denotes the dimension of the beam cross-sections in the bending plane. Accordingly, for a load of same magnitude, we obtain:

$$\frac{Z(L)}{X(L)} \approx \eta^2 \qquad [7.2]$$

On the other hand, assuming that to prevent failure under excessive load, the magnitude of the local stress must be less than some threshold σ_a everywhere within

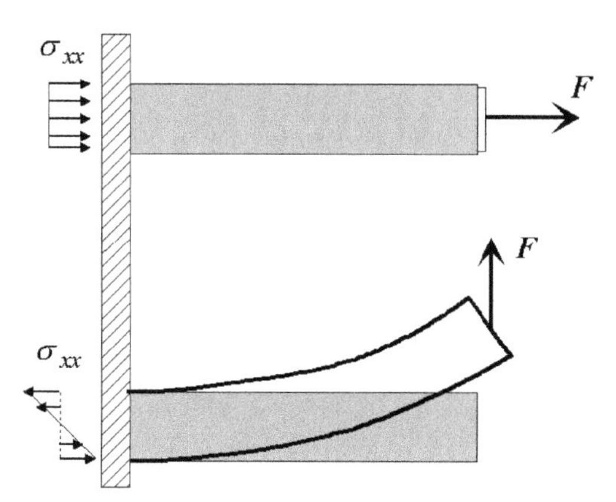

Figure 7.1. *Straight beam loaded according to either stretching or flexure mode*

the beam, the following admissible load ratio is found:

$$\frac{F_x}{F_z} \approx \eta \qquad [7.3]$$

Turning back to the roof truss of Figure 3.17, it can be viewed as a curved arch grossly discretized by using two straight beam elements. The deflection due to a transverse load applied at the top was found to be (see formula [3.113]):

$$Z_2(\alpha) = \frac{-F_z}{2(K_\ell(\sin\alpha)^2 + 3K_b(\cos\alpha)^2)}$$

As already pointed out, even a small tilt angle α, is sufficient to reduce drastically the magnitude of Z_2 (see Figure 3.18). Such a result holds qualitatively for any other loading condition.

Hence, it can be concluded that to increase the resistance of beams, it is highly preferable to let them work in traction-compression rather than in flexion and this can be achieved by using curved beams instead of straight beams. As illustrated in the next subsection, a similar conclusion arises concerning plates which are advantageously replaced by shells.

7.1.2 *Resistance of shells and plates to transverse loads*

Let us return to the example of a circular cylinder loaded by an external pressure, which has already been analysed in Chapter 5 subsection 5.4.4, as an equivalent annular plate subjected to a uniform radial force. The case of a cylinder of small thickness ratio $h/R \ll 1$ is particularly enlightening. It is recalled that the following results were obtained:

$$U = -\frac{pR^2}{Eh}; \quad \sigma_{rr} = -\frac{p(r-R)}{h}; \quad \sigma_{\theta\theta} = -\frac{pR}{h} \quad \Rightarrow \quad \frac{\sigma_{rr}}{\sigma_{\theta\theta}} \leq \frac{h}{R}$$

Since r varies between R and $R+h$, the radial stress is found to be less or equal to the external pressure and most of the radial load is balanced by the hoop stress $\sigma_{\theta\theta}$ which is obviously of membrane type. A convenient way to remove the curvature effect is to cut mentally the shell along a meridian line and to develop it as an equivalent rectangular plate, as sketched in Figure 7.2. Flexural deflection of the rectangular plate loaded by the same pressure field as the cylindrical shell and hinged along the lateral edges (AB) and $(A'B')$ can be determined with a sufficient degree of accuracy by assuming an approximate solution of the type:

$$Z(x) = Z_0 \sin\left(\frac{x}{2R}\right); \quad \text{where } x \text{ is the abscissa in the } (AA') \text{ direction.}$$

Applying the Galerkin procedure described, together with the Rayleigh–Ritz method in Chapter 5 – based here on the principle of minimum potential

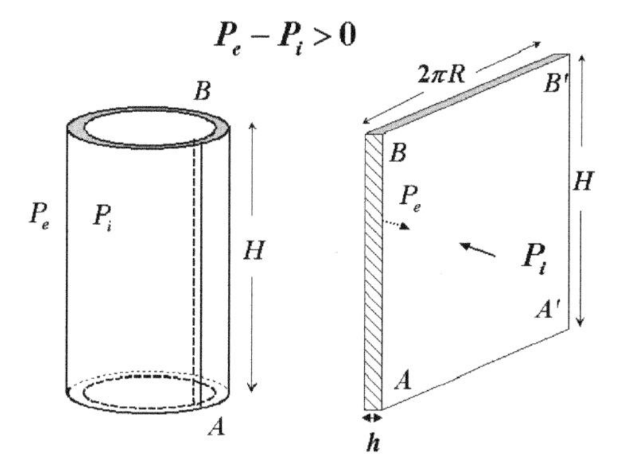

Figure 7.2. *Circular cylindrical shell developed as an equivalent plate*

energy – Z_0 is determined by solving the discretized equation:

$$Z_0 H D \left(\frac{1}{2R}\right)^4 \int_0^{2\pi R} \left(\sin\left(\frac{x}{2R}\right)\right)^2 dx$$
$$= pH \int_0^{2\pi R} \sin\left(\frac{x}{2R}\right) dx; \quad \text{where } D = \frac{Eh^3}{12(1-\nu^2)}$$

The result is

$$Z_0 = -\frac{4p(2R)^4}{\pi D} = \frac{768p(1-\nu^2)}{\pi E}\left(\frac{R}{h}\right)^3$$

So the ratio of the two displacement values is

$$\frac{U}{Z_0} \simeq \frac{\pi}{768}\left(\frac{h}{R}\right)^2$$

Hence, the advantage of using a curved instead of a plane structure to resist transverse loading is obvious.

The above examples clearly show that the basic beneficial effect of curvature is to redistribute the internal forces balancing the external transverse load, in such a way that string, or membrane, components prevail on the flexure components, as qualitatively depicted in Figure 7.3, which can stand either for a curved beam, or a shell slice. This property is of paramount importance in explaining why arches and shells are often used in structural applications – in which both stiffness and lightness are sought – and to motivate their study, to which the last two chapters of this book are devoted. The theoretical study of arches and shells has many common aspects with that of straight beams and plates respectively, but curvilinear metrics and coupling between membrane and bending effects give rise to some notable complications. For that reason, it is found appropriate to adopt a progressive

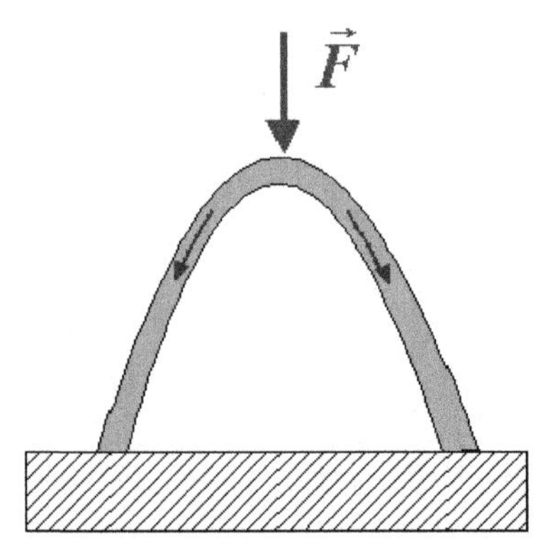

Figure 7.3. *Path of the forces resisting the external loading in a curved structure*

approach in which arches and shells are first modelled as strings and membranes respectively, the effects of bending and torsion being considered afterwards in the next and final chapter of this book.

7.2. Arches and circular rings

7.2.1 *Geometry and curvilinear metric tensor*

An arch is defined as a curved beam where curvature is in one plane only, in such a way that its neutral fibre is a plane curved line, as shown in Figure 7.4. To formulate the equilibrium equations, two coordinate systems are needed, one global and the other local, the last one being defined at any point $M(s)$ of the neutral line, where s is the curvilinear abscissa. The unit tangent vector to this line, denoted $\vec{t}(s)$, points towards increasing s values; $\vec{n}(s)$ is the unit normal vector, oriented in such a manner that the two following equations are satisfied:

$$\frac{d\vec{t}}{ds} = -\frac{\vec{n}}{R}; \quad \frac{d\vec{n}}{ds} = \frac{\vec{t}}{R} \qquad [7.4]$$

$R(s)$ is the positive curvature radius. $C(s)$ is the curvature centre and $\vec{n}(s)$ is oriented from C to M, that is, oriented towards the arch extrados. To define the metrics associated with curved structures, a set of orthonormal curvilinear coordinates is used. For further extension to the shell geometry, it is convenient to use the notations of Figure 7.5, according to which the position of a material point of the arch is defined as:

$$\vec{P}(\alpha, \varsigma) = \vec{r}(\alpha) + \varsigma\vec{n}(\alpha) \qquad [7.5]$$

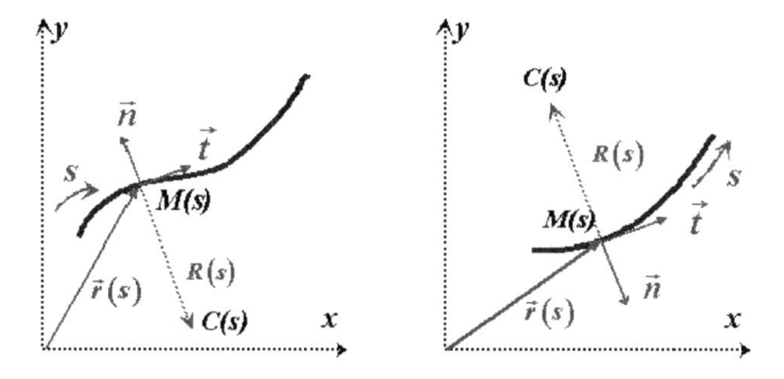

Figure 7.4. *Geometric description of an arch*

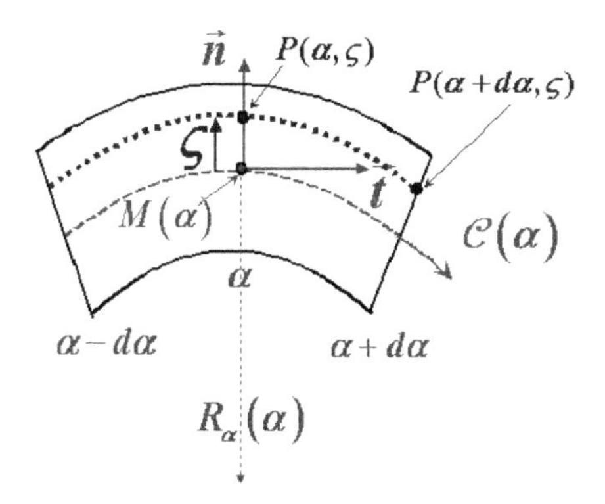

Figure 7.5. *Position of a current point P of the arch*

ς is the coordinate in the transverse direction $\vec{n}(\alpha)$ and the curvilinear coordinate α is such that:

$$s = s(\alpha) \quad \text{and} \quad ds = g_\alpha \, d\alpha \tag{7.6}$$

Then the relations defining the components G_α, G_ς of the metric tensor are:

$$d\vec{P} = d\vec{r} + d\varsigma\,\vec{n} + \varsigma\,d\vec{n} = ds\left(1 + \frac{\varsigma}{R}\right)\vec{t} + d\,\varsigma\vec{n} \;\Rightarrow$$

Whence,

$$(d\vec{P})^2 = ds^2\left(1 + \frac{\varsigma}{R}\right)^2 + (d\varsigma)^2$$

$$G_s = G_\alpha = g_\alpha\left(1 + \frac{\varsigma}{R}\right); \quad G_\varsigma = 1 \tag{7.7}$$

7.2.2 *Local and global displacements*

As this chapter is devoted to the string components exclusively, the study is restricted to the movements in the plane of the arch (i.e. the plane of the neutral line). The out of plane components will be considered in the next chapter. The local displacement of the current point is,

$$\vec{\xi}(\alpha,\varsigma) = \xi_\alpha \vec{t} + \xi_\varsigma \vec{n} \qquad [7.8]$$

The kinematic hypotheses of the Bernoulli–Euler model imply:

$$\xi_\alpha = X_\alpha - \varsigma\psi; \quad \xi_\varsigma = X_\varsigma \qquad [7.9]$$

X_α, X_ς are the components of the global displacement in the arch plane. ψ is the rotation angle around the vector $\vec{t} \times \vec{n}$.

7.2.3 *Local and global strains*

By substituting the relations [7.8] and [7.9] into the strain relationships [5.87] the following is obtained:

$$\begin{aligned}
\varepsilon_{\alpha\alpha} &= \frac{1}{g_\alpha}\left(1+\frac{\varsigma}{R}\right)^{-1}\left(\frac{\partial(X_\alpha - \varsigma\psi)}{\partial\alpha} + \frac{X_\varsigma}{R}\right)\\
\varepsilon_{\alpha\varsigma} &= \frac{1}{2}\left\{g_\alpha\left(1+\frac{\varsigma}{R}\right)\frac{\partial}{\partial\varsigma}\left((X_\alpha - \varsigma\psi)\frac{1}{g_\alpha}\left(1+\frac{\varsigma}{R}\right)^{-1}\right) + \frac{1}{g_\alpha}\left(1+\frac{\varsigma}{R}\right)^{-1}\frac{\partial X_\varsigma}{\partial\alpha}\right\}\\
\varepsilon_{\varsigma\varsigma} &= \frac{\partial X_\varsigma}{\partial\varsigma}
\end{aligned} \qquad [7.10]$$

These equations can be greatly simplified if the slenderness ratio of the arch is large enough to allow the approximation:

$$\frac{h}{R} \ll 1 \Rightarrow \left(1+\frac{\varsigma}{R}\right) \cong 1 \qquad [7.11]$$

On the other hand, keeping to the Bernoulli–Euler model, transverse shear strains $\varepsilon_{\alpha\varsigma}, \varepsilon_{\varsigma\varsigma}$ are assumed to vanish. So, the local strains reduce to:

$$\varepsilon_{\varsigma\varsigma} = \frac{\partial X_\varsigma}{\partial\varsigma} = 0$$

$$2\varepsilon_{\alpha\varsigma} = -\psi - \frac{X_\alpha}{R} + \frac{1}{g_\alpha}\frac{\partial X_\varsigma}{\partial\alpha} = 0 \Rightarrow \psi = \frac{1}{g_\alpha}\frac{\partial X_\varsigma}{\partial\alpha} - \frac{X_\alpha}{R}$$

$$\varepsilon_{\alpha\alpha} = \left(\frac{1}{g_\alpha}\frac{\partial X_\alpha}{\partial \alpha} + \frac{X_\varsigma}{R}\right) - \varsigma\frac{1}{g_\alpha}\frac{\partial \psi}{\partial \alpha}$$
$$= \left(\frac{1}{g_\alpha}\frac{\partial X_\alpha}{\partial \alpha} + \frac{X_\varsigma}{R}\right) + \varsigma\left(\frac{1}{g_\alpha}\frac{\partial}{\partial \alpha}\left(\frac{X_\alpha}{R}\right) - \frac{1}{g_\alpha}\frac{\partial}{\partial \alpha}\left(\frac{1}{g_\alpha}\frac{\partial X_\varsigma}{\partial \alpha}\right)\right) \quad [7.12]$$

NOTE. – *Formal expression of* $\varepsilon_{\alpha\varsigma}$ *and curvature terms*

To express $\varepsilon_{\alpha\varsigma}$ in terms of global quantities, derivation with respect to ς must be performed before using [7.11]. This induces an additional term to the rotation of the cross-section, proportional to the arch curvature, which of course is absent in the case of straight beams. An additional term proportional to the arch curvature is also present in the string and in the bending components of $\varepsilon_{\alpha\alpha}$. These results can be understood intuitively by considering that on the infinitesimal scale the neutral line can be assimilated to a circular arc of radius R. So, if a cross-section is displaced by the small quantity X_α along the neutral line, it rotates by the small amountX_α/R. On the other hand, the stretching strain of a fibre is the resultant of a tangential stretching and of a radial dilatation X_ζ which necessarily induces a variation of the fibre radius, from $R+\zeta$ to $R+\zeta+X_\zeta$.

7.2.4 *Equilibrium equations along the neutral line*

As in the case of slender straight beams, the rotatory inertia of the cross-sections is neglected. Then the variation of the kinetic energy density takes the form:

$$\delta[e_\kappa] = \rho S(\dot{X}_\alpha \delta \dot{X}_\alpha + \dot{X}_\varsigma \delta \dot{X}_\varsigma) \quad [7.13]$$

The variation of the strain energy density is:

$$\delta[e_s] = \mathcal{N}_{\alpha\alpha}\left(\left(\frac{1}{g_\alpha}\frac{\partial \delta X_\alpha}{\partial \alpha}\right) + \frac{\delta X_\varsigma}{R}\right) \quad [7.14]$$

Hamilton's principle leads to:

$$\int_{t_1}^{t_2} dt \int_{\alpha_1}^{\alpha_2} \left\{\rho S(\dot{X}_\alpha \delta \dot{X}_\alpha + \dot{X}_\varsigma \delta \dot{X}_\varsigma) - \mathcal{N}_{\alpha\alpha}\left(\frac{1}{g_\alpha}\delta\left(\frac{\partial X_\alpha}{\partial \alpha}\right) + \frac{\delta X_\varsigma}{R}\right)\right\} g_\alpha\, d\alpha$$
$$+ \int_{t_1}^{t_2} dt \int_{\alpha_1}^{\alpha_2} \left\{\left(F_\alpha^{(e)} \delta X_\alpha + F_\varsigma^{(e)} \delta X_\varsigma\right)\right\} g_\alpha\, d\alpha$$
$$+ \left[T_\alpha^{(e)} \delta X_\alpha + T_\varsigma^{(e)} \delta X_\varsigma\right]_{\alpha_1} + \left[T_\alpha^{(e)} \delta X_\alpha + T_\varsigma^{(e)} \delta X_\varsigma\right]_{\alpha_2} = 0 \quad [7.15]$$

where the external loading comprises the forces per unit arch length $F_\alpha^{(e)}$; $F_\varsigma^{(e)}$ and the end forces $T_\alpha^{(e)}$; $T_\varsigma^{(e)}$.

After one integration by parts the corresponding equilibrium equations are found to be:

$$\begin{aligned} \rho S\ddot{X}_\alpha - \frac{1}{g_\alpha}\frac{\partial \mathcal{N}_{\alpha\alpha}}{\partial \alpha} &= F_\alpha^{(e)} \\ \rho S\ddot{X}_\varsigma + \frac{\mathcal{N}_{\alpha\alpha}}{R} &= F_\varsigma^{(e)} \end{aligned} \qquad [7.16]$$

The first equation is very similar to the axial equation of a straight beam; the second can be easily understood by looking at Figure 7.6, which makes clear how the membrane stresses equilibrate the transverse external and inertia forces.

The boundary conditions are:

$$\begin{aligned} [(\mathcal{N}_{\alpha\alpha} - T_\alpha)\delta X_\alpha]_{\alpha_2} = 0; \quad [(\mathcal{N}_{\alpha\alpha} + T_\alpha)\delta X_\alpha]_{\alpha_1} = 0 \\ [T_\varsigma \delta X_\varsigma]_{\alpha_2} = [T_\varsigma \delta X_\varsigma]_{\alpha_1} = 0 \end{aligned} \qquad [7.17]$$

The first condition is similar to that which holds for straight beams; the second implies the disappearance of either the transverse force or of the transverse displacement at the ends of the arch. The reaction force induced at a fixed support must be balanced by bending shear stresses, as further discussed in Chapter 8, subsections 8.1.3 and 8.1.4. If the material is linear elastic, the longitudinal stress is:

$$\mathcal{N}_{\alpha\alpha} = ES\left(\frac{1}{g_\alpha}\frac{\partial X_\alpha}{\partial \alpha} + \frac{X_\varsigma}{R}\right) \qquad [7.18]$$

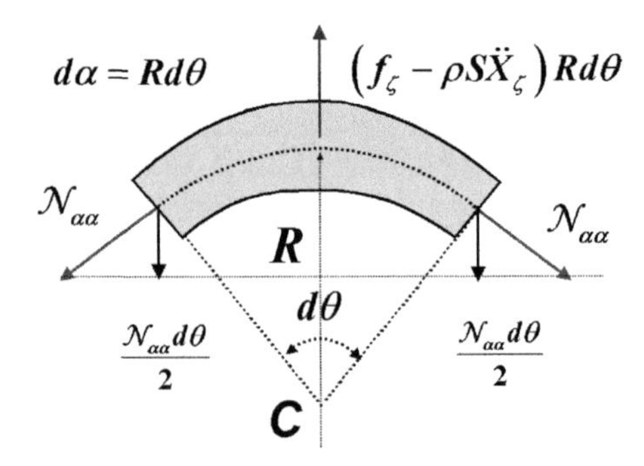

Figure 7.6. *Local transverse equilibrium of an arch*

By substituting [7.18] into [7.16], a system of two vibratory coupled equations is obtained:

$$\begin{aligned} \rho S\ddot{X}_\alpha - \frac{\partial}{g_\alpha \partial\alpha}\left(ES\left(\frac{\partial X_\alpha}{g_\alpha \partial\alpha} + \frac{X_\varsigma}{R}\right)\right) &= F_\alpha^{(e)} \\ \rho S\ddot{X}_\varsigma + \frac{ES}{R}\left(\frac{\partial X_\alpha}{g_\alpha \partial\alpha} + \frac{X_\varsigma}{R}\right) &= F_\varsigma^{(e)} \end{aligned} \qquad [7.19]$$

It is precisely through the coupling induced by the curved geometry that transverse forces are balanced by longitudinal stresses.

NOTE. – *Symmetry of the coupling operator*

The coupling operator must be self adjoint because the system is conservative. This operator is written in a matrix form as:

$$L_C[\,] = \begin{bmatrix} 0 & L_\varsigma \\ L_\alpha & 0 \end{bmatrix}$$

where

$$L_\varsigma = -\frac{\partial}{g_\alpha \partial\alpha}\left(\frac{ES[\,]}{R}\right) \quad \text{acting on } X_\varsigma$$

and

$$L_\alpha = +\frac{ES}{R}\left(\frac{\partial[\,]}{g_\alpha \partial\alpha}\right) \quad \text{acting on } X_\alpha$$

It can be verified that the energy functional of coupling is the same as its adjoint form:

$$\begin{aligned} \mathcal{E}(X_\alpha, X_\varsigma) &= \int_{s_1(\alpha_1)}^{s_2(\alpha_2)} [X_\alpha \quad X_\varsigma] \begin{bmatrix} 0 & L_\varsigma \\ L_\alpha & 0 \end{bmatrix} \begin{bmatrix} X_\alpha \\ X_\varsigma \end{bmatrix} ds \\ &= \int_{s_1(\alpha_1)}^{s_2(\alpha_2)} (X_\alpha L_\varsigma[X_\varsigma] + X_\varsigma L_\alpha[X_\alpha]) ds \\ &= \int_{s_1(\alpha_1)}^{s_2(\alpha_2)} (X_\varsigma L_\varsigma^{\#}[X_\alpha] + X_\alpha L_\alpha^{\#}[X_\varsigma]) ds \end{aligned}$$

7.2.5 *Application to a circular ring*

7.2.5.1 Simplifications inherent in axisymmetric structures

Due to the axial symmetry of the structure, a circular ring may be conveniently described by using polar coordinates, see Figure 7.7. The displacement field is defined by:

1. the tangential displacement

$$X_\alpha = V \quad \text{where } \alpha = \theta \text{ and } s = R\theta \tag{7.20}$$

2. the transverse displacement $X_\varsigma = U$.

The notation U, V for the radial and the tangential displacements respectively is used for any axisymmetric structure. The equations [7.19] are written as:

$$\begin{aligned} \rho S\ddot{V} - \frac{ES}{R^2}\left(\frac{\partial^2 V}{\partial \theta^2} + \frac{\partial U}{\partial \theta}\right) &= F_\theta^{(e)} \\ \rho S\ddot{U} + \frac{ES}{R^2}\left(\frac{\partial V}{\partial \theta} + U\right) &= F_r^{(e)} \end{aligned} \tag{7.21}$$

An immediate consequence of axisymmetry is that U and V may be expanded in a Fourier series:

$$\begin{aligned} U(\theta) &= \sum_{n=0}^{\infty} U_n^{(c)} \cos(n\theta) + U_n^{(s)} \sin(n\theta) \\ V(\theta) &= \sum_{n=0}^{\infty} V_n^{(c)} \cos(n\theta) + V_n^{(s)} \sin(n\theta) \end{aligned} \tag{7.22}$$

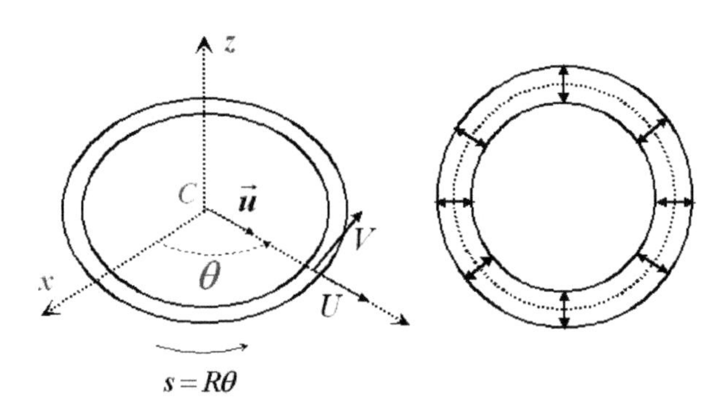

Figure 7.7. *Coordinates associated with a circular ring and breathing mode*

The Fourier coefficients can be identified term to term, once the loading is also expanded in a Fourier series. Though the procedure is mathematically valid whatever the order n of the harmonic component may be, its physical relevance is restricted here to $n = 0$ and eventually $n = 1$, because in most cases loads distributed according to higher harmonics induce bending effects which are not included in the model [7.21].

7.2.5.2 *Breathing mode of vibration of a circular ring*

Using [7.21] for $n = 0$, the modal equations are:

$$\begin{aligned} -\omega^2 \rho S V_0 &= 0 \\ -\omega^2 \rho S U_0 + \frac{ES}{R^2} U_0 &= 0 \end{aligned} \qquad [7.23]$$

There are two solutions. The first corresponds to $\omega = 0$ and $U_0 = 0$; it is immediately recognized as a rigid mode according to which the ring is revolving freely at constant and arbitrary speed around its axis. The second solution is such that $\omega \neq 0$, $V_0 = 0$. It is termed a *breathing mode* according to which the ring pulses radially at a resonance frequency equal to:

$$f_0 = \frac{1}{2\pi R}\sqrt{\frac{E}{\rho}} \qquad [7.24]$$

This natural frequency is the same as that of the first longitudinal mode of a straight beam of length $2\pi R$, either non-supported, or fixed at both ends.

7.2.5.3 *Translational modes of vibration*

The same analysis conducted with the value $n = 1$ leads to two rigid body modes which are free translations of the ring along the axes $\theta = 0$ and $\theta = \pi/2$ (see Figure 7.8).

$$\begin{aligned} &\omega_1 = 0; \quad U_1 = \cos\theta; \quad V_1 = -\sin\theta \\ &\omega_2 = 0; \quad U_2 = \sin\theta; \quad V_2 = \cos\theta \end{aligned}$$

In fact, the translation can be in any direction as the θ origin may be chosen arbitrarily; which gives rise to any admissible mode shape superposing the two orthogonal families identified just above. Finally, for $n > 1$ the mode shapes induce bending and so will be studied in the following chapter.

7.2.5.4 Cable stressed by its own weight

As already indicated, an arch modelled by equations [7.19] is equivalent to a string or a cable. Such structures must be set in tension to support loads comprising a transverse component, as illustrated in the following example where a semicircular arch is loaded by its own weight, as sketched in Figure 7.9. It is rather obvious, at least from common experience, that stable equilibrium with the apex in upper position is impossible if bending effects are not taken into account. The full problem shall be analysed in the next chapter, subsection 8.1.4.3.

Keeping here to the string model, the equilibrium equations of the arch are written as:

$$-\frac{ES}{R^2}\left(\frac{\partial^2 V}{\partial \theta^2}+\frac{\partial U}{\partial \theta}\right)=-\rho g S \sin\theta$$

$$+\frac{ES}{R^2}\left(\frac{\partial V}{\partial \theta}+U\right)=-\rho g S \cos\theta$$

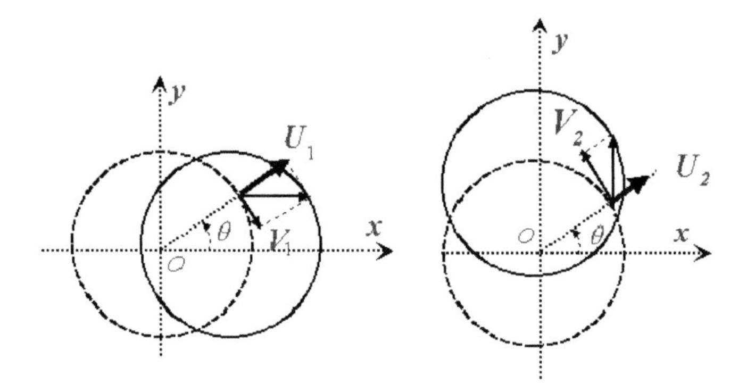

Figure 7.8. *Translational modes in the plane of the ring*

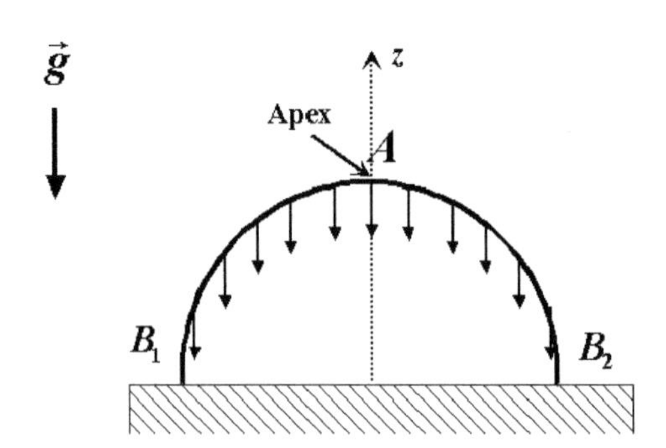

Figure 7.9. *Arch loaded by its own weight*

These two equations are found to be related to each other since the first equation is merely the derivative with respect to θ of the second. Moreover, because of the loading, they are found to be incompatible. The stable equilibrium position is a catenary curve with the apex in low position; for the solution of this problem, see Appendix A.5. Interestingly enough, the situation is however completely different for shells which can support their own weight even if the study is restricted to membrane effects, as further analysed in subsection 7.3.7.

7.3. Shells

7.3.1 *Geometry and curvilinear metrics*

Like plates, shells are three dimensional structures in which one dimension – the thickness – is very small in comparison with the two others. The midsurface is curved and a point on it is defined by a set of two curvilinear orthonormal coordinates α and β, see Figure 7.10. The local coordinate system is defined by the unit tangent vectors $\vec{t}_\alpha(\alpha,\beta)$ and $\vec{t}_\beta(\alpha,\beta)$. $\vec{t}_\alpha(\alpha,\beta)$ is tangent to a $(\mathcal{C}_\alpha)$ curve drawn at β constant and $\vec{t}_\beta(\alpha,\beta)$ is tangent to the orthogonal curve $(\mathcal{C}_\beta)$ drawn at α constant. The unit normal vector is:

$$\vec{n}(\alpha,\beta) = \vec{t}_\alpha \times \vec{t}_\beta \qquad [7.25]$$

The derivatives of these unit vectors are:

$$\begin{aligned} \frac{d\vec{t}_\alpha}{ds_\alpha} &= \frac{d\vec{t}_\alpha}{g_\alpha\, d\alpha} = -\frac{\vec{n}}{R_\alpha}; \quad \frac{d\vec{n}}{ds_\alpha} = \frac{d\vec{n}}{g_\alpha\, d\alpha} = \frac{\vec{t}_\alpha}{R_\alpha} \\ \frac{d\vec{t}_\beta}{ds_\beta} &= \frac{d\vec{t}_\beta}{g_\beta\, d\beta} = -\frac{\vec{n}}{R_\beta}; \quad \frac{d\vec{n}}{ds_\beta} = \frac{d\vec{n}}{g_\beta\, d\beta} = \frac{\vec{t}_\beta}{R_\beta} \end{aligned} \qquad [7.26]$$

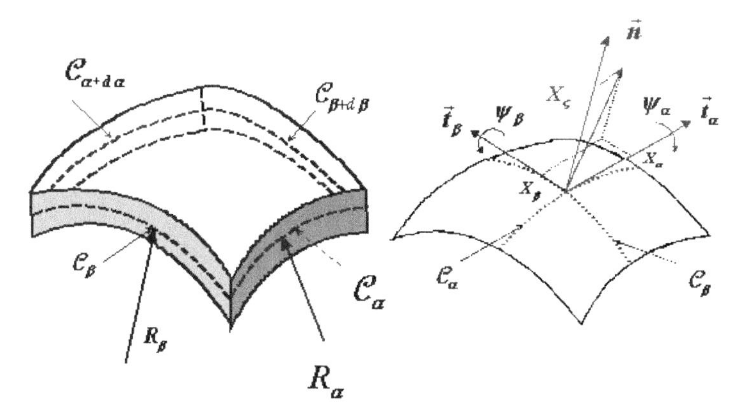

Figure 7.10. *Shell element defined by the parameters α and β*

R_α and R_β are the curvature radii of (C_α) and (C_β) respectively. The coefficients g_α, g_β define the metrics in the tangent plane of a point $P_0(\alpha,\beta)$ lying on the midsurface. The position of a point P located at the distance ζ of the midsurface is specified by the vector:

$$\vec{P}(\alpha,\beta,\varsigma) = \vec{r}(\alpha,\beta) + \varsigma\vec{n}(\alpha,\beta) \tag{7.27}$$

The components of the metric tensor at P are denoted $G_\alpha; G_\beta; G_\zeta$. They are determined as follows:

$$d\vec{P} = d\vec{r} + d\varsigma\vec{n} + \varsigma d\vec{n} \;\Rightarrow\; (d\vec{P})^2 = (d\vec{r})^2 + (d\varsigma)^2 + \varsigma^2(d\vec{n})^2 + 2\varsigma d\vec{r}\cdot d\vec{n} \tag{7.28}$$

$d\vec{r}$ is in the tangent plane so

$$(d\vec{r})^2 = g_\alpha^2(d\alpha)^2 + g_\beta^2(d\beta)^2 \tag{7.29}$$

The determination of $(d\vec{n})^2$ is obtained from the differentiation of $\vec{n}$.

$$d\vec{n} = \frac{\partial\vec{n}}{\partial\alpha}d\alpha + \frac{\partial\vec{n}}{\partial\beta}d\beta \;\Rightarrow\; (d\vec{n})^2 = \left(\frac{\partial\vec{n}}{\partial\alpha}\right)^2(d\alpha)^2 + \left(\frac{\partial\vec{n}}{\partial\beta}\right)^2(d\beta)^2 \tag{7.30}$$

or

$$d\vec{n}\cdot d\vec{n} = \frac{g_\alpha^2}{R_\alpha^2}(d\alpha)^2 + \frac{g_\beta^2}{R_\beta^2}(d\beta)^2$$

The dot product $d\vec{r}\cdot d\vec{n}$ is expressed as:

$$\begin{aligned} d\vec{r}\cdot d\vec{n} &= \left(\frac{\partial\vec{r}}{\partial\alpha}d\alpha + \frac{\partial\vec{r}}{\partial\beta}d\beta\right)\cdot\left(\frac{\partial\vec{n}}{\partial\alpha}d\alpha + \frac{\partial\vec{n}}{\partial\beta}d\beta\right) \\ &= \frac{\partial\vec{r}}{\partial\alpha}\cdot\frac{\partial\vec{n}}{\partial\alpha}(d\alpha)^2 + \frac{\partial\vec{r}}{\partial\beta}\cdot\frac{\partial\vec{n}}{\partial\beta}(d\beta)^2 + \left(\frac{\partial\vec{r}}{\partial\alpha}\cdot\frac{\partial\vec{n}}{\partial\beta} + \frac{\partial\vec{r}}{\partial\beta}\cdot\frac{\partial\vec{n}}{\partial\alpha}\right)d\alpha\, d\beta \end{aligned} \tag{7.31}$$

With [7.25] and [7.26] it follows that:

$$d\vec{r}\cdot d\vec{n} = \frac{g_\alpha^2}{R_\alpha}(d\alpha)^2 + \frac{g_\beta^2}{R_\beta}(d\beta)^2$$

The results [7.29] to [7.31] being substituted into [7.28], we arrive at:

$$(d\vec{P})^2 = \left(g_\alpha\left(1 + \frac{\varsigma}{R_\alpha}\right)\right)^2(d\alpha)^2 + \left(g_\beta\left(1 + \frac{\varsigma}{R_\beta}\right)\right)^2(d\beta)^2 + (d\varsigma)^2 \tag{7.32}$$

So, the coefficients of the curved metrics are identified as:

$$G_\alpha = g_\alpha \left(1 + \frac{\varsigma}{R_\alpha}\right); \quad G_\beta = g_\beta \left(1 + \frac{\varsigma}{R_\beta}\right); \quad G_\varsigma = 1 \qquad [7.33]$$

7.3.2 *Local and global displacements*

The local displacement is defined by:

$$\vec{\xi}(\alpha, \beta, \varsigma) = \xi_\alpha \vec{t}_\alpha + \xi_\beta \vec{t}_\beta + \xi_\varsigma \vec{n} \qquad [7.34]$$

The kinematic approximations of the Kirchhoff–Love model lead to:

$$\xi_\alpha = X_\alpha - \varsigma \psi_\beta; \quad \xi_\beta = X_\beta - \varsigma \psi_\alpha; \quad \xi_\varsigma = X_\varsigma \qquad [7.35]$$

where $X_\alpha, X_\beta, X_\varsigma, \psi_\alpha, \psi_\beta$ are the components of the global displacements depending on α and β (see Figure 7.10). In [7.34] ψ_α is the rotation angle around $\vec{t}_\alpha$ and ψ_β the angle around $\vec{t}_\beta$.

7.3.3 *Local and global strains*

The tensor of the small strains is obtained in the same way as for the arches. The calculations are rather tedious. Their results are given here for thin shells defined by the inequalities $h \ll R_\alpha, R_\beta$. The membrane components are:

$$\begin{aligned}
\eta_{\alpha\alpha} &= \frac{1}{g_\alpha}\frac{\partial X_\alpha}{\partial \alpha} + \frac{X_\beta}{g_\alpha g_\beta}\frac{\partial g_\alpha}{\partial \beta} + \frac{X_\varsigma}{R_\alpha} \\
\eta_{\beta\beta} &= \frac{1}{g_\beta}\frac{\partial X_\beta}{\partial \beta} + \frac{X_\alpha}{g_\alpha g_\beta}\frac{\partial g_\beta}{\partial \alpha} + \frac{X_\varsigma}{R_\beta} \\
2\eta_{\alpha\beta} &= \frac{g_\beta}{g_\alpha}\frac{\partial}{\partial \alpha}\left(\frac{X_\beta}{g_\beta}\right) + \frac{g_\alpha}{g_\beta}\frac{\partial}{\partial \beta}\left(\frac{X_\alpha}{g_\alpha}\right)
\end{aligned} \qquad [7.36]$$

[7.36] may be compared with [5.87] which holds for plates; the differences in normal strains originate from the curvatures, which extend the result already obtained for an arch to the 2D case.

7.3.4 *Global membrane stresses*

Global membrane stresses are sketched in Figure 7.11. As in the case of plates, $\mathcal{N}_{\alpha\alpha}, \mathcal{N}_{\beta\beta}$ are the normal and $\mathcal{N}_{\beta\alpha}, \mathcal{N}_{\alpha\beta}$ the 'in-tangential plane' shear components. Again, the dimension of these global quantities is a force per unit length.

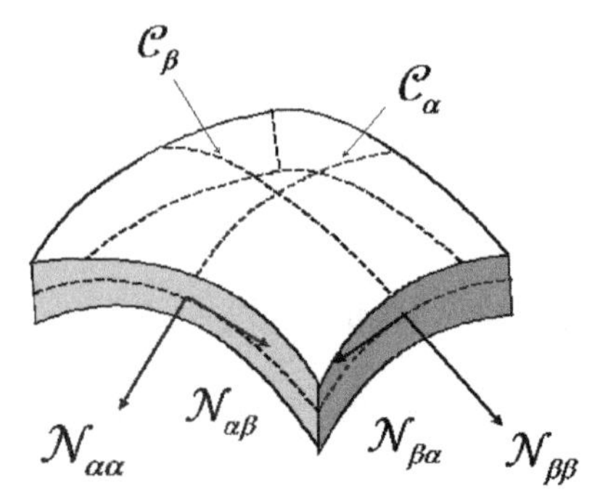

Figure 7.11. *Global membrane stresses*

7.3.5 *Membrane equilibrium*

The variation of the kinetic energy density is:

$$\delta[e_\kappa] = \rho h(\dot{X}_\alpha \delta\dot{X}_\alpha + \dot{X}_\beta \delta\dot{X}_\beta + \dot{X}_\varsigma \delta\dot{X}_\varsigma) \qquad [7.37]$$

and the variation of the strain energy density is:

$$\begin{aligned}\delta[e_s] = {} & \mathcal{N}_{\alpha\alpha}\left(\frac{1}{g_\alpha}\frac{\partial \delta X_\alpha}{\partial \alpha} + \frac{\delta X_\beta}{g_\alpha g_\beta}\frac{\partial g_\alpha}{\partial \beta} + \frac{\delta X_\varsigma}{R_\alpha}\right) \\ & + \mathcal{N}_{\beta\beta}\left(\frac{1}{g_\beta}\frac{\partial \delta X_\beta}{\partial \beta} + \frac{\delta X_\alpha}{g_\alpha g_\beta}\frac{\partial g_\beta}{\partial \alpha} + \frac{\delta X_\varsigma}{R_\beta}\right) \\ & + \mathcal{N}_{\alpha\beta}\left(\frac{g_\beta}{g_\alpha}\frac{\partial}{\partial \alpha}\left(\frac{\delta X_\beta}{g_\beta}\right) + \frac{g_\alpha}{g_\beta}\frac{\partial}{\partial \beta}\left(\frac{\delta X_\alpha}{g_\alpha}\right)\right)\end{aligned} \qquad [7.38]$$

It follows the equilibrium equations:

$$\begin{aligned}& \rho h\ddot{X}_\alpha - \frac{1}{g_\alpha g_\beta}\left\{\frac{\partial g_\beta \mathcal{N}_{\alpha\alpha}}{\partial \alpha} - \mathcal{N}_{\beta\beta}\frac{\partial g_\beta}{\partial \alpha} + \frac{1}{g_\alpha}\frac{\partial g_\alpha^2 \mathcal{N}_{\beta\alpha}}{\partial \beta}\right\} = f_\alpha^{(e)}(\alpha,\beta;t) \\ & \rho h\ddot{X}_\beta - \frac{1}{g_\alpha g_\beta}\left\{\frac{\partial g_\alpha \mathcal{N}_{\beta\beta}}{\partial \beta} - \mathcal{N}_{\alpha\alpha}\frac{\partial g_\alpha}{\partial \beta} + \frac{1}{g_\beta}\frac{\partial g_\beta^2 \mathcal{N}_{\alpha\beta}}{\partial \alpha}\right\} = f_\beta^{(e)}(\alpha,\beta;t) \\ & \rho h\ddot{X}_\varsigma + \left(\frac{\mathcal{N}_{\alpha\alpha}}{R_\alpha} + \frac{\mathcal{N}_{\beta\beta}}{R_\beta}\right) = f_\varsigma^{(e)}(\alpha,\beta;t)\end{aligned} \qquad [7.39]$$

NOTE. – *Shells and plates*

As the following relation holds:

$$\frac{\partial g_\alpha^2 \mathcal{N}_{\beta\alpha}}{g_\alpha \partial\beta} = g_\alpha \frac{\partial \mathcal{N}_{\beta\alpha}}{\partial\beta} + 2\frac{\partial g_\alpha}{\partial\beta}\mathcal{N}_{\beta\alpha} = \frac{\partial g_\alpha \mathcal{N}_{\beta\alpha}}{\partial\beta} + \frac{\partial g_\alpha}{\partial\beta}\mathcal{N}_{\beta\alpha}$$

It is noticed that the two first equations in [7.39] are the same as those which hold for plates. Thus the curvature effect inherent in the shell geometry is entirely accounted for by the third equation which links the transverse and the membrane effects. It is also worth noticing that the system [7.39] may be understood as a simple extension to the 2D case, of the 'string' equations already established for arches.

In the absence of any edge loading, the boundary conditions are:

$$\begin{aligned} &\text{Curve } (\mathcal{C}_\alpha): \quad \mathcal{N}_{\beta\beta} = 0, \text{ or } X_\beta = 0; \quad \mathcal{N}_{\beta\alpha} = 0, \text{ or } X_\alpha \\ &\text{Curve } (\mathcal{C}_\beta): \quad \mathcal{N}_{\alpha\alpha} = 0, \text{ or } X_\alpha = 0; \quad \mathcal{N}_{\alpha\beta} = 0, \text{ or } X_\beta \end{aligned} \qquad [7.40]$$

If the material is linear, isotropic and elastic,

$$\begin{aligned} &\mathcal{N}_{\alpha\alpha} = \frac{Eh}{1-\nu^2}(\eta_{\alpha\alpha} + \nu\eta_{\beta\beta}); \quad \mathcal{N}_{\beta\beta} = \frac{Eh}{1-\nu^2}(\eta_{\beta\beta} + \nu\eta_{\alpha\alpha}) \\ &\mathcal{N}_{\alpha\beta} = \mathcal{N}_{\beta\alpha} = Gh\eta_{\alpha\beta} = Gh\eta_{\beta\alpha} \end{aligned} \qquad [7.41]$$

From [7.39] and [7.41] three coupled equilibrium equations expressed in terms of displacements and their derivatives with respect to α and β can be deduced; their expressions are cumbersome and do not need to be given here.

7.3.6 *Axisymmetric shells*

7.3.6.1 Geometry and metric tensor

The midsurface is described by the parametric equations:

$$x(\alpha,\theta) = r(\alpha)\cos\theta; \quad y(\alpha,\theta) = r(\alpha)\sin\theta; \quad z(\alpha) \qquad [7.42]$$

x, y, z are the Cartesian coordinates and r, z, θ the cylindrical ones. If θ is given, the corresponding $r(z)$ curve is called the meridian or generatrix line, see Figure 7.12. The metrics is defined by:

$$d\vec{s} = \left(\frac{\partial x}{\partial\alpha}\vec{i} + \frac{\partial y}{\partial\alpha}\vec{j} + \frac{\partial z}{\partial\alpha}\vec{k}\right)d\alpha + \left(\frac{\partial x}{\partial\theta}\vec{i} + \frac{\partial y}{\partial\theta}\vec{j}\right)d\theta = \left(\frac{\partial r}{\partial\alpha}\vec{u} + \frac{\partial z}{\partial\alpha}\vec{k}\right)d\alpha + r\vec{u}_1\,d\theta$$

where $\vec{u}$ is the unit polar vector. Its derivative $\vec{u}_1$ is the unit azimuthal or circumferential vector.

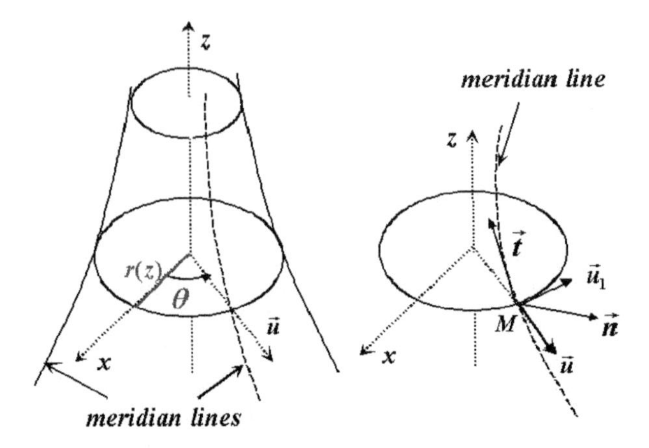

Figure 7.12. *Midsurface of an axisymmetric shell*

From [7.42] we get:

$$(d\vec{s})^2 = \left(\left(\frac{\partial r}{\partial \alpha}\right)^2 + \left(\frac{\partial z}{\partial \alpha}\right)^2\right)(d\alpha)^2 + r^2(d\theta)^2 \qquad [7.43]$$

Thus, the components of the metric tensor are found to be:

$$g_{\alpha\alpha} = \left(\left(\frac{\partial r}{\partial \alpha}\right)^2 + \left(\frac{\partial z}{\partial \alpha}\right)^2\right)^{1/2} \quad ; \quad g_{\theta\theta} = r \qquad [7.44]$$

A local coordinate system at a current point M can be also used. The unit vectors are the vector $\vec{t}$ tangential at M to the meridian line pointing towards $\alpha > 0$ and the normal vector $\vec{n} = \vec{t} \times \vec{u}_1$. Their components can be expressed in the cylindrical frame $\vec{u}, \vec{u}_1, \vec{k}$ as:

$$\vec{t} = \frac{d\vec{s}_m}{ds_m} = \frac{1}{g_{\alpha\alpha}}\left(\frac{\partial r}{\partial \alpha}\vec{u} + \frac{\partial z}{\partial \alpha}\vec{k}\right)$$
$$\vec{n} = \frac{1}{g_{\alpha\alpha}}\left(\frac{\partial r}{\partial \alpha}\vec{u} + \frac{\partial z}{\partial \alpha}\vec{k}\right) \times \vec{u}_1 = \frac{1}{g_{\alpha\alpha}}\left(\frac{\partial r}{\partial \alpha}\vec{k} - \frac{\partial z}{\partial \alpha}\vec{u}\right) \qquad [7.45]$$

ds_m is the length of an infinitesimal arc of the meridian; the second relation asserts that $\vec{n}$ lies in the meridian plane.

7.3.6.2 Curvature tensor

Let us consider a curve (C_l) obtained by cutting the midsurface by a plane which contains the normal vector $\vec{n}$ at a current point (M). An infinitesimal arc of (C_l) is

$d\vec{s} = \vec{\tau}\, ds$. Its curvature is given by:

$$\chi_1 = \frac{d\vec{\tau}}{ds} \cdot \vec{n} = -\frac{1}{R_1} \cdot \left(\frac{d\vec{\tau}}{ds} = -\frac{\vec{n}}{R_1} \right) \qquad [7.46]$$

as $\vec{\tau} \cdot \vec{n} = 0$, another equivalent form of [7.46] is:

$$\chi_1 = \frac{d\vec{\tau}}{ds} \cdot \vec{n} = -\vec{\tau} \cdot \frac{d\vec{n}}{ds} = -\frac{d\vec{s} \cdot d\vec{n}}{ds^2} \qquad [7.47]$$

or $d\vec{s} \cdot d\vec{n} = \chi_1 ds^2 = \chi_1((g_{\alpha\alpha} d\alpha)^2 + (g_{\theta\theta} d\theta)^2)$.

This quadratic form leads naturally to the following definition of the curvature tensor:

$$\begin{aligned} d\vec{s} \cdot d\vec{n} &= \begin{bmatrix} g_{\alpha\alpha}\, d\alpha & g_{\theta\theta}\, d\theta \end{bmatrix} \begin{bmatrix} \chi_{\alpha\alpha} & \chi_{\alpha\theta} \\ \chi_{\theta\alpha} & \chi_{\theta\theta} \end{bmatrix} \begin{bmatrix} g_{\alpha\alpha}\, d\alpha \\ g_{\theta\theta}\, d\theta \end{bmatrix} \\ &= \begin{bmatrix} ds_\alpha & ds_\theta \end{bmatrix} \begin{bmatrix} \chi_{\alpha\alpha} & \chi_{\alpha\theta} \\ \chi_{\theta\alpha} & \chi_{\theta\theta} \end{bmatrix} \begin{bmatrix} ds_\alpha \\ ds_\theta \end{bmatrix} \end{aligned} \qquad [7.48]$$

The components are obtained by expanding the scalar product $d\vec{s} \cdot d\vec{n}$, as follows:

$$\vec{n} = \frac{\partial r}{\partial s_\alpha} \vec{k} - \frac{\partial z}{\partial s_\alpha} \vec{u} \quad (ds_\alpha = g_{\alpha\alpha}\, d\alpha)$$

then,

$$d\vec{n} = \left(\frac{\partial^2 r}{\partial s_\alpha^2} \vec{k} - \frac{\partial^2 z}{\partial s_\alpha^2} \vec{u} \right) ds_\alpha - \frac{\partial z}{\partial s_\alpha} \vec{u}_1\, d\theta$$

$$d\vec{s} = \left(\frac{\partial r}{\partial s_\alpha} \vec{u} + \frac{\partial z}{\partial s_\alpha} \vec{k} \right) ds_\alpha + \vec{u}_1 r\, d\theta$$

and finally,

$$\begin{aligned} \chi_{\alpha\alpha} &= \frac{1}{R_\alpha} = \frac{\partial^2 z}{\partial s_\alpha^2} \frac{\partial r}{\partial s_\alpha} - \frac{\partial^2 r}{\partial s_\alpha^2} \frac{\partial z}{\partial s_\alpha} \\ \chi_{\alpha\theta} &= \chi_{\theta\alpha} = 0 \\ \chi_{\theta\theta} &= \frac{1}{R_\theta} = \frac{1}{r} \frac{\partial z}{\partial s_\alpha} \end{aligned} \qquad [7.49]$$

As expected, the curvature tensor is diagonal. $\chi_{\alpha\alpha}$ is the curvature of the meridian line. $\chi_{\theta\theta}$ is the curvature of the curve obtained by cutting the surface by the plane which contains $\vec{n}$ and $\vec{u}_1$, i.e. the (C_l) curve orthogonal to the meridian plane. These two curvatures are the principal curvatures associated with the principal curvature lines, just described. The mean curvature is defined by:

$$\bar{\chi} = \chi_{\alpha\alpha} + \chi_{\theta\theta} \qquad [7.50]$$

The Gaussian curvature, or total curvature, Ξ is defined by the product,

$$\Xi = \chi_{\alpha\alpha}\chi_{\theta\theta} \qquad [7.51]$$

The surface is said to be locally elliptic, parabolic, or hyperbolic, according to the sign of Ξ ($>0, 0, <0$) respectively. As indicated in Figure 7.13, it is also useful to introduce the angle between the vectors $\vec{u}, \vec{t}$ or $\vec{k}, \vec{n}$ which allows one to write:

$$ds_m = R_\varphi \, d\varphi \;\Rightarrow\; \frac{dz}{ds_m} = \pm \sin\varphi; \quad \frac{dr}{ds_m} = \cos\varphi \qquad [7.52]$$

The positive sign defines a negative curvature of the meridian and vice versa. With [7.49] it follows that:

$$\frac{d\varphi}{ds_m} = \frac{1}{R_\varphi}; \quad \frac{\sin\varphi}{r} = \frac{1}{R_\theta} \qquad [7.53]$$

The usual notations for the global displacements are U, V, W. U is the radial displacement in the direction of the normal vector $\vec{n}$, V the azimuth or circumferential displacement in the direction $\vec{u}_1$ tangent to a parallel circle and W is the meridian displacement in the direction $\vec{t}$ tangent to the meridian line. Finally, the global stresses $\mathcal{N}_{\theta\theta}$ and $\mathcal{N}_{\varphi\varphi}$ are named *hoop stress* and *meridian stress*, respectively.

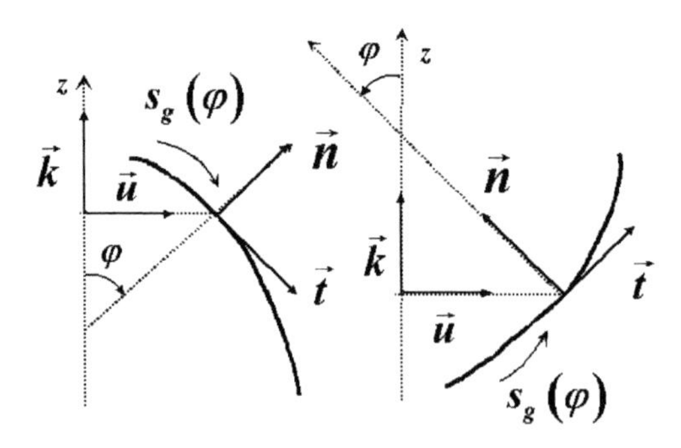

Figure 7.13. *Use of angle φ to describe the meridian line*

7.3.7 *Applications in elastostatics*

7.3.7.1 Spherical shell loaded by uniform pressure

The problem is sketched in Figure 7.14. Using spherical coordinates, a point of the surface and the metrics are defined by the following relationships:

$$r = R\sin\varphi; \quad z = R\cos\varphi$$

$$g_\theta = R\sin\varphi; \quad g_\varphi = \sqrt{\left(\frac{\partial r}{\partial\varphi}\right)^2 + \left(\frac{\partial z}{\partial\varphi}\right)^2} = R$$

$$\frac{\partial}{\partial s_\varphi} = \frac{1}{R}\frac{\partial}{\partial\varphi}; \quad \chi_{\varphi\varphi} = \frac{1}{R}(-(\cos\varphi)^2 - (\sin\varphi)^2) = -\frac{1}{R};$$

$$\chi_{\theta\theta} = \frac{1}{R\sin\varphi}\frac{-R\sin\varphi}{R} = -\frac{1}{R}$$

Because of the symmetry $V = W = 0$. In the system [7.39], the sole equation of interest is related to the normal displacement. In statics, it reduces to:

$$\mathcal{N}_{\theta\theta} + \mathcal{N}_{\varphi\varphi} = Rp \qquad [7.54]$$

where p stands for the pressure.

Relations [7.36] and [7.41] give

$$\eta_{\theta\theta} = \eta_{\varphi\varphi} = \frac{U}{R}; \quad \mathcal{N}_{\theta\theta} = \mathcal{N}_{\varphi\varphi} = \frac{Eh}{1-\nu}\frac{U}{R}$$

whence the solution:

$$U = \frac{p(1-\nu)}{2E}\frac{R^2}{h}; \quad \mathcal{N}_{\theta\theta} = \mathcal{N}_{\varphi\varphi} = \frac{pR}{2}; \quad \mathcal{N}_{\theta\varphi} = 0 \qquad [7.55]$$

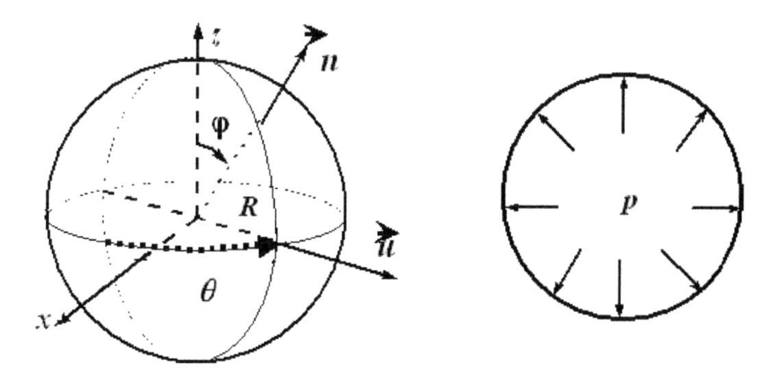

Figure 7.14. *Spherical shell under internal pressure*

As easily expected, the sphere responds to the uniform pressure according to a breathing mode of deformation and the normal stresses are tensile if $p > 0$.

7.3.7.2 *Cylindrical shell closed by hemispherical ends*

The structure sketched in Figure 7.15 is subjected to a uniform external pressure p, produced by an inbalance between internal and external pressures, P_i and P_e. The displacements and the stresses are searched using the membrane theory only. Accordingly, the hemispheres are assumed to act on the cylinder bases as end longitudinal forces only, that is any transverse shear force is neglected. The cylindrical metrics is given by:

$$ds^2 = R^2\, d\theta^2 + dz^2; \quad g_\alpha = g_z = 1; \quad g_\beta = g_\theta = R$$

On the other hand, the mechanical system is symmetric with respect to the middle plane orthogonal to the cylinder axis. So, it suffices to deal with one half of the structure. Taking the origin of the cylinder axis Oz at the plane of symmetry, W must vanish at $z = 0$. On the other hand, the solution is obviously independent of θ. The load acting on the cylinder is in the radial direction $f_\varsigma = f_U = P_i - P_e = -p$ whereas the cylinder bases are loaded by a uniform tensile force per unit length t_z whose resultant balances the pressure exerted on the hemispherical cap; so we get:

$$t_z(z = L/2) = -\frac{\pi R^2 p}{2\pi R} = -\frac{pR}{2}$$

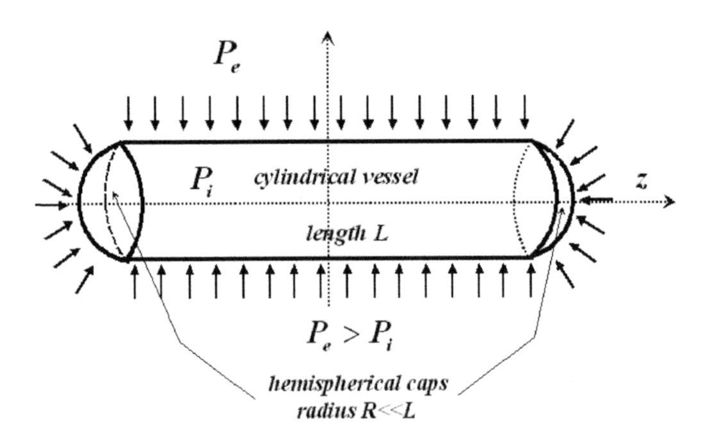

Figure 7.15. *Cylindrical shell ended by two hemispheres*

Thus, the equations to be solved are:

$$\frac{d\mathcal{N}_{zz}}{dz} = 0, \quad \mathcal{N}_{zz}(L/2) = -\frac{pR}{2}, \quad W(0) = 0$$
$$\frac{d\mathcal{N}_{\theta z}}{dz} = 0, \quad \frac{\mathcal{N}_{\theta\theta}}{R} = -p$$

The elastic stress-strain relationship is written as:

$$\begin{bmatrix} \mathcal{N}_{zz} \\ \mathcal{N}_{\theta\theta} \\ \mathcal{N}_{z\theta} \end{bmatrix} = \frac{Eh}{1-\nu^2} \begin{bmatrix} 1 & \nu & 0 \\ \nu & 1 & 0 \\ 0 & 0 & (1-\nu)/2 \end{bmatrix} \begin{bmatrix} \eta_{zz} \\ \eta_{\theta\theta} \\ 2\eta_{z\theta} \end{bmatrix}$$
$$\text{where} \quad \eta_{zz} = \frac{\partial W}{\partial z}; \quad \eta_{\theta\theta} = \frac{U}{R}; \quad \eta_{z\theta} = \frac{1}{2}\frac{\partial V}{\partial z}$$

The circumferential displacement is related to a free rotation of the structure around its symmetry axis; so it can take any arbitrarily value, zero for instance. The axial equilibrium equation gives a constant value for $\mathcal{N}_{zz}$, which is determined by using the equilibrium condition at the interface between the cylinder and the hemispherical cap. The remaining equations are:

$$\frac{d^2W}{dz^2} + \frac{\nu}{R}\frac{dU}{dz} = 0$$
$$\frac{Eh}{1-\nu^2}\left(\frac{dW}{dz} + \frac{\nu U}{R}\right)\bigg|_{L/2} = -\frac{pR}{2}; \quad W(0) = 0$$
$$\frac{\mathcal{N}_{\theta\theta}}{R} = -p \Rightarrow \frac{Eh}{1-\nu^2}\left(\frac{U}{R} + \nu\frac{dW}{dz}\right) = -pR$$

After some elementary manipulations, the following solution is obtained:

$$\frac{dW}{dz} = -\frac{pR}{Eh}\left(\frac{1}{2} - \nu\right) \Rightarrow W(z) = -\frac{pR}{Eh}\left(\frac{1}{2} - \nu\right)z$$
$$U = -\frac{pR^2}{Eh}\left(1 - \frac{\nu}{2}\right)$$

It is worth noticing that the radial displacement of the cylinder differs from that of the hemisphere, already determined in subsection 7.3.7.1:

$$U_{\text{cylinder}} = -\frac{pR^2}{Eh}\left(1 - \frac{\nu}{2}\right) \quad U_{\text{cap}} = U = -\frac{p(1-\nu)}{2E}\frac{R^2}{h}$$

As in reality U is continuous, the difference must be smoothed out by bending effects taking place at the interface. As will be shown in Chapter 8 subsection 8.3.2.4, bending is highly confined in the vicinity of the interface and outside this zone of accommodation the radial displacements are very close to the asymptotic values given here.

7.3.7.3 Pressurized toroidal shell

As sketched in Figure 7.16, we consider a toroidal shell of circular cross-section, loaded by an internal pressure p, assumed to be uniform. An approximate solution may be reasonably proposed, according to which the displacement field is the superposition of two distinct radial dilatations, one concerning the cross sectional (meridian) circles (radius a) and the other the parallel circles of radius lying between $R-a$ and $R+a$. The resulting displacement field is thus written as $\vec{\xi} = q_1\vec{i} + q_2\vec{n}$. Furthermore, if the aspect ratio R/a of the torus is sufficiently large it can be assumed that q_1 and q_2 are essentially constant. The problem can be then solved by using the Rayleigh–Ritz or Galerkin procedure, based again on the principle of minimum potential energy. However, as the trial function is constant in the local frame $\vec{i}, \vec{n}$, it may be found more expedient to write directly the equilibrium equations in terms of q_1 and q_2. Both methods are successively worked out below.

First, the metrics of the shell is given by:

$$ds^2 = g_\theta^2\, d\theta^2 + g_\varphi^2\, d\varphi^2 = r^2\, d\theta^2 + a^2\, d\varphi^2$$

$$g_\theta = r = R + a \sin\varphi, \quad g_\varphi = a$$

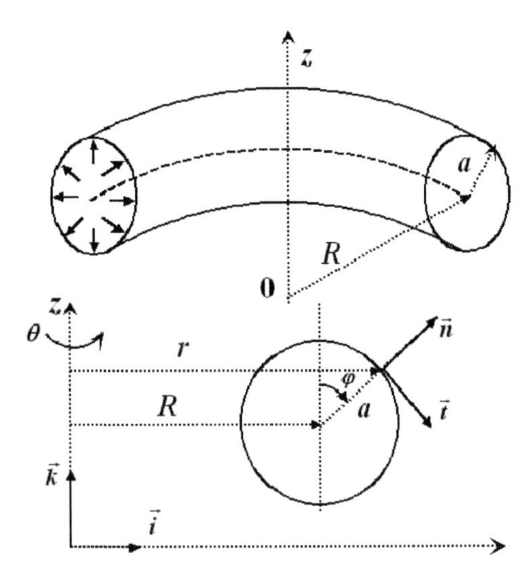

Figure 7.16. *Toroidal shell*

The principal curvature radii are:

$$\frac{1}{R_\varphi} = \frac{d\varphi}{ds_m} = \frac{1}{a} \quad \Rightarrow \quad R_\varphi = a$$

$$\frac{1}{R_\theta} = \frac{\sin\varphi}{r} \quad \Rightarrow \quad R_\theta = \frac{R + a\sin\varphi}{\sin\varphi}$$

The displacement field is:

$$\vec{\xi} = U\vec{n} + W\vec{t} = q_2\vec{n} + q_1\vec{i}$$

$$\vec{i} = \cos\varphi\vec{t} + \sin\varphi\vec{n}$$

$$U = q_2 + q_1\sin\varphi; \quad W = q_1\cos\varphi$$

The small strains are given by the relations [7.36] in which the following variables are in correspondence:

$$\alpha \to \varphi; \quad R_\alpha = R_\varphi = a; \quad X_\alpha = W$$

$$\beta \to \theta; \quad R_\theta = \frac{R + a\sin\varphi}{\sin\varphi}; \quad X_\beta = V \equiv 0$$

$$X_\varsigma \to U$$

V is a free rotation around the torus axis and may be discarded. Then:

$$\eta_{\varphi\varphi} = \frac{q_2}{a}, \quad \eta_{\theta\theta} = \frac{q_1 + q_2\sin\varphi}{R + a\sin\varphi}$$

The elastic stresses are:

$$\begin{bmatrix} \mathcal{N}_{\varphi\varphi} \\ \mathcal{N}_{\theta\theta} \end{bmatrix} = \frac{Eh}{1-\nu^2}\begin{bmatrix} 1 & \nu \\ \nu & 1 \end{bmatrix}\begin{bmatrix} \eta_{\varphi\varphi} \\ \eta_{\theta\theta} \end{bmatrix}$$

$$\mathcal{N}_{\varphi\varphi} = \frac{Eh}{1-\nu^2}\left\{\frac{q_2}{a} + \nu\frac{q_1 + q_2\sin\varphi}{R + a\sin\varphi}\right\}$$

$$\mathcal{N}_{\theta\theta} = \frac{Eh}{1-\nu^2}\left\{\frac{q_1 + q_2\sin\varphi}{R + a\sin\varphi} + \nu\frac{q_2}{a}\right\}$$

The equilibrium equations follow:

$$\frac{\mathcal{N}_{\varphi\varphi}}{a} + \frac{\mathcal{N}_{\theta\theta}\sin\varphi}{R + a\sin\varphi} = p$$

$$\frac{\partial[(R + a\sin\varphi)\mathcal{N}_{\varphi\varphi}]}{\partial\varphi} - \mathcal{N}_{\theta\theta}\frac{\partial[(R + a\sin\varphi)]}{\partial\varphi} = 0$$

$$\frac{\partial\mathcal{N}_{\varphi\varphi}}{\partial\varphi} + a\frac{\cos\varphi}{R + a\sin\varphi}(\mathcal{N}_{\varphi\varphi} - \mathcal{N}_{\theta\theta}) = 0$$

Generally, they are difficult to integrate; however if $a/R \ll 1$, the following approximations can be made:

$$\mathcal{N}_{\varphi\varphi} = \frac{Eh}{1-\nu^2}\left\{\frac{q_2}{a} + \nu\frac{q_1 + q_2\sin\varphi}{R + a\sin\varphi}\right\} \simeq \frac{Eh}{1-\nu^2}\left\{\frac{q_2}{a}\left(1 + \nu\frac{a\sin\varphi}{R}\right) + \nu\frac{q_1}{R}\right\}$$

$$\mathcal{N}_{\theta\theta} = \frac{Eh}{1-\nu^2}\left\{\frac{q_1 + q_2\sin\varphi}{R + a\sin\varphi} + \nu\frac{q_2}{a}\right\} \simeq \frac{Eh}{1-\nu^2}\left\{\frac{q_1}{R} + \nu\frac{q_2}{a}\left(1 + \frac{a\sin\varphi}{R}\right)\right\}$$

If the terms dependent on a/R are neglected, the stresses become independent of φ. The corresponding approximate values of $\mathcal{N}_{\varphi\varphi}, \mathcal{N}_{\theta\theta}$ are:

$$\mathcal{N}_{\varphi\varphi} = \mathcal{N}_{\theta\theta} \cong \frac{Eh}{1-\nu^2}\left(\frac{q_2}{a}\right)(1+\nu) = \frac{Eh}{1-\nu}\left(\frac{q_2}{a}\right)$$

$$\text{where} \quad \frac{q_2}{a} = \frac{q_1}{R} \quad \Rightarrow \quad Rq_2 = aq_1$$

whence the approximate results:

$$q_2 \cong \frac{a^2 p(1-\nu)}{Eh} \qquad q_1 \cong \frac{aRp(1-\nu)}{Eh}$$

Turning now to the more refined Galerkin procedure, the Lagrangian of the system takes the form,

$$\mathcal{L}(q_1, q_2) = -\mathcal{E}_s + \mathcal{W}_p$$

Its variation gives the equilibrium equations:

$$\delta\mathcal{L}(q_1, q_2) = -\delta\mathcal{E}_s + \delta\mathcal{W}_p = 0$$

First, the variation of the pressure work density is determined:

$$\delta w_p = -p\vec{n}\cdot\delta\vec{\xi}\,dS \;\Rightarrow\; \delta w_p = p(\delta q_2 - \delta q_1\vec{i}\cdot\vec{n})\,dS$$

$$dS = ra\,d\theta\,d\varphi, \quad r = R + a\sin\varphi, \quad \vec{i}\cdot\vec{n} = -\sin\varphi$$

then,

$$\delta \mathcal{W}_p = pa\delta q_2 \int_0^{2\pi} \int_0^{2\pi} (R + a \sin \varphi)\, d\varphi\, d\theta$$
$$+ pa\delta q_1 \int_0^{2\pi} \int_0^{2\pi} (R + a \sin \varphi) \sin \varphi\, d\varphi\, d\theta$$
$$\delta \mathcal{W}_p = 4\pi^2 a^2 Rp \left(\frac{\delta q_2}{a} + \frac{\delta q_1}{2R} \right)$$

Reduced displacements defined by $\bar{q}_1 = q_1/R$; $\bar{q}_2 = q_2/a$ may be introduced and give:

$$\delta \mathcal{W}_p = 4\pi^2 a^2 Rp \left(\delta \bar{q}_2 + \frac{\delta \bar{q}_1}{2} \right)$$

The factor $1/2$ present in the generalized force associated with the displacement $\bar{q}_1$ is associated with the term $asin\varphi$ in the curvature radius formula. This term has been neglected in the calculus made just above. Then, the strain energy is deduced from the stress and the strain expressions,

$$\eta_{\varphi\varphi} = \bar{q}_2, \quad \eta_{\theta\theta} \cong \bar{q}_1$$
$$\mathcal{N}_{\varphi\varphi} = \frac{Eh}{1 - \nu^2} (\bar{q}_1 + \nu \bar{q}_2), \quad \mathcal{N}_{\theta\theta} = \frac{Eh}{1 - \nu^2} (\bar{q}_2 + \nu \bar{q}_1)$$

which give:

$$\delta e_s = aR \frac{Eh}{1 - \nu^2} (\bar{q}_1 \delta \bar{q}_1 + \nu (\bar{q}_2 \delta \bar{q}_1 + \bar{q}_1 \delta \bar{q}_2 + \bar{q}_2 \delta \bar{q}_2)) \int_0^{2\pi} \int_0^{2\pi} d\varphi\, d\theta$$
$$= 4\pi^2 aR \frac{Eh}{1 - \upsilon^2} (\bar{q}_1 \delta \bar{q}_1 + \upsilon (\bar{q}_2 \delta \bar{q}_1 + \bar{q}_1 \delta \bar{q}_2 + \bar{q}_2 \delta \bar{q}_2))$$

Finally the equilibrium equations are:

$$\frac{Eh}{1 - \nu^2} (\bar{q}_1 + \nu \bar{q}_2) = a \frac{p}{2}$$
$$\frac{Eh}{1 - \nu^2} (\bar{q}_2 + \nu \bar{q}_1) = ap$$

leading immediately to the following approximate solution:

$$q_1 = \left(\frac{1}{2} - \nu \right) \frac{paR}{Eh}; \quad q_2 = \left(1 - \frac{\nu}{2} \right) \frac{pa^2}{Eh}$$

These results are more accurate than those obtained by the direct approximate solution.

7.3.7.4 *Spherical cap loaded by its own weight*

The problem is sketched in Figure 7.17. In local coordinates the loading is given by,

$$f_\varsigma = -\rho hg \cos\varphi; \quad f_\theta = 0; \quad f_\varphi = \rho hg \sin\varphi$$

Because of the symmetries, the solution is independent of θ and $V = 0$. The local equilibrium is then described by,

$$(\mathcal{N}_{\theta\theta} - \mathcal{N}_{\varphi\varphi})\cos\varphi - \sin\varphi\frac{d\mathcal{N}_{\varphi\varphi}}{d\varphi} = \rho hgR(\sin\varphi)^2$$
$$\mathcal{N}_{\theta\theta} + \mathcal{N}_{\varphi\varphi} = -\rho hgR\cos\varphi$$

and $\mathcal{N}_{\varphi\varphi}$ verifies:

$$2\cos\varphi\mathcal{N}_{\varphi\varphi} + \sin\varphi\frac{d\mathcal{N}_{\varphi\varphi}}{d\varphi} = -\rho hgR$$

which can be integrated (though integration is not obvious) to give:

$$\mathcal{N}_{\varphi\varphi} = \frac{-\rho hRg}{1+\cos\varphi} \Rightarrow \mathcal{N}_{\theta\theta} = \rho h\, Rg\left(\frac{1}{1+\cos\varphi} - \cos\varphi\right)$$

A simpler manner to obtain $\mathcal{N}_{\varphi\varphi}$ is to write the global equilibrium of the shell in the vertical direction. The area of an elementary strip of the shell surface delimited

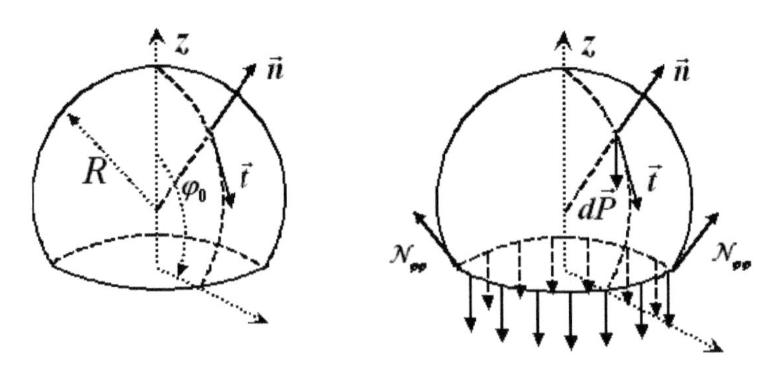

Figure 7.17. *Spherical cap loaded by its own weight*

by two horizontal planes at z and $z + dz$ is readily found to be $2\pi R \sin\varphi R\, d\varphi$; the corresponding weight is $-\rho h \vec{g} 2\pi R \sin\varphi R\, d\varphi$. So, the total weight is:

$$\vec{P} = -\rho h \vec{g} \int_0^{\varphi} 2\pi R \sin\varphi R\, d\varphi = -2\pi R^2 \rho h g (1 - \cos\varphi)$$

as it must be balanced by the vertical component of support reactions, we get:

$$2\pi R^2 \rho h g (1 - \cos\varphi) = -2\pi R (\sin\varphi)^2 \mathcal{N}_{\varphi\varphi}(\varphi) \Rightarrow$$

$$\mathcal{N}_{\varphi\varphi}(\varphi) = -\rho h g R \frac{1 - \cos\varphi}{(\sin\varphi)^2} = \frac{-\rho h g R}{1 + \cos\varphi}$$

The normal stresses are plotted in Figure 7.18 versus φ. $\mathcal{N}_{\varphi\varphi}$ is negative, hence compressive, whatever the φ value may be. However, $\mathcal{N}_{\theta\theta}$ becomes tensile in the range $\varphi > 52°$. This less intuitive result caused many flaws and even failures in masonry domes because masonry resists compressive stresses much better than tensile stresses. For instance, the dome of St Peter basilica in Rome was found largely cracked during the seventeenth century. The empirical solution implemented by the architects – which was perfectly satisfactory – has been to reinforce the dome perimeter support by a masonry ring [COT 90].

Once the stresses are known, it is possible to calculate the strains and the displacements. The analytical solution is not very simple and requires one to assume specific boundary conditions. The interest of such analytical solutions, which hold for particular boundary conditions, is to obtain approximate solutions for various

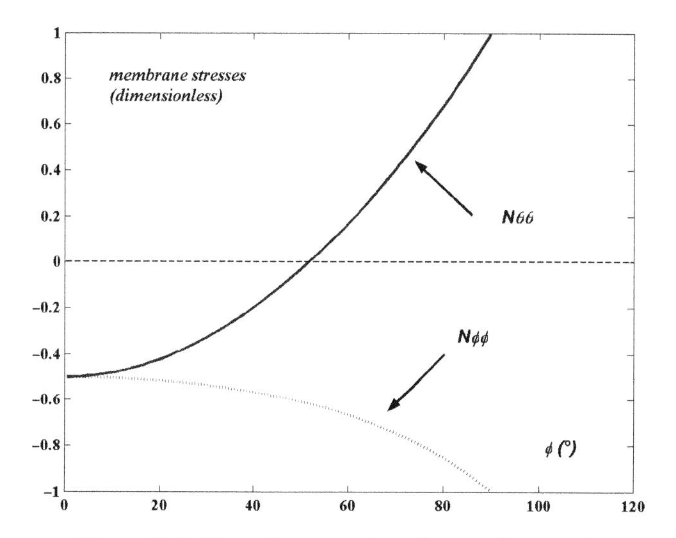

Figure 7.18. *Normal stresses plotted versus the φ angle*

actual boundary conditions which are not too far from the assumed ones, provided Saint-Venant's principle can be invoked. This point is illustrated below.

If the material is linear elastic, the strains in the spherical dome are:

$$\eta_{\theta\theta} = \frac{1}{E}(\mathcal{N}_{\theta\theta} - \nu\mathcal{N}_{\varphi\varphi}) = \frac{W}{R}\cot\varphi + \frac{U}{R}$$
$$\eta_{\varphi\varphi} = \frac{1}{E}(\mathcal{N}_{\varphi\varphi} - \nu\mathcal{N}_{\theta\theta}) = \frac{1}{R}\frac{dW}{d\varphi} + \frac{U}{R}$$

Eliminating U gives,

$$\eta_{\theta\theta} - \eta_{\varphi\varphi} = \frac{1}{R}\left(W\cot\varphi - \frac{dW}{d\varphi}\right) = \frac{\rho h g R(1+\nu)}{E}\left(\frac{2}{1+\cos\varphi} - \cos\varphi\right)$$

This equation has the general form $W\cot\varphi - dW/d\varphi = f(\varphi)$ and the solution is $W = \sin\varphi \int f(\varphi)/\sin\varphi \, d\varphi + Cste$

$$W = \frac{-\rho h g R^2(1+\nu)}{E}\sin\varphi\left\{\mathrm{Log}\left(\frac{1+\cos\varphi}{1+\cos\varphi_0}\right) + \frac{1}{1+\cos\varphi} - \frac{1}{1+\cos\varphi_0}\right\}$$

the radial displacement U is readily found to be:

$$U = \frac{-\rho h g R^2}{E}\left\{\frac{1+\nu}{2(1+\cos\varphi)} - \cos\varphi\right\} + W\mathrm{cotg}\varphi$$

In Figures 7.19, the reduced value of U is plotted as a function of φ for an hemisphere and a dome defined by a $3\pi/4$ base angle. The tangential displacement along the meridians is set to zero but the other components are left free, which is

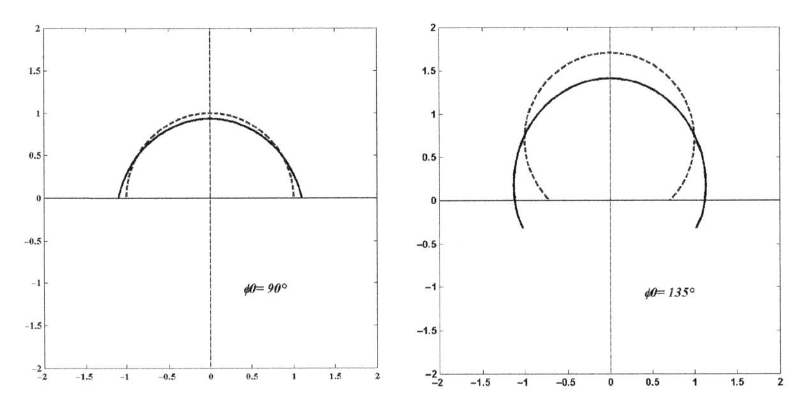

Figure 7.19. *Analytical displacements of two spherical caps*

unrealistic. However, these plots show clearly that the radius of the parallel circles is reduced in the upper zone $\varphi > 52°$ thus inducing a compressive hoop stress and enlarged in the lower zone, thus inducing a tensile hoop stress.

NOTE. – *Finite element solution*

A finite element solution which accounts for both the membrane and bending effects shows that a clamped boundary condition does not deeply modify the analytical solution except in the vicinity of the clamping. In Figures 7.20a, 7.20b and 7.21a, 7.21b, the deformed shapes of a masonry dome with the base either on sliding or clamped support can be compared. The radius is 20 m, the thickness 10 cm and the weight 400 tons. To make the deformed shapes clearly visible, the actual displacements have been multiplied by the factor 20000. The left-hand side plot shows the deflection of a meridian and the right-hand side plot shows that of an

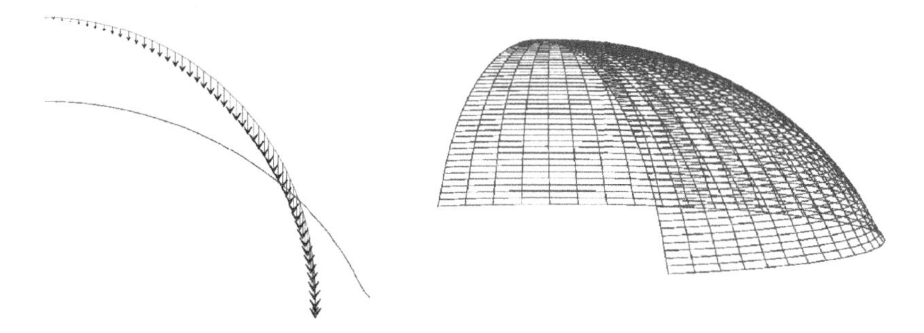

Figure 7.20a. *Hemispherical cap provided with sliding supports at the base*

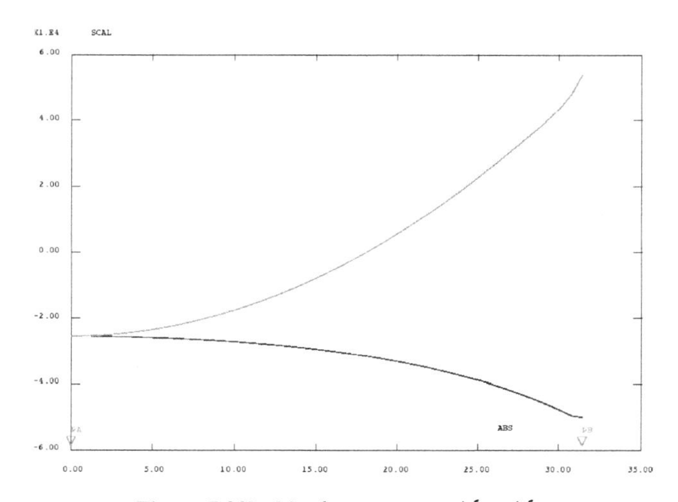

Figure 7.20b. *Membrane stresses* $N_{\varphi\varphi}, N_{00}$

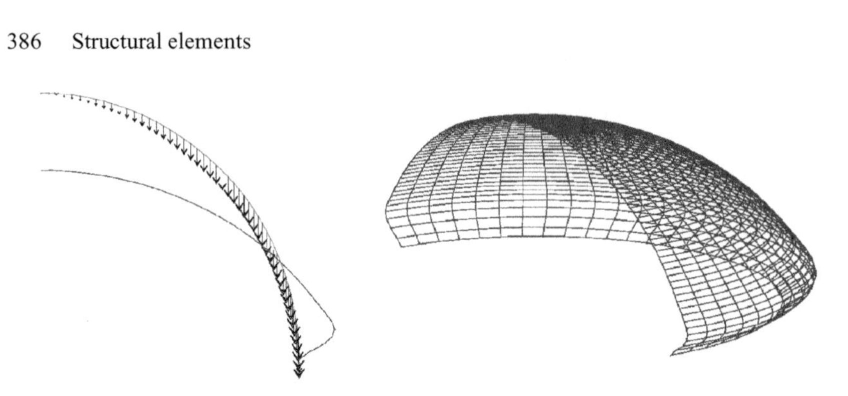

Figure 7.21a. *Hemispherical cap clamped at the base*

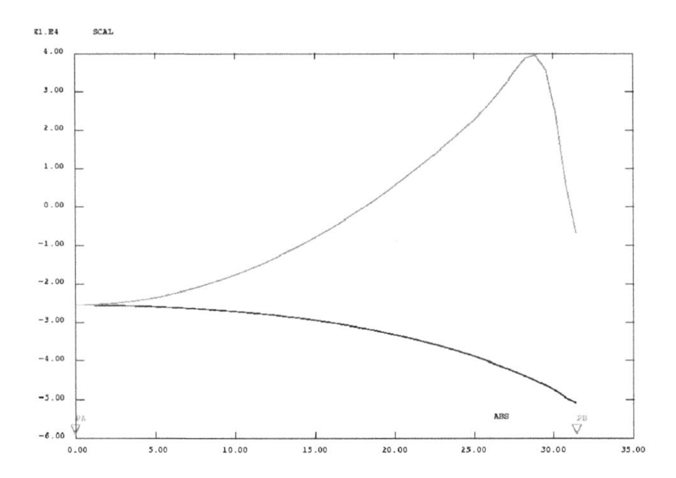

Figure 7.21b. *Membrane stresses* $N_{\varphi\varphi}, N_{\theta\theta}$

angular sector. Computed deflection in the case of a sliding support is found to be very close to the analytical solution which discards bending effects, except at the immediate vicinity of the base, see Figure 7.20a. This close agreement also holds so far as the membrane stresses are concerned, except near the top and the base of the dome, where bending stresses are present, see Figure 7.20b. As could be expected, the effect of clamping the dome at the base is to increase further the importance of bending near the base which becomes significant over a much larger zone than in the case of a sliding base, see Figure 7.21a. Outside this perturbed zone, the dome response remains essentially the same in both support configurations.

7.3.7.5 Conical shell of revolution loaded by its own weight

The structure is shown in Figure 7.22. The shell is shaped as a conical frustum of revolution around the axis Oz; the cone half angle is denoted ψ, the shell thickness

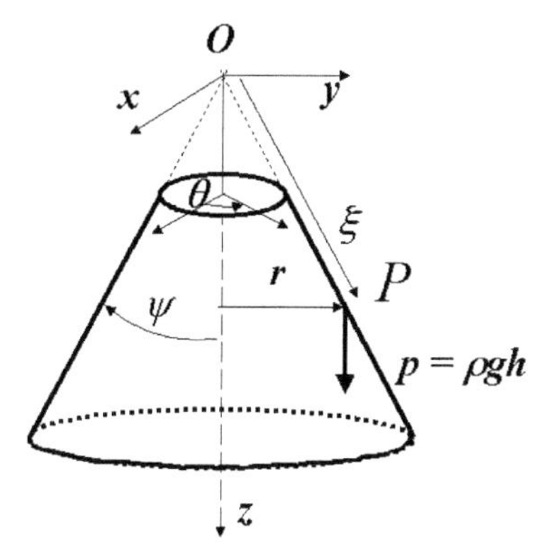

Figure 7.22. *Truncated cone loaded by its own weight*

is h and $p = \rho g h$ is the weight per unit area of the shell. The bottom of the frustum is supported in the Oz direction. We are interested in determining the membrane stresses.

The position of a current point is defined by the two parameters θ and $\xi = \overline{OP}$. The top and the bottom of the cone are defined by $\xi = \xi_0$ and $\xi = \xi_1$ respectively. The coordinates of a point are:

$$r = \xi \sin\psi; \quad x = \xi \sin\psi \cos\theta; \quad y = \xi \sin\psi \cos\theta; \quad z = \xi \cos\psi$$

So, the coefficients of the metrics are given by:

$$\begin{aligned}(ds)^2 &= (dx)^2 + (dy)^2 + (dz)^2 \\ (ds)^2 &= \left\{(\sin\psi\cos\theta)^2 + (\sin\psi\sin\theta)^2 + (\cos\psi)^2\right\}(d\xi)^2 \\ &\quad + \left\{(\sin\theta)^2 + (\cos\theta)^2\right\}(\xi\sin\psi\, d\theta)^2 \\ (ds)^2 &= (d\xi)^2 + (\xi\sin\psi\, d\theta)^2\end{aligned}$$

then,

$$g_\xi = 1; \quad g_\theta = \xi \sin\psi$$

The principal curvatures are given by:

$$\chi_{\xi\xi} = \frac{1}{R_\xi} = \frac{\partial^2 z}{\partial \xi^2}\frac{\partial r}{\partial \xi} - \frac{\partial^2 r}{\partial \xi^2}\frac{\partial z}{\partial \xi} = 0;$$

$$\chi_{\theta\theta} = \frac{1}{R_\theta} = \frac{1}{r}\frac{\partial z}{\partial \xi} = \frac{\cos\psi}{\xi\sin\psi} \quad \Rightarrow \quad R_\theta = \xi tg\psi$$

The meridian is a straight line, so its curvature is zero. It has to be stressed that the parallel circles are not the principal curvature lines; the latter are defined by the intersection of the frustum surface by planes which are orthogonal to the meridian lines. As determined just above, their curvature radius is $R_\theta = \xi tg\psi$. The equilibrium equations are obtained from [7.39]. Because the problem does not depend on θ they can be written as:

$$-\frac{1}{\xi \sin\psi}\left\{\frac{\partial(\xi \sin\psi \mathcal{N}_{\xi\xi})}{\partial \xi} - \mathcal{N}_{\theta\theta}\frac{\partial(\xi \sin\psi)}{\partial \xi}\right\} = p\cos\psi, \qquad \frac{\mathcal{N}_{\theta\theta}}{\xi tg\psi} = -p\sin\psi$$

So the hoop stress is immediately obtained as:

$$\mathcal{N}_{\theta\theta} = -p\xi \sin\psi \, tg\psi$$

The meridian stress is governed by the differential equation:

$$-\frac{\mathcal{N}_{\xi\xi}}{\xi} - \frac{\partial \mathcal{N}_{\xi\xi}}{\partial \xi} = p\cos\psi + p\sin\psi \, tg\psi = \frac{p}{\cos\psi} \Rightarrow \frac{\partial(\xi\mathcal{N}_{\xi\xi})}{\partial \xi} = -\frac{p\xi}{\cos\psi}$$

Integration is also immediate and the top base being assumed to be free, we get:

$$\mathcal{N}_{\xi\xi} = -\frac{p(\xi^2 - \xi_0^2)}{2\xi \cos\psi}$$

Accordingly, the shell is found to be everywhere in a compressive state, both in the meridian and the circumferential directions. As $\psi \to 0$ the structure tends to a cylindrical shell, so $\mathcal{N}_{\theta\theta} \to 0$, as suitable. In contrast, if $\psi \to \pi/2$ the structure tends to a plate loaded by its own weight (bending only) and the membrane solution is obviously meaningless. Finally, it can be noted that, as for the dome, $\mathcal{N}_{\xi\xi}$ can be calculated by balancing directly the meridian stresses and the weight.

7.3.7.6 Conical container

A container full of liquid is shown in Figure 7.23. It is uniformly supported at its top $z = H = L\cos\psi$ and full of liquid whose density is ρ; the cone half-angle is ψ. The equilibrium equations are:

$$-\frac{1}{\xi \sin\psi}\left\{\frac{\partial(\xi \sin\psi \mathcal{N}_{\xi\xi})}{\partial\xi} - \mathcal{N}_{\theta\theta}\sin\psi\right\} = f_\xi = 0$$

$$-\frac{1}{\xi \sin\psi}\left\{\frac{1}{\xi \sin\psi}\frac{\partial((\xi \sin\psi)^2\mathcal{N}_{\theta\xi})}{\partial\xi}\right\} = f_\theta = 0$$

$$\frac{\mathcal{N}_{\theta\theta}}{\xi tg\psi} = p(\xi) = \rho g(L - \xi)\cos\psi$$

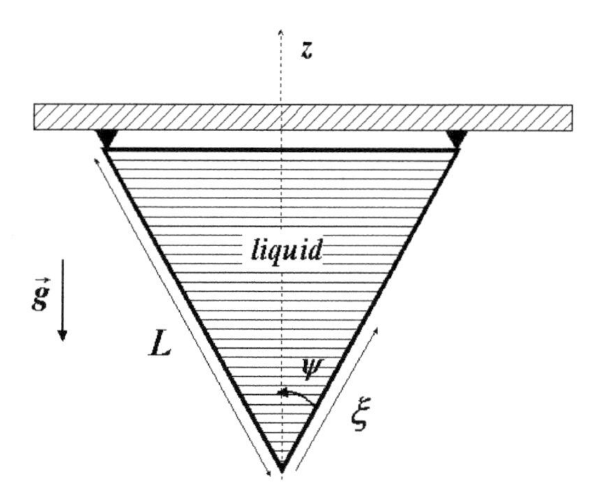

Figure 7.23. *Conical container full of liquid*

or,

$$\frac{\partial(\xi\mathcal{N}_{\xi\xi})}{\partial\xi} - \mathcal{N}_{\theta\theta} = 0, \qquad \frac{\partial(\xi^2\mathcal{N}_{\theta\xi})}{\partial\xi} = 0, \quad \mathcal{N}_{\theta\theta} = \rho g\xi(L-\xi)\sin\psi$$

From the second equation, $\xi^2\mathcal{N}_{\theta\xi} = a \Rightarrow \mathcal{N}_{\theta\xi} = a/\xi^2$

At the top, the stresses must verify the condition $\mathcal{N}_{\theta\xi}(H) = 0 \Rightarrow a = 0 \Rightarrow \mathcal{N}_{\theta\xi} \equiv 0$ everywhere within the shell, as could be anticipated.

The meridian stress is given by:

$$\frac{\partial(\xi\mathcal{N}_{\xi\xi})}{\partial\xi} = \rho g\xi(L-\xi)\sin\psi \;\Rightarrow\; \xi\mathcal{N}_{\xi\xi} = \rho g\sin\psi\left(\frac{L\xi^2}{2} - \frac{\xi^3}{3}\right) + b,$$

$$\mathcal{N}_{\xi\xi} = \rho g\sin\psi\,\xi\left(\frac{L\xi}{2} - \frac{\xi^2}{3}\right) + \frac{b}{\xi}$$

The constant b must be zero because the stress must remain finite at the apex, that is as $\xi \to 0$. So we get:

$$\mathcal{N}_{\xi\xi} = \rho g\sin\psi\left(\frac{L\xi}{2} - \frac{\xi^2}{3}\right)$$

Finally, the stress maxima are found to be:

$$\mathcal{N}_{\theta\theta} = \rho g \xi (L - \xi) \sin\psi \;\Rightarrow\; (\mathcal{N}_{\theta\theta})_{\max} = \mathcal{N}_{\theta\theta}(L/2) = \frac{\rho g L^2 \sin\psi}{4}$$

$$\mathcal{N}_{\xi\xi} = \rho g \sin\psi \left(\frac{L\xi}{2} - \frac{\xi^2}{3} \right) \;\Rightarrow\; (\mathcal{N}_{\xi\xi})_{\max} = \mathcal{N}_{\xi\xi}(3L/4) = \frac{3\rho g L^2 \sin\psi}{16}$$

$$\mathrm{Max} \left(\frac{\mathcal{N}_{\theta\theta}}{\mathcal{N}_{\xi\xi}} \right) = \frac{4}{3}$$

Thus the shell is in a compressive state everywhere, both in the meridian and the circumferential directions. On the other hand, here again, the hoop stresses are found to be larger than the meridian stresses, though by about 30% only.

Chapter 8

Bent and twisted arches and shells

To deal with general loading conditions, it is necessary to include bending and torsion into the equilibrium equations of arches and shells; which leads to interesting coupling effects between various elementary modes of deformation. In the case of arches, in-plane bending is found to be coupled with tangential stretching and out-of-plane bending with torsion. In the case of shells, all the elementary modes of deformation are found to be coupled together, except in a few particular loading cases. The problem of shell vibrations was first attacked by Sophie Germain in the early nineteenth century. However, the basic development of the thin shell theory is due to Love in 1888. As already indicated in the preceding chapter, Love's model is based on simplifying assumptions which extend in a natural manner those already used to model straight beams and plates. The thin shell theory was the object of various refinements during the twentieth century. As a result, there exists a wide variety of shell equations. However, all of them are basically of the Love type, differing only by the approximations made to deal with the metric coefficients G_α, G_β. Furthermore, such differences turn out to be of little practical importance. Therefore, presentation is restricted here to the Love model. Particularization to cylindrical shells of revolution gives us the opportunity to discuss a few problems of practical interest and the validity of various simplifications of Love's equations, in relation to the specificities of the loading considered.

8.1. Arches and circular rings

8.1.1 *Local and global displacement fields*

The relations [7.4] to [7.9] of Chapter 7 are extended to describe the motions out or in the arch plane, see Figure 8.1. For this purpose, a local frame defining a direct Cartesian coordinate system is needed. The unit vectors are $\vec{t}$, tangent to the neutral fibre, $\vec{n}_1$ normal to the neutral fibre, in the arch plane and pointing towards the extrados, and $\vec{n}_2 = \vec{t} \times \vec{n}_1$. The local displacement vector is written as:

$$\begin{aligned} \vec{\xi} = \vec{X} + \vec{\Psi} \times \vec{\varsigma} \quad &\text{where } \vec{\varsigma} = \varsigma_1 \vec{n}_1 + \varsigma_2 \vec{n}_2 \\ \vec{X} = X_s \vec{t} + X_1 \vec{n}_1 + X_2 \vec{n}_2 \quad &\text{and} \quad \vec{\Psi} = \psi_s \vec{t} + \psi_1 \vec{n}_1 + \psi_2 \vec{n}_2 \end{aligned} \qquad [8.1]$$

where $\vec{X}$ and $\vec{\Psi}$ given by the second row of [8.1], define the global displacement field. Whence:

$$\xi_s = X_s + \varsigma_2 \psi_1 - \varsigma_1 \psi_2; \quad \xi_1 = X_1 - \varsigma_2 \psi_s; \quad \xi_2 = X_2 + \varsigma_1 \psi_s \qquad [8.2]$$

The coefficients of the metric tensor are obtained from:

$$\begin{aligned} P(s, \varsigma_1, \varsigma_2) &= \vec{r}(s) + \varsigma_1 \vec{n}_1 + \varsigma_2 \vec{n}_2 \\ dP(s, \varsigma_1, \varsigma_2) &= d\vec{r}(s) + d\varsigma_1 \vec{n}_1 + d\varsigma_2 \vec{n}_2 + \varsigma_1 d\vec{n}_1 \\ dP(s, \varsigma_1, \varsigma_2) &= ds \left(1 + \frac{\varsigma_1}{R}\right) \vec{t} + d\varsigma_1 \vec{n}_1 + d\varsigma_2 \vec{n}_2 \end{aligned} \qquad [8.3]$$

Analytical expressions are made somewhat less cumbersome by using the curvilinear abscissa s of a current point defined along the curved neutral fibre. The local radius of curvature of the arch is $R(s)$. From [8.3], the following metric coefficients are readily found to be:

$$G_s = \left(1 + \frac{\varsigma_1}{R}\right); \quad G_1 = G_2 = 1 \qquad [8.4]$$

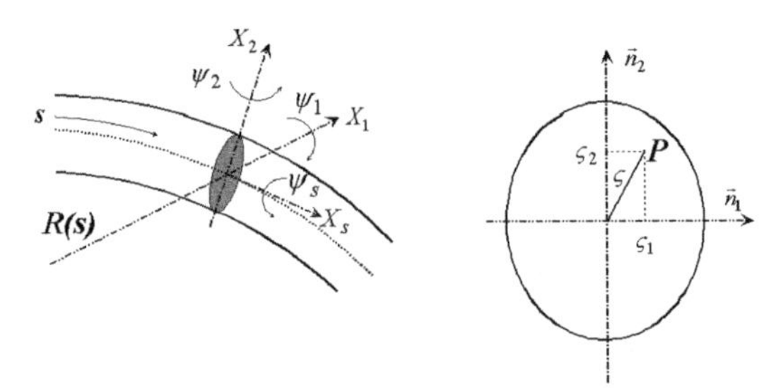

Figure 8.1. *Arch: local frame and global displacements*

8.1.2 *Tensor of small local strains*

The results [8.2] and [8.4] are substituted into the general expression [5.87] of the strain tensor. Here, the non-identically zero components are first written as:

$$
\begin{aligned}
\varepsilon_{ss} &= \frac{1}{G_s}\left(\frac{\partial \xi_s}{\partial s} + \frac{\xi_1}{G_1}\frac{\partial G_s}{\partial \varsigma_1}\right) \\
\varepsilon_{s1} &= \frac{1}{2}\left\{\frac{G_s}{G_1}\frac{\partial}{\partial \varsigma_1}\left(\frac{\xi_s}{G_s}\right) + \frac{G_1}{G_s}\frac{\partial}{\partial s}\left(\frac{\xi_1}{G_1}\right)\right\} \\
\varepsilon_{s2} &= \frac{1}{2}\left\{\frac{G_s}{G_2}\frac{\partial}{\partial \varsigma_2}\left(\frac{\xi_s}{G_s}\right) + \frac{G_2}{G_s}\frac{\partial}{\partial s}\left(\frac{\xi_2}{G_2}\right)\right\}
\end{aligned}
\qquad [8.5]
$$

and then:

$$
\begin{aligned}
\varepsilon_{ss} &= \left(1+\frac{\varsigma_1}{R}\right)^{-1}\left(\frac{\partial(X_s+\varsigma_2\psi_1-\varsigma_1\psi_2)}{\partial s} + \frac{X_1-\varsigma_2\psi_s}{R}\right) \\
\varepsilon_{s1} &= \frac{1}{2}\left(\left(1+\frac{\varsigma_1}{R}\right)\frac{\partial}{\partial \varsigma_1}\left(\frac{X_s+\varsigma_2\psi_1-\varsigma_1\psi_2}{(1+\varsigma_1/R)}\right)\right. \\
&\quad \left.+\left(1+\frac{\varsigma_1}{R}\right)^{-1}\left(\frac{\partial(X_1-\varsigma_2\psi_s)}{\partial s}\right)\right) \\
\varepsilon_{s2} &= \frac{1}{2}\left(\left(1+\frac{\varsigma_1}{R}\right)\frac{\partial}{\partial \varsigma_2}\left(\frac{\partial X_s+\varsigma_2\psi_1-\varsigma_1\psi_2}{(1+\varsigma_1/R)}\right)\right. \\
&\quad \left.+\left(1+\frac{\varsigma_1}{R}\right)^{-1}\left(\frac{\partial(X_2+\varsigma_1\psi_s)}{\partial s}\right)\right)
\end{aligned}
\qquad [8.6]
$$

In the case of slender arches the strains [8.6] can be simplified, reducing to:

$$
\begin{aligned}
\varepsilon_{ss} &= \left(\frac{\partial X_s}{\partial s} + \varsigma_2\frac{\partial \psi_1}{\partial s} - \varsigma_1\frac{\partial \psi_2}{\partial s} + \frac{X_1-\varsigma_2\psi_s}{R}\right) \\
\varepsilon_{s1} &= \frac{1}{2}\left(-\psi_2 - \frac{(X_s+\varsigma_2\psi_1-\varsigma_1\psi_2)}{R} + \frac{\partial X_1}{\partial s} - \varsigma_2\frac{\partial \psi_s}{\partial s}\right) \\
\varepsilon_{s2} &= \frac{1}{2}\left(\psi_1 + \frac{\partial X_2}{\partial s} + \varsigma_1\frac{\partial \psi_s}{\partial s}\right)
\end{aligned}
\qquad [8.7]
$$

It is noted that all the deformation modes are coupled with respect to all the global displacement components.

8.1.3 *Pure bending in the arch plane*

8.1.3.1 Equilibrium equations

It is possible to formulate the equilibrium equations of an arch by taking into account all the terms present in the strains [8.7]. Nevertheless, it is more convenient and instructive to start by investigating each basic mode of deformation separately, namely the in-plane flexure, the out-of-plane flexure and the torsion. The models derived by this manner have the advantage of simplicity and can be used as a guideline for formulating afterwards more refined coupled models.

Let us consider first the in-plane flexure. The relevant components of the global displacement field are X_s, X_1, ψ_2. Therefore, the local strain field is reduced to:

$$\begin{aligned} \varepsilon_{ss} &= \left(\frac{\partial X_s}{\partial s} - \varsigma_1 \frac{\partial \psi_2}{\partial s} + \frac{X_1}{R}\right) \\ \varepsilon_{s1} &= \frac{1}{2}\left(-\frac{X_s}{R} - \psi_2 + \frac{\partial X_1}{\partial s}\right) = 0 \end{aligned} \qquad [8.8]$$

These expressions are further simplified by making two additional assumptions. First, transverse shear deformation is neglected, as in the straight beam Bernoulli–Euler model; so ε_{s1} is assumed to be identically zero. Further, if coupling with the axial mode of deformation is discarded, the axial local strains are simplified since the axial global strain must vanish:

$$\eta_{ss} = \varepsilon_{ss}(\varsigma_1 = 0) = 0 \qquad [8.9]$$

These conditions give two independent holonomic conditions:

$$\frac{\partial X_s}{\partial s} + \frac{X_1}{R} = 0; \quad \varepsilon_{ss} = -\varsigma_1 \frac{\partial \psi_2}{\partial s}; \quad \psi_2 = \frac{\partial X_1}{\partial s} - \frac{X_s}{R} \qquad [8.10]$$

which can be readily used to express the axial local strain in terms of the sole transverse displacement X_1, just as for straight beams:

$$\varepsilon_{ss} = -\varsigma_1 \left(\frac{\partial^2 X_1}{\partial s^2} + \frac{X_1}{R^2}\right) \qquad [8.11]$$

In [8.11], the term arising from the differentiation of $R(s)$ is neglected, which is justified in the case of slender arches with moderate curvatures. The global bending strain is thus found to be:

$$\chi_{ss} = -\left(\frac{\partial^2 X_1}{\partial s^2} + \frac{X_1}{R^2}\right) \qquad [8.12]$$

The bending strain [8.12] differs from that obtained in straight beams by an additional term due to the finite curvature of the neutral fibre in the non-deformed state. It is also important to be aware that the arch axial displacement X_s does not vanish. The equation of equilibrium and the boundary conditions can now be easily established. The variation of kinetic energy is:

$$\delta[\mathcal{E}_\kappa] = \int_{s_1}^{s_2} \rho S(\dot{X}_s \delta\dot{X}_s + \dot{X}_1 \delta\dot{X}_1)\, ds \qquad [8.13]$$

and the variation of strain energy is:

$$-\delta[\mathcal{E}_s] = \int_{s_1}^{s_2} \mathcal{M}_2 \left(\frac{\partial^2 \delta X_1}{\delta s^2} + \frac{\delta X_1}{R^2} \right) ds = \int_{s_1}^{s_2} \left(\frac{\partial^2 \mathcal{M}_2}{\partial s^2} + \frac{\mathcal{M}_2}{R^2} \right) \delta X_1\, ds$$
$$+ \left[\mathcal{M}_2 \delta \left(\frac{\partial X_1}{\partial s} \right) - \frac{\partial \mathcal{M}_2}{\partial s} \delta X_1 \right]_{s_1}^{s_2} \qquad [8.14]$$

in which $\mathcal{M}_2$ designates the bending moment about the flexure axis $\vec{n}_2$.

The equation of transverse equilibrium and the related boundary conditions are found to be:

$$\rho S \ddot{X}_1 - \frac{\partial^2 \mathcal{M}_2}{\partial s^2} - \frac{\mathcal{M}_2}{R^2} = F_1^{(e)}(s;t) - \frac{\partial \mathfrak{M}_2^{(e)}(s;t)}{\partial s}$$
$$\left[\mathcal{M}_2 + \mathcal{M}^{(e)} \right]_{s=s_1} = 0; \quad \left[-\mathcal{M}_2 + \mathcal{M}^{(e)} \right]_{s=0} = 0 \qquad [8.15]$$
$$\left[-\frac{\partial \mathcal{M}_2}{\partial s} + T_1^{(e)} \right]_{s=s_1} = 0; \quad \left[+\frac{\partial \mathcal{M}_2}{\partial s} + T_1^{(e)} \right]_{s=0} = 0$$

The external loading comprises a force field acting in the transverse and in-plane direction $\vec{n}_1$ and a moment field about the out-of-plane direction $\vec{n}_2$. The force and moment densities per unit arch length are denoted $F_1^{(e)}(s;t)$ and $\mathfrak{M}_2^{(e)}(s;t)$, respectively. These components are assumed to vanish at the arch ends. The latter can be loaded by a force $T_1^{(e)}$ and a moment $\mathcal{M}^{(e)}$.

The tangential equation reduces to:

$$\rho S \ddot{X}_s = F_s^{(e)}(s;t) \qquad [8.16]$$

The left-hand side of [8.16] comprises an inertial term only since no axial global stress results according to the 'pure bending' model. Finally, if the arch material is linear elastic, the stresses are found to be:

$$\sigma_{ss} = -E\varsigma_1 \left(\frac{\partial^2 X_1}{\partial s^2} + \frac{X_1}{R^2} \right) \Rightarrow \mathcal{M}_2 = -EI_2 \left(\frac{\partial^2 X_1}{\partial s^2} + \frac{X_1}{R^2} \right) \qquad [8.17]$$

The equation of transverse vibration is:

$$\rho S \ddot{X}_1 + \frac{\partial^2}{\partial s^2}\left(EI_2\left(\frac{\partial^2 X_1}{\partial s^2} + 2\left(\frac{X_1}{R^2}\right)\right)\right) + \frac{EI_2}{R^4}X_1 = F_1^{(e)} - \frac{\partial \mathfrak{M}_2^{(e)}}{\partial s} \quad [8.18]$$

Equation [8.18] differs from the straight beam case by two additional stiffness terms which arise as a consequence of the finite curvature of the neutral fibre.

8.1.3.2 Vibration modes of a circular ring

In agreement with the usual notations used for structures of revolution, the displacement field and the curvilinear coordinate are rewritten as:

$$X_s = V; \quad X_1 = U; \quad X_2 = W; \quad ds = R\,d\theta \quad [8.19]$$

R being the radius of the ring neutral fibre. For mathematical convenience, ring cross-sections are assumed to be circular; their radius being denoted a. The equations of free vibration are:

$$\rho S \ddot{U} + \frac{EI}{R^4}\left(\frac{\partial^4 U}{\partial \theta^4} + 2\frac{\partial^2 U}{\partial \theta^2} + U\right) = 0; \quad \rho S \ddot{V} = 0 \quad [8.20]$$

If the ring is free to move in its own plane, the sole condition to be fulfilled by U and V is a 2π periodicity. Therefore, it is natural to search for modal solutions such as $U_n(\theta;t) = \cos(n\theta)e^{i\omega_n t}$ or $U_n(\theta;t) = \sin(n\theta)e^{i\omega_n t}$, or more generally any linear superposition of these two orthogonal families. Adopting the cosine family is equivalent to setting the axis $\theta = 0$ at an antinode of vibration. According to this choice, the first equation [8.10] implies necessarily the following mode shapes:

$$\begin{aligned} U_n^{(1)} &= a_n \cos(n\theta); \quad V_n^{(1)} = -\frac{a_n}{n}\sin(n\theta) \\ U_n^{(2)} &= a_n \sin(n\theta); \quad V_n^{(2)} = +\frac{a_n}{n}\cos(n\theta) \end{aligned} \quad [8.21]$$

Again, a_n is an arbitrary constant used to normalize the mode shapes. Hereafter it is set to one. Substituting [8.21] into the radial equation [8.20] leads to:

$$\left(-\omega_n^2 \rho S + \frac{EI}{R^4}(n^2-1)^2\right)\cos(n\theta) = 0 \quad [8.22]$$

From [8.22] the natural frequencies of the in-plane bending modes are given by:

$$f_n = \frac{n^2-1}{2\pi R^2}\sqrt{\frac{EI}{\rho S}} = \frac{(n^2-1)}{2\pi R}\frac{c}{\eta}, \quad n = 1, 2, \ldots$$

$$\text{where} \quad \eta = \sqrt{\frac{SR^2}{I}} \quad \text{and} \quad c = \sqrt{\frac{E}{\rho}} \qquad [8.23]$$

η is interpreted as a slenderness ratio. For a ring of circular cross-section of radius a it is readily found that $\eta = 2R/a$. It is noticed that the modes $n = 1$ refer to a free in-plane translation of the ring, as already discussed in Chapter 7 section 7.5. In Figure 8.2 two mode shapes of the first family (antinode at $\theta = 0$) are plotted and the corresponding frequencies are specified for a steel ring $R = 1\text{m}$, $a = 1\text{cm}$. Full and dashed dotted heavy lines refer to two vibration patterns separated by half a period. It can be verified that the radial and tangential nodes are not located at the same angular positions.

The ratio between the frequencies [8.23] and the breathing mode frequency (formula [7.23]) is:

$$\frac{f_n}{f_0} = (n^2-1)\eta^{-1} \qquad [8.24]$$

As η is usually much larger than one, the result [8.24] indicates that the bending stiffness of the ring is much less than the axial one. On the other hand, the tangential equation produces another rigid body mode. It is immediately identified as a free rotation around the ring axis.

NOTE. – *Correction of the natural frequencies due to the tangential vibration*

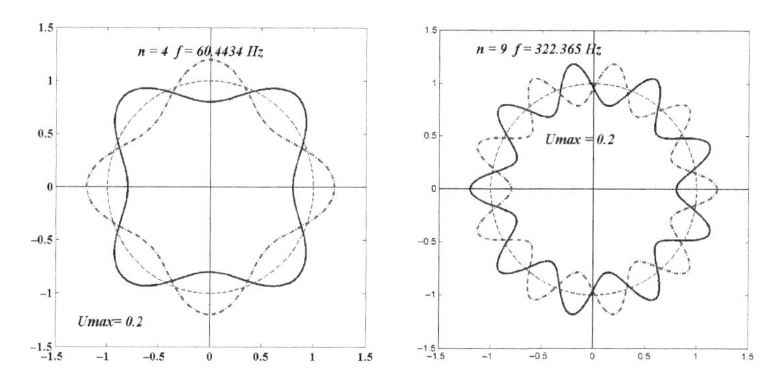

Figure 8.2. *Natural modes of vibration of a circular ring: in-plane bending*

The model of pure in-plane bending can be corrected by taking into account the tangential inertia. Including the tangential kinetic energy into the Rayleigh quotient we get:

$$f_n = \frac{n(n^2-1)c}{2\pi R\eta\sqrt{n^2+1}}, \qquad n = 1, 2, \ldots \qquad [8.25]$$

8.1.4 *Model coupling in-plane bending and axial vibrations*

8.1.4.1 Coupled equations

The model described in the last subsection needs to be improved for dealing with most of the problems concerned with transverse loading; the static equilibrium of an arch loaded by its own weight for instance. A better model can be built by taking into account coupling between longitudinal and normal in-plane displacements. Accordingly, the simplifying assumption [8.9] is relaxed. As a consequence, strain and kinetic energies are modified, now including terms which depend both on normal X_1 and axial X_s displacements and two coupled vibration equations are obtained. The calculation is detailed below.

Local strains are still given by [8.8]. However, as the condition [8.9] is withdrawn, the axial strain ε_{ss} is written as:

$$\varepsilon_{ss} = \eta_{ss} + \varsigma_1\chi_{ss} = \left(\frac{\partial X_s}{\partial s} + \frac{X_1}{R} - \varsigma_1\left(\frac{\partial^2 X_1}{\partial s^2} - \frac{\partial}{\partial s}\left(\frac{X_s}{R}\right)\right)\right)$$
$$\text{where} \quad \chi_{ss} = -\frac{\partial \psi_2}{\partial s} \quad \text{and} \quad \varepsilon_{s1} = 0 \Rightarrow \psi_2 = \frac{\partial X_1}{\partial s} - \frac{X_s}{R} \qquad [8.26]$$

The variation of strain energy is:

$$\delta[\mathcal{E}_s] = -\int_{s_1}^{s_2} \{\mathcal{M}_2\delta[\chi_{ss}] - \mathcal{N}_{ss}\delta[\eta_{ss}]\}\, ds$$

$$\delta[\mathcal{E}_s] = -\int_{s_1}^{s_2} \left\{\mathcal{M}_2\left(\frac{\partial^2(\delta X_1)}{\partial s^2} - \frac{\partial}{\partial s}\left(\frac{\delta X_s}{R}\right)\right) - \mathcal{N}_{ss}\left(\frac{\partial(\delta X_s)}{\partial s} + \frac{\delta X_1}{R}\right)\right\} ds \qquad [8.27]$$

From the variations of strain and kinetic energies plus external work, the equilibrium equations in the normal and the axial directions are obtained in the usual way as:

$$\rho S\ddot{X}_1 - \frac{\partial^2 \mathcal{M}_2}{\partial s^2} + \frac{\mathcal{N}_{ss}}{R} = F_1^{(e)}(s;t) - \frac{\partial \mathfrak{M}_2^{(e)}}{\partial s}$$
$$\rho S\ddot{X}_s - \frac{1}{R}\frac{\partial}{\partial s}(\mathcal{M}_2) - \frac{\partial \mathcal{N}_{ss}}{\partial s} = F_s^{(e)}(s;t) \quad [8.28]$$

Particularization to elastodynamics is straightforward. Elastic stresses relevant to the problem are:

$$\sigma_{ss} = E\varepsilon_{ss} \Rightarrow \mathcal{M}_2 = -EI_2\chi_{ss} \Rightarrow \mathcal{M}_2 = -EI_2\left(\frac{\partial^2 X_1}{\partial s^2} - \frac{\partial}{\partial s}\left(\frac{X_s}{R}\right)\right)$$
$$\mathcal{N}_{ss} = ES\eta_{ss} \Rightarrow \mathcal{N}_{ss} = ES\left(\frac{\partial X_s}{\partial s} + \frac{X_1}{R}\right) \quad [8.29]$$

whence the following coupled vibration equations:

$$\rho S\ddot{X}_1 + \frac{\partial^2}{\partial s^2}\left\{EI_2\left(\frac{\partial^2 X_1}{\partial s^2} - \frac{\partial}{\partial s}\left(\frac{X_s}{R}\right)\right)\right\} + \frac{ES}{R}\left\{\frac{\partial X_s}{\partial s} + \frac{X_1}{R}\right\}$$
$$= F_1^{(e)} - \frac{\partial \mathfrak{M}_2^{(e)}}{\partial s}$$
$$\rho S\ddot{X}_s + \frac{1}{R}\frac{\partial}{\partial s}\left\{EI_2\left(\frac{\partial^2 X_1}{\partial s^2} - \frac{\partial}{\partial s}\left(\frac{X_s}{R}\right)\right)\right\} - \frac{\partial}{\partial s}\left\{ES\left(\frac{\partial X_s}{\partial s} + \frac{X_1}{R}\right)\right\} = F_s^{(e)} \quad [8.30]$$

NOTE. – *Symmetry of the coupling operator*

As the system is conservative, the coupling stiffness operator has to be formally self-adjoint (see Chapter 3, subsection 3.3.5.). Such a property is not conspicuous in the present case since differential operators are of odd order. In the first equation (the so called transverse equation) the coupling differential operator is:

$$L_1[\,] = -\frac{\partial^2}{\partial s^2}\left(EI_2\frac{\partial}{\partial s}\frac{1}{R}[\,]\right) + \frac{ES}{R}\frac{\partial}{\partial s}[\,] \quad [8.31]$$

It superposes a bending term and an axial (or string) term. In the second equation (the axial equation) the coupling operator is:

$$L_2[\,] = +\frac{1}{R}\frac{\partial}{\partial s}\left(EI_2\frac{\partial^2}{\partial s^2}[\,]\right) - \frac{\partial}{\partial s}\left(\frac{ES}{R}\right)[\,] \quad [8.32]$$

Actually, the property of self-adjointness does not mean that $L_2[\,] = L_1[\,]$, but that the following symmetry condition of the energy functional holds:

$$\begin{aligned}\mathcal{W}(X_1, X_s) &= \int_{s_1}^{s_2} (X_1 L_1[X_s] + X_s L_2[X_1]) ds \\ &= \int_{s_1}^{s_2} \left(X_s L_1^{\#}[X_1] + X_1 L_2^{\#}[X_s] \right) ds \end{aligned} \qquad [8.33]$$

A necessary and sufficient condition for [8.33] to be valid is that L_1 and L_2 form a pair of adjoint operators $L_1 = L_2^{\#}; L_2 = L_1^{\#}$. It is left to the reader to check that this is the case here.

8.1.4.2 *Vibration modes of a circular ring*

The system [8.30] leads to the modal equations:

$$\begin{aligned}&\frac{EI}{R^4}\left(\frac{\partial^4 U}{\partial \theta^4} - \frac{\partial^3 V}{\partial \theta^3}\right) + \frac{ES}{R^2}\left(\frac{\partial V}{\partial \theta} + U\right) - \rho S \omega^2 U = 0 \\ -&\frac{ES}{R^2}\left(\frac{\partial^2 V}{\partial \theta^2} + \frac{\partial U}{\partial \theta}\right) + \frac{EI}{R^4}\left(\frac{\partial^3 U}{\partial \theta^3} - \frac{\partial^2 V}{\partial \theta^2}\right) - \rho S \omega^2 V = 0\end{aligned} \qquad [8.34]$$

which are rewritten in the following dimensionless matrix form:

$$\begin{gathered}\begin{bmatrix} L_{11} & L_{12} \\ L_{21} & L_{22} \end{bmatrix} \begin{bmatrix} u_n \\ v_n \end{bmatrix} = \begin{bmatrix} 0 \\ 0 \end{bmatrix} \\ L_{11} = 1 + \eta^{-2}\frac{\partial^4}{\partial \theta^4} - \varpi^2; \quad L_{12} = \frac{\partial}{\partial \theta} - \varpi^2 \frac{\partial^3}{\partial \theta^3} \\ L_{22} = -(1+\eta^{-2})\frac{\partial^2}{\partial \theta^2} - \varpi^2; \quad L_{21} = -\frac{\partial}{\partial \theta} + \varpi^2 \frac{\partial^3}{\partial \theta^3} \\ \varpi = \frac{\omega R}{c}\end{gathered} \qquad [8.35]$$

The mode shapes are of the following admissible type:

$$\begin{aligned}u_n(\theta) &= \alpha_n \cos n\theta + \beta_n \sin n\theta \\ v_n(\theta) &= a_n \cos n\theta + b_n \sin n\theta\end{aligned} \qquad [8.36]$$

The energy functional can be written as:

$$\mathcal{E} = \int_0^{2\pi} [u_n v_n] \begin{bmatrix} L_{11}[u_n] + L_{12}[v_n] \\ L_{21}[u_n] + L_{22}[v_n] \end{bmatrix} d\theta$$

$$\begin{aligned}
&L_{11}[u_n] = \{1 + \eta^{-2} n^4 - \varpi^2\} u_n \\
&L_{12}[v_n] = n(1 + \eta^{-2} n^2)(b_n \cos(n\theta) - a_n \sin(n\theta)) \\
&L_{21}[u_n] = n(1 + \eta^{-2} n^2)(\alpha_n \sin(n\theta) - \beta_n \cos(n\theta)) \\
&L_{22}[v_n] = \{n^2(1 + \eta^{-2}) - \varpi^2\} v_n
\end{aligned} \qquad [8.37]$$

which leads to the quadratic and symmetrical form:

$$\mathcal{E} = A_n \left(\alpha_n^2 + \beta_n^2\right) + B_n \left(a_n^2 + b_n^2\right) + 2C_n (b_n \alpha_n - \beta_n a_n)$$

$$A_n = 1 + \eta^{-2} n^4 - \varpi^2; \quad B_n = n^2(1 + \eta^{-2}) - \varpi^2; \quad C_n = n(1 + \eta^{-2} n^2) \qquad [8.38]$$

from which the following modal matrix system is obtained:

$$\begin{bmatrix} A_n & 0 & 0 & C_n \\ 0 & A_n & -C_n & 0 \\ 0 & -C_n & B_n & 0 \\ C_n & 0 & 0 & B_n \end{bmatrix} \begin{bmatrix} \alpha_n \\ \beta_n \\ a_n \\ b_n \end{bmatrix} = \begin{bmatrix} 0 \\ 0 \\ 0 \\ 0 \end{bmatrix} \qquad [8.39]$$

The characteristic equation is:

$$A_n B_n \left(A_n B_n - 2C_n^2\right) + C_n^4 = 0 \qquad [8.40]$$

In the particular case $n = 1$, [8.40] reduces to:

$$\varpi^4 \left(\varpi^2 - 2C_1\right)^2 = 0 \qquad [8.41]$$

The modes of free and rigid in-plane translations correspond to the multiple root $\varpi = 0$, which constitute a vector manifold spanned by the two orthonormed mode shapes:

$$u_1^{(1)} = \cos\theta; \quad v_1^{(1)} = -\sin\theta; \quad u_1^{(2)} = \sin\theta; \quad v_1^{(2)} = \cos\theta \qquad [8.42]$$

The vector manifold spanned by the two orthonormal mode shapes corresponds to the multiple root $\varpi = \sqrt{2C_1} = \sqrt{2(1+\eta^2)}$:

$$u_1^{(3)} = \cos\theta; \quad v_1^{(3)} = +\sin\theta; \quad u_1^{(4)} = \sin\theta; \quad v_1^{(4)} = -\cos\theta \qquad [8.43]$$

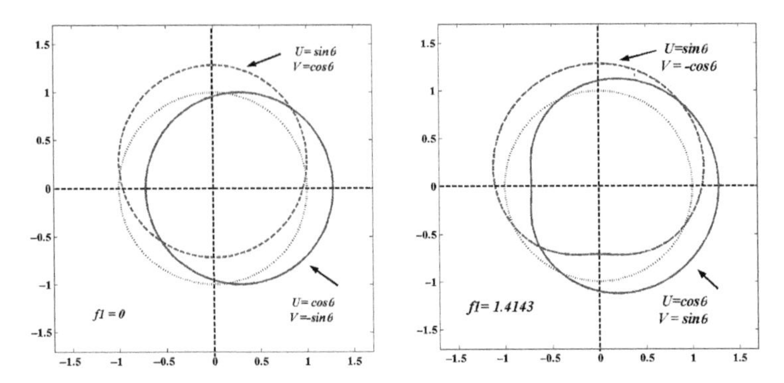

Figure 8.3. *In-plane modes shapes* $n = 1$ *of a circular ring*

Mode shapes [8.42] and [8.43] are shown in Figure 8.3. The value of the dimensionless natural frequency is also specified on the plots.

The natural pulsations of the modes $n > 1$ are given by the roots of the equation:

$$\varpi_n^4 - \Omega_1^2 \varpi_n^2 + \Omega_2^4 = 0$$

$$\text{where} \quad \Omega_1^2 = (n^2 + 1)(n^2 \eta^{-2} + 1); \quad \Omega_2^4 = \eta^{-2} n^2 (n^2 - 1)^2 \qquad [8.44]$$

As expected, there are two distinct families or branches defined as:

$$\varpi_{n1}^2 = \frac{\Omega_1^2}{2}\left\{1 - \sqrt{1 - 4\frac{\Omega_2^4}{\Omega_1^4}}\right\}; \quad \varpi_{n2}^2 = \frac{\Omega_1^2}{2}\left\{1 + \sqrt{1 - 4\frac{\Omega_2^4}{\Omega_1^4}}\right\} \qquad [8.45]$$

As indicated in Figure 8.4, the lower modal branch corresponds to predominant bending vibrations, such that $B_n \simeq -1/n$, and the upper branch corresponds to predominant longitudinal vibrations, such that $B_n \simeq n$. The large frequency gap between the two families is a consequence of the large value of the slenderness ratio $\eta = 2R/a$. Figure 8.5 shows the shapes of the first few modes (antinode assumed at $\theta = 0$), and specifies also their dimensionless natural frequencies.

8.1.4.3 Arch loaded by its own weight

A half-circular arch erected in a vertical plane is loaded by its own weight and supported on a rigid horizontal floor, see Figure 8.6. The curvilinear coordinate is defined as the direct angle θ, counted from the radius passing through the apex A.

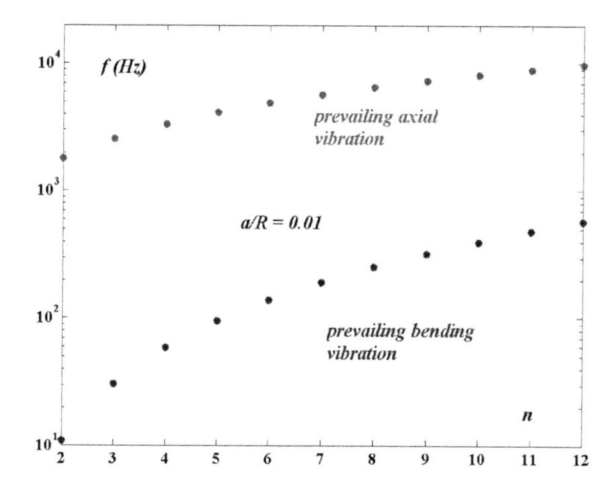

Figure 8.4. *In-plane modes coupling bending and axial vibration*

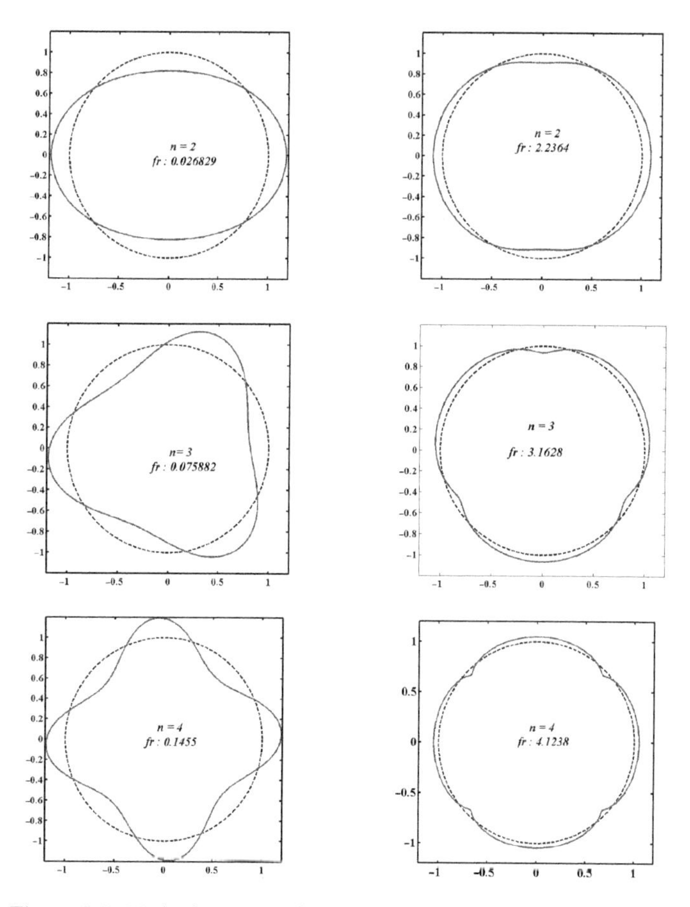

Figure 8.5. *Mode shapes coupling in-plane bending and axial vibration*

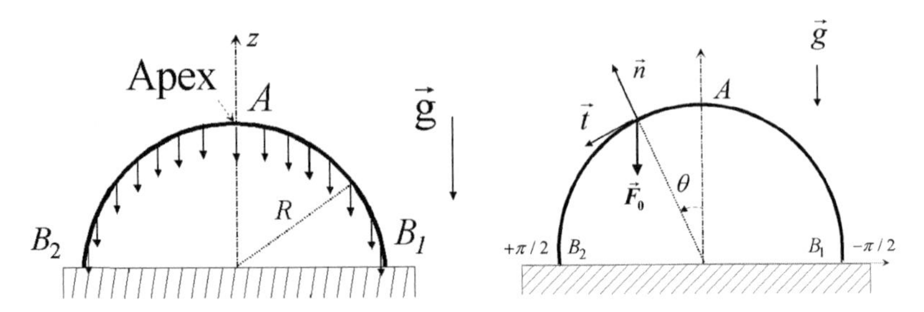

Figure 8.6. *Half-circular arch loaded by its own weight*

The load is defined by:

$$\vec{F}_0 = \rho S \vec{g} \Rightarrow \begin{cases} F_r = \vec{F}_0 \cdot \vec{n} = -F_0 \cos\theta \\ F_t = \vec{F}_0 \cdot \vec{t} = F_0 \sin\theta \end{cases} \quad \text{where } F_0 = \rho S g \qquad [8.46]$$

$S = \pi a^2$ is the cross-sectional area. The static equilibrium equations are written in the following dimensionless form:

$$\begin{aligned} \eta^{-2}\left(\frac{d^4 u}{d\theta^4} - \frac{d^3 v}{d\theta^3}\right) + \left(\frac{dv}{d\theta} + u\right) &= -f_0 \cos\theta \\ \eta^{-2}\left(\frac{d^3 u}{d\theta^3} - \frac{d^2 v}{d\theta^2}\right) - \left(\frac{d^2 v}{d\theta^2} + \frac{du}{d\theta}\right) &= +f_0 \sin\theta \end{aligned} \qquad [8.47]$$

where $u = U/a; v = V/a\ \eta = 2R/a$. The reduced loading is defined as:

$$f_0 = \frac{\rho g R^2}{Ea} \qquad [8.48]$$

As the arch is assumed to slide freely on the horizontal ground, the boundary conditions at B_1, B_2 are:

$$\begin{aligned} & v(\pm\pi/2) = 0 \\ & \left.\frac{d^2 v}{d\theta^2} - \frac{du}{d\theta}\right|_{\pm\pi/2} = 0; \quad \left.\frac{d^3 v}{d\theta^3} - \frac{d^2 u}{d\theta^2}\right|_{\pm\pi/2} = 0 \end{aligned} \qquad [8.49]$$

As already indicated in Chapter 7 subsection 7.2.5, the problem may be solved by expanding the displacement and the load in Fourier series. As these fields are of interest in the interval $-\pi/2 \leq \theta \leq +\pi/2$ solely, they can be assumed to be π periodic. In agreement with the coordinate system specified in Figure 8.6, the series are written as:

$$u(\theta)=\sum_{n=0}^{\infty} q_n \cos(n\theta); \quad v(\theta)=\sum_{n=1}^{\infty} p_n \sin(n\theta)$$

$$f_r(\theta)=-f_0\cos\theta=\sum_{n=0}^{\infty} r_n\cos(n\theta); \quad f_t(\theta)=f_0\sin\theta=\sum_{n=1}^{\infty} t_n \sin(n\theta) \tag{8.50}$$

The generalized radial and tangential forces r_n, t_n are given by the following integrals:

$$r_0=-\frac{f_0}{2\pi}\int_{-\pi/2}^{+\pi/2}\cos\theta\, d\theta=-\frac{f_0}{\pi}; \quad r_1=-\frac{f_0}{\pi}\int_{-\pi/2}^{+\pi/2}(\cos\theta)^2 d\theta=-\frac{f_0}{2}$$

$$r_{n>1}=-\frac{f_0}{\pi}\int_{-\pi/2}^{+\pi/2}\cos(n\theta)\cos\theta\, d\theta=\begin{cases}\dfrac{2f_0(-1)^k}{\pi(4k^2-1)} & \text{if } n=2k\\ 0 & \text{if } n=2k+1\end{cases} \tag{8.51}$$

$$t_1=\frac{f_0}{\pi}\int_{-\pi/2}^{+\pi/2}(\sin\theta)^2\, d\theta=\frac{f_0}{2}$$

$$t_{n>1}=-\frac{f_0}{\pi}\int_{-\pi/2}^{+\pi/2}\sin(n\theta)\sin\theta\, d\theta=\begin{cases}\dfrac{-4f_0k(-1)^k}{\pi(4k^2-1)} & \text{if } n=2k\\ 0 & \text{if } n=2k+1\end{cases} \tag{8.52}$$

In principle, the generalized displacements q_n, p_n are obtained by identifying term by term the harmonic coefficients appearing on the left-hand and the right-hand sides of the equations. However, if the Fourier series just defined above are directly substituted into the system [8.47], the method fails. Indeed, for $n=1$ the system has no solution, as it is of the following form:

$$\begin{bmatrix}1+\eta^{-2} & 1+\eta^{-2}\\ 1+\eta^{-2} & 1+\eta^{-2}\end{bmatrix}\begin{bmatrix}q_1\\ p_1\end{bmatrix}=\frac{f_0}{2}\begin{bmatrix}-1\\ +1\end{bmatrix} \tag{8.53}$$

The difficulty can be avoided by writing the Fourier series in a slightly distinct, but equivalent form, as follows:

$$-f_0\cos\theta=-f_0\left\{\frac{1}{\pi}+\frac{\cos\theta}{2}-\frac{2}{\pi}\sum_{k=1}^{\infty}\frac{(-1)^k\cos(2k\theta)}{4k^2-1}\right\} \Rightarrow$$

$$-f_0\cos\theta=-2f_0\left\{\frac{1}{\pi}-\frac{2}{\pi}\sum_{k=1}^{\infty}\frac{(-1)^k\cos(2k\theta)}{4k^2-1}\right\}$$

$$f_0 \sin\theta = f_0 \left\{ \frac{\sin\theta}{2} - \frac{4}{\pi} \sum_{k=1}^{\infty} \frac{k(-1)^k \sin(2k\theta)}{4k^2 - 1} \right\} \Rightarrow$$

$$f_0 \sin\theta = -\frac{8F_0}{\pi} \left\{ \sum_{k=1}^{\infty} \frac{k(-1)^k \sin(2k\theta)}{4k^2 - 1} \right\}$$

[8.54]

Using the series [8.54], which comprises even harmonics only, the following results are found:

$$n = 0 \Rightarrow q_0 = -\frac{2f_0}{\pi} \qquad [8.55]$$

$$n = 2k \neq 0$$

$$\begin{bmatrix} 16\eta^{-2}k^4 + 1 & 2k(4\eta^{-2}k^2 + 1) \\ 2k(4\eta^{-2}k^2 + 1) & 4k^2(\eta^{-2} + 1) \end{bmatrix} \begin{bmatrix} q_k \\ p_k \end{bmatrix} = \frac{4f_0(-1)^k}{\pi(4k^2 - 1)} \begin{bmatrix} 1 \\ -2k \end{bmatrix} \qquad [8.56]$$

The deflection evaluated with ten harmonics, for a steel arch $R = 10$ m and $a = 20$ cm, is shown in Figure 8.7. The apex is lowered by about $Z_m \simeq 30$ cm and the sliding ends separate horizontally by $X_m \simeq 48$ cm. As a conclusion of this example, it is worth emphasizing that when the in-plane load applied to an arch is not purely normal, it is balanced by stresses comprising necessarily string and bending components. This result is in contrast to that obtained in the case of shells in Chapter 7 subsection 7.3.7, where the most important component to resist external loading was found to be the hoop stress.

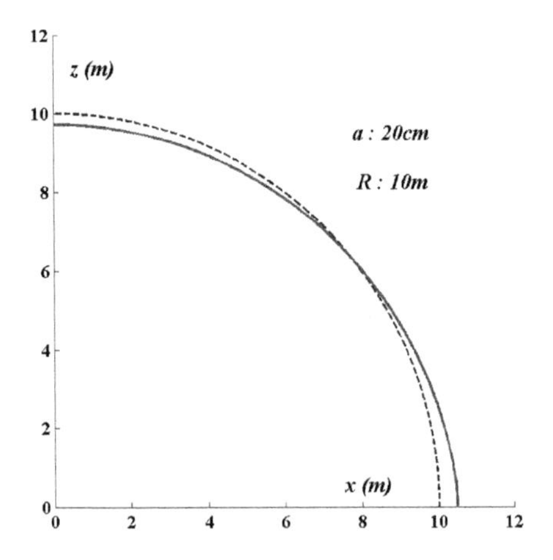

Figure 8.7. *Static deflection of a half circular arch loaded by its own weight*

8.1.5 *Model coupling torsion and out-of-plane bending*

8.1.5.1 Coupled equations of vibration

Coupling between torsion and out-of-plane bending can be put in evidence, in a rather intuitive way, based on Figures 8.8–8.10. Figure 8.8 shows a ring element and a fibre passing through a point P of the cross-section. Torsion is described by the small axial rotation ψ_s whereas out-of-plane bending is described by the small rotation about the in-plane normal axis $\vec{n}_1$ and the related translation X_2 of cross-sections in the $\vec{n}_2$ out-of-plane direction. The displacement, induced by ψ_s, of a pair of diametrically opposite points P and Q is shown in Figure 8.9. The polar coordinates of P, Q in the non-deformed state are noted r, α and r, $\alpha + \pi$ (on the figure $\alpha = \pi/2$). The two fibre elements subtended by the angle $d\theta$ which pass through P and Q respectively, are of the same length $\ell = R(1 + r\cos\alpha)\,d\theta$. Due to the rotation ψ_s, P is mapped into P' which is closer to the ring centre than P.

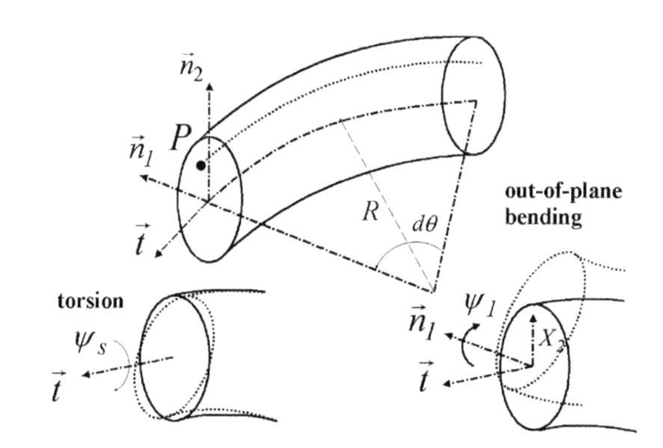

Figure 8.8. *Out-of-plane bending and torsion of a circular ring*

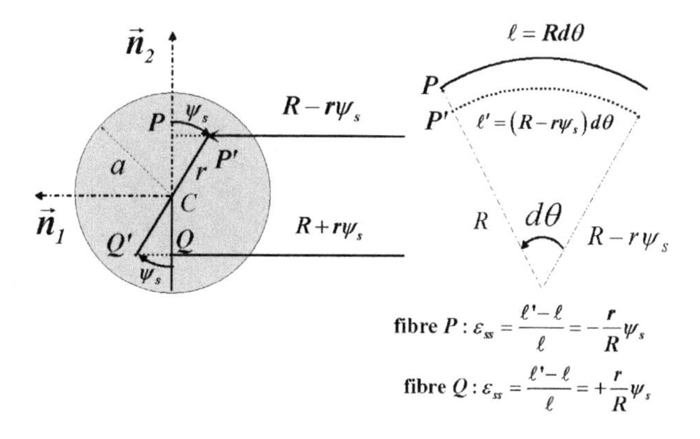

Figure 8.9. *Axial strains induced by torsion*

So the fibre length is modified, becoming

$$\ell' = (R(1 + r\cos\alpha) - r\sin\psi_s)\, d\theta \simeq (R(1 + r\cos\alpha) - r\psi_s)\, d\theta$$

Q is transformed into Q' which is further from the ring centre than Q. The fibre length becomes $\ell'' \cong (R(1 + r\cos\alpha) + r\psi_s)\, d\theta$. Accordingly, it is found that torsion induces a longitudinal local strain given by:

$$\varepsilon_{ss} = -\frac{r}{R}\psi_s \operatorname{sign}(\sin\alpha) \qquad [8.57]$$

The local longitudinal stress field for an elastic material is sketched in Figure 8.10. Thus, the resulting global stress reduces to a bending moment about $\vec{n}_1$ in the ring plane, which is proportional to the torsion angle and to the geometrical curvature of the ring. To formulate the problem, the following components of displacement are assumed to vanish:

$$X_1 = X_s = 0; \quad \psi_2 = 0 \qquad [8.58]$$

It follows that the local strains are:

$$\begin{aligned} \varepsilon_{ss} &= \varsigma_2 \left(\frac{\partial \psi_1}{\partial s} - \frac{\psi_s}{R} \right) \\ \varepsilon_{s1} &= -\frac{\varsigma_2}{2} \left(\frac{\psi_1}{R} + \frac{\partial \psi_s}{\partial s} \right) \\ \varepsilon_{s2} &= \frac{1}{2} \left(\psi_1 + \frac{\partial X_2}{\partial s} + \varsigma_1 \frac{\partial \psi_s}{\partial s} \right) \end{aligned} \qquad [8.59]$$

The rotation angle of the cross-sections is related to the out-of-plane displacement by:

$$\psi_1 = -\frac{\partial X_2}{\partial s} \qquad [8.60]$$

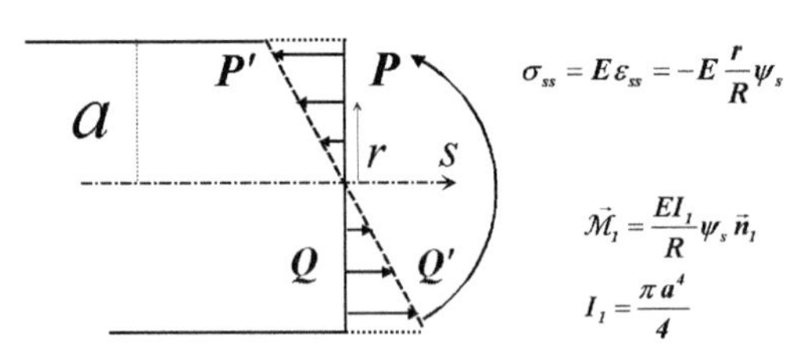

Figure 8.10. *Local bending stresses induced by torsion*

Substituting [8.60] into [8.59], we obtain:

$$\varepsilon_{ss} = -\varsigma_2 \left(\frac{\partial^2 X_2}{\partial s^2} + \frac{\psi_s}{R} \right); \quad \varepsilon_{s1} = \frac{\varsigma_2}{2} \left(\frac{1}{R} \frac{\partial X_2}{\partial s} - \frac{\partial \psi_s}{\partial s} \right); \quad \varepsilon_{s2} = \frac{\varsigma_1}{2} \frac{\partial \psi_s}{\partial s} \tag{8.61}$$

The variation in strain energy is written as:

$$\delta \left[\mathcal{E}_s \right] = -\int_{t_1}^{t_2} dt \int_{s_1}^{s_2} \left\{ \mathcal{M}_1 \left(\left(\frac{\partial^2 \delta X_2}{\partial s^2} + \frac{\delta \psi_s}{R} \right) \right) - \mathcal{M}_{ss} \frac{\partial \delta \psi_s}{\partial s} - \frac{\mathcal{M}_{s_1}}{R} \frac{\partial \delta X_2}{\partial s} \right\} ds \tag{8.62}$$

where $\mathcal{M}_1$ is the bending moment about $\vec{n}_1$. $\mathcal{M}_{ss}$ is the torsion moment. Finally, $\mathcal{M}_{s1}$ designates the moment of the shear stress σ_{s1} which does not induces torsion. Hamilton's principle is written as:

$$\delta[\mathcal{A}] = -\int_{t_1}^{t_2} dt \int_{s_1}^{s_2} \left\{ \begin{array}{l} \rho S \dot{X}_2 \delta \dot{X}_2 + \rho J \dot{\psi}_s \delta \dot{\psi}_s \\ + \mathcal{M}_1 \left(\left(\dfrac{\partial^2 \delta X_2}{\partial s^2} + \dfrac{\delta \psi_s}{R} \right) \right) \\ - \mathcal{M}_{ss} \dfrac{\partial \delta \psi_s}{\partial s} - \dfrac{\mathcal{M}_{s_1}}{R} \dfrac{\partial \delta X_2}{\partial s} \end{array} \right\} ds = 0 \tag{8.63}$$

S is the area and J the area polar moment of inertia of the cross-sections.

After a few standard manipulations, the following equilibrium equations are obtained, which are coupled in torsion and out-of-plane bending:

$$\begin{aligned} \rho S \ddot{X}_2 - \frac{\partial^2 \mathcal{M}_1}{\partial s^2} - \frac{\partial}{\partial s} \left(\frac{\mathcal{M}_{s1}}{R} \right) &= F_2^{(e)} - \frac{\partial \mathfrak{M}_1^{(e)}}{\partial s} \\ \rho J \ddot{\psi}_s - \frac{\partial \mathcal{M}_{ss}}{\partial s} - \frac{\mathcal{M}_1}{R} &= \mathfrak{M}_{ss}^{(e)} \end{aligned} \tag{8.64}$$

In [8.64] external loading comprises the densities per unit length of a transverse force, denoted $F_2^{(e)}$, of a bending moment denoted $\mathfrak{M}_1^{(e)}$ and of a torsion moment denoted $\mathfrak{M}_{ss}^{(e)}$. The elastic stresses are:

$$\sigma_{ss} = E \varepsilon_{ss} \Rightarrow \mathcal{M}_1 = -E \int_{(s)} \varsigma_2^2 \left(\frac{\partial^2 X_2}{\partial s^2} + \frac{\psi_s}{R} \right) dS = -E I_1 \left(\frac{\partial^2 X_2}{\partial s^2} + \frac{\psi_s}{R} \right) \tag{8.65}$$

where I_1 is the area moment about the principal axis $\vec{n}_1$.

On the other hand, the shear elastic stresses are:

$$\sigma_{s1} = 2G\varepsilon_{s1}; \quad \sigma_{s2} = 2G\varepsilon_{s2} \Rightarrow$$

$$\mathcal{M}_{ss} = G\frac{\partial \psi_s}{\partial s}\iint_{(S)}\left(\varsigma_1^2 + \varsigma_2^2\right) dS = GJ\frac{\partial \psi_s}{\partial s};$$

$$\mathcal{M}_{s1} = \frac{G}{R}\frac{\partial X_2}{\partial s}\iint_{(S)}\frac{\varsigma_2^2}{R}dS = \frac{GI_1}{R}\frac{\partial X_2}{\partial s} \qquad [8.66]$$

Substituting [8.65] and [8.66] into the equilibrium equations [8.64], the vibration equations are written as:

$$\rho S\ddot{X}_2 + \frac{\partial^2}{\partial s^2}EI_1\left(\frac{\partial^2 X_2}{\partial s^2} + \frac{\psi_s}{R}\right) - \frac{\partial}{\partial s}\left(\frac{GI_1}{R^2}\frac{\partial X_2}{\partial s}\right) = F_2^{(e)} - \frac{\partial \mathfrak{M}_1^{(e)}}{\partial s}$$

$$\rho J\ddot{\psi}_s + \frac{EI_1}{R}\left(\frac{\partial^2 X_2}{\partial s^2} + \frac{\psi_s}{R}\right) - \frac{\partial}{\partial s}\left(GJ\frac{\partial \psi_s}{\partial s}\right) = \mathfrak{M}_{ss}^{(e)} \qquad [8.67]$$

Here also, it could be checked that the coupling operator is self-adjoint.

8.1.5.2 Natural modes of vibration of a circular ring

Denoting by ψ the torsion angle and by W the out-of-plane translation, the system [8.67] takes the particular form:

$$\rho S\ddot{W} + \frac{EI_1}{R^4}\left(\frac{\partial^4 W}{\partial \theta^4} + R\frac{\partial^2 \psi}{\partial \theta^2}\right) - \frac{GI_1}{R^4}\frac{\partial^2 W}{\partial \theta^2} = 0$$

$$\rho J\ddot{\psi} + \frac{EI_1}{R^3}\left(\frac{\partial^2 W}{\partial \theta^2} + R\psi\right) - \frac{GJ}{R^2}\frac{\partial^2 \psi}{\partial \theta^2} = 0 \qquad [8.68]$$

We proceed in the same manner as in subsection 8.1.4.2. Mathematical manipulations are alleviated by selecting a priori mode shapes with an antinode at $\theta = 0$. Admissible functions are found to be $W_n = \alpha_n \cos(n\theta)$ and $\psi_n = \beta_n \cos(n\theta)$. The energy functional can be written as:

$$\mathcal{E} = \int_0^{2\pi} [W_n \quad R\psi_n]\begin{bmatrix} L_{11}[W_n] + L_{12}[R\psi_n] \\ L_{21}[W_n] + L_{22}[R\psi_n] \end{bmatrix} d\theta$$

$$L_{11}[W_n] = \left\{\eta^{-2}n^4 + \frac{n^2}{2(1+\nu)} - \varpi^2\right\} W_n; \quad L_{12}[\nu_n] = -\eta^{-2}n^2 R\psi_n \qquad [8.69]$$

$$L_{22}[R\psi_n] = 2\left\{\left(\frac{1}{2} + \frac{n^2}{2(1+\nu)}\right) - \varpi^2\right\} R\psi_n; \quad L_{12}[u_n] = -nW_n$$

So, the energy functional takes the form:

$$\mathcal{E} = \left\{\Omega_1^2 - \varpi^2\right\}\alpha_n^2 + 2\left\{\Omega_2^2 - \varpi^2\right\}\beta_n^2 - 2\Omega_c^2\alpha_n\beta_n$$

$$\Omega_1^2 = \eta^{-2}n^4 + \frac{n^2}{2(1+\nu)}; \quad \Omega_2^2 = \left(\frac{1}{2} + \frac{n^2}{2(1+\nu)}\right); \quad \Omega_c^2 = \frac{n^2\left(1+\eta^{-2}\right)}{2} \qquad [8.70]$$

which leads to the following modal system:

$$\begin{bmatrix} \Omega_1^2 - \omega^2 & -\Omega_c^2 \\ -\Omega_c^2 & 2\left(\Omega_2^2 - \omega^2\right) \end{bmatrix} \begin{bmatrix} \alpha_n \\ \beta_n \end{bmatrix} = \begin{bmatrix} 0 \\ 0 \end{bmatrix} \qquad [8.71]$$

Again there exist two modal branches, such that:

$$\varpi_{(1)n}^2 = \frac{\Omega_1^2 + \Omega_2^2 + \sqrt{\Delta}}{2}; \quad \varpi_{(2)n}^2 = \frac{\Omega_1^2 + \Omega_2^2 - \sqrt{\Delta}}{2};$$

$$\Delta = \left(\Omega_1^2 - \Omega_2^2\right)^2 + 2\Omega_c^4 \qquad [8.72]$$

In Figure 8.11, the natural frequencies of the out-of-plane modes are plotted versus n, for a steel ring $R = 1\,\mathrm{m}$ $a/R = 0.1$. The modes of the lower branch vibrate more in bending than in torsion up to a given modal index n_i beyond which torsion prevails, and the modes of the upper branch behaves the opposite. n_i is proportional to R/a. Such a behaviour is due to the fact that the bending stiffness operator comprises a θ derivative of order four, while that of the torsion stiffness comprises

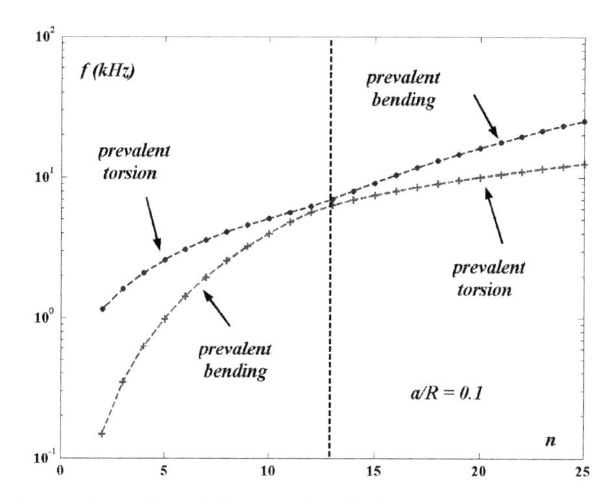

Figure 8.11. *Out-of-plane modes of vibration of a circular ring*

a θ derivative of the order two only. So, as n increases, the ring becomes stiffer in bending than in torsion. On the other hand, it is worth noting that the two frequency branches of Figure 8.11 are very close to each other in the vicinity of n_i ($n_i = 13$ in Figure 8.11), without however crossing each other. This is because the discriminant $\Delta(n)$ is strictly positive whatever the value of n may be. Such behaviour is a common feature to any coupled conservative system. This can be easily justified by noting that in the case when Δ could cross a zero value, the natural frequencies of the system would become complex in the range $\Delta < 0$; which is characteristic of nonconservative systems.

8.2. Thin shells

8.2.1 *Local and global tensor of small strains*

The calculation procedure of small strains in a thin shell is similar to that already described for the arch case. Though the mathematical manipulations are heavier, no new difficulties are encountered.

8.2.1.1 Local displacement field

It is recalled (cf. Chapter 7, subsection 7.3.2) that the local displacement field is written as:

$$\xi_\alpha = X_\alpha - \varsigma\psi_\beta; \quad \xi_\beta = X_\beta - \varsigma\psi_\alpha; \quad \xi_\varsigma = X_\varsigma \qquad [8.73]$$

where ζ designates the coordinate in the direction $\vec{n}$, normal to the mid-surface of the shell.

The coefficients of the metric tensor are:

$$G_\alpha = g_\alpha\left(1 + \frac{\varsigma}{R_\alpha}\right); \quad G_\beta = g_\beta\left(1 + \frac{\varsigma}{R_\beta}\right); \quad G_\varsigma = 1 \qquad [8.74]$$

Furthermore, in this book, study is restricted to thin shells, so h/R_α and h/R_β are much less than one. As a consequence, the ratios $\varsigma/R_\alpha, \varsigma/R_\beta$ are interpreted as infinitesimal dimensionless quantities of first order.

8.2.1.2 Expression of the local and global strain components

Making use of the results established in Appendix A.3, it is easily found that the small strains of a 2D structure can be expressed in orthogonal curvilinear coordinates as:

$$\varepsilon_{\alpha\alpha} = \frac{1}{G_\alpha}\left\{\frac{\partial \xi_\alpha}{\partial \alpha} + \frac{\xi_\beta}{G_\beta}\frac{\partial G_\alpha}{\partial \beta}\right\}; \quad \varepsilon_{\beta\beta} = \frac{1}{G_\beta}\left\{\frac{\partial \xi_\beta}{\partial \beta} + \frac{\xi_\alpha}{G_\alpha}\frac{\partial G_\beta}{\partial \alpha}\right\}$$

$$\varepsilon_{\alpha\beta} = \frac{1}{2}\left\{\frac{G_\alpha}{G_\beta}\frac{\partial}{\partial \beta}\left(\frac{\xi_\alpha}{G_\alpha}\right) + \frac{G_\beta}{G_\alpha}\frac{\partial}{\partial \alpha}\left(\frac{\xi_\beta}{G_\beta}\right)\right\} \qquad [8.75]$$

Substituting [8.73] and [8.74] into [8.75], and focusing on flexure and torsion components only, since membrane components were described in the last chapter

$$\varepsilon_{\alpha\alpha} = -\varsigma\chi_{\alpha\alpha}; \quad \varepsilon_{\beta\beta} = -\varsigma\chi_{\beta\beta}; \quad \varepsilon_{\alpha\beta} = -\varsigma\chi_{\alpha\beta} \qquad [8.76]$$

is obtained.

Small bending and torsional strains are given up to the first order by:

$$\chi_{\alpha\alpha} = -\left(\frac{1}{g_\alpha}\frac{\partial \psi_\beta}{\partial \alpha} + \frac{\psi_\alpha}{g_\alpha g_\beta}\frac{\partial g_\alpha}{\partial \beta}\right); \quad \chi_{\beta\beta} = -\left(\frac{1}{g_\beta}\frac{\partial \psi_\alpha}{\partial \beta} + \frac{\psi_\beta}{g_\alpha g_\beta}\frac{\partial g_\beta}{\partial \alpha}\right)$$

$$\chi_{\alpha\beta} = -\frac{1}{2}\left\{\frac{g_\alpha}{g_\beta}\frac{\partial}{\partial \beta}\left(\frac{\psi_\beta}{g_\alpha}\right) + \frac{g_\beta}{g_\alpha}\frac{\partial}{\partial \alpha}\left(\frac{\psi_\alpha}{g_\beta}\right)\right\}$$

[8.77]

According to the simplifying assumptions of the Kirchhoff–Love model, the transverse shear strains are neglected. As a consequence, the flexure rotations can be expressed in terms of the global translation components. The formal expression of the transverse shear strains is:

$$\varepsilon_{\alpha\varsigma} = \frac{1}{2}\left\{G_\alpha\frac{\partial}{\partial \varsigma}\left(\frac{\xi_\alpha}{G_\alpha}\right) + \frac{1}{G_\alpha}\frac{\partial X_\varsigma}{\partial \alpha}\right\}$$

$$= \frac{1}{2}\left\{G_\alpha\left(\frac{G_\alpha\xi'_\alpha - G'_\alpha\xi_\alpha}{G_\alpha^2}\right) + \frac{1}{G_\alpha}\frac{\partial X_\varsigma}{\partial \alpha}\right\} \equiv 0 \qquad [8.78]$$

$$\xi'_\alpha = \frac{\partial \xi_\alpha}{\partial \varsigma} = \psi_\beta; \quad G'_\alpha = \frac{\partial G_\alpha}{\partial \varsigma} = \frac{g_\alpha}{R_\alpha}$$

Retaining the terms of first order only, [8.78] becomes:

$$\varepsilon_{\alpha\varsigma} = \frac{1}{2}\left\{-\psi_\beta - \frac{X_\alpha}{R_\alpha} + \frac{1}{g_\alpha}\frac{\partial X_\varsigma}{\partial \alpha}\right\} = 0; \quad \varepsilon_{\beta\varsigma} = \frac{1}{2}\left\{\psi_\alpha - \frac{X_\beta}{R_\beta} + \frac{1}{g_\beta}\frac{\partial X_\varsigma}{\partial \beta}\right\} = 0$$

[8.79]

The flexure rotations are thus similar to those established for an arch:

$$\psi_\alpha = \frac{1}{g_\beta}\frac{\partial X_\varsigma}{\partial \beta} - \frac{X_\beta}{R_\beta}; \quad \psi_\beta = \frac{1}{g_\alpha}\frac{\partial X_\varsigma}{\partial \alpha} - \frac{X_\alpha}{R_\alpha} \qquad [8.80]$$

Relations [8.80] differ from those which hold for plates, by a single additional term related to the geometric curvature of the shell. Substituting [8.80] into [8.77], global strains in flexure and torsion are expressed as:

$$\chi_{\alpha\alpha} = \frac{-1}{g_\alpha}\left[\left\{\frac{\partial}{\partial\alpha}\left(\frac{X_\alpha}{R_\alpha}\right)+\frac{X_\beta}{R_\beta}\frac{\partial g_\alpha}{g_\beta\partial\beta}\right\}-\left\{\frac{\partial}{\partial\alpha}\left(\frac{\partial X_\varsigma}{g_\alpha\partial\alpha}\right)+\frac{\partial X_\varsigma}{g_\beta^2\partial\beta}\frac{\partial g_\alpha}{\partial\beta}\right\}\right]$$

$$\chi_{\beta\beta} = \frac{-1}{g_\beta}\left[\left\{\frac{\partial}{\partial\beta}\left(\frac{X_\beta}{R_\beta}\right)+\frac{X_\alpha}{R_\alpha}\frac{\partial g_\beta}{g_\alpha\partial\alpha}\right\}-\left\{\frac{\partial}{\partial\beta}\left(\frac{\partial X_\varsigma}{g_\beta\partial\beta}\right)+\frac{\partial g_\beta}{g_\alpha^2\partial\alpha}\frac{\partial X_\varsigma}{\partial\alpha}\right\}\right]$$

$$\chi_{\alpha\beta} = \frac{-1}{2}\left[\left\{\frac{g_\alpha}{g_\beta}\frac{\partial}{\partial\beta}\left(\frac{X_\alpha}{g_\alpha R_\alpha}\right)+\frac{g_\beta}{g_\alpha}\frac{\partial}{\partial\alpha}\left(\frac{X_\beta}{g_\beta R_\beta}\right)\right\}\right.$$
$$\left.-\left\{\left(\frac{g_\alpha}{g_\beta}\frac{\partial}{\partial\beta}\left(\frac{\partial X_\varsigma}{g_\alpha^2\partial\alpha}\right)+\frac{g_\beta}{g_\alpha}\frac{\partial}{\partial\alpha}\left(\frac{\partial X_\varsigma}{g_\beta^2\partial\beta}\right)\right)\right\}\right]$$

[8.81]

In [8.81], the terms related to the shell's curvatures are within the first pair of braces while the plate terms are within the second pair.

8.2.2 *Love's equations of equilibrium*

The expression [8.81] indicates clearly that the shell equations differ from the plate equations by additional terms related to the geometrical curvatures into two principal and orthogonal directions. Furthermore, in a shell, coupling between the radial and the tangential displacements arises as a consequence of curvature effects. The plate equations are recalled here, to make easier the comparison with the shell equations, which shall be established afterwards. The in-plane motions are governed by the two membrane equations [5.90]:

$$\rho h\ddot{X}_\alpha - \frac{1}{g_\alpha g_\beta}\left\{\frac{\partial\left(g_\beta \mathcal{N}_{\alpha\alpha}\right)}{\partial\alpha}-\mathcal{N}_{\beta\beta}\frac{\partial g_\beta}{\partial\alpha}+\frac{1}{g_\alpha}\frac{\partial\left(g_\alpha^2\mathcal{N}_{\beta\alpha}\right)}{\partial\beta}\right\} = f_\alpha^{(e)}(\alpha,\beta;t)$$

$$\rho h\ddot{X}_\beta - \frac{1}{g_\alpha g_\beta}\left\{\frac{\partial\left(g_\alpha \mathcal{N}_{\beta\beta}\right)}{\partial\beta}-\mathcal{N}_{\alpha\alpha}\frac{\partial g_\alpha}{\partial\beta}+\frac{1}{g_\beta}\frac{\partial\left(g_\beta^2\mathcal{N}_{\alpha\beta}\right)}{\partial\alpha}\right\} = f_\beta^{(e)}(\alpha,\beta;t)$$

The out-of-plane motions are governed by the flexure-torsion equation [6.115]:

$$\rho h\ddot{Z} - \frac{1}{g_\alpha g_\beta}\left(\frac{\partial\left(g_\beta Q_{\alpha z}\right)}{\partial\alpha}+\frac{\partial\left(g_\alpha Q_{\beta z}\right)}{\partial\beta}\right) = f_z^{(e)}(\alpha,\beta;t)+\frac{\partial\mathfrak{M}_\alpha^{(e)}}{g_\alpha\partial\alpha}+\frac{\partial\mathfrak{M}_\beta^{(e)}}{g_\beta\partial\beta}$$

where the transverse shear forces are expressed as:

$$Q_{\alpha\varsigma} = \frac{1}{g_\alpha g_\beta}\left(\frac{\partial\left(g_\beta \mathcal{M}_{\alpha\alpha}\right)}{\partial\alpha} + \frac{1}{g_\alpha}\frac{\partial\left(g_\alpha^2 \mathcal{M}_{\alpha\beta}\right)}{\partial\beta} - \mathcal{M}_{\beta\beta}\frac{\partial g_\beta}{\partial\alpha}\right)$$

$$Q_{\beta\varsigma} = \frac{1}{g_\alpha g_\beta}\left(\frac{\partial\left(g_\alpha \mathcal{M}_{\beta\beta}\right)}{\partial\beta} + \frac{1}{g_\beta}\frac{\partial\left(g_\beta^2 \mathcal{M}_{\beta\alpha}\right)}{\partial\alpha} - \mathcal{M}_{\alpha\alpha}\frac{\partial g_\alpha}{\partial\beta}\right)$$

Turning now to the case of shells, the variational calculus concerning the strain energy due to the shell curvature terms is straightforward. Summing the plate and the shell curvature terms the equations of thin shells, termed Love's equations, are finally obtained. They are written as:

$$\rho h\ddot{X}_\alpha - \frac{1}{g_\alpha g_\beta}\left\{\frac{\partial\left(g_\beta \mathcal{N}_{\alpha\alpha}\right)}{\partial\alpha} - \mathcal{N}_{\beta\beta}\frac{\partial g_\beta}{\partial\alpha} + \frac{1}{g_\alpha}\frac{\partial\left(g_\alpha^2 \mathcal{N}_{\beta\alpha}\right)}{\partial\beta}\right\} - \frac{Q_{\alpha\varsigma}}{R_\alpha}$$
$$= f_\alpha^{(e)}(\alpha,\beta;t)$$

$$\rho h\ddot{X}_\beta - \frac{1}{g_\alpha g_\beta}\left\{\frac{\partial\left(g_\alpha \mathcal{N}_{\beta\beta}\right)}{\partial\beta} - \mathcal{N}_{\alpha\alpha}\frac{\partial g_\alpha}{\partial\beta} + \frac{1}{g_\beta}\frac{\partial\left(g_\beta^2 \mathcal{N}_{\alpha\beta}\right)}{\partial\alpha}\right\} - \frac{Q_{\beta\varsigma}}{R_\beta}$$
$$= f_\beta^{(e)}(\alpha,\beta;t)$$

$$\rho h\ddot{X}_\varsigma - \frac{1}{g_\alpha g_\beta}\left(\frac{\partial\left(g_\beta Q_{\alpha\varsigma}\right)}{\partial\alpha} + \frac{\partial\left(g_\alpha Q_{\beta\varsigma}\right)}{\partial\beta}\right) + \left(\frac{\mathcal{N}_{\alpha\alpha}}{R_\alpha} + \frac{\mathcal{N}_{\beta\beta}}{R_\beta}\right)$$
$$= f_\varsigma^{(e)}(\alpha,\beta;t) + \frac{\partial\mathfrak{M}_\alpha^{(e)}}{g_\alpha\partial\alpha} + \frac{\partial\mathfrak{M}_\beta^{(e)}}{g_\beta\partial\beta} \qquad [8.82]$$

The equations for elastic vibrations can be easily deduced from [8.82]; however, the expressions are cumbersome. So it is generally preferred to specify them by using the peculiarities of the problem to be solved. This will be done in the next subsection taking the case of circular cylindrical shells.

8.3. Circular cylindrical shells

8.3.1 *Equilibrium equations*

8.3.1.1 Love's equations in cylindrical coordinates

As shown in Figure 8.12, the mid-surface is described by the axial coordinate z and the polar angle θ. The radius of curvature of the meridian lines is infinite and

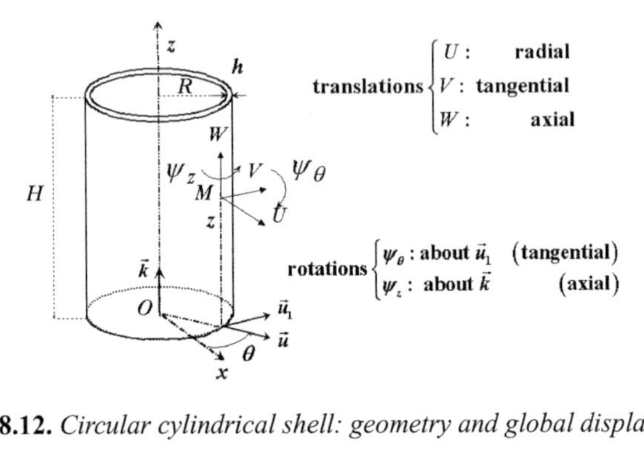

Figure 8.12. *Circular cylindrical shell: geometry and global displacements*

$R_\theta = R$ is constant. The metrics is defined by:

$$(ds)^2 = (dz)^2 + (R\,d\theta)^2 \Rightarrow g_z = 1; \quad g_\theta = R \qquad [8.83]$$

The global displacement field comprises the radial U, the tangential V and the axial W translations and two rotations ψ_θ (about the tangential axis $\vec{u}_1$) and ψ_z (about the axial axis $\vec{k}$). The equilibrium equations have the form:

$$\rho h \ddot{W} - \left\{ \frac{\partial \mathcal{N}_{zz}}{\partial z} + \frac{\partial \mathcal{N}_{\theta z}}{R\partial \theta} \right\} = f_z^{(e)}(\theta, z; t)$$

$$\rho h \ddot{V} - \left\{ \frac{\partial \mathcal{N}_{\theta\theta}}{R\partial \theta} + \frac{\partial \mathcal{N}_{z\theta}}{\partial z} + \frac{\mathcal{Q}_{\theta r}}{R} \right\} = f_\theta^{(e)}(\theta, z; t) \qquad [8.84]$$

$$\rho h \ddot{U} - \left(\frac{\partial \mathcal{Q}_{zr}}{\partial z} + \frac{\partial \mathcal{Q}_{\theta r}}{R\partial \theta} \right) + \frac{\mathcal{N}_{\theta\theta}}{R} = f_r^{(e)}(z, \theta; t) + \frac{\partial \mathfrak{M}_z^{(e)}}{\partial z} + \frac{\partial \mathfrak{M}_\theta^{(e)}}{R\partial \theta}$$

$$\mathcal{Q}_{zr} = \frac{\partial \mathcal{M}_{zz}}{\partial z} + \frac{\partial \mathcal{M}_{z\theta}}{R\partial \theta}; \quad \mathcal{Q}_{\theta r} = \frac{\partial \mathcal{M}_{\theta\theta}}{R\partial \theta} + \frac{\partial \mathcal{M}_{\theta z}}{\partial z} \qquad [8.85]$$

The Figures 8.13a, 8.13b, and 8.13c help to make concrete the meaning of these force balances, in a similar manner as already done for beams and plates.

8.3.1.2 Boundary conditions

Boundary terms which result from integration by parts performed in the variational calculus provide the boundary conditions at the cylinder bases. They are expressed as relations which associate the relevant pairs of conjugated displacement

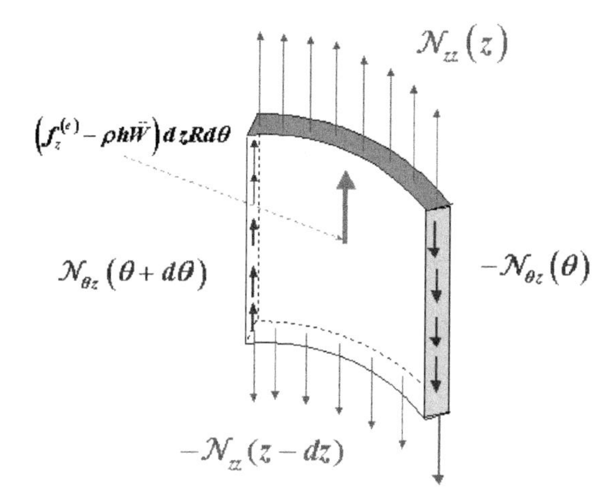

Figure 8.13a. *Axial force balance*

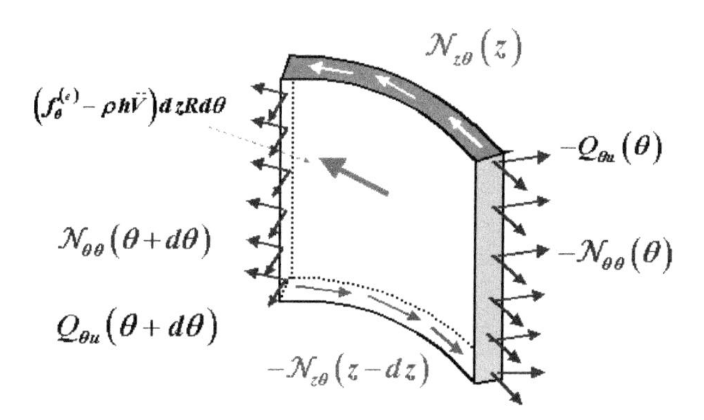

Figure 8.13b. *Tangential force balance*

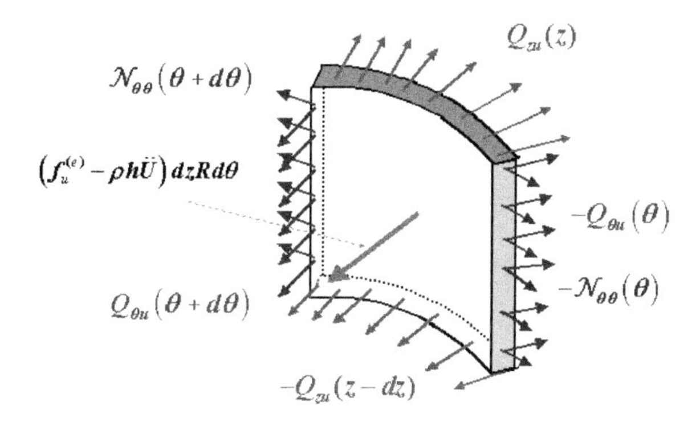

Figure 8.13c. *Radial force balance*

and stress variables:

$$W \longleftrightarrow \mathcal{N}_{zz}$$

$$V \longleftrightarrow \mathcal{N}_{z\theta} + \frac{\mathcal{M}_{z\theta}}{R}$$

$$U \longleftrightarrow \frac{\partial \mathcal{M}_{zz}}{\partial z} + 2\frac{\partial \mathcal{M}_{z\theta}}{R\partial \theta} \qquad [8.86]$$

$$\varphi_\theta \longleftrightarrow \mathcal{M}_{zz}$$

Either the displacement or the stress variable is set to zero according to whether the degree of freedom concerned is fixed or free. As in the case of plates, the transverse shear forces at the boundaries differ from their expression at a current point within the shell:

$$\mathcal{V}_{zu} = \mathcal{Q}_{zu} + \frac{\partial \mathcal{M}_{z\theta}}{R\partial \theta}; \quad \mathcal{V}_{z\theta} = \mathcal{N}_{z\theta} + \frac{\mathcal{M}_{z\theta}}{R} \qquad [8.87]$$

$\mathcal{V}_{zu}$ and $\mathcal{V}_{z\theta}$ are known as the Kirchhoff effective shear stress resultants of the first and of the second kind, respectively.

8.3.2 *Elastic vibrations*

8.3.2.1 Small elastic strain and stress fields

The membrane components of global strains are:

$$\eta_{zz} = \frac{\partial W}{\partial z}; \quad \eta_{\theta\theta} = \frac{\partial V}{R\partial \theta} + \frac{U}{R}; \quad \eta_{z\theta} = \frac{1}{2}\left(\frac{\partial W}{R\partial \theta} + \frac{\partial V}{\partial z}\right) \qquad [8.88]$$

The global strains in flexure and torsion are:

$$\chi_{zz} = -\frac{\partial \psi_\theta}{\partial z}; \quad \chi_{\theta\theta} = -\frac{\partial \psi_z}{R\partial \theta}; \quad \chi_{z\theta} = \frac{-1}{2}\left(\frac{\partial \psi_\theta}{R\partial \theta} + \frac{\partial \psi_z}{\partial z}\right) \qquad [8.89]$$

Flexure angles are related to translation displacements by:

$$\psi_\theta = +\frac{\partial U}{\partial z}; \quad \psi_z = -\frac{V}{R} + \frac{\partial U}{R\partial \theta} \qquad [8.90]$$

Whence the following components of the elastic stress field:

$$\mathcal{N}_{zz} = \frac{Eh}{1-\nu^2}\left\{\frac{\partial W}{\partial z} + \nu\left(\frac{\partial V}{R\partial \theta} + \frac{U}{R}\right)\right\}$$

$$\mathcal{N}_{\theta\theta} = \frac{Eh}{1-\nu^2}\left\{\frac{\partial V}{R\partial \theta} + \frac{U}{R} + \nu\frac{\partial W}{\partial z}\right\} \qquad [8.91]$$

$$\mathcal{N}_{z\theta} = Gh\left(\frac{\partial W}{R\partial \theta} + \frac{\partial V}{\partial z}\right)$$

$$\mathcal{M}_{zz} = -D\left\{\frac{\partial^2 U}{\partial z^2} + \frac{\nu}{R^2}\left(\frac{\partial^2 U}{\partial \theta^2} - \frac{\partial V}{\partial \theta}\right)\right\}$$

$$\mathcal{M}_{\theta\theta} = -D\left\{\frac{1}{R^2}\left(\frac{\partial^2 U}{\partial \theta^2} - \frac{\partial V}{\partial \theta}\right) + \nu\frac{\partial^2 U}{\partial z^2}\right\}$$

$$\mathcal{M}_{\theta z} = \frac{-D(1-\nu)}{R}\left(\frac{\partial^2 U}{\partial \theta \partial z} - \frac{1}{2}\frac{\partial V}{\partial z}\right) \qquad [8.92]$$

$$\text{where} \quad D = \frac{Eh^3}{12(1-\nu^2)}$$

8.3.2.2 Equations of vibrations

By substituting the elastic stresses [8.91], [8.92] into the Love equations [8.84], we arrive at the following system of vibration equations:

1. *Axial equation:*

$$\rho h\ddot{W} - \frac{Eh}{1-\nu^2}\left\{\frac{\partial^2 W}{\partial z^2} + \frac{1-\nu}{2}\frac{\partial^2 W}{R^2\partial \theta^2} + \frac{1+\nu}{2}\frac{\partial^2 V}{R\partial \theta \partial z} + \nu\frac{\partial U}{R\partial z}\right\}$$
$$= f_z^{(e)}(\theta, z; t) \qquad [8.93]$$

2. *Tangential equation:*

$$\rho h\ddot{V} - \frac{Eh}{1-\nu^2}\left\{\begin{array}{l}\left(1 + \frac{h^2}{12R^2}\right)\left(\frac{\partial^2 V}{R^2\partial \theta^2} + \frac{1-\nu}{2}\frac{\partial^2 V}{\partial z^2}\right) + \frac{1+\nu}{2}\frac{\partial^2 W}{R\partial \theta \partial z} \\ + \frac{\partial U}{R^2\partial \theta} - \frac{h^2}{12R^2}\left(\frac{\partial^3 U}{R^2\partial \theta^3} + \frac{\partial^3 U}{\partial z^2 \partial \theta}\right)\end{array}\right\}$$
$$= f_\theta^{(e)}(\theta, z; t) \qquad [8.94]$$

3. *Radial equation:*

$$\rho h\ddot{U} + \frac{Eh}{1-\nu^2}\left\{\begin{array}{l}\frac{U}{R^2} + \frac{h^2}{12R^2}\left(\frac{\partial^4 U}{R^2\partial \theta^4} + 2\frac{\partial^4 U}{\partial z^2\partial \theta^2} + R^2\frac{\partial^4 U}{\partial z^4}\right) \\ + \nu\frac{\partial W}{R\partial z} + \frac{\partial V}{R^2\partial \theta} - \frac{h^2}{12R^2}\left(\frac{\partial^3 V}{\partial z^2\partial \theta} + \frac{\partial^3 V}{R^2\partial \theta^3}\right)\end{array}\right\}$$
$$= f_r^{(e)}(\theta, z; t) + \frac{\partial \mathfrak{M}_\theta^{(e)}}{R\partial \theta} + \frac{\partial \mathfrak{M}_z^{(e)}}{\partial z} \qquad [8.95]$$

In these equations, the terms proportional to $h^2/12R^2$ are due to bending and twisting. More generally, to understand the relative importance of the various terms appearing in the coupled system [8.93] to [8.95], it is advisable to discuss first a few particular problems which can be solved by using further simplifying assumptions.

8.3.2.3 Pure bending model

As for a circular ring, the flexure terms can be singled out by neglecting the membrane strains. So, turning back to equations [8.88], it is assumed that:

$$\frac{\partial W}{\partial z} = 0; \qquad \frac{\partial V}{R\partial \theta} = -\frac{U}{R}; \qquad \frac{\partial W}{R\partial \theta} = -\frac{\partial V}{\partial z} \qquad [8.96]$$

Accordingly, the bending and torsion stresses are:

$$\begin{aligned} \mathcal{M}_{zz} &= -D\left\{\frac{\partial^2 U}{\partial z^2} + \frac{\nu}{R^2}\left(\frac{\partial^2 U}{\partial \theta^2} + U\right)\right\} \\ \mathcal{M}_{\theta\theta} &= -D\left\{\frac{1}{R^2}\left(\frac{\partial^2 U}{\partial \theta^2} + U\right) + \nu\frac{\partial^2 U}{\partial z^2}\right\} \\ \mathcal{M}_{\theta z} &= \frac{-D(1-\nu)}{R}\left(\frac{\partial^2 U}{\partial \theta \partial z} + \frac{\partial W}{R\partial \theta}\right) \end{aligned} \qquad [8.97]$$

The shear forces are:

$$\begin{aligned} Q_{zr} &= -D\left\{\frac{\partial^3 U}{\partial z^3} + \frac{1}{R^2}\left(\frac{\partial^3 U}{\partial z \partial \theta^2} + \left(1 + \frac{\nu}{2}\right)\frac{\partial U}{\partial z}\right)\right\} \\ Q_{\theta r} &= -D\left\{\frac{1}{R^3}\frac{\partial^3 U}{\partial \theta^3} + \frac{1}{R}\frac{\partial^3 U}{\partial \theta \partial z^2} + \frac{1}{R^3}\frac{\partial U}{\partial \theta}\right\} \end{aligned} \qquad [8.98]$$

Substituting [8.98] into the radial equation [8.84], an equation dealing with the sole radial displacement U is obtained. In the absence of external load, it is written as:

$$\begin{aligned} \rho h\ddot{U} + D\Bigg\{&\left(\frac{\partial^4 U}{\partial z^4} + \frac{2}{R^2}\frac{\partial^4 U}{\partial z^2 \partial \theta^2} + \frac{1}{R^4}\frac{\partial^4 U}{\partial \theta^4}\right) \\ &+ \left(\frac{(1+0.5\nu)}{R^2}\frac{\partial^2 U}{\partial z^2} + \frac{\partial^2 U}{R^4 \partial \theta^2}\right)\Bigg\} = 0 \end{aligned} \qquad [8.99]$$

As expected, the stiffness operator comprises plate and shell components which appear within the first and second brackets, respectively. A convenient way to assess the relative importance of these components is to calculate the natural modes of vibration. Analytical solution is straightforward if the cylinder bases are hinged. The radial displacement can be written as:

$$U_{n,m} = \sin\left(\frac{m\pi z}{H}\right)\sin(n\theta) \tag{8.100}$$

where H denotes the cylinder height; n is the circumferential (or azimuth) index and m is the axial index.

Substituting [8.100] into [8.99], the following natural pulsations are found:

$$\omega_{n,m} = \frac{c}{R\sqrt{12(1-\nu^2)}}\frac{h}{R}\sqrt{\left(\left(\frac{m\pi R}{H}\right)^2 + n^2\right)\left(\left(\frac{m\pi R}{H}\right)^2 + n^2 - 1\right) - \frac{\nu}{2}\left(\frac{m\pi R}{H}\right)^2} \tag{8.101}$$

where $c = \sqrt{E/\rho}$

The ratio of the contribution of the shell terms over that of the plate terms to the modal stiffness is found to be:

$$\kappa_{n,m} = \frac{\left|K_{n,m}^{(S)}\right|}{K_{n,m}^{(P)}}\left(n^2 + \left(\frac{m\pi R}{H}\right)^2\right)^{-1} \tag{8.102}$$

Hence, $\kappa_{n,m}$ decreases rather quickly as the modal order increases, as a mere consequence of the wavelength shortening. If $\kappa_{n,m}$ is sufficiently small [8.101] may be replaced by:

$$\omega_{n,m} = \frac{c}{R\sqrt{12(1-\nu^2)}}\frac{h}{R}\left(\left(\frac{m\pi R}{H}\right)^2 + n^2\right) \tag{8.103}$$

which is identical to the result obtained for a rectangular plate $L_x = \pi R$ and $L_y = H$ hinged at the four edges, see formula [6.96]. The reason for $L_x = \pi R$ instead of $L_x = 2\pi R$, as could be assumed at first sight, is the 2π periodicity which holds in the case of the cylinder instead of the π periodicity which holds in the case of the rectangular plate.

8.3.2.4 *Constriction of a circular cylindrical shell*

Let us consider a circular cylindrical shell loaded by a constant radial force density $f_0 = F/2\pi R$ along a parallel line located at height z_0:

$$f_0\delta(z - z_0) \tag{8.104}$$

The present problem is clearly independent of θ. Furthermore, the tangential displacement must be zero, because of the axial symmetry. Nevertheless, the hoop stress $\mathcal{N}_{\theta\theta}$ is of paramount importance. Thus the membrane strains reduce to:

$$\eta_{zz} = \frac{\partial W}{\partial z}; \quad \eta_{\theta\theta} = \frac{U}{R}; \quad \eta_{z\theta} = 0 \qquad [8.105]$$

The non-zero stress components are:

$$\begin{aligned} \mathcal{N}_{zz} &= \frac{Eh}{1-\nu^2}\left\{\frac{\partial W}{\partial z} + \nu\frac{U}{R}\right\}; \quad \mathcal{N}_{\theta\theta} = \frac{Eh}{1-\nu^2}\left\{\frac{U}{R} + \nu\frac{\partial W}{\partial z}\right\} \\ \mathcal{M}_{zz} &= -D\frac{\partial^2 U}{\partial z^2}; \quad \mathcal{M}_{\theta\theta} = -D\nu\frac{\partial^2 U}{\partial z^2}; \quad \mathcal{Q}_{zr} = -D\frac{\partial^3 U}{\partial z^3} \end{aligned} \qquad [8.106]$$

In the static (or quasi-static) domain, the equilibrium equations simplify to:

$$\frac{\partial \mathcal{N}_{zz}}{\partial z} = 0 \Rightarrow \frac{Eh}{1-\nu^2}\left\{\frac{\partial^2 W}{\partial z^2} + \frac{\nu}{R}\frac{\partial U}{\partial z}\right\} = 0 \Rightarrow \frac{\partial W}{\partial z} + \frac{\nu}{R}U = C \qquad [8.107]$$

The radial equation is written as:

$$\begin{aligned} &\left(\frac{\partial \mathcal{Q}_{zr}}{\partial z} +\right) + \frac{\mathcal{N}_{\theta\theta}}{R} = f_0\delta(z-z_0) \Rightarrow \\ &D\frac{\partial^4 U}{\partial z^4} + \frac{Eh}{1-\nu^2}\left\{\frac{U}{R^2} + \frac{\nu}{R}\frac{\partial W}{\partial z}\right\} = f_0\delta(z-z_0) \end{aligned} \qquad [8.108]$$

In equation [8.107], the constant Ccan be set to zero, which means that the shell is assumed to be free axially at both ends. The bending equation [8.108] is thus reduced to:

$$D\frac{d^4 U}{dz^4} + \frac{Eh}{R^2}U = f_0\delta(z-z_0) \qquad [8.109]$$

It can be noticed that the left-hand side of the equilibrium equation is the same as that of a bended straight beam resting on a uniform elastic foundation, see Figure 8.14, the beam rigidity being set to $D = EI$ and the support stiffness coefficient per unit being set to $K_S = Eh/R^2$.

As the problem is symmetric with respect to the plane $z = z_0$, it can be formulated as:

$$\begin{aligned} &D\frac{d^4 U}{dz^4} + \frac{Eh}{R^2}U = 0 \\ &D\left.\frac{d^3 U}{dz^3}\right|_{z=z_0} = \frac{1}{2}f_0; \quad \left.\frac{dU}{dz}\right|_{z=z_0} = 0 \end{aligned} \qquad [8.110]$$

where half a shell loaded by half the actual load is considered.

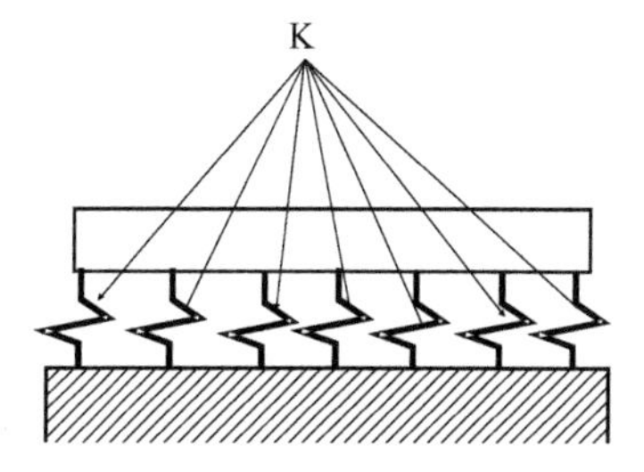

Figure 8.14. *Bending beam on an elastic foundation*

Another consequence of symmetry is the disappearance of the first derivative of U along the loaded parallel. Solving first for the characteristic equation, we arrive at:

$$\lambda^4 = -\frac{12(1-\nu^2)}{(Rh)^2} \Rightarrow$$

$$\lambda_1 = k(1+i); \quad \lambda_2 = k(1-i); \quad \lambda_3 = -\lambda_1; \quad \lambda_4 = -\lambda_2 \qquad [8.111]$$

$$\text{where} \quad k = \sqrt[4]{\frac{3(1-\nu^2)}{(Rh)^2}}$$

After some elementary trigonometric manipulations, the general solution of the differential equation [8.110] is expressed as:

$$U(z) = e^{-kz}(A_1 \cos(kz) + B_1 \sin(kz)) + e^{+kz}(A_2 \cos(kz) + B_2 \sin(kz)) \qquad [8.112]$$

A necessary condition is $A_2 = B_2 = 0$, otherwise the solution would increase exponentially when moving off the loaded parallel, which is unrealistic. The two other constants are determined by using the conditions at the loaded parallel. Finally, the shell radial deflection is found to be:

$$U(z) = \frac{f_0 e^{-kz}}{8Dk^3} \cos(k(z - z_0)) \qquad [8.113]$$

Figure 8.15 shows the shell constricted by a radial inward load. The results of the finite element computation are found to agree closely with the analytical result [8.113]. As a remarkable result it is found that the deflected portion of the shell is highly confined around the loaded parallel line, as U vanishes exponentially with the short characteristic length $\ell_c \simeq \sqrt{Rh}$. Accordingly, as soon as $H \gg \ell_c$, constriction is quite insensitive to the support conditions at the cylinder bases.

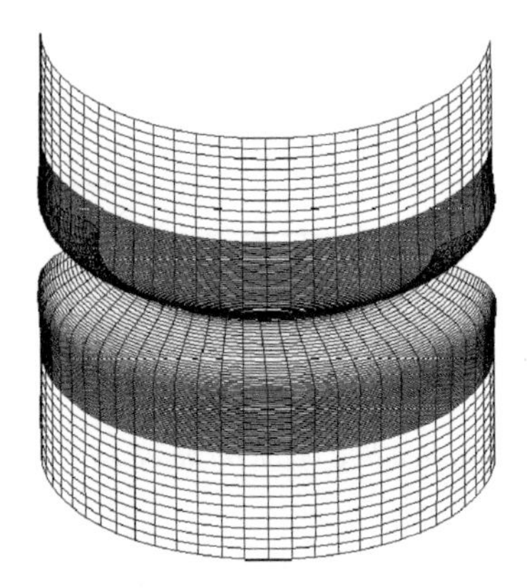

Figure 8.15. *Deformed shell (180° sector)*

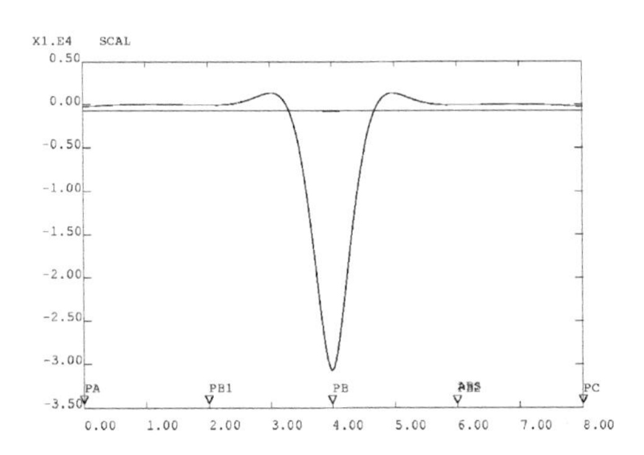

Figure 8.16a. *Membrane stress $\mathcal{N}_{\theta\theta}$, $\mathcal{N}_{zz}$, see text for indentification.*

The stress distributions are plotted along a meridian in Figures 8.16a and 8.16b. It is noted that the profile of $\mathcal{N}_{\theta\theta}$ is marked by a sharp compressive peak, whose maximum is centred at the loaded parallel. Moving off this parallel, a secondary traction peak of much smaller magnitude is first observed, then the stress vanishes in an exponential way. As expected, $\mathcal{N}_{zz}$ is practically zero. On the other hand, the most prominent component of the bending moment is $\mathcal{M}_{zz}$; however $\mathcal{M}_{\theta\theta}$ is not negligible.

Finally, the present problem gives us the opportunity to turn back to that of the pressurized vessel already studied in Chapter 7 subsection 7.3.7.2. Indeed, it is now possible to discuss the end effects at the interface between the cylindrical body and the hemispherical caps. Based on pure membrane effects, the radial displacement of the caps was found to be less than that of the cylinder by the ratio $(1-\nu)/(2-\nu)$. As a consequence, the cylinder ends are constricted and bending is induced at the interface, which accounts for the condition of displacement continuity. According

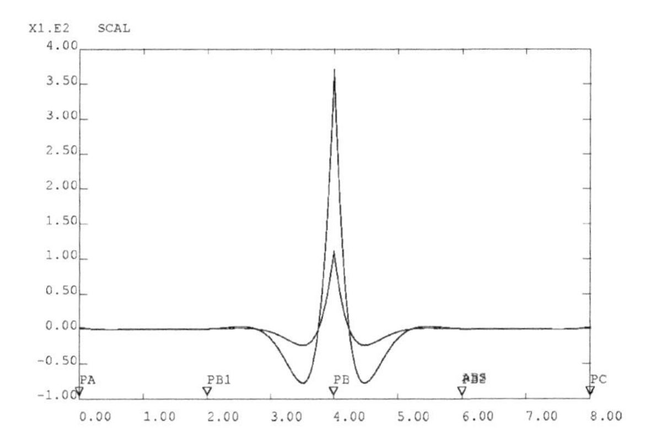

Figure 8.16b. *Bending moments $\mathcal{M}_{\theta\theta}$, $\mathcal{M}_{zz}$, see text for indentification.*

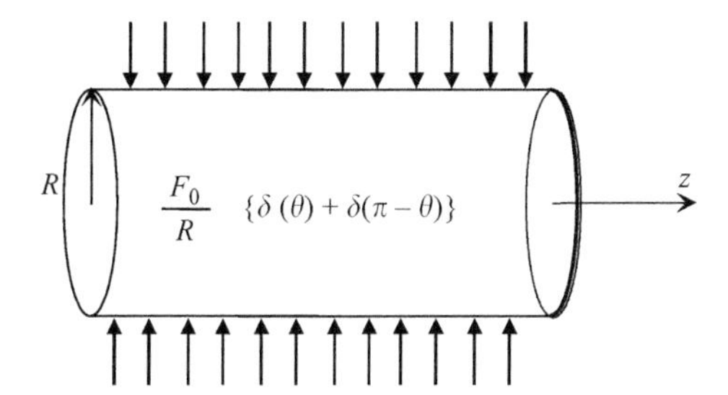

Figure 8.17. *Cylindrical shell pinched by radial forces along two opposite meridian lines*

to the present analysis, it can be asserted that bending remains a local effect, closely confined within a characteristic length scale $\sqrt{Rh}$ from the interface.

8.3.2.5 Bending about the meridian lines

A circular cylindrical shell is loaded along two opposite meridian lines by a radial force density, which is assumed to be uniform, as sketched in Figure 8.17. Because of the loading distribution, the deflection of the shell is independent from the axial coordinate z. It can be also expected that bending is the main deformation about the loaded meridian lines. Accordingly, the membrane effects are discarded, and the elastic energy is reduced to the bending term:

$$\mathcal{E}_e = \frac{HR}{2}\int_0^{2\pi} \mathcal{M}_{\theta\theta}\chi_{\theta\theta}\, d\theta \qquad [8.114]$$

The n-th harmonic component of the displacement field is first written as:

$$U_n \cos(n\theta);\ V_n \sin(n\theta);\ W_n \equiv 0 \qquad [8.115]$$

However, the disappearance of the membrane hoop strain $\eta_{\theta\theta}$, implies:

$$V_n = -\frac{U_n}{n} \qquad [8.116]$$

Substituting the field [8.115], [8.116] into equations [8.89], [8.92], the elastic energy contained in the harmonic n is found to be:

$$\mathcal{E}_e^{(n)} = \frac{\pi DH(n^2-1)^2}{2R^3} U_n^2 \qquad [8.117]$$

The load is expanded as a Fourier series:

$$F(\theta) = F_0\{\delta(\theta) + \delta(\pi - \theta)\} = \frac{2F_0}{\pi}\left\{\frac{1}{2} + \sum_{k=1}^{\infty} \cos(2k\theta)\right\} \qquad [8.118]$$

The equilibrium equation for the harmonic $n = 2k$ is:

$$\frac{\pi DH(4k^2-1)^2}{R^3} U_{2k} = \frac{2F_0}{\pi}; \quad k \geq 1 \qquad [8.119]$$

Thus the lateral deflection is given by the Fourier series:

$$U(\theta) = \frac{24(1-\nu^2)R^3 F_0}{\pi^2 E h^3 H} \sum_{k=1}^{\infty} \frac{\cos(2k\theta)}{(4k^2-1)^2} \qquad [8.120]$$

The series converges very fast to about 0.11. Most of the deflection is produced by the ovaling mode $n = 2$. However, the contribution of harmonics of a few higher ranks is also perceptible, see Figure 8.18. Finally, it is worth stressing that the solution [8.120] accounts for bending only. The harmonic $n = 0$ of the external loading is expected to deform the shell according to a breathing mode of the type already described in Chapter 7 subsection 7.2.5.2 for a circular ring. The ring model agrees with equation [8.108], provided the z dependency is removed. However, owing to the large difference between the membrane and bending rigidities of a shell, the contribution of the $n = 0$ mode to the shell deformation can be safely neglected in comparison with the contribution of the first bending modes $n = 2, 4, \ldots$

8.3.2.6 *Natural modes of vibration $n = 0$*

In order to analyse the breathing modes, the Love equations are simplified by neglecting the membrane stresses $\mathcal{N}_{zz}$ and $\mathcal{N}_{z\theta}$, which are much smaller than the

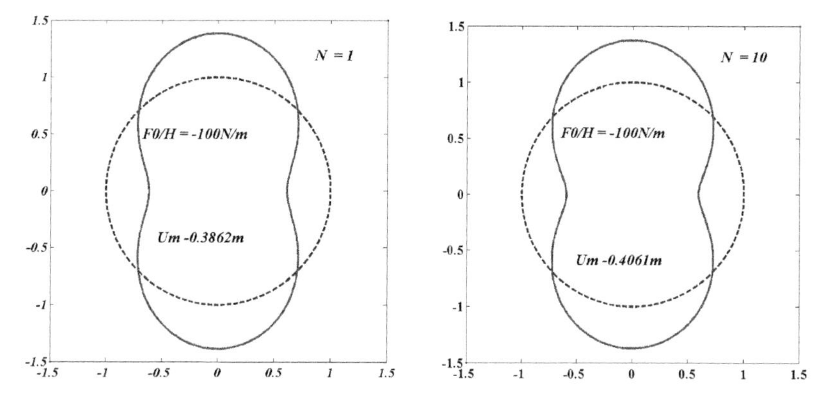

Figure 8.18. *Deflection of the shell, harmonic truncation:* $2N = 20$ *and* $2N = 2$

hoop stress $\mathcal{N}_{\theta\theta}$. Accordingly, the admissible field of displacement is:

$$V = 0; \quad W = -\frac{R}{\nu}\frac{\partial U}{\partial z} \qquad [8.121]$$

As the breathing modes are independent of θ, the strains of interest can be simplified by using [8.121] as:

$$\begin{aligned} \eta_{\theta\theta} &= \frac{U}{R}; \quad \chi_{zz} = -\frac{d^2U}{dz^2} \\ \chi_{\theta\theta} &= \frac{\partial\psi_\theta}{R\partial\theta} = 0; \quad \chi_{z\theta} = \frac{V}{R} - \frac{\partial U}{R\partial\theta} = 0 \end{aligned} \qquad [8.122]$$

The variation of elastic energy is:

$$\delta\mathcal{E}_e = \int_0^H \int_0^{2\pi} \{\mathcal{N}_{\theta\theta}\delta\eta_{\theta\theta} + \mathcal{M}_{zz}\delta\chi_{zz}\} R\, d\theta\, dz \qquad [8.123]$$

That of kinetic energy is:

$$\delta\mathcal{E}_\kappa = \rho h \int_0^H \int_0^{2\pi} \dot{U}\delta\dot{U} R\, d\theta\, dz \qquad [8.124]$$

Whence the modal equation:

$$\frac{Eh}{R^2(1-\nu^2)}\left(U + \frac{h^2}{12R^2}\frac{\partial^4 U}{\partial z^4}\right) - \omega^2\rho h U = 0 \qquad [8.125]$$

The shell stiffness comprises a membrane term similar to the term present in the breathing mode of a circular ring and a flexure term, similar to the term present in the bending modes of a straight beam. Going a step further in the analysis, it is realized that the bending energy is far smaller than the membrane energy provided the shell is assumed to be thin ($h/R \ll 1$). Therefore, it is expected that the breathing modes of a circular cylindrical shell are characterized by a high modal density. This is because the contribution of bending to the modal stiffness increases very slowly with the axial index m. A simple application suffices to confirm this point. Let us consider a shell hinged at both bases; the natural frequencies are immediately found to be:

$$f_{0,m} = \frac{c}{2\pi R}\sqrt{1 + \frac{1}{12}\left(\frac{h}{R}\right)^2 \left(\frac{m\pi R}{H}\right)^4}, \quad m = 1, 2, \ldots$$

$$\text{where} \quad c = \sqrt{\frac{E}{\rho(1-\nu^2)}} \qquad [8.126]$$

The result [8.126] puts in evidence the large preponderance of the ring breathing term, which is corrected by a beam bending term, whose relative importance increases with m. However, for a thin shell, m must take on large values for obtaining a significant contribution of bending to the shell energy. The Figure 8.19 shows a few breathing modes for a steel shell, $R = 0.535$ m, $H = 4R, h = 1$ mm, clamped at the lower base and free at the other.

8.3.3 *Bending coupled in z and θ*

8.3.3.1 Simplified model neglecting the hoop and shear stresses

To built a simplified model to account for bending about the tangential and the axial directions – coupling thus the z and θ dependent terms – several simplifications of Love's equations can be proposed. Here as a first assumption, the membrane strains $\eta_{\theta\theta}$ and $\eta_{\theta z}$ are neglected as in subsection 8.3.2. This allows one to express the displacement field in terms of the radial component only:

$$\eta_{\theta\theta} = 0 \Rightarrow V_n = -\frac{U_n}{n}$$
$$\eta_{\theta z} = 0 \Rightarrow W_n = -\frac{R}{n^2}\frac{\partial U_n}{\partial z} \qquad [8.127]$$

On the other hand, the axial membrane component η_{zz} cannot be neglected since it plays an important role as soon as the radial displacement varies with z. Using again the simplified expressions [8.121], the global strain and stress components,

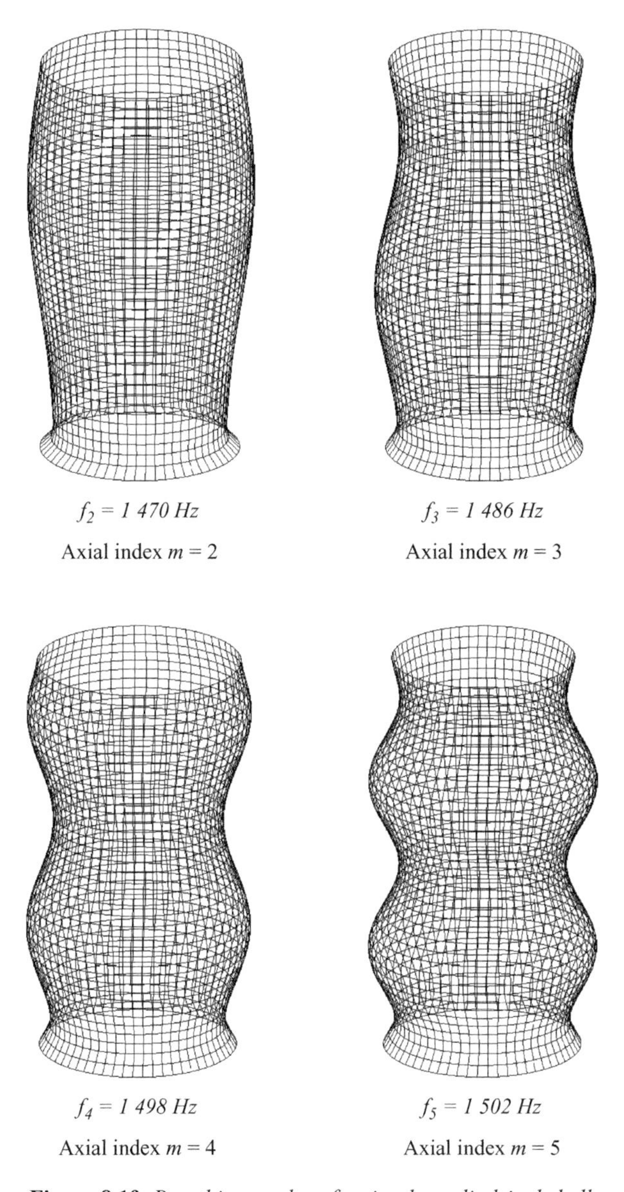

Figure 8.19. *Breathing modes of a circular cylindrical shell*

for a given harmonic n, reduce to:

$$\eta_{zz}^{(n)} = \frac{dW_n}{dz} = -\frac{R}{n^2}\frac{d^2U_n}{dz^2}; \quad \mathcal{N}_{zz}^{(n)} = \frac{Eh}{1-\nu^2}\left(-\frac{R}{n^2}\frac{d^2U_n}{dz^2}\right)$$

$$\chi_{zz}^{(n)} = -\frac{d\psi_\theta^{(n)}}{dz} = -\frac{d^2U_n}{dz^2}; \quad \mathcal{M}_{zz}^{(n)} = -D\left\{\frac{d^2U_n}{dz^2} + \frac{\nu(1-n^2)U_n}{R^2}\right\}$$

$$\chi_{\theta\theta}^{(n)} = -\frac{n\psi_z^{(n)}}{R} = \frac{(n^2-1)}{R^2}U_n; \quad \mathcal{M}_{\theta\theta}^{(n)} = -D\left\{-\frac{(n^2-1)}{R^2}U_n + \nu\frac{d^2U_n}{dz^2}\right\}$$

$$\chi_{\theta z}^{(n)} = -\frac{1}{R}\left(\frac{1}{n} - n\right)\frac{dU_n}{dz}; \quad \mathcal{M}_{\theta z}^{(n)} = -\frac{D(1-\nu)}{R}\left(-n + \frac{1}{2n}\right)\frac{dU_n}{dz}$$

[8.128]

8.3.3.2 Membrane and bending-torsion terms of elastic energy

The elastic energy of the shell includes a membrane term together with bending and twisting terms. It is important to study first their relative magnitude. This is worked out just below, starting from the expressions [8.128].

1. *Membrane energy*

$$\mathcal{E}_{m(zz)}^{(n)} = \frac{\pi R}{2}\int_0^H \mathcal{N}_{zz}^{(n)}\eta_{zz}^{(n)}dz = \frac{\pi EhR^3}{2(1-\nu^2)n^4}\int_0^H \left(\frac{d^2U_n}{dz^2}\right)^2 dz \quad [8.129]$$

2. *Bending energy about a tangential direction*

$$\mathcal{E}_{b(zz)}^{(n)} = \frac{\pi R}{2}\int_0^H \mathcal{M}_{zz}^{(n)}\chi_{zz}^{(n)}dz$$
$$= \frac{\pi RD}{2}\int_0^H \left\{\left(\frac{d^2U_n}{dz^2}\right)^2 + \nu\left(\frac{1-n^2}{R^2}\right)\left(\frac{d^2U_n}{dz^2}\right)U_n\right\}dz$$

[8.130]

3. *Bending energy about a meridian line*

$$\mathcal{E}_{b(\theta\theta)}^{(n)} = \frac{\pi R}{2}\int_0^H \mathcal{M}_{\theta\theta}^{(n)}\chi_{\theta\theta}^{(n)}dz$$
$$= \frac{\pi RD}{2}\int_0^H \left\{\left(\frac{1-n^2}{R^2}\right)^2 U_n^2 + \nu\left(\frac{1-n^2}{R^2}\right)\left(\frac{d^2U_n}{dz^2}\right)U_n\right\}dz$$

[8.131]

4. *Torsional energy*

$$\mathcal{E}^{(n)}_{t(z\theta)} = \pi R \int_0^H \mathcal{M}^{(n)}_{z\theta} \chi^{(n)}_{z\theta}\, dz$$
$$= \frac{\pi RD(1-\nu)}{R^2} \int_0^H \left(\frac{1-n^2}{n}\right)\left(\frac{1-2n^2}{2n}\right)\left(\frac{dU_n}{dz}\right)^2 dz \qquad [8.132]$$

To assess the relative importance of the terms [8.129] to [8.132], the following dimensionless parameters are introduced:

$$\eta = \frac{H}{R}; \quad \varepsilon = \frac{h}{R} \qquad [8.133]$$

η is the aspect ratio and ε the thickness ratio of the shell. The following relative orders of magnitude are arrived at:

$$\frac{\mathcal{E}^{(n)}_{b(zz)}}{\mathcal{E}^{(n)}_{m(zz)}} \cong \frac{\varepsilon^2}{6}\{1+\nu(1-n^2)\eta^2\}$$
$$\frac{\mathcal{E}^{(n)}_{b(\theta\theta)}}{\mathcal{E}^{(n)}_{m(zz)}} \cong \frac{\varepsilon^2}{6}\{(1-n^2)^2\eta^4 + (1-n^2)\eta^2\} \qquad [8.134]$$
$$\frac{\mathcal{E}^{(n)}_{t(z\theta)}}{\mathcal{E}^{(n)}_{m(zz)}} \cong \frac{\varepsilon^2}{6}\frac{(1-n^2)(1-2n^2)}{2n^2}\eta^2$$

where $\mathcal{E}^{(n)}_{m(zz)} \cong \dfrac{\pi EhU_n^2}{2(1-\nu^2)\eta^3}\left(\dfrac{1}{n^4}\right)$

These ratios show that values of η, ε being fixed, the relative contribution of the membrane energy decreases as n increases. Furthermore, as soon as n is sufficiently large, the bending term about the meridian lines becomes preponderant:

$$\mathcal{E}^{(n)}_{b(\theta\theta)}/\mathcal{E}^{(n)}_{m(zz)} \cong \frac{\varepsilon^2 n^4 \eta^4}{6} \qquad [8.135]$$

Consequently, provided the geometrical parameters η, ε are not too large, in such a way that $\varepsilon\eta^2 < 1$, the total elastic energy is largely dominated either by the membrane component if n is small enough, or by the bending component about the meridian lines if n is sufficiently large. Thus, the interesting result arises that the elastic potential of the shell is marked by a minimum value which occurs at a certain circumferential index $n_{\min}$, governed by the aspect and thickness ratios of the shell, as illustrated in Figures 8.20a and 8.20b.

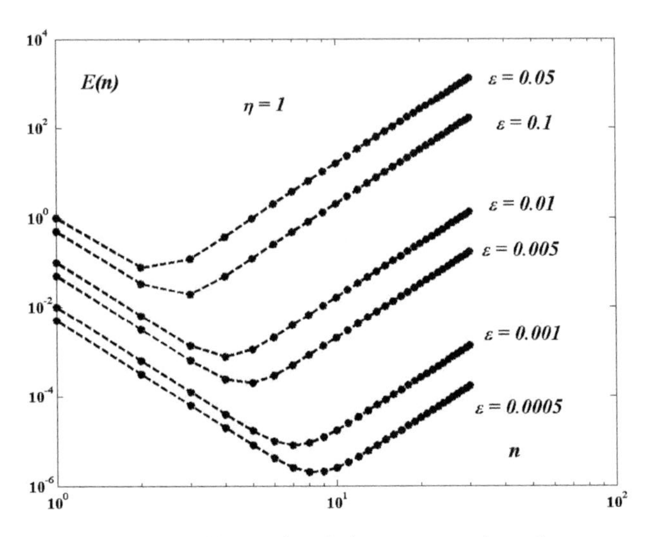

Figure 8.20a. *Reduced elastic potential* $\eta = 1$

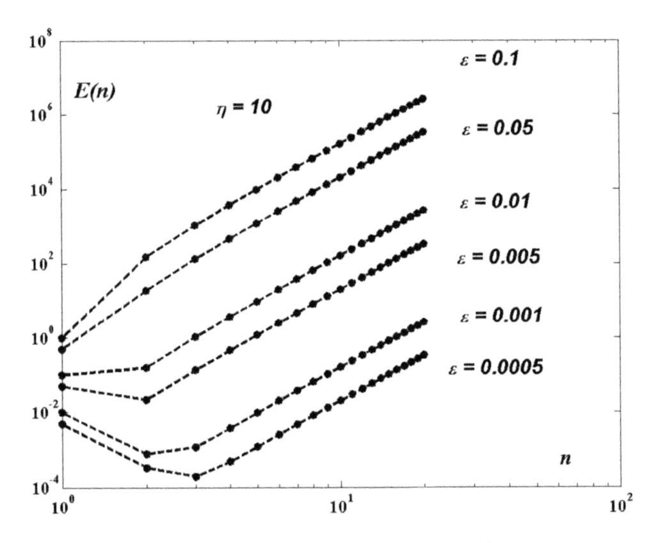

Figure 8.20b. *Reduced elastic potential* $\eta = 10$

Furthermore, by retaining the elastic terms arising from the two preponderant membrane and flexure terms only, the radial vibration equation is written as:

$$\frac{EhR^2}{1-\nu^2}\frac{1}{n^4}\frac{\partial^4 U_n}{\partial z^4} + \frac{Eh^3(n^2-1)^2 U_n}{12(1-\nu^2)R^4} + \rho h\left(1+\frac{1}{n^2}\right)\ddot{U}_n = Q_n^{(e)} \qquad [8.136]$$

8.3.3.3 Point-wise punching of a circular cylindrical shell

As a first application of the very simplified model [8.136], let us consider a circular cylindrical shell loaded by a pair of radial forces applied at two diametrically opposite points, as shown in Figure 8.21.

The displacement field is represented by the Fourier series:

$$U(\theta,z) = \sum_{n=0}^{\infty} U_n(z)\cos(n\theta);$$

$$V(\theta,z) = \sum_{n=1}^{\infty} V_n(z)\sin(n\theta);$$

$$W(\theta,z) = \sum_{n=0}^{\infty} W_n(z)\cos(n\theta)$$

Starting from [8.136], the following radial equation is obtained:

$$\frac{d^4U_n}{dz^4} + 4\beta_n^4 U_n = \frac{(1-\nu^2)n^4}{\pi E h R^3} F_n\delta(z)$$
$$\beta_n^4 = \frac{h^2 n^4 (n^2-1)^2}{48R^6} \qquad [8.137]$$

which shows that the axial evolution of the radial displacement is governed, for each harmonic, by the bending equation of a straight beam resting on a uniform elastic foundation, whose stiffness increases rapidly with the harmonic rank. The

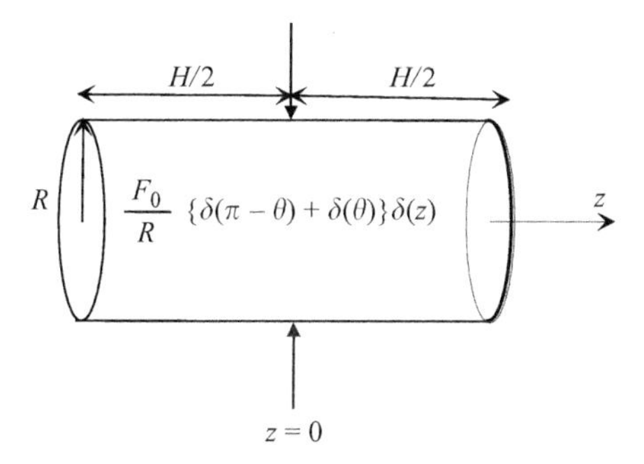

Figure 8.21. *Point-wise punching of a cylindrical cylinder*

Fourier series of the loading may be written as:

$$f^{(e)}(\theta, z) = 2F_0 \left\{ \frac{1}{2} + \sum_{k=1}^{\infty} \cos(2k\theta) \right\} \delta(z) \qquad [8.138]$$

According to this form, the angular element $d\theta$ is used to integrate the force density along the azimuth, instead of the elementary arc length $R\,d\theta$. Then the solution is found to be:

$$n = 2k; \quad z \geq 0$$

$$U_n(z) = \frac{3^{3/4}(1-\nu^2) n F_0}{\pi E (n^2-1)^{3/2}} \left(\frac{R}{h^2}\right) \sqrt{\frac{2R}{h}} e^{-\beta_n z} \sin\left(\beta_n z + \frac{\pi}{4}\right) \qquad [8.139]$$

The shell deflection at the loaded points is:

$$U(0,0) = \frac{3^{3/4}(1-\nu^2) F_0}{\pi E} \left(\frac{R}{h^2}\right) \sqrt{\frac{R}{h}} \sum_{k=1}^{\infty} \frac{2k}{\left(4k^2-1\right)^{3/2}} \qquad [8.140]$$

The equivalent shell stiffness for point-wise punching can be defined as:

$$K_e = \frac{F_0}{U(0,0)} = \alpha E h \left(\frac{h}{R}\right)^{3/2} \quad \text{with } \alpha = \frac{\pi}{3^{3/4}} \cong 1.45, \quad \nu = 0.3 \qquad [8.141]$$

This approximate result is found to be very close to that established in [SEID 75], starting from a more refined model than the present one.

8.3.3.4 Natural modes of vibration

According to the simplified model established in subsection 8.3.3, the modal equation is:

$$\frac{ER^2}{1-\nu^2} \frac{1}{n^4} \frac{d^4 U_n}{dz^4} + \frac{Eh^2(n^2-1)^2 U_n}{12(1-\nu^2)R^4} - \omega^2 \rho \left(1 + \frac{1}{n^2}\right) U_n = 0 \qquad [8.142]$$

The mode shapes are the same as for straight beams and the natural frequencies can be easily determined by using the Rayleigh quotient; for instance, if the shell is hinged at its bases they are found to be:

$$f_{n,m} = \frac{c}{2\pi R} \sqrt{\frac{1}{(1-\nu^2)} \left(\frac{n^2}{n^2+1}\right) \left\{ \left(\frac{m\pi}{n\eta}\right)^4 + \frac{\varepsilon^2 (1-n^2)^2}{12} \right\}}; \quad c = \sqrt{\frac{E}{\rho}} \qquad [8.143]$$

From this expression it can be pointed out that:

1. The mode shapes $n = 1$ are independent of the shell bending term about the meridian lines and are similar to the bending modes of the beam equivalent to the cylinder:

$$f_{1,m} = \frac{1}{2\pi}\left(\frac{m\pi}{H}\right)^2\sqrt{\frac{ER^2}{2\rho(1-\nu^2)}} \cong \frac{1}{2\pi}\left(\frac{m\pi}{H}\right)^2\sqrt{\frac{EI}{\rho S}}$$

   where

$$S = \pi((R+h)^2 - R^2) \cong 2\pi Rh; \quad I = \frac{\pi((R+h)^4 - R^4)}{4} \cong \pi R^3 h \qquad [8.144]$$

2. If the shell aspect ratio is large enough, the modes defined by $n > 1$ have a generalized stiffness, and thus a natural frequency, larger than the corresponding beam value associated with the same axial index m.
3. If the shell aspect ratio is small enough, the modal stiffness, and thus the natural frequency has a minimum for $n = n_{\min}$, which depends on the shell aspect and thickness ratios.

   The simplified model adopted here gives satisfactory results for a few problems such as the point-wise punching of circular cylindrical shells. It is also useful to describe qualitatively the major features of the bending modes of circular cylindrical shells. Nevertheless, comparison of results [8.143] with those produced by using the finite element method indicate quite significant discrepancies.

8.3.3.5 Donnel–Mushtari–Vlasov model

Circa 1938, L.H. Donnel and K.M. Mushtari developed independently a simplified model for axisymmetric cylindrical shells. The approach was then generalized for any geometry by V.Z. Vlasov, see for instance [KRA 67], [DON 76]. The model results in a radial vibration equation significantly more complicated than equation [8.136], which is widely used in shell vibrations. The first assumption is to neglect the contributions of in-plane components in the bending strains but to retain them in the membrane strain components. Accordingly, the harmonic components of the strains are reformulated as:

$$\begin{aligned}
&\eta_{zz}^{(n)} = \frac{\partial W_n}{\partial z}; \quad \eta_{\theta\theta}^{(n)} = \frac{nV_n}{R}; \quad \eta_{z\theta}^{(n)} = \frac{1}{2}\left(\frac{\partial V_n}{\partial z} - \frac{nW_n}{R}\right)\\
&\chi_{zz}^{(n)} = -\frac{\partial^2 U_n}{\partial z^2}; \quad \chi_{\theta\theta}^{(n)} = \frac{n^2}{R^2}U_n; \quad \chi_{\theta z}^{(n)} = +\frac{n}{R}\frac{\partial U_n}{\partial z}
\end{aligned} \qquad [8.145]$$

As a second assumption, the inertia term is neglected in the axial and tangential Love equations [8.84]. Finally, as a third assumption, the contribution of the shear term $Q_{\theta r}^{(n)}/R$ in the tangential Love equation is also neglected.

The remaining task is to eliminate the V_n and W_n components of motion between the simplified Love equations. It turns out that the elimination process is not straightforward. It is described in particular in [NOV 64], [SOE 93]. The final result is:

$$D\left(\frac{n^2}{R^2}-\frac{\partial^2}{\partial z^2}\right)^4 U_n+\frac{Eh}{R^2}\frac{\partial^4 U_n}{\partial z^4}+\rho h\left(\frac{n^2}{R^2}-\frac{\partial^2}{\partial z^2}\right)^2\ddot{U}_n=\left(\frac{n^2}{R^2}-\frac{\partial^2}{\partial z^2}\right)^2 f_n^{(e)} \qquad [8.146]$$

Approximate solutions of the modal problem associated with equation [8.146] can be obtained by using the Rayleigh quotient method. As a reasonable approximation of the radial mode shapes $u_n(z)$, it is natural to select the beam mode shapes which agree with the boundary conditions applied to the shell. For instance, if the shell is hinged at its bases they are found to be:

$$f_{n,m}=\frac{c}{2\pi R}\sqrt{\frac{(m\pi)^4}{((n\eta)^2+(m\pi)^2)^2}+\frac{\varepsilon^2((n\eta)^2+(m\pi)^2)^2}{12(1-\nu^2)}} \qquad [8.147]$$

8.3.4 *Modal analysis of Love's equations*

The Rayleigh–Ritz method can be used to solve the modal problem arising from Love's equations. A set of analytical shapes is first assumed to describe the actual mode shape of rank (n,m). The solution is then searched for as a linear superposition of the admissible shapes whose multiplicative coefficients are the new unknown of the problem, acting as a set of generalized displacements (cf. Chapter 5, subsection 5.3.6.3). In the case of circular cylindrical shells, we are led naturally to use a set of harmonic sine and cosine functions in the azimuth direction and to use bending mode shapes of straight beams, which agree with the boundary conditions applied to the shell, at least those concerning the displacement field.

The simplest case of application is a cylinder simply supported at its two ends, that is hinged in the radial direction, but free to move in the axial direction. The following trial functions are used:

$$\begin{cases} U_{n,m}=u_{n,m}\sin\left(\dfrac{m\pi z}{H}\right)\cos(n\theta)\\ V_{n,m}=v_{n,m}\sin\left(\dfrac{m\pi z}{H}\right)\sin(n\theta)\\ W_{n,m}=w_{n,m}\cos\left(\dfrac{m\pi z}{H}\right)\cos(n\theta)\end{cases} \qquad [8.148]$$

Projecting Love's equations on the functions [8.148], we obtain the following linear and algebraic system of the canonical form:

$$\left[c^2\begin{bmatrix} K_{11} & K_{12} & K_{13} \\ K_{21} & K_{22} & K_{23} \\ K_{31} & K_{32} & K_{33} \end{bmatrix} - \omega_{n,m}^2 \begin{bmatrix} 1 & 0 & 0 \\ 0 & 1 & 0 \\ 0 & 0 & 1 \end{bmatrix}\right] \begin{bmatrix} w_{n,m} \\ v_{n,m} \\ u_{n,m} \end{bmatrix} = \begin{bmatrix} 0 \\ 0 \\ 0 \end{bmatrix} \qquad [8.149]$$

$$\text{where} \quad c^2 = \frac{E}{\rho(1-\nu^2)}$$

The stiffness coefficients appearing in [8.149] are given by:

$$K_{11} = \left(\frac{m\pi}{H}\right)^2 + \left(\frac{1-\nu}{2}\right)\left(\frac{n}{R}\right)^2$$

$$K_{22} = (1+\kappa)\left(\left(\frac{n}{R}\right)^2 + \left(\frac{1-\nu}{2}\right)\left(\frac{m\pi}{H}\right)^2\right)$$

$$K_{33} = \frac{1}{R^2} + \kappa R^2\left(\left(\frac{n}{R}\right)^2 + \left(\frac{m\pi}{H}\right)^2\right)^2$$

$$K_{12} = K_{21} = -\left(\frac{1+\nu}{2}\right)\left(\frac{n}{R}\right)\left(\frac{m\pi}{H}\right) \qquad [8.150]$$

$$K_{13} = K_{31} = -\left(\frac{\nu}{R}\right)\left(\frac{m\pi}{H}\right)$$

$$K_{23} = K_{32} = +n\left(\frac{1}{R^2} + \kappa\left(\left(\frac{n}{R}\right)^2 + \left(\frac{m\pi}{H}\right)^2\right)\right)$$

$$\text{where} \quad \kappa = \frac{h^2}{12R^2}$$

Numerical solution of the modal system [8.149] can be conveniently carried out by using a software such as MATLAB. The results thus obtained are found to be in close agreement with those arising from a finite element model. As expected, there exist three distinct families of modes, which can be identified by looking at the relative amplitude of the generalized displacements. As a particular case, the $n = 0$ modes can be suitably split into the three following families of predominant deformation,

(1) Torsion modes: $f_t = \dfrac{mc}{2H}\sqrt{\dfrac{1-\nu}{2}\left(1+\dfrac{h^2}{12R^2}\right)}, \quad v \neq 0; \quad u \equiv w \equiv 0$

(2) Axial modes: $f_a \simeq \dfrac{mc}{2H}, \quad w \gg u; \quad v \equiv 0$

(3) Breathing modes: $f_r \simeq \dfrac{c}{2\pi R}\sqrt{1+\dfrac{h^2}{12R^2}\left(\dfrac{m\pi R}{H}\right)^4}, \quad u \gg w; \quad v \equiv 0$

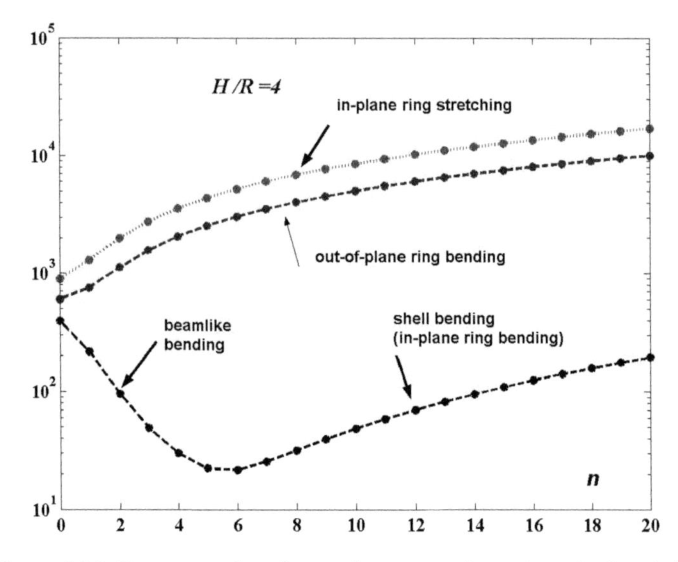

Figure 8.22. *Frequency plot of a simply supported circular cylindrical shell*

The expression given here for f_r is identical to the result [8.126]. It is also noticed that to describe the breathing modes the following trial functions are chosen:

$$U_{n,m} = u_{n,m} \sin\left(\frac{m\pi z}{H}\right); \quad V_{n,m} = 0; \quad W_{n,m} = w_{n,m} \cos\left(\frac{m\pi z}{H}\right)$$

When n differs from zero, mode shapes can be qualitatively interpreted by referring to the straight flexure beam and to the circular ring cases. Considering a typical frequency plot such as shown in Figure 8.22, the lowest branch corresponds to the so called shell bending modes, which combine beamlike bending and in-plane bending of rings. The two other branches present similarities with the out-of-plane bending and the in-plane stretching modes of rings. Results obtained by such an approximate procedure agree satisfactorily with those arising from the finite element method, provided a sufficient number of terms are retained in the trial functions.

In Figure 8.23, a few mode shapes of the lowest branch obtained by using the finite element method are shown, ($m = 1, n = 1$ to 7). It refers to a steel shell ($R = 0.53$ m, $H = 1$ m, $h = 1$ mm) clamped at the lower base and free at the other.

8.3.5 *Axial loading: global and local responses*

As a final example of coupling effects in cylindrical shells of revolution, it is of interest to discuss briefly the elastic deflection induced by axial forces of equal

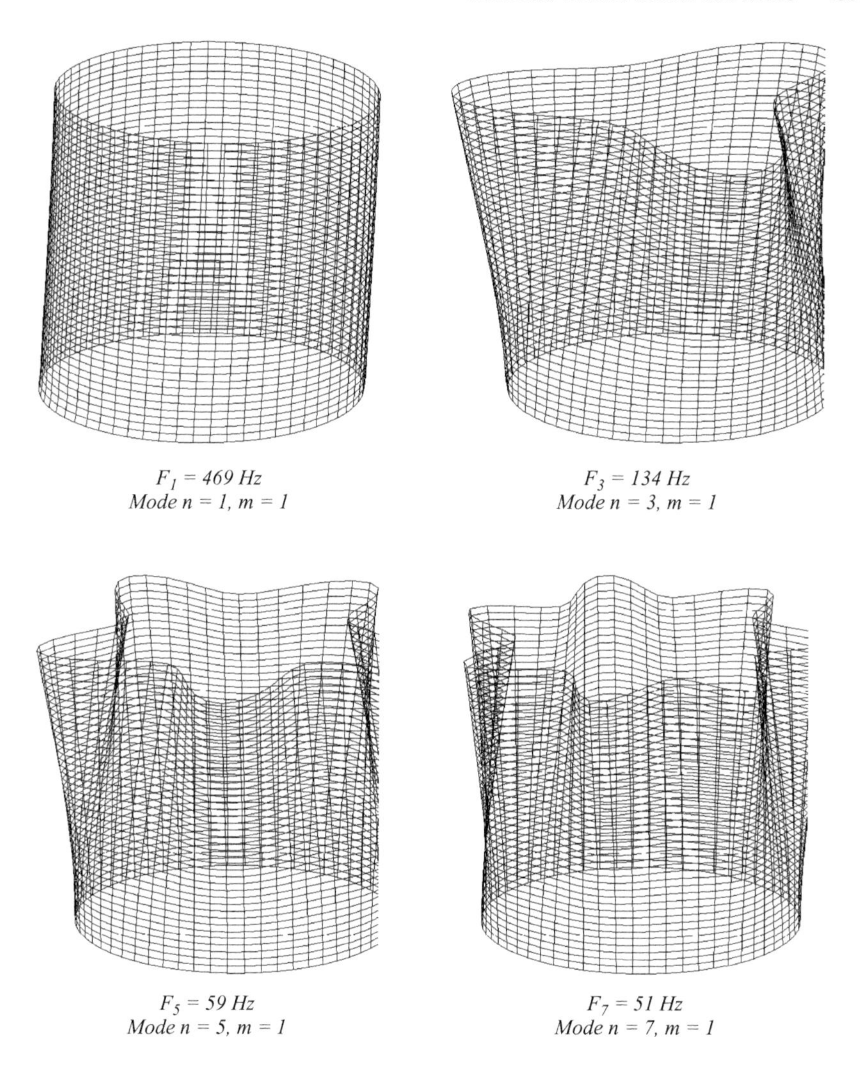

Figure 8.23. *Natural vibration modes of a circular cylindrical shell of small aspect ratio; the shell is clamped at the lower base and free at the other*

magnitude which are applied to a pair of diametrically opposed points, as sketched in Figure 8.24. Besides its practical interest, this problem gives us the opportunity to revisit the Saint-Venant principle, which was already seen to be the cornerstone for modelling structural elements, in Chapter 1 section 1.5. As evidenced in colour plate 2, the elastic response of a circular cylindrical shell to an axial load can present local features which are more or less pronounced, depending upon the space distribution of the loading. As a major result, it is found that if the shell is sufficiently thin (typically $h/R<0.01$), the Saint-Venant principle is not verified.

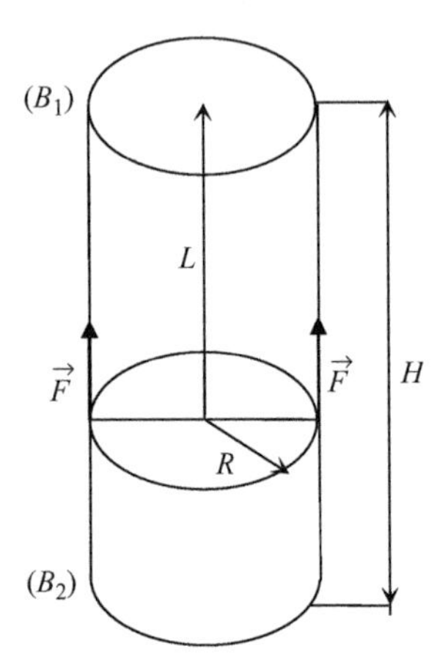

Figure 8.24. *Cylindrical shell loaded by a pair of concentrated axial forces*

The problem sketched in Figure 8.24 was investigated in detail in [CAR 01], based on Love's equations and an expansion in the Fourier series of the load. According to this work, the contribution of a harmonic of rank n of the load to the total response of the shell is transmitted along the shell practically without attenuation according to the approximate law:

$$\frac{H}{R} \simeq \frac{2}{n^2}\sqrt{\frac{R}{h}} \qquad [8.151]$$

For instance in the case of a shell $R = 1\,\text{m}, h = 1\,\text{cm}$, the contribution of the second harmonic remains practically the same up to $H = 5\,\text{m}$. Furthermore, it is also possible to establish that the axial distance up to which the contribution diminishes by a factor ten is five time larger. Such a result, illustrated in colour plates 1 and 2, precludes any practical relevance of the Saint-Venant principle for thin shells of revolution, subjected to concentrated loads. The fairly large length scales necessary to attenuate the so called local responses is clearly a consequence of the shell curvature, since the same problem when investigated on a plate leads to a much smaller length scale as already evidenced in Chapter 5 subsection 5.3.6.4.

Finally, colour plates 13 and 14 show the response of a thin cylindrical shell of revolution, with an aspect ratio $\eta = 4$ to a pair of axial and tangential forces respectively. The forces are applied to two diametrically opposite points at mid-height of the shell, which is clamped at a base and free at the other.

Appendices

A.1. Vector and tensor calculus

A.1.1 *Definition and notations of scalar, vector and tensor fields*

The object of this appendix is to provide the reader with a short presentation of the major concepts and results concerning vector and tensor calculus, which are necessary for the applications discussed in this book.

First, it is recalled that physical entities depending on the position within a continuous medium are categorized, either as scalar, vector, or tensor fields. To identify them independently of the coordinate system used to describe their components, a symbolic notation is used. The notation is not uniform; in this book a quantity denoted q is a scalar, $\vec{q}$ is a vector and $\overline{\overline{q}}$ is a tensor. Symbolic notation is useful to write out the physical laws in an intrinsic form, that is independent from the coordinate system. For instance, the dynamic equilibrium equation,

$$\rho(\vec{r})\ddot{\vec{X}}(\vec{r};t) - \operatorname{div}\overline{\overline{\sigma}}(\vec{r};t) = \vec{F}^{(e)}(\vec{r};t),$$

where $\vec{r}$ is the position vector, holds in any coordinate system, Cartesian or not. $\rho(\vec{r})$ is a scalar field, $\ddot{\vec{X}}(\vec{r};t)$ and $\vec{F}^{(e)}(\vec{r};t)$ are fluctuating (time varying) vector fields and $\overline{\overline{\sigma}}(\vec{r};t)$ is a fluctuating tensor field.

If calculus is performed using symbolic notation, operations such as addition, product, differentiation etc. have to be made according to distinct rules depending on the nature of the quantities involved. Moreover, symbolic notation provides only abstract results not suited to produce numerical results. For this last purpose, it is necessary to describe the physical entities by their components in a specific coordinate system which, in our applications, is always orthogonal. Indicial notation is a convenient way to identify the physical entities by their components. As a scalar, q has only one component, so no index is needed. A vector $\vec{q}$ of the Euclidean 3D space is denoted by its generic component q_i where the index i takes on the values $1, 2, 3$. A tensor is said to be ‘Cartesian’ when its components are defined in an orthogonal coordinate system. The generic component of a Cartesian tensor $\overline{\overline{q}}$ of the Nth rank is written as $q_{ijk...}$ with N indices which take on the values $1, 2, 3$, independently from each other. Note that all the indices are written as subscripts because for Cartesian tensors the concept of covariant and contravariant indices is not required. As such components are scalar variables, or fields depending on the components r_1, r_2, r_3 of $\vec{r}$, we have to deal only with the operation rules applying to scalars, when using indicial notation.

The touchstone to determine the nature of a field is not only based on the number of its components, but, more importantly, on the manner in which they transform under a change of orthogonal coordinate system, as defined by the orthonormal matrix $[A]$ built with the components of the unit vectors of the new orthonormal frame as expressed in the first. The generic term of $[A]$ is denoted a_{ij}, where it is stated that i is the line index and j the column index. By definition, a_{ij} is the i-th component of the j-th unit vector.

A scalar field is unchanged through the transformation:

$$q'\left(r'_1, r'_2, r'_3\right) = q(r_1, r_2, r_3) \qquad [A.1.1]$$

The components of a vector field are changed according to the formula:

$$q'_j\left(r'_1, r'_2, r'_3\right) = a_{ij} q_i(r_1, r_2, r_3) \qquad [A.1.2]$$

The components of a tensor field are changed according to the formula:

$$q'_{ijk...}\left(r'_1, r'_2, r'_3\right) = a_{i\ell} a_{jm} a_{kn} \dots q_{\ell mn...}(r_1, r_2, r_3) \qquad [A.1.3]$$

where the primed quantities refer to the new coordinate system and where the convention of implicit summation applies to the repeated indices. Thus, if the same index appears twice in a tensor formula, one should automatically sum over that index. Such a repeated index is often called a dummy index, since its name can be changed at will, without changing the tensor formula.

Note that indicial notation is not used to identify the coordinates of the position vector $\vec{r}$. They are enclosed within parenthesis to indicate a function, that is the indexed field component is a function of the space coordinates r_1, r_2, r_3, which vary within the domain of definition of the functions, independently from the indices of the field component.

On the other hand, comparing [A.1.2] and [A.1.3], it can be concluded that a vector is a tensor of first rank and comparing [A.1.1] and [A.1.2], a scalar can be interpreted as a tensor of zero rank. Hence, according to indicial notation we have to deal with a single fundamental entity which is a tensor.

A.1.2 *Tensor algebra*

There are a great many different ways to multiply the components of two tensors together. As indicated in the last subsection, it is convenient to perform such operations by using the indicial notation and then to convert it into a symbolic form to produce an intrinsic expression.

A.1.2.1 Contracted product

It is recalled that the scalar product of two vectors is written in symbolic notation as the 'dot product':

$$s = \vec{u} \cdot \vec{v} \qquad [A.1.4]$$

It can also be written by using the matrix notation as:

$$s = [u]^T[v] = [u_1, u_2, u_3]\begin{bmatrix} v_1 \\ v_2 \\ v_3 \end{bmatrix} \qquad [A.1.5]$$

The contracted product of two tensors extends the scalar, or inner product, of two vectors. For tensors of the first rank u_i and v_j the contracted product is defined as:

$$s = u_1v_1 + u_2v_2 + u_3v_3 = u_iv_i \qquad [A.1.6]$$

To prove that the contracted product of two tensors of the first rank (vectors) is a scalar, it is necessary to consider first the contracted products of two tensors of the second rank. The products formed by using a single dummy index result in another tensor of the second rank. Depending on whether the dummy index is a line, or a column index, four distinct tensors are obtained:

$$c_{ij} = a_{ik}b_{kj}; \quad d_{ij} = a_{ki}b_{kj}; \quad e_{ij} = a_{ik}b_{jk}; \quad f_{ij} = a_{ki}b_{jk} \qquad [A.1.7]$$

It is easily verified that these products correspond to the following matrix products:

$$[C] = [A][B]; \quad [D] = [A]^T[B]; \quad [E] = [A][B]^T; \quad [F] = [A]^T[B]^T \qquad [A.1.8]$$

The upper script (^{T}) stands for a matrix transposition, which allows the lines of a matrix to transform to the columns of its transpose. In index notation, the line and column indices are exchanged.

If $[A]$ is an orthonormal matrix, the column vectors of components a_{ij} with j fixed, form an orthonormal set. Therefore, it follows that:

$$a_{ki}a_{kj} = \delta_{ij} \quad \text{where } \delta_{ij} = 1 \ \text{ if } i = j \ \text{ and } \ 0 \text{ otherwise} \qquad [A.1.9]$$

or in matrix notation:

$$[A]^T[A] = [I] \Longleftrightarrow [A]^T = [A]^{-1} \qquad \text{[A.1.10]}$$

where $[I]$ is the identity matrix built with the coefficients δ_{ij}.

Now, we are in position to prove that s, as given by [A.1.6], is a scalar quantity. According to the transformation law [A.1.2] it follows that:

$$u_i' = a_{ij}u_j, \quad v_i' = a_{ik}v_k \Rightarrow u_i'v_i' = a_{ij}u_j a_{ik}v_k = u_j a_{ij} a_{ik} v_k \qquad \text{[A.1.11]}$$

Using [A.1.9], [A.1.11] is further transformed into:

$$u_i'v_i' = u_j a_{ij} a_{ik} v_k = u_j \delta_{jk} v_k = u_j v_j = u_k v_k = s \qquad \text{[A.1.12]}$$

In matrix notation, the same coordinate transformation reads as:

$$\begin{aligned} &[u'] = [A][u];\ [v'] = [A][v] \\ &[u']^T[v'] = [u]^T[A]^T[A][v] = [u]^T[v] = s \end{aligned} \qquad \text{[A.1.13]}$$

For using symbolic notation in the case of tensors of the second rank, the symbol (^{T}) is appropriate to indicate a transposition, as in matrix notation, and a dot is used to indicate a contracted product. So, the products [A.1.7] and [A.1.8] are written as:

$$\overline{\overline{C}} = \overline{\overline{A}} \cdot \overline{\overline{B}}; \overline{\overline{D}} = \left(\overline{\overline{A}}\right)^T \cdot \overline{\overline{B}}; \overline{\overline{E}} = \overline{\overline{A}} \cdot \left(\overline{\overline{B}}\right)^T; \overline{\overline{F}} = \left(\overline{\overline{A}}\right)^T \cdot \left(\overline{\overline{B}}\right)^T \qquad \text{[A.1.14]}$$

On the other hand, a tensor product can be contracted with respect to several indices, which gives a new tensor, the rank of which is equal to the number of the non-repeated indices. For instance, if the product of two tensors of the second rank is contracted twice, the result is a scalar:

$$\begin{aligned} &\text{indicial notation:} && s = a_{ij}b_{ij} \\ &\text{symbolic notation:} && s = \overline{\overline{A}} : \overline{\overline{B}} \end{aligned} \qquad \text{[A.1.15]}$$

As in [A.1.15], both i and j are dummy indices, they can be exchanged at will producing the same scalar. In the same way, if the symbolic notation is used the tensors can be replaced by their transpose.

Finally, it is worth emphasizing that the contacted product can be used as a very convenient touchstone to prove whether an indexed quantity $q_{ijkl\ldots}$ is a tensor or not. Indeed, if $q_{ijkl\ldots}$ is a tensor, the contracted product $q_{ijkl\ldots}q_{ijkl\ldots} = (q_{ijkl\ldots})^2 = s$, where all the indices are repeated, must give a scalar. To alleviate the mathematical manipulations, let us consider the case of a tensor of the second rank. According to the transformation law [A.1.3]:

$$\left(q'_{ij}\right)^2 = a_{ik}a_{j\ell}q_{k\ell}a_{im}a_{jn}q_{mn} = (a_{ik}a_{im})(a_{j\ell}a_{jn})q_{k\ell}q_{mn}$$

Using again the relation [A1.9], it follows that:

$$\left(q'_{ij}\right)^2 = \delta_{km}\delta_{\ell n}q_{k\ell}q_{mn} = (q_{mn})^2 = (q_{k\ell})^2 \quad [A.1.16]$$

Conversely, if the results of the contracted product $(q_{ijkl\ldots})^2$ is found to be independent from the coordinate system (i.e. a scalar), it can be stated that $q_{ijkl\ldots}$ is a tensor. The criterion can be extended to the case of the product of quantities indexed twice. For instance q_{ij} and p_{ij} are two tensors, provided $p_{ij}q_{ij}$ is a scalar.

A.1.2.2 Non-contracted product

The non-contracted product, or dyadic product, of two tensors of the first rank u_i and v_j is a tensor of the second rank defined as:

$$p_{ij} = u_i v_j \quad [A.1.17]$$

It can be expressed in matrix notation as:

$$[p] = \begin{bmatrix} u_1v_1 & u_1v_2 & u_1v_3 \\ u_2v_1 & u_2v_2 & u_2v_3 \\ u_3v_1 & u_3v_2 & u_3v_3 \end{bmatrix} \quad [A.1.18]$$

where, as a convention, i stands for the line index and j for the column index.

In symbolic notation, such a non-contracted product is written as:

$$\overline{\overline{p}} = \overline{\overline{u}}\,\overline{\overline{v}} \quad [A.1.19]$$

A.1.2.3 Cross-product of two vectors in indicial notation

The cross-product of two vectors is written in symbolic notation as:

$$\vec{c} = \vec{a} \times \vec{b} \quad [A.1.20]$$

It is recalled that the Cartesian components of the resulting vector $\vec{c}$ are:

$$c_1 = a_2b_3 - a_3b_2; \quad c_2 = a_3b_1 - a_1b_3; \quad c_3 = a_1b_2 - a_2b_1$$

a_1, a_2, a_3 and b_1, b_2, b_3 are the Cartesian components of $\vec{a}$ and $\vec{b}$, respectively.

The cross-product can also be written in index notation by using the alternate unit tensor of third rank ε_{ijk} whose components are defined as follows:

$$\varepsilon_{ijk} = \begin{cases} 0 & \text{if at least one index is repeated} \\ +1 & \text{if } i, j, k \text{ form a clockwise circulatory sequence} \\ -1 & \text{if } i, j, k \text{ form an anticlockwise circulatory sequence} \end{cases} \qquad \text{[A.1.21]}$$

It allows one to write the cross-product of two vectors as:

$$c_i = \varepsilon_{ijk} a_j b_k \qquad \text{[A.1.22]}$$

A.2. Differential operators

A.2.1 *The Nabla differential operator*

In symbolic notation, differential operators such as the divergence, the gradient the curl etc. can be identified after their specific name, as for instance div $\vec{X}$ reads as the divergence of the vector field $\vec{X}$, $\overrightarrow{\text{grad}\,\Phi}$ as the gradient of the scalar field Φ, or $\overline{\overline{\text{grad}}}\,\vec{X}$ as the gradient of the vector field $\vec{X}$ etc. Note that the upper arrow is used here simply to stress that the gradient of a scalar is a vector and the double bar to stress that the gradient of a vector is a tensor. However, for the sake of clarity and as a preliminary step toward an appropriate index notation, a more convenient symbolic notation makes use of the so called 'Nabla' differential operator denoted here $\vec{\nabla}$ to mark that it can be used as a vector. In Cartesian coordinates $\vec{\nabla}$ is written as:

$$\vec{\nabla} = \frac{\partial}{\partial x}\vec{i} + \frac{\partial}{\partial y}\vec{j} + \frac{\partial}{\partial z}\vec{k} \qquad \text{[A.2.1]}$$

Going a step further, in index notation $\vec{\nabla}$ is written as ∇_i, $i = 1, 2, 3$.

A.2.2 *The divergence operator*

Let $\vec{A}$ be a vector of Cartesian coordinates, A_x, A_y, A_z. Its divergence can be written as:

$$\text{div}\,\vec{A} = \vec{\nabla}\cdot\vec{A} = \frac{\partial A_x}{\partial x} + \frac{\partial A_y}{\partial y} + \frac{\partial A_z}{\partial z} \iff \text{div}\,\vec{A} = \nabla_i A_i = \frac{\partial A_i}{\partial x_i} \qquad \text{[A.2.2]}$$

As a classical application, it is shown that the rate of change of an infinitesimal volume subjected to a displacement field $\vec{X}$ is given by div$\vec{X}$. Indeed, to the first order, the lengths of the cubical element $dx\,dy\,dz$ are changed into

$(1 + \partial X/\partial x)\,dx$ etc. Thus the change of volume is given to the first order by:

$$d\mathcal{V} = \left(\left(1 + \frac{\partial X}{\partial x}\right)\left(1 + \frac{\partial Y}{\partial y}\right)\left(1 + \frac{\partial Z}{\partial z}\right) - 1\right) dx\,dy\,dz \simeq \operatorname{div}\vec{X} dx\,dy\,dz \qquad [A.2.3]$$

In the same way, the divergence of a tensor $\overline{\overline{A}}$ of rank higher than one can be written as:

$$\operatorname{div}\overline{\overline{A}} = \vec{\nabla}\cdot\overline{\overline{A}} = \nabla_i A_{ijk\ldots} = \frac{\partial A_{ijk\ldots}}{\partial x_i} \qquad [A.2.4]$$

The divergence theorem allows one to transform the volume integral of $\operatorname{div}\overline{\overline{A}}$ into the integral of the flux of $\overline{\overline{A}}$ over the surface bounding the volume, which is written as:

$$\int_{(\mathcal{V})} \operatorname{div}\overline{\overline{A}}\, d\mathcal{V} = \int_{(\mathcal{S})} \overline{\overline{A}}\cdot\vec{n}\, d\mathcal{S} \qquad [A.2.5]$$

where $\vec{n}$ is the unit vector normal to the elementary surface $d\mathcal{S}$ of the boundary of the volume $(\mathcal{V})$.

This very important theorem can be proved by using indicial notation:

$$\int_{(\mathcal{V})} \frac{\partial A_{ijk\ldots}}{\partial x_i}\, dx_1\,dx_2\,dx_3 = \int_{(\mathcal{S})} A_{1jk\ldots} dx_2\,dx_3 + \int_{(\mathcal{S})} A_{2jk\ldots} dx_1\,dx_3 + \int_{(\mathcal{S})} A_{3jk\ldots} dx_1\,dx_2$$

It is noticed that $dx_2\,dx_3$, $dx_1\,dx_3$, $dx_1\,dx_2$ are the elementary areas normal to the unit vectors $\vec{i}, \vec{j}, \vec{k}$ respectively whereas $A_{1jk\ldots}, A_{2jk\ldots}, A_{3jk\ldots}$ are the components of $\overline{\overline{A}}$ in the directions $\vec{i}, \vec{j}, \vec{k}$ respectively. Accordingly, the three surface integrals can be merged into the single surface integral which appears on the right-hand side of [A.2.5].

A.2.3 *The gradient operator*

Let Φ be a scalar field. Its gradient is a vector which may be written by using one of the following possible notations:

$$\overrightarrow{\text{grad}\,\Phi} = \vec{\nabla}\Phi = \nabla_i \Phi = \frac{\partial \Phi}{\partial x_i} \qquad [A.2.6]$$

$\overrightarrow{\text{grad}\,\Phi}$ allows one to measure the variation of Φ when the position is shifted by the infinitesimal quantity $d\vec{r} \cdot d\Phi$ is given by:

$$d\Phi = d\vec{r} \cdot \overrightarrow{\text{grad}\,\Phi} = \frac{\partial \Phi}{\partial x_i} dx_i \qquad [A.2.7]$$

where dx_i is the generic Cartesian component of $d\vec{r}$.

In a similar way, the rate of change of Φ along the $\vec{\ell}$ direction, where $\vec{\ell}$ is a unit vector, is given by the scalar product:

$$\vec{\ell} \cdot \overrightarrow{\text{grad}\,\Phi} = \frac{\partial \Phi}{\partial x_i} \ell_i \qquad [A.2.8]$$

In terms of geometry, [A.2.8] is interpreted as the orthogonal projection of the gradient on the $\vec{\ell}$ direction.

In the same way the gradient of a tensor can be written as the non-contracted product, written in symbolic notation as:

$$\text{grad}\,\overline{\overline{A}} = \vec{\nabla}\overline{\overline{A}} \qquad [A.2.9]$$

The gradient of a vector is a tensor of the second rank, written as:

$$\overline{\overline{\text{grad}\,\vec{A}}} = \vec{\nabla}\vec{A} = \frac{\partial A_j}{\partial x_i} = \begin{bmatrix} \frac{\partial A_x}{\partial x} & \frac{\partial A_y}{\partial x} & \frac{\partial A_z}{\partial x} \\ \frac{\partial A_x}{\partial y} & \frac{\partial A_y}{\partial y} & \frac{\partial A_z}{\partial y} \\ \frac{\partial A_x}{\partial z} & \frac{\partial A_y}{\partial z} & \frac{\partial A_z}{\partial z} \end{bmatrix} \qquad [A.2.10]$$

A.2.4 *The curl operator*

$$\text{curl}\,(\vec{A}) = \vec{\nabla} \times \vec{A} = \varepsilon_{ijk} \frac{\partial A_k}{\partial x_j} \qquad [A.2.11]$$

where ε_{ijk} is the alternate unit tensor defined by [A.1.21].

Let $\vec{\Omega}\,dt$ be an infinitesimal rotation applied to a rigid body. The displacement field is given by $\vec{X} = (\vec{\Omega} \times \vec{r})\,dt$, whose Cartesian components are:

$$X = (z\Omega_y - y\Omega_z)\,dt; \quad Y = (x\Omega_z - z\Omega_x)\,dt; \quad Z = (y\Omega_x - x\Omega_y)\,dt$$

Forming the curl of this displacement field, it can be verified that:

$$\text{curl}\,\vec{X} = 2\vec{\Omega} \qquad [A.2.12]$$

A.2.5 *The Laplace operator*

$$\vec{\nabla} \cdot (\vec{\nabla}\Phi) = \Delta\Phi = \frac{\partial^2 \Phi}{\partial x^2} + \frac{\partial^2 \Phi}{\partial y^2} + \frac{\partial^2 \Phi}{\partial z^2} \qquad [A.2.13]$$

$$\vec{\nabla} \cdot (\vec{\nabla}\vec{A}) = \Delta\vec{A} = (\Delta A_x)\vec{i} + (\Delta A_y)\vec{j} + (\Delta A_z)\vec{k} \qquad [A.2.14]$$

A.2.6 *Other useful formulas*

$$\text{curl}\left(\overrightarrow{\text{grad}\,\Phi}\right) = 0 \qquad [A.2.15]$$

$$\text{div}\left(\text{curl}\,\vec{A}\right) = 0 \qquad [A.2.16]$$

$$\text{curl}\left(\text{curl}\,\vec{A}\right) = \overrightarrow{\text{grad}\left(\text{div}\vec{A}\right)} - \Delta\vec{A} \qquad [A.2.17]$$

$$\text{div}\left(\alpha\vec{A}\right) = \vec{A} \cdot \overrightarrow{\text{grad}\,\alpha} + \alpha\,\text{div}\left(\vec{A}\right) \qquad [A.2.18]$$

$$\text{curl}\left(\alpha\vec{A}\right) = \overrightarrow{\text{grad}\,\alpha} \times \vec{A} + \alpha\,\text{curl}\,\vec{A} \qquad [A.2.19]$$

$$\overrightarrow{\text{grad}\left(\vec{A} \cdot \vec{B}\right)} = \vec{A} \cdot \overline{\overline{\text{grad}\,\vec{B}}} + \vec{B} \cdot \overline{\overline{\text{grad}\,\vec{A}}} + \vec{A} \times \text{curl}\,\vec{B} + \vec{B} \times \text{curl}\,\vec{A} \qquad [A.2.20]$$

$$2\vec{A} \cdot \overline{\overline{\text{grad}\,\vec{B}}} = -\text{curl}\left(\vec{A} \times \vec{B}\right) + \overrightarrow{\text{grad}\left(\vec{A} \cdot \vec{B}\right)} + \vec{A}\,\text{div}\vec{B} - \vec{B}\,\text{div}\vec{A} - \vec{B} \times \text{curl}\,\vec{A} - \vec{A} \times \text{curl}\,\vec{B} \qquad [A.2.21]$$

$$\text{div}\left(\vec{A} \times \vec{B}\right) = \vec{A} \cdot \text{curl}\,\vec{B} - \vec{B} \cdot \text{curl}\,\vec{A} \qquad [A.2.22]$$

$$\text{curl}\left(\vec{A} \times \vec{B}\right) = \vec{B} \cdot \overline{\overline{\text{grad}\,\vec{A}}} - \vec{A} \cdot \overline{\overline{\text{grad}\,\vec{B}}} + \vec{A}\,\text{div}\vec{B} - \vec{B}\,\text{div}\vec{A} \qquad [A.2.23]$$

A.3. Differential operators in curvilinear and orthonormal coordinates

A.3.1 *Metrics*

The curvilinear coordinate system (Figure A.3.1) is defined by two families of orthogonal curves denoted $(\mathcal{C}_\alpha)$ and $(\mathcal{C}_\beta)$ respectively, which are described in the Cartesian system Oxy, by the parametric equations:

$$x = x(\alpha, \beta);\; y = y(\alpha, \beta) \qquad [A.3.1]$$

The curves $(\mathcal{C}_\alpha)$ are such that α is constant and the curves $(\mathcal{C}_\beta)$ are such that β is constant. Accordingly, $\vec{e}_\alpha = \left(\partial x/\partial\alpha\vec{i} + \partial y/\partial\alpha\,\vec{j}\right)d\alpha$ is tangent to $(\mathcal{C}_\beta)$ and normal

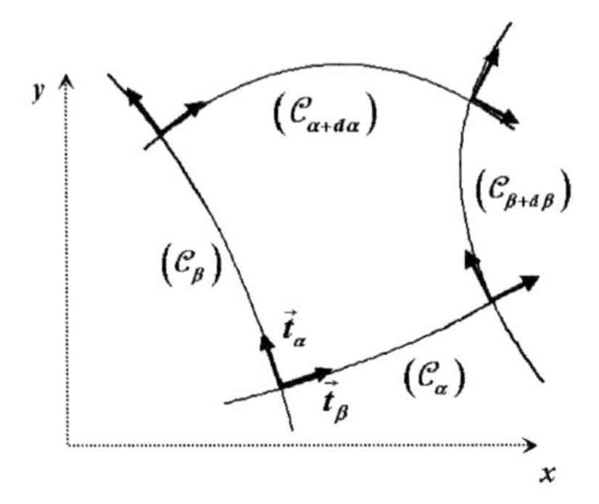

Figure A.3.1. *Curvilinear and orthonormal coordinate system*

to $(\mathcal{C}_{\alpha})$. The corresponding unit vector is $\vec{t}_{\alpha}$, such that:

$$\vec{e}_{\alpha} = g_{\alpha}\, d\alpha \vec{t}_{\alpha} \quad \text{where } g_{\alpha} = \sqrt{\left(\frac{\partial x}{\partial \alpha}\right)^2 + \left(\frac{\partial y}{\partial \alpha}\right)^2} \qquad \text{[A.3.2]}$$

In the same way $\vec{e}_{\beta} = \left(\partial x/\partial\beta \vec{i} + \partial y/\partial\beta \vec{j}\right) d\beta$ is tangent to $(\mathcal{C}_{\alpha})$ and normal to $(\mathcal{C}_{\beta})$. The corresponding unit vector is $\vec{t}_{\beta}$, such that:

$$\vec{e}_{\beta} = g_{\beta} d\beta \vec{t}_{\beta} \quad \text{where } g_{\beta} = \sqrt{\left(\frac{\partial x}{\partial \beta}\right)^2 + \left(\frac{\partial y}{\partial \beta}\right)^2} \qquad \text{[A.3.3]}$$

From the orthogonality of $\vec{t}_{\alpha}$ and $\vec{t}_{\beta}$ it follows that:

$$\vec{t}_{\alpha} \cdot \vec{t}_{\beta} = \frac{1}{g_{\alpha} g_{\beta}\, d\alpha\, d\beta} \left(\frac{\partial x}{\partial \alpha}\vec{i} + \frac{\partial y}{\partial \alpha}\vec{j}\right) \cdot \left(\frac{\partial x}{\partial \beta}\vec{i} + \frac{\partial y}{\partial \beta}\vec{j}\right) \Rightarrow \frac{\partial x}{\partial \alpha}\frac{\partial x}{\partial \beta} + \frac{\partial y}{\partial \alpha}\frac{\partial y}{\partial \beta} = 0 \qquad \text{[A.3.4]}$$

Invariance of the length of an infinitesimal segment implies:

$$\begin{aligned} ds^2 = dx^2 + dy^2 &= \left(\frac{\partial x}{\partial \alpha} d\alpha + \frac{\partial x}{\partial \beta} d\beta\right)^2 + \left(\frac{\partial y}{\partial \alpha} d\alpha + \frac{\partial y}{\partial \beta} d\beta\right)^2 \\ &= \left(\left(\frac{\partial x}{\partial \alpha}\right)^2 + \left(\frac{\partial y}{\partial \alpha}\right)^2\right) d\alpha^2 + \left(\left(\frac{\partial x}{\partial \beta}\right)^2 + \left(\frac{\partial y}{\partial \beta}\right)^2\right) d\beta^2 = g_{\alpha}^2\, d\alpha^2 + g_{\beta}^2\, d\beta^2 \end{aligned} \qquad \text{[A.3.5]}$$

The result is suitably written in matrix form as:

$$ds^2 = dx^2 + dy^2 = [dx\ dy]\begin{bmatrix}1 & 0\\ 0 & 1\end{bmatrix}\begin{bmatrix}dx\\ dy\end{bmatrix}$$
$$ds^2 = g_\alpha^2\, d\alpha^2 + g_\beta^2\, d\beta^2 = [d\alpha\ d\beta]\begin{bmatrix}g_\alpha^2 & 0\\ 0 & g_\beta^2\end{bmatrix}\begin{bmatrix}d\alpha\\ d\beta\end{bmatrix} \qquad \text{[A.3.6]}$$

where the matrices are identified with the metric tensor in Cartesian and in orthonormal curvilinear coordinates, respectively.

Other useful relationships concern the rate of change of the unit vectors along the curves $(\mathcal{C}_\alpha)$ and $(\mathcal{C}_\beta)$. They may be gathered into the single matrix equality:

$$\begin{bmatrix}\dfrac{\partial \vec{t}_\alpha}{\partial \alpha} & \dfrac{\partial \vec{t}_\alpha}{\partial \beta}\\ \dfrac{\partial \vec{t}_\beta}{\partial \alpha} & \dfrac{\partial \vec{t}_\beta}{\partial \beta}\end{bmatrix} = \begin{bmatrix}-\dfrac{1}{g_\beta}\dfrac{\partial g_\alpha}{\partial \beta}\vec{t}_\beta & +\dfrac{1}{g_\alpha}\dfrac{\partial g_\beta}{\partial \alpha}\vec{t}_\beta\\ +\dfrac{1}{g_\beta}\dfrac{\partial g_\alpha}{\partial \beta}\vec{t}_\alpha & -\dfrac{1}{g_\alpha}\dfrac{\partial g_\beta}{\partial \alpha}\vec{t}_\alpha\end{bmatrix} \qquad \text{[A.3.7]}$$

Proof of [A.3.7] is outlined as follows:

$$\vec{t}_\alpha \cdot \vec{t}_\alpha = 1 \ \Rightarrow\ \vec{t}_\alpha \cdot \frac{\partial \vec{t}_\alpha}{\partial \alpha} = 0 \ \Rightarrow\ \frac{\partial \vec{t}_\alpha}{\partial \alpha} = a\vec{t}_\beta$$
$$\vec{t}_\alpha \cdot \vec{t}_\alpha = 1 \ \Rightarrow\ \vec{t}_\alpha \cdot \frac{\partial \vec{t}_\alpha}{\partial \beta} = 0 \ \Rightarrow\ \frac{\partial \vec{t}_\alpha}{\partial \beta} = b\vec{t}_\beta$$
$$\vec{t}_\alpha \cdot \vec{t}_\beta = 0 \ \Rightarrow\ \vec{t}_\alpha \cdot \frac{\partial \vec{t}_\beta}{\partial \alpha} = -\frac{\partial \vec{t}_\alpha}{\partial \alpha}\cdot \vec{t}_\beta = -a$$

Let $\vec{r}$ be a position vector, so:

$$d\vec{r} = dx\vec{i} + dy\vec{j} = g_\alpha\, d\alpha\, \vec{t}_\alpha + g_\beta\, d\beta\, \vec{t}_\beta$$
$$\Rightarrow\ \frac{d\vec{r}}{d\alpha} = g_\alpha \vec{t}_\alpha;\quad \frac{d\vec{r}}{d\beta} = g_\beta \vec{t}_\beta$$
$$\Rightarrow\ \frac{d^2\vec{r}}{d\alpha d\beta} = \frac{\partial g_\beta}{\partial \alpha}\vec{t}_\beta + g_\beta \frac{\partial \vec{t}_\beta}{\partial \alpha} = \frac{\partial g_\alpha}{\partial \beta}\vec{t}_\alpha + g_\alpha \frac{\partial \vec{t}_\alpha}{\partial \beta}$$

Projection of the last expression on $\vec{t}_\alpha$ gives $g_\beta \partial\vec{t}_\beta/\partial\alpha \cdot \vec{t}_\alpha = -a g_\beta = \partial g_\alpha/\partial\beta$, whence the desired result $\partial\vec{t}_\alpha/\partial\alpha = -(\partial g_\alpha/g_\beta \partial_\beta)\vec{t}_\beta$. The other derivatives in [A.3.7] are obtained in the same way.

A.3.2 *Differential operators in curvilinear and orthogonal coordinates*

A.3.2.1 Gradient of a scalar and the Nabla operator

$$\overrightarrow{\text{grad}\,\Phi} = \frac{\partial \Phi}{\partial x}\vec{i} + \frac{\partial \Phi}{\partial y}\vec{j} = \vec{\nabla}\Phi = G_\alpha \vec{t}_\alpha + G_\beta \vec{t}_\beta \qquad \text{[A.3.8]}$$

$$G_\alpha = \frac{\partial \Phi}{\partial x}\vec{i}\cdot\vec{t}_\alpha + \frac{\partial \Phi}{\partial y}\vec{j}\cdot\vec{t}_\alpha = \frac{1}{g_\alpha}\left(\frac{\partial x}{\partial \alpha}\frac{\partial \Phi}{\partial x} + \frac{\partial y}{\partial \alpha}\frac{\partial \Phi}{\partial y}\right) = \frac{1}{g_\alpha}\frac{\partial \Phi}{\partial \alpha}$$

$$G_\beta = \frac{\partial \Phi}{\partial x}\vec{i}\cdot\vec{t}_\beta + \frac{\partial \Phi}{\partial y}\vec{j}\cdot\vec{t}_\beta = \frac{1}{g_\beta}\left(\frac{\partial x}{\partial \beta}\frac{\partial \Phi}{\partial x} + \frac{\partial y}{\partial \beta}\frac{\partial \Phi}{\partial y}\right) = \frac{1}{g_\beta}\frac{\partial \Phi}{\partial \beta}$$

whence:

$$\vec{\nabla} = \vec{t}_\alpha \frac{1}{g_\alpha}\frac{\partial}{\partial \alpha} + \vec{t}_\beta \frac{1}{g_\beta}\frac{\partial}{\partial \beta} \qquad \text{[A.3.9]}$$

$$\vec{\nabla}\Phi = \vec{t}_\alpha \frac{1}{g_\alpha}\frac{\partial \Phi}{\partial \alpha} + \vec{t}_\beta \frac{1}{g_\beta}\frac{\partial \Phi}{\partial \beta} \qquad \text{[A.3.10]}$$

A.3.2.2 Gradient of a vector

$$\overline{\overline{\text{grad}\,\vec{V}}} = \vec{\nabla}\vec{V} = \left(\vec{t}_\alpha \frac{1}{g_\alpha}\frac{\partial}{\partial \alpha} + \vec{t}_\beta \frac{1}{g_\beta}\frac{\partial}{\partial \beta}\right)\left(X_\alpha \vec{t}_\alpha + X_\beta \vec{t}_\beta\right) \qquad \text{[A.3.11]}$$

As $\vec{t}_\alpha, \vec{t}_\beta$ are unit and orthogonal vectors, in matrix notation, the non-contracted product $\vec{t}_\alpha\vec{t}_\beta$ produces a matrix in which all elements are zero, except those located at the α-th row and the β-th column which is equal to one. The components of [A.3.11] are:

$$\begin{aligned}
\vec{t}_\alpha \frac{1}{g_\alpha}\frac{\partial X_\alpha \vec{t}_\alpha}{\partial \alpha} &= (\vec{t}_\alpha\vec{t}_\alpha)\frac{1}{g_\alpha}\frac{\partial X_\alpha}{\partial \alpha} + \left(\vec{t}_\alpha \frac{\partial \vec{t}_\alpha}{\partial \alpha}\right)\frac{X_\alpha}{g_\alpha} \\
\vec{t}_\alpha \frac{1}{g_\alpha}\frac{\partial X_\beta \vec{t}_\beta}{\partial \alpha} &= (\vec{t}_\alpha\vec{t}_\beta)\frac{1}{g_\alpha}\frac{\partial X_\beta}{\partial \alpha} + \left(\vec{t}_\alpha \frac{\partial \vec{t}_\beta}{\partial \alpha}\right)\frac{X_\beta}{g_\alpha} \\
\vec{t}_\beta \frac{1}{g_\beta}\frac{\partial X_\alpha \vec{t}_\alpha}{\partial \beta} &= (\vec{t}_\beta\vec{t}_\alpha)\frac{1}{g_\beta}\frac{\partial X_\alpha}{\partial \beta} + \left(\vec{t}_\beta \frac{\partial \vec{t}_\alpha}{\partial \beta}\right)\frac{X_\alpha}{g_\beta} \\
\vec{t}_\beta \frac{1}{g_\beta}\frac{\partial X_\beta \vec{t}_\beta}{\partial \beta} &= (\vec{t}_\beta\vec{t}_\beta)\frac{1}{g_\beta}\frac{\partial X_\beta}{\partial \beta} + \left(\vec{t}_\beta \frac{\partial \vec{t}_\beta}{\partial \beta}\right)\frac{X_\beta}{g_\beta}
\end{aligned} \qquad \text{[A.3.12]}$$

The derivatives of the unit vectors are given by the formula [A.3.7], whence:

$$\vec{\nabla}\vec{V} = \begin{bmatrix} \dfrac{1}{g_\alpha}\dfrac{\partial X_\alpha}{\partial \alpha} + \dfrac{X_\beta}{g_\alpha g_\beta}\dfrac{\partial g_\alpha}{\partial \beta} & \dfrac{1}{g_\alpha}\dfrac{\partial X_\beta}{\partial \alpha} - \dfrac{X_\alpha}{g_\alpha g_\beta}\dfrac{\partial g_\alpha}{\partial \beta} \\ \dfrac{1}{g_\beta}\dfrac{\partial X_\alpha}{\partial \beta} - \dfrac{X_\beta}{g_\alpha g_\beta}\dfrac{\partial g_\beta}{\partial \alpha} & \dfrac{1}{g_\beta}\dfrac{\partial X_\beta}{\partial \beta} + \dfrac{X_\alpha}{g_\alpha g_\beta}\dfrac{\partial g_\beta}{\partial \alpha} \end{bmatrix} \qquad \text{[A.3.13]}$$

A.3.2.3 Divergence of a vector

$$\text{div}\vec{V} = \vec{\nabla}\cdot\vec{V} = \left(\vec{t}_\alpha \frac{1}{g_\alpha}\frac{\partial}{\partial \alpha} + \vec{t}_\beta \frac{1}{g_\beta}\frac{\partial}{\partial \beta}\right)\cdot\left(X_\alpha \vec{t}_\alpha + X_\beta \vec{t}_\beta\right) \qquad \text{[A.3.14]}$$

Whence:

$$\begin{aligned} \vec{\nabla}\cdot\vec{V} &= \vec{t}_\alpha \cdot \frac{1}{g_\alpha}\frac{\partial\left(X_\alpha \vec{t}_\alpha\right)}{\partial \alpha} + \vec{e}_\alpha \cdot \frac{1}{g_\alpha}\frac{\partial\left(X_\beta \vec{t}_\beta\right)}{\partial \alpha} + \vec{t}_\beta \cdot \frac{1}{g_\beta}\frac{\partial\left(X_\alpha \vec{t}_\alpha\right)}{\partial \alpha} \\ &\quad + \vec{t}_\beta \cdot \frac{1}{g_\beta}\frac{\partial\left(X_\beta \vec{t}_\beta\right)}{\partial \beta} \\ &= \left(\vec{t}_\alpha\cdot\vec{t}_\alpha\right)\frac{1}{g_\alpha}\frac{\partial X_\alpha}{\partial \alpha} + \frac{X_\alpha}{g_\alpha}\left(\vec{t}_\alpha\cdot\frac{\partial \vec{t}_\alpha}{\partial \alpha}\right) + \left(\vec{t}_\alpha\cdot\vec{t}_\beta\right)\frac{1}{g_\alpha}\frac{\partial X_\beta}{\partial \alpha} + \frac{X_\beta}{g_\alpha}\left(\vec{t}_\beta\cdot\frac{\partial \vec{t}_\beta}{\partial \alpha}\right) \\ &\quad + \left(\vec{t}_\beta\cdot\vec{t}_\beta\right)\frac{1}{g_\beta}\frac{\partial X_\beta}{\partial \beta} + \frac{X_\beta}{g_\beta}\left(\vec{t}_\beta\cdot\frac{\partial \vec{t}_\beta}{\partial \beta}\right) + \left(\vec{t}_\beta\cdot\vec{t}_\alpha\right)\frac{1}{g_\beta}\frac{\partial X_\alpha}{\partial \beta} \\ &\quad + \frac{X_\alpha}{g_\beta}\left(\vec{t}_\alpha\cdot\frac{\partial \vec{t}_\alpha}{\partial \beta}\right) \end{aligned}$$

Using the orthogonality conditions and the relations [A3.7], we get:

$$\vec{\nabla}\cdot\vec{V} = \frac{1}{g_\alpha}\frac{\partial X_\alpha}{\partial \alpha} + \frac{X_\beta}{g_\alpha g_\beta}\frac{\partial g_\alpha}{\partial \beta} + \frac{1}{g_\beta}\frac{\partial X_\beta}{\partial \beta} + \frac{X_\alpha}{g_\beta g_\alpha}\frac{\partial g_\beta}{\partial \alpha}$$

written in a compact form as:

$$\vec{\nabla}\cdot\vec{V} = \frac{1}{g_\alpha g_\beta}\left[\frac{\partial(g_\beta X_\alpha)}{\partial \alpha} + \frac{\partial(g_\alpha X_\beta)}{\partial \beta}\right] \qquad \text{[A.3.15]}$$

A.3.2.4 Divergence of a tensor of the second rank

$$\text{div}\,\overline{\overline{T}} = \begin{bmatrix} \dfrac{\partial(g_\beta T_{\alpha\alpha})}{\partial \alpha} - T_{\beta\beta}\dfrac{\partial g_\beta}{\partial \alpha} + \dfrac{1}{g_\alpha}\dfrac{\partial(g_\alpha^2 T_{\beta\alpha})}{\partial \beta} \\ \dfrac{\partial(g_\alpha T_{\beta\beta})}{\partial \beta} - T_{\alpha\alpha}\dfrac{\partial g_\alpha}{\partial \beta} + \dfrac{1}{g_\beta}\dfrac{\partial(g_\beta^2 T_{\alpha\beta})}{\partial \alpha} \end{bmatrix} \qquad \text{[A.3.16]}$$

A.3.2.5 Curl of a vector

$$\vec{\nabla} \times \vec{V} = \left(\vec{t}_\alpha \frac{1}{g_\alpha}\frac{\partial}{\partial \alpha} + \vec{t}_\beta \frac{1}{g_\beta}\frac{\partial}{\partial \beta}\right) \times (X_\alpha \vec{t}_\alpha + X_\beta \vec{t}_\beta)$$

$$\begin{aligned}\vec{\nabla} \times \vec{V} = {} & \vec{t}_\alpha \times \left(\frac{1}{g_\alpha}\frac{\partial (X_\alpha \vec{t}_\alpha)}{\partial \alpha}\right) + \vec{t}_\alpha \times \left(\frac{1}{g_\beta}\frac{\partial (X_\beta \vec{t}_\beta)}{\partial \beta}\right) \\ & + \vec{t}_\beta \times \left(\frac{1}{g_\beta}\frac{\partial (X_\alpha \vec{t}_\alpha)}{\partial \beta}\right) + \vec{t}_\beta \times \left(\frac{1}{g_\beta}\frac{\partial (X_\beta \vec{t}_\beta)}{\partial \beta}\right) \\ & - (\vec{t}_\alpha \times \vec{t}_\beta)\left(\frac{X_\alpha}{g_\alpha g_\beta}\frac{\partial g_\alpha}{\partial \beta}\right) + (\vec{t}_\alpha \times \vec{t}_\beta)\frac{1}{g_\alpha}\frac{\partial X_\beta}{\partial \alpha} \\ & - (\vec{t}_\beta \times \vec{t}_\alpha)\left(\frac{X_\beta}{g_\alpha g_\beta}\frac{\partial g_\beta}{\partial \alpha}\right) + (\vec{t}_\alpha \times \vec{t}_\beta)\frac{1}{g_\beta}\frac{\partial X_\alpha}{\partial \beta}\end{aligned}$$

and finally:

$$\vec{\nabla} \times \vec{V} = \frac{1}{g_\alpha g_\beta}\left(\frac{\partial (g_\beta X_\beta)}{\partial \alpha} - \frac{\partial (g_\alpha X_\alpha)}{\partial \beta}\right)\vec{k} \qquad [A.3.17]$$

A.3.2.6 Laplacian of a scalar

$$\Delta \Phi = \vec{\nabla} \cdot \vec{\nabla} \Phi = \frac{1}{g_\alpha g_\beta}\left(\frac{\partial}{\partial \alpha}\left(\frac{g_\beta}{g_\alpha}\frac{\partial \Phi}{\partial \alpha}\right) + \frac{\partial}{\partial \beta}\left(\frac{g_\alpha}{g_\beta}\frac{\partial \Phi}{\partial \beta}\right)\right) \qquad [A.3.18]$$

A.3.2.7 Polar coordinates

Polar coordinates are defined as the radial distance r from the origin and the azimuth angle θ, see Figure A.3.2. The unit radial vector is denoted $\vec{u}$ and the unit tangential vector is denoted $\vec{u}_1$. The displacement vector is written as:

$$\vec{X} = U\vec{u} + V\vec{u}_1$$

The metrics is given by:

$$ds^2 = dr^2 + r^2 d\theta^2 = [dr \quad d\theta]\begin{bmatrix}1 & 0 \\ 0 & r^2\end{bmatrix}\begin{bmatrix}dr \\ d\theta\end{bmatrix} \Rightarrow g_r = 1; \quad g_\theta = r;$$

$$\vec{\nabla} = \vec{u}\frac{\partial}{\partial r} + \vec{u}_1 \frac{1}{r}\frac{\partial}{\partial \theta}$$

$$\vec{\nabla}\Phi = \vec{u}\frac{\partial \Phi}{\partial r} + \vec{u}_1 \frac{1}{r}\frac{\partial \Phi}{\partial \theta}$$

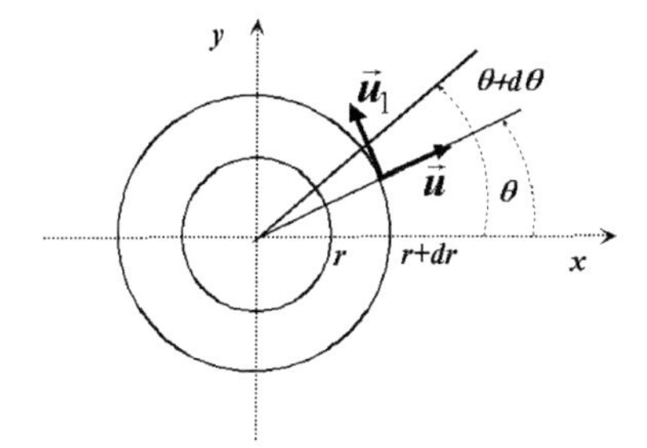

Figure A.3.2. *Polar coordinates system*

$$\vec{\nabla}\vec{X} = \left(\vec{u}\frac{\partial}{\partial r} + \vec{u}_1\frac{1}{r}\frac{\partial}{\partial\theta}\right)(U\vec{u} + V\vec{u}_1)$$

$$\begin{bmatrix} \dfrac{\partial\vec{u}}{\partial r} & \dfrac{\partial\vec{u}}{\partial\theta} \\ \dfrac{\partial\vec{u}_1}{\partial r} & \dfrac{\partial\vec{u}_1}{\partial\theta} \end{bmatrix} = \begin{bmatrix} 0 & +\vec{u}_1 \\ 0 & -\vec{u} \end{bmatrix}$$

Whence:

$$\vec{\nabla}\vec{X} = \begin{bmatrix} \dfrac{\partial U}{\partial r} & \dfrac{\partial V}{\partial r} \\ \dfrac{1}{r}\dfrac{\partial U}{\partial\theta} - \dfrac{V}{r} & \dfrac{1}{r}\dfrac{\partial V}{\partial\theta} + \dfrac{U}{r} \end{bmatrix}$$

The small strain tensor is thus:

$$\begin{bmatrix} \varepsilon_{rr} & \varepsilon_{r\theta} \\ \varepsilon_{\theta r} & \varepsilon_{\theta\theta} \end{bmatrix} = \frac{\vec{\nabla}\vec{X} + (\vec{\nabla}\vec{X})^T}{2}$$
$$= \begin{bmatrix} \dfrac{\partial U}{\partial r} & \dfrac{1}{2}\left(\dfrac{\partial V}{\partial r} + \dfrac{1}{r}\dfrac{\partial U}{\partial\theta} - \dfrac{V}{r}\right) \\ \dfrac{1}{2}\left(\dfrac{\partial V}{\partial r} + \dfrac{1}{r}\dfrac{\partial U}{\partial\theta} - \dfrac{V}{r}\right) & \dfrac{1}{r}\dfrac{\partial V}{\partial\theta} + \dfrac{U}{r} \end{bmatrix}$$

A.3.2.8 Cylindrical coordinates

Extension of the above formulas to the cylindrical case is straightforward. Cylindrical coordinates are r, θ, defined as above, plus the abscissa z along the cylinder axis. The unit vectors are $\vec{u}, \vec{u}_1$ and $\vec{k}$. The displacement vector is defined as $\vec{X} = U\vec{u} + V\vec{u}_1 + W\vec{k}$. Coefficients of the metrics are given by:

$$ds^2 = dr^2 + r^2 d\theta^2 + dz^2 = [dr \quad d\theta \quad dz]\begin{bmatrix} 1 & 0 & 0 \\ 0 & r^2 & 0 \\ 0 & 0 & 1 \end{bmatrix}\begin{bmatrix} dr \\ d\theta \\ dz \end{bmatrix};$$

$$\vec{\nabla} = \vec{u}\frac{\partial}{\partial r} + \vec{u}_1\frac{1}{r}\frac{\partial}{\partial \theta} + \vec{k}\frac{\partial}{\partial z}$$

$$\begin{bmatrix} \dfrac{\partial \vec{u}}{\partial r} & \dfrac{\partial \vec{u}}{\partial \theta} & \dfrac{\partial \vec{u}}{\partial z} \\ \dfrac{\partial \vec{u}_1}{\partial r} & \dfrac{\partial \vec{u}_1}{\partial \theta} & \dfrac{\partial \vec{u}_1}{\partial z} \\ \dfrac{\partial \vec{k}}{\partial r} & \dfrac{\partial \vec{k}}{\partial \theta} & \dfrac{\partial \vec{k}}{\partial z} \end{bmatrix} = \begin{bmatrix} 0 & +\vec{u}_1 & 0 \\ 0 & -\vec{u} & 0 \\ 0 & 0 & 0 \end{bmatrix}$$

$$\vec{\nabla}\cdot(U\vec{u} + V\vec{u}_1 + W\vec{k}) = \frac{1}{r}\frac{\partial(rU)}{\partial r} + \frac{\partial V}{r\partial \theta} + \frac{\partial W}{\partial z}$$

$$\vec{\nabla}\Phi = \vec{u}\frac{\partial \Phi}{\partial r} + \vec{u}_1\frac{1}{r}\frac{\partial \Phi}{\partial \theta} + \vec{k}\frac{\partial \Phi}{\partial z}$$

$$\Delta\Phi = \vec{\nabla}\cdot\vec{\nabla}\Phi = \frac{1}{r}\left(\frac{\partial}{\partial r}\left(r\frac{\partial \Phi}{\partial r}\right) + \frac{\partial}{\partial \theta}\left(\frac{1}{r}\frac{\partial \Phi}{\partial \theta}\right)\right) + \frac{\partial^2 \Phi}{\partial z^2}$$

$$\vec{\nabla}\vec{X} = \begin{bmatrix} \dfrac{\partial U}{\partial r} & \dfrac{\partial V}{\partial r} & \dfrac{\partial W}{\partial r} \\ \dfrac{1}{r}\dfrac{\partial U}{\partial \theta} - \dfrac{V}{r} & \dfrac{1}{r}\dfrac{\partial V}{\partial \theta} + \dfrac{U}{r} & \dfrac{1}{r}\dfrac{\partial W}{\partial \theta} \\ \dfrac{\partial U}{\partial z} & \dfrac{\partial V}{\partial z} & \dfrac{\partial W}{\partial z} \end{bmatrix}$$

$$\begin{bmatrix} \varepsilon_{rr} & \varepsilon_{r\theta} & \varepsilon_{rz} \\ \varepsilon_{\theta r} & \varepsilon_{\theta\theta} & \varepsilon_{\theta z} \\ \varepsilon_{zr} & \varepsilon_{z\theta} & \varepsilon_{zz} \end{bmatrix} =$$

$$\begin{bmatrix} \dfrac{\partial U}{\partial r} & \dfrac{1}{2}\left(\dfrac{\partial V}{\partial r} + \dfrac{1}{r}\dfrac{\partial U}{\partial \theta} - \dfrac{V}{r}\right) & \dfrac{1}{2}\left(\dfrac{\partial W}{\partial r} + \dfrac{\partial U}{\partial z}\right) \\ \dfrac{1}{2}\left(\dfrac{\partial V}{\partial r} + \dfrac{1}{r}\dfrac{\partial U}{\partial \theta} - \dfrac{V}{r}\right) & \dfrac{1}{r}\dfrac{\partial V}{\partial \theta} + \dfrac{U}{r} & \dfrac{1}{2}\left(\dfrac{\partial W}{r\partial \theta} + \dfrac{\partial V}{\partial z}\right) \\ \dfrac{1}{2}\left(\dfrac{\partial W}{\partial r} + \dfrac{\partial U}{\partial z}\right) & \dfrac{1}{2}\left(\dfrac{\partial W}{r\partial \theta} + \dfrac{\partial V}{\partial z}\right) & \dfrac{\partial W}{\partial z} \end{bmatrix}$$

A.4. Plate bending in curvilinear coordinates

A.4.1 *Formulation of Hamilton's principle*

In the absence of external loading, the variation of the action of the Lagrangian is expressed as:

$$\delta\mathcal{A} = \int_{t_1}^{t_2} dt \int_{\beta_1}^{\beta_2} \int_{\alpha_1}^{\alpha_2} (\rho h \dot{Z} \delta \dot{Z} - \mathcal{M}_{\alpha\alpha}\delta\chi_{\alpha\alpha}$$
$$-2\mathcal{M}_{\alpha\beta}\delta\chi_{\alpha\beta} - \mathcal{M}_{\beta\beta}\delta\chi_{\beta\beta})\, g_\alpha g_\beta \, d\alpha\, d\beta = 0 \qquad \text{[A.4.1]}$$

It is necessary to carry out the integrations by parts which allows one to express the variations in terms of the displacement variables Z, $\partial Z/\partial\alpha, \partial Z/\partial\beta$.

$$-\int_{\beta_1}^{\beta_2}\int_{\alpha_1}^{\alpha_2} (\mathcal{M}_{\alpha\alpha}\delta\chi_{\alpha\alpha}) g_\alpha g_\beta\, d\alpha\, d\beta$$
$$= \int_{\beta_1}^{\beta_2}\int_{\alpha_1}^{\alpha_2} \mathcal{M}_{\alpha\alpha}\left\{\frac{1}{g_\alpha}\left(\frac{\partial}{\partial\alpha}\left(\frac{1}{g_\alpha}\delta\left(\frac{\partial Z}{\partial\alpha}\right)\right)\right)\right.$$
$$\left. + \frac{1}{g_\alpha g_\beta^2}\left(\frac{\partial}{\partial\beta}\left(g_\alpha\delta\left(\frac{\partial Z}{\partial\beta}\right)\right)\right)\right\} g_\alpha\, g_\beta\, d\alpha\, d\beta$$
$$= \int_{\beta_1}^{\beta_2}\int_{\alpha_1}^{\alpha_2}\left(\mathcal{M}_{\alpha\alpha} g_\beta \frac{\partial}{\partial\alpha}\left(\frac{1}{g_\alpha}\delta\left(\frac{\partial Z}{\partial\alpha}\right)\right)\right) d\alpha\, d\beta$$
$$+ \int_{\beta_1}^{\beta_2}\int_{\alpha_1}^{\alpha_2}\left(\frac{\mathcal{M}_{\alpha\alpha}}{g_\beta}\frac{\partial g_\alpha}{\partial\beta}\right)\delta\left(\frac{\partial Z}{\partial\beta}\right) d\alpha\, d\beta$$

$$\int_{\beta_1}^{\beta_2}\int_{\alpha_1}^{\alpha_2}\left(\mathcal{M}_{\alpha\alpha} g_\beta \frac{\partial}{\partial\alpha}\left(\frac{1}{g_\alpha}\delta\left(\frac{\partial Z}{\partial\alpha}\right)\right)\right) d\alpha\, d\beta$$
$$= \int_{\beta_1}^{\beta_2}\left[\frac{g_\beta}{g_\alpha}\mathcal{M}_{\alpha\alpha}\delta\left(\frac{\partial Z}{\partial\alpha}\right)\right]_{\alpha_1}^{\alpha_2} d\beta$$
$$- \int_{\beta_1}^{\beta_2}\int_{\alpha_1}^{\alpha_2}\frac{1}{g_\alpha}\left(\frac{\partial(g_\beta\mathcal{M}_{\alpha\alpha})}{\partial\alpha}\delta\left(\frac{\partial Z}{\partial\alpha}\right)\right) d\alpha\, d\beta$$
$$= \int_{\beta_1}^{\beta_2}\left[\frac{g_\beta}{g_\alpha}\mathcal{M}_{\alpha\alpha}\delta\left(\frac{\partial Z}{\partial\alpha}\right)\right]_{\alpha_1}^{\alpha_2} d\beta - \int_{\beta_1}^{\beta_2}\left[\frac{1}{g_\alpha}\frac{\partial(g_\beta\mathcal{M}_{\alpha\alpha})}{\partial\alpha}\delta Z\right]_{\alpha_1}^{\alpha_2} d\beta$$
$$+ \int_{\beta_1}^{\beta_2}\int_{\alpha_1}^{\alpha_2}\frac{\partial}{\partial\alpha}\left(\frac{1}{g_\alpha}\left(\frac{\partial\left(g_\beta\mathcal{M}_{\alpha\alpha}\right)}{\partial\alpha}\right)\right)\delta Z\, d\alpha\, d\beta$$

$$\int_{\beta_1}^{\beta_2}\int_{\alpha_1}^{\alpha_2}\left(\frac{\mathcal{M}_{\alpha\alpha}}{g_\beta}\frac{\partial g_\alpha}{\partial\beta}\right)\delta\left(\frac{\partial Z}{\partial\beta}\right)d\alpha\, d\beta$$

$$= \int_{\alpha_1}^{\alpha_2}\left[\frac{\mathcal{M}_{\alpha\alpha}}{g_\beta}\frac{\partial g_\alpha}{\partial\beta}\delta Z\right]_{\beta_1}^{\beta_2} d\alpha - \int_{\beta_1}^{\beta_2}\int_{\alpha_1}^{\alpha_2}\frac{\partial}{\partial\beta}\left(\frac{\mathcal{M}_{\alpha\alpha}}{g_\beta}\frac{\partial g_\alpha}{\partial\beta}\right)\delta Z\, d\alpha\, d\beta$$

$$-\int_{\beta_1}^{\beta_2}\int_{\alpha_1}^{\alpha_2}\mathcal{M}_{\alpha\alpha}\delta\,\chi_{\alpha\alpha}\, g_\alpha\, g_\beta\, d\alpha\, d\beta$$

$$= \int_{\beta_1}^{\beta^2}\frac{1}{g_\alpha}\left[g_\beta\mathcal{M}_{\alpha\alpha}\delta\left(\frac{\partial Z}{\partial\alpha}\right) - \frac{\partial(g_\beta\mathcal{M}_{\alpha\alpha})}{\partial\alpha}\partial Z\right]_{\alpha_1}^{\alpha^2} d\beta$$

$$+\int_{\alpha_1}^{\alpha^2}\left[\frac{\mathcal{M}_{\alpha\alpha}}{g_\beta}\frac{\partial g_\alpha}{\partial\beta}\delta Z\right]_{\beta_1}^{\beta_2} d\alpha$$

$$+\int_{\beta_1}^{\beta_2}\int_{\alpha_1}^{\alpha_2}\left(\frac{\partial}{\partial\alpha}\left(\frac{\partial(g_\beta\mathcal{M}_{\alpha\alpha})}{g_\alpha\partial\alpha}\right) - \frac{\partial}{\partial\beta}\left(\frac{\mathcal{M}_{\alpha\alpha}}{g_\beta}\frac{\partial g_\alpha}{\partial_\beta}\right)\right)\delta Z\, d\alpha\, d\beta \quad \text{[A.4.2]}$$

In the same way:

$$-\int_{\beta_1}^{\beta_2}\int_{\alpha_1}^{\alpha_2}\mathcal{M}_{\beta\beta}\delta\chi_{\beta\beta}g_\alpha g_\beta\, d\alpha\, d\beta$$

$$= \int_{\alpha_1}^{\alpha^2}\frac{1}{g_\beta}\left[g_\alpha\mathcal{M}_{\beta\beta}\delta\left(\frac{\partial Z}{\partial\alpha}\right) - \frac{\partial(g_\beta\mathcal{M}_{\beta\beta})}{\partial\beta}\partial Z\right]_{\beta_1}^{\beta^2} d\alpha$$

$$+\int_{\beta_1}^{\beta^2}\left[\frac{\mathcal{M}_{\beta\beta}}{g_\alpha}\frac{\partial g_\beta}{\partial\alpha}\delta Z\right]_{\alpha_1}^{\alpha_2} d\beta$$

$$+\int_{\beta_1}^{\beta_2}\int_{\alpha_1}^{\alpha_2}\left(\frac{\partial}{\partial\beta}\left(\frac{\partial(g_\alpha\mathcal{M}_{\beta\beta})}{g_\beta\partial\beta}\right) - \frac{\partial}{\partial\alpha}\left(\frac{\mathcal{M}_{\beta\beta}}{g_\alpha}\frac{\partial g_\beta}{\partial_\alpha}\right)\right)\delta Z\, d\alpha\, d\beta \quad \text{[A.4.3]}$$

$$2\int_{\beta_1}^{\beta_2}\int_{\alpha_1}^{\alpha_2}(\mathcal{M}_{\alpha\beta}\delta\chi_{\alpha\beta})g_\alpha g_\beta\, d\alpha\, d\beta$$

$$= \int_{\beta_1}^{\beta_2}\int_{\alpha_1}^{\alpha_2}\left\{\mathcal{M}_{\alpha\beta}\frac{g_\alpha}{g_\beta}\frac{\partial}{\partial\beta}\left(\frac{1}{g_\alpha^2}\delta\left(\frac{\partial Z}{\partial\alpha}\right)\right)\right\}g_\alpha\, g_\beta\, d\alpha\, d\beta$$

$$+\int_{\beta_1}^{\beta_2}\int_{\alpha_1}^{\alpha_2}\left\{\mathcal{M}_{\beta\alpha}\frac{g_\beta}{g_\alpha}\frac{\partial}{\partial\alpha}\left(\frac{1}{g_\beta^2}\delta\left(\frac{\partial Z}{\partial\beta}\right)\right)\right\}g_\alpha\, g_\beta\, d\alpha\, d\beta$$

$$\int_{\beta_1}^{\beta_2}\int_{\alpha_1}^{\alpha_2}\mathcal{M}_{\alpha\beta}\left\{\frac{g_\alpha}{g_\beta}\frac{\partial}{\partial\beta}\left(\frac{1}{g_\alpha^2}\delta\left(\frac{\partial Z}{\partial\alpha}\right)\right)\right\}g_\alpha\, g_\beta\, d\alpha\, d\beta$$

$$= \int_{\alpha_1}^{\alpha_2}\left[\mathcal{M}_{\alpha\beta}\delta\left(\frac{\partial Z}{\partial\alpha}\right)\right]_{\beta_1}^{\beta_2} d\alpha - \int_{\beta_1}^{\beta_2}\int_{\alpha_1}^{\alpha_2}\frac{1}{g_\alpha^2}\frac{\partial}{\partial\beta}\left(g_\alpha^2\mathcal{M}_{\alpha\beta}\right)\delta\left(\frac{\partial Z}{\partial\alpha}\right)d\alpha\, d\beta$$

$$= \int_{\alpha_1}^{\alpha_2} \left[\mathcal{M}_{\alpha\beta} \delta \left(\frac{\partial Z}{\partial \alpha} \right) \right]_{\beta_1}^{\beta_2} d\alpha - \int_{\beta_1}^{\beta_2} \left[\frac{1}{g_\alpha^2} \frac{\partial}{\partial \beta} (g_\alpha^2 \mathcal{M}_{\alpha\beta}) \delta Z \right]_{\alpha_1}^{\alpha_2} d\beta$$
$$+ \int_{\beta_1}^{\beta_2} \int_{\alpha_1}^{\alpha_2} \frac{\partial}{\partial \alpha} \left(\frac{1}{g_\alpha^2} \frac{\partial}{\partial \beta} (g_\alpha^2 \mathcal{M}_{\alpha\beta}) \right) \delta Z \, d\alpha \, d\beta$$

Finally:

$$\int_{\beta_1}^{\beta_2} \int_{\alpha_1}^{\alpha_2} \mathcal{M}_{\beta\alpha} \frac{g_\beta}{g_\alpha} \frac{\partial}{\partial \alpha} \left(\frac{1}{g_\beta^2} \delta \left(\frac{\partial Z}{\partial \beta} \right) \right) g_\alpha \, g_\beta \, d\alpha \, d\beta$$
$$= \int_{\beta_1}^{\beta_2} \left[\mathcal{M}_{\beta\alpha} \delta \left(\frac{\partial Z}{\partial \alpha} \right) \right]_{\alpha_1}^{\alpha_2} d\beta - \int_{\alpha_1}^{\alpha_2} \left[\frac{1}{g_\beta^2} \frac{\partial \left(g_\alpha^2 \mathcal{M}_{\beta\alpha} \right)}{\partial \alpha} \delta Z \right]_{\beta_1}^{\beta_2} d\alpha \qquad \text{[A.4.4]}$$
$$+ \int_{\beta_1}^{\beta_2} \int_{\alpha_1}^{\alpha_2} \frac{\partial}{\partial \beta} \left(\frac{1}{g_\beta^2} \frac{\partial \left(g_\beta^2 \mathcal{M}_{\beta\alpha} \right)}{\partial \alpha} \right) \delta Z \, d\alpha \, d\beta$$

A.4.2 *Equation of local equilibrium in terms of shear forces*

Gathering together the surface terms of [A.4.1] and changing their sign, we get:

$$\delta \mathcal{A} = \int_{t_1}^{t_2} dt \int_{\beta_1}^{\beta_2} \int_{\alpha_1}^{\alpha_2} \delta Z \left\{ \rho h \ddot{Z} g_\alpha g_\beta - \frac{\partial}{\partial \alpha} \left(\frac{1}{g_\alpha} \frac{\partial (g_\beta \mathcal{M}_{\alpha\alpha})}{\partial \alpha} \right. \right.$$
$$\left. + \frac{1}{g_\alpha^2} \frac{\partial (g_\alpha^2 \mathcal{M}_{\alpha\beta})}{\partial \beta} - \frac{\mathcal{M}_{\beta\beta}}{g_\alpha} \frac{\partial g_\beta}{\partial \alpha} \right) - \frac{\partial}{\partial \beta} \left(\frac{1}{g_\beta} \frac{\partial \left(g_\alpha \mathcal{M}_{\beta\beta} \right)}{\partial \beta} \right.$$
$$\left. \left. + \frac{1}{g_\beta^2} \frac{\partial (g_\beta^2 \mathcal{M}_{\beta\alpha})}{\partial \alpha} - \frac{\mathcal{M}_{\alpha\alpha}}{g_\beta} \frac{\partial g_\alpha}{\partial \beta} \right) \right\} d\alpha \, d\beta = 0 \qquad \text{[A.4.5]}$$

Transverse shear forces acting in the normal direction to the plate midplane are defined as:

$$Q_{\alpha z} = \frac{1}{g_\alpha g_\beta} \left(\frac{\partial (g_\beta \mathcal{M}_{\alpha\alpha})}{\partial \alpha} + \frac{1}{g_\alpha} \frac{\partial \left(g_\alpha^2 \mathcal{M}_{\alpha\beta} \right)}{\partial \beta} - \mathcal{M}_{\beta\beta} \frac{\partial g_\beta}{\partial \alpha} \right)$$
$$Q_{\beta z} = \frac{1}{g_\alpha g_\beta} \left(\frac{\partial (g_\alpha \mathcal{M}_{\beta\beta})}{\partial \beta} + \frac{1}{g_\beta} \frac{\partial \left(g_\beta^2 \mathcal{M}_{\beta\alpha} \right)}{\partial \alpha} - \mathcal{M}_{\alpha\alpha} \frac{\partial g_\alpha}{\partial \beta} \right) \qquad \text{[A.4.6]}$$

Hamilton's principle is written in a more compact form than [4.1] as:

$$\delta\mathcal{A} = \int_{t_1}^{t_2} dt \int_{\beta_1}^{\beta_2} \int_{\alpha_1}^{\alpha_2} \delta Z \left\{ \rho h \ddot{Z} g_\alpha g_\beta - \frac{\partial (g_\beta Q_{\alpha z})}{\partial \alpha} - \frac{\partial (g_\alpha Q_{\beta z})}{\partial \beta} \right\} d\alpha \, d\beta = 0$$

The equation of local equilibrium follows as:

$$\rho h \ddot{Z} - \frac{1}{g_\alpha g_\beta} \left\{ \frac{\partial (g_\beta Q_{\alpha z})}{\partial \alpha} + \frac{\partial (g_\alpha Q_{\beta z})}{\partial \beta} \right\} = 0 \qquad \text{[A.4.7]}$$

A.4.3 *Boundary conditions: effective Kirchhoff's shear forces and corner forces*

Gathering together the boundary terms of [A.4.1] associated with the moments and rotations, we obtain:

$$\begin{aligned} &\int_{\beta_1}^{\beta_2} \left[\frac{g_\beta}{g_\alpha} \mathcal{M}_{\alpha\alpha} \delta \left(\frac{\partial Z}{\partial \alpha} \right) + \mathcal{M}_{\beta\alpha} \delta \left(\frac{\partial Z}{\partial \beta} \right) \right]_{\alpha_1}^{\alpha_2} d\beta = 0 \\ &\int_{\alpha_1}^{\alpha_2} \left[\frac{g_\alpha}{g_\beta} \mathcal{M}_{\beta\beta} \delta \left(\frac{\partial Z}{\partial \beta} \right) + \mathcal{M}_{\alpha\beta} \delta \left(\frac{\partial Z}{\partial \alpha} \right) \right]_{\alpha_1}^{\alpha_2} d\alpha = 0 \end{aligned} \qquad \text{[A.4.8]}$$

The important point is that, in such expressions, the rotations must be considered as dependent variables. Accordingly, the torsion term is integrated to express the variation in terms of δZ solely, so [A.4.11] becomes:

$$\begin{aligned} &\int_{\beta_1}^{\beta_2} \left[\frac{g_\beta}{g_\alpha} \mathcal{M}_{\alpha\alpha} \delta \left(\frac{\partial Z}{\partial \alpha} \right) d\beta \right]_{\alpha_1}^{\alpha_2} + \left[[\mathcal{M}_{\beta\alpha} \delta Z]_{\alpha_1}^{\alpha_2} \right]_{\beta_1}^{\beta_2} - \int_{\beta_1}^{\beta_2} \left[\frac{\partial \mathcal{M}_{\beta\alpha}}{\partial \beta} \delta Z \right]_{\alpha_1}^{\alpha_2} d\beta = 0 \\ &\int_{\alpha_1}^{\alpha_2} \left[\frac{g_\alpha}{g_\beta} \mathcal{M}_{\beta\beta} \delta \left(\frac{\partial Z}{\partial \beta} \right) \right]_{\alpha_1}^{\alpha_2} d\alpha + \left[[\mathcal{M}_{\alpha\beta} \delta Z]_{\alpha_1}^{\alpha_2} \right]_{\beta_1}^{\beta_2} - \int_{\alpha_1}^{\alpha_2} \left[\frac{\partial \mathcal{M}_{\alpha\beta}}{\partial \alpha} \delta Z \right]_{\beta_1}^{\beta_2} d\alpha = 0 \end{aligned} \qquad \text{[A.4.9]}$$

Whence, the homogeneous boundary conditions:

along the edges α_1 and α_2: $\mathcal{M}_{\alpha\alpha} = 0$; or $\partial Z / \partial \alpha = 0$

along the edges β_1 and β_2: $\mathcal{M}_{\beta\beta} = 0$; or $\partial Z / \partial \beta = 0$ [A.4.10]

at the corners: $\mathcal{M}_{\alpha\beta} + \mathcal{M}_{\beta\alpha} = 2\mathcal{M}_{\alpha\beta} = 0$; or $Z = 0$

Gathering the boundary terms associated with the shear forces and displacements we get:

$$\int_{\beta_1}^{\beta_2}\left[-\frac{1}{g_\alpha}\frac{\partial(g_\beta\mathcal{M}_{\alpha\alpha})}{\partial\alpha}-\frac{1}{g_\alpha^2}\frac{\partial}{\partial\beta}\left(g_\alpha^2\mathcal{M}_{\alpha\beta}\right)+\frac{\mathcal{M}_{\beta\beta}}{g_\alpha}\frac{\partial g_\beta}{\partial\alpha}-\frac{\partial\mathcal{M}_{\beta\alpha}}{\partial\beta}\right]_{\alpha_1}^{\alpha_2}\delta Z\,d\beta=0$$

$$\int_{\alpha_1}^{\alpha_2}\left[-\frac{1}{g_\beta}\frac{\partial(g_\alpha\mathcal{M}_{\beta\beta})}{\partial\beta}-\frac{1}{g_\beta^2}\frac{\partial}{\partial\alpha}\left(g_\beta^2\mathcal{M}_{\beta\alpha}\right)+\frac{\mathcal{M}_{\alpha\alpha}}{g_\beta}\frac{\partial g_\alpha}{\partial\beta}-\frac{\partial\mathcal{M}_{\alpha\beta}}{\partial\alpha}\right]_{\beta_1}^{\beta_2}\delta Z\,d\alpha$$

[A.4.11]

Dividing the first row by g_β and the second row by g_α, we get:

$$\int_{\beta_1}^{\beta_2}\left[-\mathcal{Q}_{\alpha z}-\frac{\partial\mathcal{M}_{\beta\alpha}}{\partial\beta}\right]_{\alpha_1}^{\alpha_2}\delta Z\,d\beta=0$$
$$\int_{\alpha_1}^{\alpha_2}\left[-\mathcal{Q}_{\beta z}-\frac{\partial\mathcal{M}_{\alpha\beta}}{\partial\alpha}\right]_{\beta_1}^{\beta_2}\delta Z\,d\alpha=0$$

[A.4.12]

Then, to the shear forces already defined by [A.4.9], the term arising from the integration of [A.4.11] must be added, resulting in the effective Kirchhoff shear force per unit length, expressed as:

$$\begin{aligned}
\mathcal{V}_{\alpha z}&=\mathcal{Q}_{\alpha z}+\frac{\partial\mathcal{M}_{\beta\alpha}}{g_\beta\partial\beta}\\
&=\frac{1}{g_\alpha g_\beta}\left\{\frac{\partial(g_\beta\mathcal{M}_{\alpha\alpha})}{\partial\alpha}+\frac{1}{g_\alpha}\frac{\partial}{\partial\beta}\left(g_\alpha^2\mathcal{M}_{\alpha\beta}\right)-\mathcal{M}_{\beta\beta}\frac{\partial g_\beta}{\partial\alpha}\right\}+\frac{\partial\mathcal{M}_{\beta\alpha}}{g_\beta\partial\beta}\\
\mathcal{V}_{\beta z}&=\mathcal{Q}_{\beta z}+\frac{\partial\mathcal{M}_{\alpha\beta}}{g_\alpha\partial\alpha}\\
&=\frac{1}{g_\alpha g_\beta}\left\{\frac{\partial(g_\alpha\mathcal{M}_{\beta\beta})}{\partial\beta}+\frac{1}{g_\beta}\frac{\partial}{\partial\alpha}(g_\beta^2\mathcal{M}_{\beta\alpha})-\mathcal{M}_{\alpha\alpha}\frac{\partial g_\alpha}{\partial\beta}\right\}+\frac{\partial\mathcal{M}_{\alpha\beta}}{g_\alpha\partial\alpha}
\end{aligned}$$

[A.4.13]

The free and fixed boundary conditions are:

$$\begin{aligned}
&\text{along the edges } \alpha_1 \text{ and } \alpha_2: \quad \mathcal{V}_{\alpha z}=0;\quad \text{or } Z=0\\
&\text{along the edges } \beta_1 \text{ and } \beta_2: \quad \mathcal{V}_{\beta z}=0;\quad \text{or } Z=0
\end{aligned}$$

[A.4.14]

A.5. Static equilibrium of a sagging cable loaded by its own weight

A cable of length L_0, is stretched between two points A and B, of coordinates $x_A=-L/2, z_A=0, x_B=+L/2, z_B=0$, where $L<L_0$, see Figure A.5.1. The problem is to determine the static equilibrium configuration $z(x)$ of the cable in

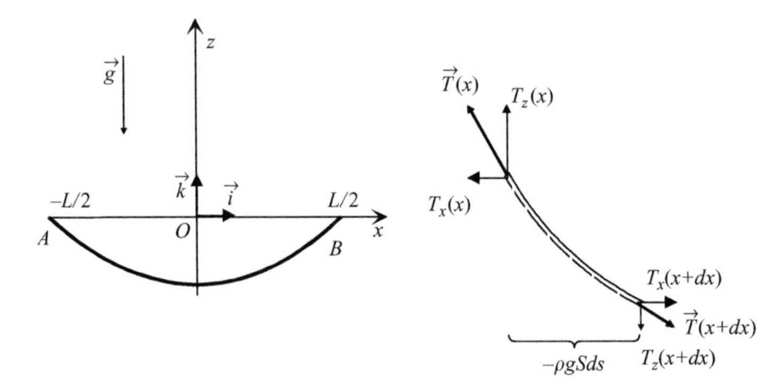

Figure A.5.1. *Cable subjected to its own weight*

a uniform gravity field $-g\vec{k}$. As soon as the tensile stress in the cable is sufficiently large, the elastic deflection of the cable can be safely neglected. The equilibrium equation will be derived by using successively two distinct methods. The first follows the Newtonian approach, which consists in writing down directly the force balance for a cable element of infinitesimal length $ds(x)$. The second method consists in deriving the Lagrange equation of the cable constrained by the condition of length invariance. It turns out that the last condition can be prescribed either in the global scale of the whole structure or in the local scale of a cable element.

A.5.1 *Newtonian approach*

The tensile force acting on the cable at the Cartesian abscissa x is denoted $\vec{T}(x) = T_x(x)\vec{i} + T_z(x)\vec{k}$. It is interpreted as the string stress related to the length invariance of the cable. The force balances are written as:

$$\begin{cases} \dfrac{dT_x}{dx} = 0 \;\Rightarrow\; T_x = T_0 \\ \dfrac{dT_z}{dx} = \rho g S \dfrac{ds}{dx} = \rho g S \sqrt{1 + \left(\dfrac{dz}{dx}\right)^2} \end{cases} \qquad [A.5.1]$$

On the other hand, the following relationship holds:

$$\frac{T_z}{T_x} = \frac{dz}{dx} \;\Rightarrow\; T_z = T_0 \frac{dz}{dx} \qquad [A.5.2]$$

whence the nonlinear differential equation:

$$\frac{d^2 z}{dx^2} - \gamma \sqrt{1 + \left(\frac{dz}{dx}\right)^2} = 0, \quad \text{where } \gamma = \frac{\rho g S}{T_0} \qquad [A.5.3]$$

The solution is found to be of the type:

$$\frac{dz}{dx} = \sinh(\gamma x) \quad \Rightarrow \quad z(x) = \frac{\cosh(\gamma x)}{\gamma} + a \qquad \text{[A.5.4]}$$

The boundary conditions imply the solution:

$$\begin{aligned} \frac{z(x)}{L} &= \frac{T_0}{\rho g S L}\left(\cosh\left(\frac{\rho g S x}{T_0}\right) - \cosh\left(\frac{\rho g S L}{2T_0}\right)\right) \\ T_z(x) &= T_0 \sinh\left(\frac{\rho g S x}{T_0}\right) \end{aligned} \qquad \text{[A.5.5]}$$

The horizontal component of the tensile force is obtained by stating that the resultant of the vertical components of the support reactions must balance the weight of the cable; whence the following transcendental equation:

$$\sinh\left(\frac{\rho g S L}{2T_0}\right) = \frac{\rho g S L_0}{2T_0} \qquad \text{[A.5.6]}$$

It can be easily checked that [A.5.6] has only one solution if $L \leq L_0$. It is also possible to express the Cartesian components of the deflected cable in terms of the arc length s:

$$\begin{aligned} &ds = \cosh(\gamma x) \quad \Rightarrow \quad s = \frac{1}{\gamma}\sinh(\gamma x) \quad \Rightarrow \quad x = \frac{1}{\gamma}(\sinh(\gamma s))^{-1} \\ &z = \frac{1}{\gamma}\left\{\cosh\left((\sinh(\gamma s))^{-1}\right) - \cosh\left(\frac{\gamma L}{2}\right)\right\} \end{aligned} \qquad \text{[A.5.7]}$$

A.5.2 *Constrained Lagrange's equations, invariance of the cable length*

The constraint condition is written as:

$$\int_{-L/2}^{L/2} ds = \int_{-L/2}^{L/2} \sqrt{1 + \left(\frac{dz}{dx}\right)^2}\, dx = L_0 \qquad \text{[A.5.8]}$$

The constrained Lagrangian is:

$$\mathcal{L}' = -\int_{-L/2}^{L/2} \{\rho g S z(x) + \lambda\} \sqrt{1 + \left(\frac{dz}{dx}\right)^2}\, dx + \lambda L_0 \qquad \text{[A.5.9]}$$

The Euler–Lagrange equations are:

$$\begin{cases} \rho g S\sqrt{(z')^2+1} - \dfrac{d}{dx}\left(\dfrac{z'(\rho g S z+\lambda)}{\sqrt{(z')^2+1}}\right) = 0 \\ -\dfrac{d}{dx}\left(\dfrac{(\rho g S z+\lambda)}{\sqrt{(z')^2+1}}\right) = 0 \;\Rightarrow\; \dfrac{(\rho g S z+\lambda)}{\sqrt{(z')^2+1}} = C \end{cases} \quad \text{where } z' = \frac{dz}{dx}$$

[A.5.10]

The Lagrange multiplier λ can be eliminated between the two equations [5.10] to obtain the nonlinear differential equation:

$$\rho g S\sqrt{(z')^2+1} - Cz'' = 0 \iff \sqrt{(z')^2+1} - \frac{z''}{\gamma} = 0$$
$$\text{where} \quad \gamma = \frac{\rho g S}{C} \quad \text{and} \quad z'' = \frac{d^2z}{dx^2} \qquad \text{[A.5.11]}$$

The solution is found to be of the type:

$$\frac{dz}{dx} = \sinh(\gamma x) \;\Rightarrow\; z(x) = \frac{\cosh(\gamma x)}{\gamma} + a \qquad \text{[A.5.12]}$$

The boundary condition determines the constant a:

$$a = -\frac{\cosh(\gamma L/2)}{\gamma} \qquad \text{[A.5.13]}$$

On the other hand, λ is given by:

$$\frac{(\rho g S z+\lambda)}{\sqrt{(z')^2+1}} = C \;\Rightarrow\; C + \frac{\lambda - C\cosh(\gamma L/2)}{\cosh(\gamma x)} = C \qquad \text{[A.5.14]}$$

and finally:

$$\lambda = C\cosh(\gamma L/2) \qquad \text{[A.5.15]}$$

The constant C is a force per unit length determined by using the condition [A.5.8], which implies that:

$$L_0 = \int_{-L/2}^{L/2} \cosh(\gamma x)\,dx = \frac{2}{\gamma}\sinh\left(\frac{\gamma L}{2}\right) \qquad \text{[A.5.16]}$$

γ is a solution of the transcendental equation:

$$\frac{\gamma L_0}{2} = \sinh\left(\frac{\gamma L}{2}\right) \qquad \text{[A.5.17]}$$

[A.5.17], which is of the same type as [5.6], gives the physical meaning of C. Indeed, the horizontal T_x and the vertical T_z components of the support reaction must verify the following relation:

$$\frac{T_z}{T_x} = \frac{\rho g S L_0}{2T_x} = \left.\frac{dz}{dx}\right|_{L/2} = \sinh\left(\frac{\gamma L}{2}\right) = \frac{\gamma L_0}{2} = \frac{\rho g S L_0}{2C} \Rightarrow T_x = C \qquad \text{[A.5.18]}$$

This final result can be used to check that [A.5.17] is identical to [A.5.6].

A.5.3 *Constrained Lagrange's equations: length invariance of a cable element*

The constraint condition is now written as:

$$\left(\frac{dx}{ds}\right)^2 + \left(\frac{dy}{ds}\right)^2 = 1 \qquad \text{[A.5.19]}$$

The constrained Lagrangian is:

$$\mathcal{L}' = -\int_{-L/2}^{L/2} \left\{\rho g S z(s) + \lambda(s)\left(\left(\frac{dx}{ds}\right)^2 + \left(\frac{dz}{ds}\right)^2 - 1\right)\right\} ds \qquad \text{[A.5.20]}$$

Here, Lagrange's multiplier depends on s since the constraint condition is written at the local scale of a cable element. The Euler–Lagrange equations are:

$$\begin{cases} \dfrac{d}{ds}\left(\lambda\dfrac{dx}{ds}\right) = 0 \\ \dfrac{d}{ds}\left(\lambda\dfrac{dz}{ds}\right) - \rho g S = 0 \end{cases} \qquad \text{[A.5.21]}$$

The first equation [A.5.21] gives:

$$\lambda = \alpha\frac{ds}{dx} \qquad \text{[A.5.22]}$$

Substituting [A.5.22] into the second equation [A.5.21], one gets:

$$\frac{d}{ds}\left(\frac{dz}{dx}\right) = \frac{\rho g S}{\alpha} \qquad \text{[A.5.23]}$$

Equation [A.5.23] can be identified with equation [A.5.3] by expressing z as a function of the variable x. Starting from the relation,

$$ds = dx\sqrt{\left(\frac{dz}{dx}\right)^2 + 1} \qquad \text{[A.5.24]}$$

equation [A.5.23] is written as:

$$\frac{d^2z/dx^2}{\sqrt{1+(dz/dx)^2}} = \frac{\rho g S}{\alpha} \qquad \text{[A.5.25]}$$

Accordingly, the solution can be obtained in the same way as in subsection A.5.2, α being identified with T_0.

A.6. Mechanical properties of some solids in common use

Metals

Stainless steel	$E \cong 1.94 10^{11}$ Pa, $\nu = 0.265$, $\rho = 7970\,\text{kgm}^{-3}$
Aluminium	$E \cong 6.9 10^{10}$ Pa, $\nu = 0.33$, $\rho = 2700\,\text{kgm}^{-3}$
Brass	$E \cong 1.2 10^{11}$ Pa, $\rho = 8800\,\text{kgm}^{-3}$
Chromium	$E \cong 4.8 10^{9}$ Pa, $\rho = 7200\,\text{kgm}^{-3}$
Copper	$E \cong 1.1 10^{11}$ Pa, $\rho = 8970\,\text{kgm}^{-3}$
Iron	$E \cong 2.1 10^{11}$ Pa, $\rho = 7860\,\text{kgm}^{-3}$
Magnesium	$E \cong 4.5 10^{10}$ Pa, $\rho = 1740\,\text{kgm}^{-3}$
Gold	$E \cong 7.4 10^{10}$ Pa, $\rho = 19\,300\,\text{kgm}^{-3}$
Nickel	$E \cong 2.1 10^{11}$ Pa, $\rho = 8910\,\text{kgm}^{-3}$
Lead	$E \cong 1.8 10^{10}$ Pa, $\rho = 11300\,\text{kgm}^{-3}$
Titane	$E \cong 1.2 10^{11}$ Pa, $\rho = 4540\,\text{kgm}^{-3}$
Tungstene	$E \cong 3.4 10^{11}$ Pa, $\rho = 19\,300\,\text{kgm}^{-3}$
Zinc	$E \cong 8.3 10^{10}$ Pa, $\rho = 7140\,\text{kgm}^{-3}$
Zirconium	$E \cong 7.6 10^{10}$ Pa, $\rho = 6370\,\text{kgm}^{-3}$

Composites material using carbon fibers

$$1.5 10^{11} \le E \le 8 10^{11}\ \text{Pa},$$

$$1500 \le \rho \le 2000\,\text{kgm}^{-3}$$

Thermoplastics

$$4 10^8 \leq E \leq 10^9\,\text{Pa},$$
$$1000 \leq \rho \leq 2000\,\text{kgm}^{-3}$$

Glass

$$E \cong 7 10^{10}\,\text{Pa},$$
$$2300 \leq \rho \leq 2600\,\text{kgm}^{-3}$$

Wood

$$3.5 10^9(\text{balsa}) \leq E \leq 2 10^{10}\,\text{Pa}(\text{ebony}),$$
$$100(\text{balsa}) \leq \rho \leq 1000\,\text{kgm}^{-3}(\text{ebony})$$

References

[ACH 73] ACHENBACH J.D., *Wave Propagation in Elastic solids*, North-Holland American Series, 1987

[ANT 90a] ANTUNES J., AXISA F., BEAUFILS B., GUILBAUD D., Coulomb Friction Modelling in Numerical Simulations of Vibration and Wear Work Rate of Multispan Tube Bundles, *Journal of Fluid and Structures*, **Vol. 4**, pp. 287–304, 1990

[ANT 90b] ANTUNES J., AXISA F., VENTO M.A., Experiments on Vibro-Impact Dynamics under Fluidelastic Instability, *ASME Journal of Pressure Vessel Technology*, **Vol. 114**, pp. 23–32, 1992

[AXI 84] AXISA F., DESSEAUX A., GIBERT R.J., Experimental Study of Tube/Support Impact Forces in Multi-Span PWR Steam Generator Tubes, *ASME Winter Annual Meeting, December 9–14, New Orleans, Louisiana, Symposium on Flow Induced Vibration, PVP*, **Vol. 3**, pp. 139–148, 1984

[AXI 88] AXISA F., ANTUNES J., VILLARD B., Overview of Numerical Methods for Predicting Flow-Induced Vibration and Wear of Heat Exchanger Tubes, *ASME Journal of Pressure Vessel Technology*, **Vol. 110(1)**, pp. 6–14, 1988

[AXI 92] AXISA F., IZQUIERDO P., Experiments on Vibro-Impact Dynamics of Loosely Supported Tubes under Harmonic Excitation, *ASME Winter Annual Meeting, Symposium on Flow Induced Vibration and Noise, November 8–13, Anaheim California*, 1992

[AXI 04] AXISA F., *Modelling of Mechanical Systems, Vol 1 Discrete Systems*, Kogan Page Science, 2004

[BAN 97] BANSAL A.S., Free Waves in Periodically Disordered Systems: Natural and Bounding Frequencies of Unsymmetric Systems and Normal Node Localization, *Journal of Sound and Vibration*, **Vol. 207(3)**, pp. 365–382, 1997

[BEC 52] BECK M., Die Knicklast des einseitig eingespannten, tangential gedrückten Stabes, *ZAMP*, **Vol. 3**, pp. 225–288, 1952

[BLE 79] BLEVINS R.D., *Natural Frequencies and Modal Shapes*, Van Nostrand Reinhold, New York, 1979

[BOU 00] BOUZIT D., CHRISTOPHE P., Wave Localization and Conversion Phenomena in Multi-coupled Multi-span Beams, *Chaos, Solitons and Fractals*, **Vol. 11**, pp. 1575–1596, 2000

[CAR 01] CARVAL S., Atténuation des surflux de contraintes dans les assemblages de coques de révolution, Thèse de doctorat, Ecole Centrale de Paris, December 2001

[CAS 92] CASTEM 2000, *Manuel de référence*, CEA/DMT/LAMS, July 1992

[CET 99] CETINKAYA C., Localization of Longitudinal Waves in Bi-periodic Elastic Structures with Disorder, *Journal of Sound and Vibration*, **Vol. 221(1)**, pp. 49–66, 1999

[CHA 87] CHAKRABARTY J., *Theory of Plasticity*, McGraw-Hill Engineering Mechanics Series, 1987

[COL 63] COLLATZ L., *Eigenwertaufgaben mit technischen anwendungen*, Akademischen Verlags-gesellschaft Geest & Portig K.-G., Leipzig, 1963

[COT 90] COTTEREL B., KAMMINGA J., *Mechanics of Pre-industrial Technology*, Cambridge University Press, 1990

[COW 66] COWPER G.R., The Shear Coefficient in Timoshenko's Beam Theory, *Journal of Applied Mechanics*, **Vol. 33**, pp. 335–340, 1966

[CRIS 86] CRISFIELD M.A., *Finite Elements and Solution Procedures for Structural Analysis*, Pineridge Press, 1986

[CRIS 96] CRISFIELD M.A., *Non linear Finite Element Analysis of Solids and Structures*, Vol. 1 & 2, J. Wiley, 1996

[DON 76] DONNELL L.H., *Beams, Plates, and Shells*, McGraw-Hill, 1976

[EWI 00] EWINS D.J., *Modal Testing: Theory, Practice and Application*, Research Studies Press, 2000

[FLE 98] FLETCHER N.H., ROSSING T.D., *The Physics of Musical Instruments*, Springer-Verlag, 1998

[FUN 68] FUNG Y.C., *Foundations of Solid Mechanics*, Prentice-Hall, 1968

[FUN 01] FUNG Y.C., *Classical and Computational Solid Mechanics*, World Scientific Publishing Company, 2001

[JON 89] JONES N., *Structural Impact*, Cambridge University Press, 1989

[KIS 91] KISSEL G.J., Localization Factor for Multichannel Disordered Systems, *Phys. Rev. A*, **Vol. 44(2)**, pp. 1008–1014, 1991

[KRA 67] KRAUS H., *Thin Elastic Shells*, John Wiley and Sons, Inc., 1967

[LAN 94] LANGLEY R.S., Wave Transmission Through One-dimensional Near-periodic Structures: Optimum and Random Disorder, *Journal of Sound and Vibration*, **Vol. 178**, pp. 411–428, 1994

[MIK 78] MIKLOWITZ J., *The Theory of Elastic Waves and Waveguides*, North Holland, 1978

[MOU 04] MOUMNI Z., AXISA F., Simplified Modelling of Vehicle Frontal Crashworthiness Using a Modal Approach, *International Journal of Crash*, **Vol. 9(3)**, pp. 285–297, 2004

[NEU 85] NEUMANN F., *Vorlesungen über die Theorie der Elasticität der festen Körper und des Lichttäthers*, B.G. Teubner, Leipzig, 1885

[NOV 64] NOVOZHILOV V.V., *The Theory of thin elastic shells*, P. Noordhoff, 1964

[PIL 02] PILKEY W.D., *Analysis and Design of Elastic Beams*, John Wiley & Sons, 2002

[SAL 01] SALENCON J.L., *Handbook of Continuum Mechanics*, Springer Verlag, 2001

[SEI 75] SEIDE P., *Small Elastic Deformations of Thin shells*, Noordhoff, 1975

[SOE 93] SOEDEL W., *Vibrations of Shells and Plates*, Marcel Dekker, Inc., 1993

[SOM 50] SOMMERFELD A., *Mechanics of Deformable Bodies*, Academic Press Inc. Publishers, 1950

[STA 70] STAKGOLD I., *Boundary Value Problems of Mathematical Physics*, The Macmillan Company, 1970

[STR 93] STRONGE W.J., YU T.X., *Dynamic Models for Structural Plasticity*, Springer-Verlag, 1993

[TIM 51] TIMOSHENKO S., GOODIER J.N., *Theory of Elasticity*, 3rd edn, McGraw-Hill, 1951

Index

acoustic mode 58
action 152
  integral 19, 20, 153, 154, 156
  and reaction 172
    reaction principle 105
actual configuration 10
additional inertia 238
  stiffness 231
additive interferences 26
adjoint
  form 156, 363
  operator 156, 157, 172, 112
admissible
  load 365
  shape 436
  variation 19
angle
  of incidence 43, 45
  of reflection 45
annular plate 307
anticlastic
  bending 343
  deformation 331
antinode 52, 246
  of vibration 396
apex 366
apparent
  phase speed 51, 52, 63
  wave number 49, 50, 53
arch 358
  and circular rings 358, 392
  coupled in-plane model 398
  curvature 358
  deformation modes 394
  extrados 382
  global bending strain (in-plane) 394
  global displacements 392
  (half circular) 406
  kinetic energy (in-plane bending) 395
  local displacements 392
  local strains 394
  metric tensor 392
  neutral fibre 392
  out-of-plane bending and torsion 407
  out-of-plane local strains 394, 408
  out-of-plane vibration equations (bending and torsion) 409
  pure bending model (in-plane) 395
  and shells 354 et seq
    (bent and twisted) 391 et seq
  strain energy (in-plane bending) 395
  vibration equation (in-plane bending) 396
area inertia moment 74
  polar inertia moment 75, 94
aspect ratio 99, 283
assembled
  matrix 167, 173
  vector 167
assembling finite elements 179
asymmetrical cross-section 77
axial
  displacement 69
  elastic vibration 78
  force 73
  mode 78
  moment 89
  preload 144
  rotation 89
  strain 70
axisymetric shell 371
axisymmetry 364

balance of moments 50
bandwidth 224
Barré de Saint-Venant 60, 92

beam 66
clamped-clamped 112
elastic-plastic
element 115, 117, 174
geometry 67
pinned-pinned 112
pinned supported 111
traction-compression 167
transverse loads 355
beats 39
bending 77
angle 69
axis 118
of beam 141, 207
element 174, 212
equations 316
mode 51, 212
mode of vibration 200–201, 212
moment 73, 125, 222
operator 159
plane 71
and shear modes 205
stiffness coefficient 327
strain 70
and transverse shear motion 135
Bernoulli–Euler 83
branch 194
model 99, 118, 135, 175, 262
Betti, E. 162
bilinear form 158
body force 11, 14
boundary, branch dispersion equation 50
boundary condition 2, 13, 19, 23, 83, 114–115, 172, 319
homogeneous 14
inhomogeneous 15
loading 27
value problem 83, 85, 95
vibration equation 143
breathing mode 365, 376
buckling 147
instability 94, 145, 214–215
load 216, 237, 337
mode 214–215
mode shape 215, 232, 238
sagging 46

cable 73, 148, 366
canonical form 274
cantilevered beam 141, 161, 208, 280
configuration 104
Cartesian coordinate 4
CASTEM 2000 179, 292
catenary curve 367
Cauchy kinematic tensor 6
stress tensor 12
central
differences scheme 253
symmetry 68, 95, 107, 278
central axis 67
centre-of-mass 77
centroid 67, 117
line 67, 100
Chladni, E. 241, 319, 340
circular cylinder 307
circular cylindrical shell 415
breathing modes 427, 429
constriction of 421
coupled modes of vibration 434
elastic stress field 418
equilibrium equations 415
point-wise punching 433
pure bending model 420
simplified model for bending 428
vibration equations 419
circular plate 350
circular ring 364
breathing mode 365, 397
in-plane vibration modes (coupled model) 402
modal branches 411
modes of vibration (out-of-plane) 400, 410
slenderness ratio 397
vibration equations (in-plane bending) 396
circumferential vector 371
clamped-clamped configuration 108
clamped end 102
Clapeyron's formula 26
complex
amplitude 34, 41
field 33
wave number 50
composite materials, carbon fibres 466
compound wave 36
compression, solid body 61
concentrated
force 223
load 82, 88, 103, 151, 152, 270, 303
mass 238
concentred non-linearity 247
concomitant 158, 159, 161
conical
container 388
frustum 386
conjugate quantities 73
connecting element 247
force 232
conservation of mechanical energy 28, 44

conservative
  boundary condition 2
  force 156
  mechanics 162
  medium 32
  operator 162
  systems 15
constrained
  Lagrangian 24
  medium 24
  model 248
  system 233
constraint
  condition 61
  reaction 14
constricted shell 423
contact
  force 11, 14, 38, 249
  time 251, 258
continuous system 56
continuum, equilibrium
    conditions 3
contour integral 68
contracted product 10
contraction 70
convergence rate 28
conversion coefficient 6
coordinate
  system 20
  transformation 77
corner
  edge 260
  force 321, 323, 331, 460
critical load 147, 149, 337
  for buckling 217
cross-section 67
cross-sectional rigidity 79
cuboid 54
curl operator 448
curvature 309
  centre 358
  curvature effect 371
  radius 368
  tensor 373
curved
  metrics 369
  structure 309, 355
curvilinear
  coordinate 212, 303, 304
  metrics 358
cut-off frequency 32, 224
cylindrical
  coordinates 3, 307, 348
  rod 38, 59, 133
cylindrical shell 355, 376
  aspect ratio 431
  pinching of 425

damping operator 219
deformable medium 2
  solid, saving DOF 186
deformation mode
  longitudinal 78
  pure bending 99
  shear 86
  several modes 128
  torsion 86
deformation rate 2
degree of freedom 186, 262
deterministic excitation 220
differential operators 449
dilatation, mode 54, 55
  dilatation wave 32, 54, 290
dipole action 154
Dirac delta distribution 151, 152, 156
Dirac dipole 154, 302
discrete model 143
  spectrum 53
  system 19, 156
discretization procedure 295
dispersion equation 50, 53, 56
dispersive wave 35, 36, 50
displacement
  amplitude 113
  field 19
  potential 33, 56
  and strains 3
  vector 4
distribution 154, 156
divergence 28, 215
  operator 446
  theorem 94
Donnel–Mushtari–Vlasov model 435
dynamical instability 149

edge
  force 267
  supports 274
  surface 260
effective Kirchhoff shear force 321, 350
effective stiffness 147
eigenvalue 13, 157, 295
eigenvector 13, 157
elastic
  bar 61
  collision 257
  core 83

elastic model 248
energy 131
energy density 273
foundation 209, 422
impact 247
impedance 76, 233, 236
limit 83, 84
material 16
motion 78
solid 3
solid rod, waves in 53
stress 74, 272
structures 189
support 15, 173
vibrations 27, 327
wave 31
wave vibrations 17, 27
elastic-plastic law 81
elasticity
constant 11
constant tensor 16
law 272, 273, 307
elastodynamic problem 59
elastostatics 275
electromagnetic wave 1
element
functional 163, 165, 169, 172
mass matrix 161, 169
matrix 174
stiffness matrix 161, 169
elementary wave 48
energy
balance 29
conservation 29
functional 151, 162
engineering strain 7
equation of dispersion 290
equations of motion 29, 218, 349
equilibrium 13, 23
forces, general stress 75
moments 77
equivalent stiffness 229
equivoluminal motion 33
essential boundary condition 295
Euclidean space 2, 20
Eulerian description 2
excitation spectrum 64, 227
expansion
joint 285
third law 119
explicit algorithm 248, 253
extended Lagrangian 18
extrados 358

face 260
facet 11
fibre 71
neutral 71
finite
difference method 248
discontinuity 83, 151, 157
finite element approximation 210
finite element discretization 163
finite element method 59, 98, 163, 167, 168, 171, 293, 303
model 108, 109, 118
finite jump 153, 156
fixed boundary 20
fixed end 53
flexible bridge 221, 225
flexural axis 118
flexure
angle 69
curvature 313
mode 201
strain 70, 313
stress tensor 13
wave 93
fly wheel 238, 239
follower
force 148, 161
load 97, 148, 216
force, balance 11, 75
transducer 74
Fourier series 97, 197, 230, 300, 364
Fourier transform 35, 37, 226
free
boundary 43
end 80
motion 57
oscillation 162
rigid mode 236
free-free modal basis 235
frequency spectrum 50, 51
function of dispersion 40
functional
scalar product 21
space 21, 150
vector 14

Galerkin method 293, 356, 378
Gaussian curvature 374
generalized displacement 19
force 19
mechanical impedance 116
stress 74
vector 294
generatrix curve 371

geometric support 164
geometrical
 defects 224
 nonlinearity 112, 114, 217
Germain S. 319
Gibbs oscillations 301
Global
 coordinate system 167
 displacement 115, 262
 displacement field 70
 elastic-plastic law 128
 equilibrium 14
 force balance 14
 frame 68, 174
 law of plasticity 128
 response 60, 61, 300
 strain 70
 stress 72, 314
 tensor 265
gradient 5, 182
 operator 447
gradient transformation matrix 6
grazing incidence 5, 47, 52
Green–Lagrange strain tensor 7, 8, 10, 141
group velocity 38, 40
guided wave 48, 49

half-circular arch 404
Hamilton's principle 3, 18, 130 et seq, 265, 267, 305
 three dimensional solid 20
hammer impact 255
harmonic
 oscillation 34
 plane wave 38
 sequence 197, 199, 214
 solution 197
Heaviside step function 221, 250
hemispherical cap 377
Hilbert space 20, 21, 156, 189, 227
holonomic constraint 24
 relation 100
homogeneous
 boundary condition 14
 condition 114–117
 system 18
Hooke's
 elasticity tensor 16
 law 16, 74
hoop
 strain 426
 stress 356, 374
hyperboloid 331

identity
 matrix 6
 tensor 5
impact
 problem 248
 stiffness 249, 252
impulsive loading 151
incidence angle 48
incident wave 46, 51
incompressible fluid 24
indicial notation 9
indifferent equilibrium 81
inertia
 operator 159
 tensor 78
inertial connection 238
inhomogeneous
 boundary condition 15
 condition 116
 system 15
initial
 condition 218
 configuration 4
 load 141, 143
 stress 141, 143, 261
 stress operator 161
in-phase mode 292
in-plane, motion 261
 bending and axial vibrations 398
 shear force 261
in-shear strain 264
 shear stress 261
 stress 205
instantaneous power 28
interaction force 242
intermediate support 68, 84, 155
interpolated displacement field 108
intrinsic form 5, 267
irrotational
 motion 33
 wave 33
isotropic elastic material, stress–strain relationships 16
isotropic material 16

Jacobian matrix 6, 8

kinematical constraint 19
kinetic energy 19, 21
 density 21, 267
Kirchhoff, G.R. 320
Kirchhoff–Love
 assumptions 278

hypotheses 312
model 262, 312, 313, 314, 315, 369
Kirchoff shear forces 460
Kirchoff's theorem 30
knife edge support 102

Lagrange's
equations 19, 24, 463
multiplier 19, 24, 74, 80, 172, 173, 315
Lagrangian 12, 19, 20, 184, 265
description 2
displacement 3
Lamé's parameters 3, 16, 304, 307, 348, 350
Laplace
operator 28, 449
series 250
transform 223, 250
Laplacian 58
lateral
displacement 69
elongation 264
limit load 127
line of centroids 67
linear
differential operator 157
elastic material law 16
elasticity 51
manifold 294
momentum 251, 256
operator 102
spring 80
strain tensor 304
load factor 216
loading step 124
loadings, surface and concentrated 324
local
displacement field 68, 312
elastic-plastic stress 123–124
equilibrium 14, 149
frame 68, 167, 174
local quantity 14
local response 59, 60, 299, 302
shear strain 91
strain 70, 313
strain tensor 46, 70
stress 72, 314
local stress field 303
system of coordinates 166, 174
locking condition 23, 80
longitudinal
displacement 69
elongation 264
mode 196, 198
mode of vibration 55, 197, 212
wave 78, 190
Love–Rayleigh
equation 134, 192
model 133
Love's equations 414
equations in cylindrical coordinates 415
lumped mass matrix 210

main cross-section 262
manifold 295
masonry dome 383
mass
matrix 274
operator 150, 157
material
law 21, 24, 25, 167
line 71
point 2, 3, 436
oscillations 31
wave 2, 32
MATLAB 39, 435
matrix notation 6
mean curvature 374
mean square value 227
mechanical
energy 28, 30
impedance 116
membrane 261
component 264
displacement 263
equation 414
equilibrium 265
mode 289, 290, 293, 300
strain tensor 305, 313
strains 263
stress 65, 302
wave 289
meridian 371
line 374
plane 373
stress 374
mesh 163, 168
method of variables separation 32
metric tensor 257
midplane 261, 312
midsurface 260, 367
mixed formulation of equilibrium
equations 16
modal
analysis 100, 188 et seq, 189, 332
basis 190, 217 (truncation of) 222
coordinate system 217
density 245, 292, 338, 428
displacement 190, 219

modal (*continued*)
equation 196, 289
expansion method 230, 294
force 219, 220
frequency 199
mass 196, 210
model 248
oscillator 219
projection method 189, 217, 233, 238
series 190, 219
stiffness 196, 210
truncation 222, 233
vector 58
wavelength 132, 133
mode
conversion 41, 53
of deformation 70
of propagation 50
shape 55, 57, 190, 197, 290
shape expansion method 197
shape localization 247
shape, natural, vibration 53
wavelength 210
model coupling torsion 407
moment balance 77
moment-curvature law 84, 126
motion
equations of 29
or shear 33
irrotational or potential 33
multispan beam 245, 249
musical instrument 74, 148, 144, 214

Nabla differential operator 446
Napoleon Bonaparte 319
natural
frequency 53, 196–197, 224
mode of vibration 3, 53, 58, 189, 196
pulsation 57
Navier's equations 17, 32
Neumann, F. 29
neutral fibre 71, 125
Newtonian approach 144
nodal
displacement 165, 167
force 166
line irrotational or potential 340
loading 171
node 52, 164
nonaxial 174
nonconservative
force 32, 156
system 149

nondispersive wave 37
nonlinear strain 141
norm 196
normal
coordinate 190, 217
fibre 262
force 73
incidence 41, 524
stress 13, 54, 55, 265

operator 150
formally self-adjoint 158
positive 106
positive definite 162
symmetry 150
orthogonality rule for mode shapes 199
orthonormal basis 189, 197, 218
curvilinear coordinates 358
orthotropic material 316
oscillations, material 31
out-of-phase mode 244, 257, 292
out-of-plane load 261
motion 261
ovalization 61

P wave 42
reflection 43, 44
Parallelepiped 57
penalty
coefficient 173
method 173
phase
angle 34, 42
function 38
in-mode motion 259 et seq
shift 34
velocity 38, 52
physical coordinate 217
pinned support 102
plane
of constant phase 42, 50
harmonic wave 32
layer 52
mode 50, 64
strain 273
plane stress 272, 273
stress 92
wave 41, 50, 58
plastic
deformation 124
failure 125, 126
flow 125
strain 124
zone 126

plate
  bending 457
  contour 267, 322
  geometry 260
  loaded 299
  out-of-plane motion 311
  prestressed 236
Pochhammer 53, 191
Poisson effect 79, 132, 274, 288
Poisson ratio 16, 47
polar coordinates 95, 307
polarization 43, 481
polynomial interpolation 166, 171
portal frame 174, 179, 211–212
position vector 2, 5
post-buckling behaviour 214
potential
  energy 18
  motion 33
power density spectrum 224
preload 144
prescribed
  displacement 15, 173
  motion 173
pressurized
  toroidal shell 378
  vessel 424
prestress
  energy 141, 334
  operator 238
  stiffness operator 214
  stiffness term 143
prestressed
  beam 141, 148, 238
  equilibrium 261
  system 141
prime unknown 169
principal
  axis of inertia 68
  curvature line 374, 379
    principal curvature 374, 379
  directions 13
  stress 13
principle
  of causality 34
  of superposition 29
  of virtual work 151
projection 151, 171, 217
propagation
  delay 31, 35
  speed 34
pseudo-mode 228
pure bending mode 99
pure shear model 86, 87

quadratic form 21, 162, 171, 267
quasi-inertial range 225, 228
quasi-static
  correction 228, 235
  mode 228
  range 225, 228

Rayleigh minimum principle 294, 299
Rayleigh–Love model 199
Rayleigh–Ritz method 242, 293, 295, 335, 345, 436
Rayleigh–Timoshenko model 195, 205
Rayleigh's quotient 207, 211, 295, 335, 434
reciprocity theorem 162
rectangular plate 262, 265, 275, 286
rectilinear mode 291, 300
reflected wave 43, 44, 46, 54
reflecting
  plane 43
  surface 43
reflection coefficient 46
  law 44, 46
  wave 40
relation of constraint 2
resonant range 225, 226
  response 226
response spectrum 227
resultant stress 72
rigid body 24, 262
  mode 236
  motion 8
rigid connection
  mode 256
  spring 244
  support 173
roof truss 180
rotation
  angle 100
  matrix 8, 178
rotational inertia 132
rotatory inertia 194, 361
rotor 238

saddle point 321
St Peter basilica 383
Saint Venant's principle 38, 59, 278, 282, 299, 303, 439
Saint Venant's theory of warping 91
scalar displacement potential 3
scalar field 2, 441
self-adjoint
  matrix 157
  operator 156, 158, 172, 294, 399

self-adjointness 274
series expansion method 65
SH, guided wave 45, 48
  wave 43, 48
  wave reflection 43
shape
  criterion 61
  shape function 212
shear
  angle 183
  centre 117, 118, 137
  deformation 63
  force 73, 101, 145, 225, 315
  modal stress 57
  mode 78, 86
  modulus 16
  motion 33
  operator 190
  shear plane mode of vibration 58
  stiffness coefficient 118
shear-centre 117, 118
shear stress 13
  strain 8
  vibration equation 86
  wave 33, 38
  weighting factor 138
shearing 62
shell 358
  bending energy 428
  element 367
  flexure rotations 413
  force balances 416
  global flexure and torsion strains 413, 414
  local displacement field 412
  local strains 412
  membrane energy 430
  metric tensor 412
  ovaling mode 426
  transverse shear forces 415
  transverse shear strains 413
shells
  Kirchhoff effective shear stress 418
  and plates, transverse loads 356
shock
  duration 249
  force 252
singular distribution 270
singularity 158
skin
  stress 125
  structure 261
slender
  arch 284
  beam 139
slenderness ratio 67, 100, 114, 118
sliding
  boundary 48
  edge 275, 276
  support 102, 106, 284
small
  curvature 100
  deformation 263
  elastic motion 78
  rotation 68
  strain tensor 9, 161, 264
solid
  layer 48
  solid mechanics 1
solid body 2
  compression 61
solids, mechanical properties 466
spatially evanescent wave 50
spectral
  criterion 224, 228
  domain 225
spherical
  cap 382
  coordinates 375
  shell 375
square plate 284, 298
stability 149
staircase signal 255
standard boundary condition 269, 276, 277
standard boundary conditions (beams) 196, 201
standing wave 3, 53, 54, 196, 201
static buckling 214
stationary principle 295
stiffness
  additional 231
  coefficient 15, 79, 855
  matrix 169, 174, 177, 274
  operator 15, 150, 155
straight beam 68, 70, 116, 189
  Hamilton's principle 130 et seq
  longitudinal motion 132
  Newtonian approach 66 et seq
  variational formulation of equations 132
  vibration modes 188, 190
strain
  density 268
  energy 21, 25, 131, 142, 305
  hardening 123, 128
  isotherm tensors 284
  rate 124
  tensor 4, 24
  vector 272, 305
strain/stress relationship 272

stress 11
  coefficient 137
  discontinuity 83, 84, 271
  elastic 74
  energy density 189
  isotherm tensor 284
  local and global 72
  modal expansion 229
  principal 13
  singularity 302
  tensor 24, 242
  vector 11, 72, 265, 272
  wave 251
stress–strain relationship 16, 55
stretched
  membrane 65, 47
  plate 278, 340
stretching 70
  operator 274
string 73, 148
  components 159 et seq
  force 73
  instrument 214
  string model 66
structural
  component 3, 48
  element 1, 40, 59, 64
structure of revolution 396
substructure 240
substructuring method 190, 256
support
  condition 14, 167
  reaction 79, 171
surface force 11
SV wave 43, 45
symbolic notation 6, 9
symmetrical matrix 157

tangential stress 13
tensile
  force 60, 74
  stress 261
  test machine 59
tension structure 261
tensor 5
  algebra 443
  field 2, 441
  of Lagrange multipliers 16
  notation 267
  Saint Venant's principle 3, 60
  small local strains 393
  test function 157
thermal
  dilatation coefficient 118, 283
  expansion 249
  expansion law 118
  gradient 287
  strain 118, 283
  stress 119, 122, 284, 285
thermal buckling 146, 217, 337
thermoelastic
  behaviour, straight beam 118
  bending of beam 121
  equilibrium 120
  law 283
  stress 119
thermoelasticity 283
thickness 260
  ratio 356
thin shells 412
thin walled beam 91
time-history 220, 227, 248
time step 248
Timoshenko model 136, 139
toroidal shell 378
torsion 77
  angle 69
  constant 94, 200
  mode of vibration 133
    torsion mode 51, 200
  mode without warping 89
  moment 73, 314, 324
  spring 90
  strain 70
torsional
  mode 89
  strain 313
total
  conversion 47
  reflection 53
traction element 212
transfer function 340
transport velocity 40
transverse
  displacement 69
  shear force 99
  shear strain 70
  strain 117
  wave 31
travelling
  load 221
  wave 31
trial function 56, 294
truncated
  modal basis 190
  model 228
  truncated series 219

truncation
  criterion 219
  flexibility 235
  mass 229
  order 224
  stiffness 228, 235, 256
twist
  angle 69, 92
  centre 117, 118

uniaxial
  plate expansion 286
  stress 62
uniform plate stretching 278
uniqueness of solution 29
unloading step 124

variables separation 32
variation operator 18
variational
  approach 131
  method 141
  principle 18, 294
Vaslov, V. Z. 435
Vector
  displacement potential 33
  field 2, 441
vibration
  antinode 198
  equation 17, 28, 78, 143, 190
  modes 53, 55, 58, 335
  node 198
    in-plane or membrane 289
virtual
  displacement field 22
  variation 185, 200
  work 22, 151, 268
viscous damping operator 150, 159

warping 91, 92, 94, 97, 197
  beam 347
  function 92, 97, 347
water in rigid tube 24
wave
  dispersive and nondispersive 35
  energy 38
  frequency 50
  number 35, 49, 56, 290
  packet 40
  plane 50
  reflection 3, 40
  refraction 41
  shape 49
  vector 56
wavelength 35, 545
weighted
  equation of motion 151
  function 156
  functional vector 151
  integral 149, 151, 156, 157, 238
work 23
  density 267
  rate 28

yield stress 123
Young's modulus 16

zig zag function 97

Printed and bound by CPI Group (UK) Ltd, Croydon, CR0 4YY

08/05/2025

01864811-0001